Fundamentals of RISK ANALYSIS *and* RISK MANAGEMENT

Fundamentals of RISK ANALYSIS and RISK MANAGEMENT

Edited by
VLASTA MOLAK
President
GAIA UNLIMITED, Inc.
Cincinnati, Ohio

LEWIS PUBLISHERS
Boca Raton New York London Tokyo

Publisher:	Joel Stein
Project Editor:	Carole Sweatman
Marketing Manager:	Greg Daurelle
Direct Marketing Manager:	Arline Massey
Cover Design:	Denise Craig
PrePress:	Carlos Esser
Manufacturing:	Sheri Schwartz

Library of Congress Cataloging-in-Publication Data

Molak, Vlasta.
Fundamentals of risk analysis and risk management / Vlasta Molak.
p. cm.
Includes bibliographical references and index.
ISBN 1-56670-130-9 (alk. paper)
1. Technology—Risk assessment. I. Title.
T174.5.M64 1996
363.1—dc20 96-19681
CIP

This book contains information obtained from authentic and highly regarded sources. Reprinted material is quoted with permission, and sources are indicated. A wide variety of references is listed. Reasonable efforts have been made to publish reliable data and information, but the authors, editor, and the publisher cannot assume responsibility for the validity of all materials or for the consequences of their use.

Direct all inquiries to CRC Press, Inc., 2000 Corporate Blvd., N.W., Boca Raton, Florida 33431.

Lewis Publishers is an imprint of CRC Press

No claim to original U.S. Government works
International Standard Book Number 1-56670-130-9
Library of Congress Card Number 96-19681
Printed in the United States of America 1 2 3 4 5 6 7 8 9 0
Printed on acid-free paper

Foreword

My Uncle Steve, who worked on one of the government's first computers, had his own mathematical system wherein he calculated the probability of a horse winning a race. Sometimes Uncle Steve won money on the horses. Sometimes he lost money on the horses. All of his winning and losing was done very scientifically: studying *The Daily Racing Digest*, calculating the odds according to such dependent variables (such as the track records of the stable, the trainer, the jockey, the horse, and the length of the race), and assigning proper weight to intervening variables (such as the condition of the track and weather at the time of the race). He did well. My Aunt Betty, who also did well at the track, used the time-honored "Hunch System of Equine Competition," also known as intuition. "I've just got a feeling that this horse is due," she would say to me during our frequent summer visits to Thistledown. All this risk taking with money, whether through science or intuition, can be best summed by the immortal tout who once said: "Ya places yer bets and ya takes yer chances." And then there was Betty and Steve's younger brother, Frank, (my father) who never bet on the horses because he believed all horse races were fixed.

Risk analysis and risk management are, for most people, much more lofty and consequential than the outcome of a horse race. Nevertheless, Uncle Steve and Aunt Betty's track assessment styles came to my mind when a nuclear scientist testifying before our Ohio Senate Energy and Environment committee claimed a planned multistate radioactive waste dump would be of little risk to Ohio. I thought of Uncle Steve and how he would have demanded the track record of the industry of containment of nuclear waste in the past. I thought of Aunt Betty and what her instincts would have told her about whether it was the right time to bet on a long shot named *Glows in the Dark*. I thought of my father and his wariness about the fix being in. Thus I came to vote against Senate Bill 19.

Informed opinions by the highly educated and much lettered are available to support nearly every point of view. Human decision-making is a terribly complicated matter. We all want to make the best decision. We would hope that the best decision is made on the basis of the best available information. Often it is. Sometimes it is not. In the chain reaction of real world decision-making, science collides with economics, which collide with politics, and the decision rests with that body of knowledge, which is (accidentally) left standing.

Vlasta Molak has gathered together the works of some of the most impressive authors of papers on risk analysis and risk management in the world. Her writings and her compilation of the work of so many leading scientists in one complete volume is a public service, in that it enables both novice and expert to ponder the many and diverse factors that are at work in assessing, analyzing, and managing risk.

This book will be useful to both legislators (local, state, and federal) and their staff to help devise better laws to protect the public, encourage responsible business development, and increase profits – rather than using risk analysis to promote *status quo* or reduce environmental safeguards.

Several chapters that deal with economics and risk analysis have convinced me that being PRO-working average person and PRO-environmental protection is NOT being ANTI-business. On the contrary, responsible and effective business organizations profit from a loyal, well-trained work force and reasonable, smart environmental regulations that encourage efficiency and nonpollution. ***Numerous studies, cited in this book, demonstrate that application of most enlighted environmental management increases profits (since pollution is equivalent to wasted resources) and thus fiscal conservatism and emphasis on private property rights also mean increased environmental protection. Only in an unenlightened society are environmental safeguards mistakenly considered as opposed to business interests and free markets.*** Better business with cleaner environment is the paradigm for the 21st century. The old paradigm "business vs. environment" needs to be retired. *Fundamentals of Risk Analysis and Risk Management* will help raise this awareness and finally bury the old nonproductive paradigm, which has been one of the major sources of controversy in our legislative process.

I would recommend this book to my colleagues, who are often involved in designing very complex environmental and occupational protection laws, as a reference and as a useful book to increase their analytical skills in dealing with the complexity of legislation, regulations, risk-benefit analysis, and risk management. Also, the wealth of references provided in this book can help us better understand how our laws affect our environmental and occupational safety and health, and ultimately our quality of life.

Senator Dennis Kucinich
Ohio State Senator

Preface

The idea for this book started as a consequence of my directing and teaching a one-day course on "Fundamentals of Risk Analysis" at the annual meetings of the Society for Risk Analysis (1991, 1992, and 1994). Also, teaching a course at the United Nations Division for Sustainable Development, New York, on "Use of Risk Analysis in Sustainable Development", and teaching a course on "Environmental Risk Assessment and Management" at the University of São Paulo and University of Mato Grosso, Cuiába, Brazil, madẹ me aware of the need for a reference that I could give to students to get a comprehensive overview of the field and lead them to valuable references if they wanted to increase their knowledge in specific aspects of risk analysis. Moreover, my position as Secretary of the Society for Risk Analysis (from 1989–1994) convinced me that there is a great need for integrating the rapidly expanding field of risk analysis and risk management, and for providing a common language for all the practitioners and members of this varied interdisciplinary professional group.

The last few years have witnessed the concepts of risk analysis and risk management permeating public discussion, often confusing decision makers and the public. When Lewis Publishers called me in 1995, after having seen the title of the course I taught at the SRA Annual Meeting in December 1994, and asked me to write a book on the subject of risk analysis and risk managment, I decided that the need for such a book was overwhelming, and that providing such a book would be a worthwhile project. Since no single person could accomplish such a monumental task of integrating the diverse fields of risk analysis and risk management, I asked my colleagues to help me write the chapters for which they were recognized experts in their particular practice of risk analysis and risk management. Most of them graciously agreed, or gave up under my incessant prodding. Some of them cancelled at the last moment, but I was fortunate to find new authors who were not intimidated by the task. With the miracle of Internet, I was able to bring in several authors from different parts of the world to help expand our understanding of how risk analysis is practiced around the world.

After almost two years of work, we have completed the task of producing this book of 26 chapters, in which we cover the fundamentals of what is known as risk analysis and risk management in the contemporary western world. Most chapters also provide a summary, questions and answers to be used as tools in teaching courses in risk analysis. The glossary should also be helpful both to students and practitioners of risk analysis. Finally, the index should make it easier to focus on a particular area of the reader's interest. The addresses of co-authors are given as an easy access for those readers and students of risk analysis who may have some questions. The E-mail addresses of some of the authors should be particularly useful for further communication.

I want to thank all of the 20 co-authors who have graciously accepted the task of making their chapters understandable to an educated general reader, while at the same time providing references and in-depth discussion for those who want more detailed understanding. My work and discussions with them were very enlightening and fun. They have done an excellent job in educating me of the aspects of risk analysis of which I was not aware, and helping to deepen my understanding of different applications of risk analysis. Also, I want to thank Brian Lewis, who asked me to do this book before selling his company, Lewis Publishers, to CRC Press. My thanks go to the professionals at CRC Press, who have been very helpful in explaining the "nuts and bolts" of publishing and have been encouraging in finishing this work. Finally, I want to thank my daughter, Yelena, and Ohio State Senator, Dennis Kucinich for their review of some of my chapters and useful discussions and suggestions. They brought to my attention broader implications of the topics in this book of real life and political functioning in which risk analysis and risk management have become household words, frequently used without ever being properly defined and understood. Any mistakes found in this book are mine and unintentional, and I would appreciate if the reader brings them to my attention.

We hope that this book will be a useful guide to all who want to improve their knowledge in confronting dangers of living, and particularly to those who make decisions that affect public safety and the general safety of this planet. The increased awareness and application of risk analysis and risk management can improve our understanding of the dangers that we face on our life journey and help us make better choices.

Vlasta Molak

The Editor

Dr. Vlasta Molak is the International Coordinator and former Secretary of the Society for Risk Analysis (SRA). In 1989 she convened an international communication network to promote uses of risk analysis in solving some of the environmental problems resulting from misuse of technology. On her several trips to Eastern Europe and the former Soviet Union, Dr. Molak initiated activities to start chapters of the SRA in Prague (Republic of Czech), Zagreb (Croatia), Osijek (Croatia), Warsaw (Poland), Budapest (Hungary), Moscow (Russia), and Kharkov (Ukraine) with interested scientists, engineers, and policy makers in those countries. Dr. Molak represented the U.S. at a four-day workshop on "How to improve environmental awareness of local decision makers in Eastern Europe," sponsored by the European Commission. Dr. Molak taught in a training program in Brazil, which was organized by Taft's University Environmental Management Program, at the University of Cuiába and the University of São Paulo. The subject was "Environmental Risk Assessment and Risk Management" for professionals involved in Brazilian environmental management. She also taught a course at the United Nations headquarters (New York) on "The Use of Risk Analysis in Sustainable Development."

Dr. Molak is the founder and president of the Biotechnology Forum, Inc. in Cincinnati and chairs the Subcommittee for Technical Interpretation of the Local Emergency Planning Committee for Hamilton County, Ohio. Under her leadership, the Biotechnology Forum has organized series of lectures and workshops. One of the workshops, "The Alaska Story: In the Context of Oil Spill Problems in the Marine Environments," with special emphasis on the biological cleanup efforts, resulted in the proceedings edited by Dr. Molak. As a chair of the Subcommittee for Technical Interpretation, Dr. Molak initiated the efforts for hazard analysis in Hamilton County, Ohio and formulated the strategy for hazard analysis. She was a member of the Planning Committee for Comparative Risk Analysis for Hamilton County (Cincinnati, Ohio) and a member of the Quality of Life Committee of the Ohio Comparative Risk Analysis Project. She presently is coordinating the efforts to deal with more complex aspects of chemical safety: process safety in manufacturing, transportation of hazardous materials, and adverse effects of routine chronic releases of toxic chemicals.

Dr. Molak has worked at the U.S. Environmental Protection Agency and the National Institute for Occupational Safety and Health (NIOSH) on developing methodologies for risk analysis of toxic chemicals. These methodologies are used to derive various environmental and occupational criteria. Dr. Molak also worked for a private environmental consulting company and now is the founder and president of GAIA UNLIMITED, Inc., her own consulting company dealing with environmental and occupational risk assessment, risk management, and general environmental problems including strategies for pollution prevention. She is teaching various courses for risk analysis (including courses for local and state governments). She is also developing the AGENDA 21 PROGRAM as a dean at the Athena University, based entirely on the Internet. It is intended to be a fully accredited program promoting ideas and operational skills necessary for sustainable development. Her training is interdisciplinary: she has a B.S. in physical engineering, an M.S. in chemistry, a Ph.D. in biochemistry, and postdoctoral training in molecular genetics. Dr. Molak is a Diplomat of the American Board of Toxicology (DABT).

Contributors

Joseph Alvarez, Ph.D.
Auxier & Associates
Parker, Colorado 80134
E-mail: jalverez@rmii.com

Vicki M. Bier, Ph.D.
Department of Industrial Engineering
Department of Nuclear Engineering and Engineering Physics
University of Wisconsin–Madison
Madison, Wisconsin 53706
E-mail: bier@ie.wisc.edu

William E. Dean, Ph.D.
Private Consultant
Sacramento, California 95814
E-mail: billdean@calweb.com

Paul F. Deisler, Ph.D.
Private Consultant
Austin, Texas 78703

Jeffrey H. Driver, Ph.D.
Technology Sciences Group, Inc.
Washington, D.C. 20036
E-mail: TSG@TSGUSA.com

Paul K. Freeman, J.D.
The ERIC Group, Inc.
Englewood, Colorado 80112

B. John Garrick, Ph.D.
PLG, Inc.
Newport Beach, California 92660
E-mail: garrick@plg.com

Herman J. Gibb, Ph.D.
National Center for Environmental Assessment
U.S. Environmental Protection Agency
Washington, D.C. 20460
E-mail: gibb.herman@epamail.epa.gov

P. J. (Bert) Hakkinen, Ph.D.
Department of Risk, Policy, and Regulatory Sciences
The Procter and Gamble Company
Ivorydale Technical Center
Cincinnati, Ohio 45224
E-mail: hakkinen.pj@pg.com

Barbara Harper, Ph.D., DABT
Department of Health Risk
Pacific Northwest Laboratory
Richland, Washington 99352
E-mail: bl_harper@pnl.gov

Peter Barton Hutt, LL.M.
Covington & Burling
Washington, D.C. 20044

Howard Kunreuther, Ph.D.
Center for Risk Management and Decision Processing
Wharton School
University of Pennsylvania
Philadelphia, Pennsylvania 19104
E-mail: kunreuther@wharton.upenn.edu

Robert T. Lackey, Ph.D.
Environmental Research Laboratory
U.S. Environmental Protection Agency
Corvallis, Oregon 97333
E-mail: lackey.robert@epamail.epa.gov

Howard Latin, J.D.
John J. Francis Scholar
Rutgers University School of Law at Newark
Newark, New Jersey 07102
E-mail: latin@andromeda.rutgers.edu

Terence L. Lustig, Ph.D.
Environmental Management Pty. Ltd.
Kensington, NSW, Australia
E-mail: TLUSTIG@ZETA.ORG.AU

Stuart C. MacDiarmid, Ph.D.
Department of Regulatory Authority
Ministry of Agriculture
Wellington, New Zealand
E-mail: macdiarmids@ra.maf.govt.nz

Vlasta Molak, Ph.D.
GAIA UNLIMITED, Inc.
Cincinnati, Ohio 45231
E-mail: vlasta@tso.cin.ix.net

Alexander Shlyakhter, Ph.D.
Department of Physics
Harvard Center for Risk Analysis
Harvard University
Cambridge, Massachusetts 02138
E-mail: shlyakhter@huhepl.harvard.edu

Paul Slovic, Ph.D.
Decision Research
Eugene, Oregon 97401
E-mail: pslovic@oregon.uoregon.edu

James A. Swaney, Ph.D.
Department of Economics
Wright State University
Dayton, Ohio 45435
E-mail: jswaney@desire.wright.edu

David Vose, M.Sc.
Risk Analysis Services
Wincanton, Somerset
United Kingdom BA9 9AP
E-mail: 100616.320@compuserve.com

Gary K. Whitmyre, M.A.
Technology Sciences Group, Inc.
Washington, D.C. 20036
E-mail: tsg@cais.com

Richard Wilson, Ph.D.
Department of Physics
Harvard Center for Risk Analysis
Harvard University
Cambridge, Massachusetts 02138
E-mail: wilson@huhepl.harvard.edu

Rae Zimmerman, Ph.D.
New York University
Robert F. Wagner Graduate
School of Public Service
New York, New York 10003
E-mail: zimmerman@is2.nyu.edu

Dedication

This book is dedicated to my dear husband, Peter and our children, Yelena, Ina, and Allen, and to my friends who have helped expand my view of the universe and of the impending dangers we all must confront to make our world a better place in which to live.

Special gratitude is extended to Yelena and my friend, Dennis, whose help came when it was most needed.

Contents

Foreword by Ohio State Senator Dennis Kucinich
Preface
The Editor
Contributors
Dedication

Introduction and Overview 1
Vlasta Molak

I. THEORETICAL BACKGROUND OF RISK ANALYSIS

Chapter I.1
Toxic Chemicals Noncancer Risk Analysis and U.S. Institutional Approaches to Risk Analysis 13
Vlasta Molak

Chapter I.2
Epidemiology and Cancer Risk Assessment 23
Herman J. Gibb

Chapter I.3
Uncertainty and Variability of Risk Analysis 33
Richard Wilson and Alexander Shlyakhter

Chapter I.4
Monte Carlo Risk Analysis Modeling 45
David Vose

Chapter I.5
An Overview of Probabilistic Risk Analysis for Complex Engineered Systems 67
Vicki M. Bier

Chapter I.6
Ecological Risk Analysis 87
Robert T. Lackey

Chapter I.7
The Basic Economics of Risk Analysis 99
James A. Swaney

II. APPLICATIONS OF RISK ANALYSIS

Chapter II.1
Assessment of Residential Exposures to Chemicals 125
Gary K. Whitmyre, Jeffrey H. Driver, and P. J. (Bert) Hakkinen

Chapter II.2
Pesticide Regulation and Human Health: The Role of Risk Assessment 143
Jeffrey H. Driver and Gary K. Whitmyre

Chapter II.3
Ionizing Radiation Risk Assessment 163
Joseph L. Alvarez

Chapter II.4
Use of Risk Analysis in Pollution Prevention 177
Vlasta Molak

Chapter II.5
Integrated Risk Analysis of Global Climate Change 187
Alexander Shlyakhter and Richard Wilson

Chapter II.6
Computer Software Programs, Databases, and the Use of the Internet, World Wide Web, and Other Online Systems 203
P. J. (Bert) Hakkinen

III. RISK PERCEPTION, LAW, POLITICS, AND RISK COMMUNICATION

Chapter III.1
Risk Perception and Trust 233
Paul Slovic

Chapter III.2
The Insurability of Risks 247
Howard Kunreuther and Paul K. Freeman

Chapter III.3
Setting Environmental Priorities Based on Risk 259
Paul F. Deisler, Jr.

Chapter III.4
Comparative Risk Analysis: A Panacea or Risky Business? 273
Vlasta Molak

Chapter III.5
Environmental Justice 281
Rae Zimmerman

Chapter III.6
Law and Risk Assessment in the United States 293
Peter Barton Hutt

Chapter III.7
Science, Regulation, and Toxic Risk Assessment 303
Howard Latin

IV. RISK MANAGEMENT

Chapter IV.1
Risk Management of the Nuclear Power Industry 327
B. John Garrick

Chapter IV.2
Seismic Risk and Management in California 341
William E. Dean

Chapter IV.3
Sustainable Management of Natural Disasters in Developing Countries 355
Terrence L. Lustig

Chapter IV.4
Risk Analysis, International Trade, and Animal Health 377
Stuart C. MacDiarmid

Chapter IV.5
Incorporating Tribal Cultural Interests and Treaty-Reserved
Rights in Risk Management 389
Barbara L. Harper

Chapter IV.6
Global Use of Risk Analysis for Sustainable Development 407
Vlasta Molak

Conclusion 423
Vlasta Molak

Answers to Questions 427

Glossary 451

Index 459

Introduction and Overview

Vlasta Molak

We are all more or less successful risk assessors and managers if we are still alive. Life is intrinsically filled with dangers, real or perceived. Planes may explode and go down either because of terrorist activities or safety rules violations, a nuclear power plant may blow up (Chernobyl) or release radioactive clouds (Three Mile Island), a chemical plant may release toxic gas (Bhopal), or a natural disaster (hurricane, flood, tornado, volcano, landslide) can strike the area in which we live. We may get acute food poisoning from either bacterial or chemical contamination, or we may suffer from chronic diseases that are in part caused by the food choices we make. Whether we are crossing the street, making investments, deciding what to eat, how to get from one place to another, choosing our profession, or getting married, we are making our decisions based on evaluating risks and benefits that a particular activity or avoidance would bring us. The subject of this book is to improve our analytical techniques in evaluating dangers and develop skills in confronting them.

1. DEFINITION OF RISK ANALYSIS

We can define risk analysis as a body of knowledge (methodology) that evaluates and derives a probability of an adverse effect of an agent (chemical, physical, or other), industrial process, technology, or natural process. Definition of an "adverse effect" is a value judgement. It could be defined as death or disease (in most cases of human health risk analysis); it could be a failure of a nuclear power plant, or a chemical plant accident, or a loss of invested money. In some recent cases of risk analysis, even vaguely defined terms such as "quality of life" or "sense of community" have been evaluated using risk analysis. Traditionally, most risk assessments (risk analysis applied in a particular situation) deal with health effects or, more recently, with the ecological health or economic well-being (in case of business risk

1-56670-130-9/97/$0.00+$.50

analysis). Although there are many types of risk analysis, some common elements are necessary to qualify the process as risk analysis, particularly when dealing with the potential health effects of toxic chemicals. Those elements are (NAS 1983)

1. Hazard (agent) identification
2. Dose-response relationship (how is quantity, intensity, or concentration of a hazard related to adverse effect)
3. Exposure analysis (who is exposed? to what and how much? how long? other exposures?)
4. Risk characterization (reviews all of the previous items and makes calculations based on data, with all the assumptions clearly stated; often the conclusion is that more data and/or improvement in methodology is needed and that no numerical risk number can be derived to express accurately the magnitude of risk)

Deciding WHAT is an adverse effect (and to some extent hazard identification) is a value judgment that can be made by well-informed citizens. The consideration of other components of risk analysis is a complex process, which in order to be properly conducted requires extensive training. Just as one would not want to have a surgery performed by an untrained layman, risk analysis may be a risky business if performed by untrained people. Because of its interdisciplinary nature and complexity, risk analysis requires an appropriate amount of time to evaluate all pertinent data, even when one deals with problems of lesser complexity. We are constantly performing risk analysis and risk management in everyday situations, such as observing traffic when planning to cross the street or driving. However, in more complex situations where we may be exposed to toxic substances, radiation, or the possibility of a nuclear power plant disaster, formal risk analysis may be necessary in order to derive reasonable (and sometimes optimal) recommendations for the most appropriate risk management.

2. PURPOSE OF THIS BOOK

This book provides a comprehensive overview of risk analysis and its applications to a broad range of human activities. The editor and co-authors seek to bridge the gap between theory and application and to create a common basic language of risk analysis. They hope that the material in this book will provide a common knowledge base for risk analysts, which can be expanded according to their specific interests and fields of study by using the references provided in each chapter. The co-authors are experienced and recognized practitioners in the various types of risk analysis and risk management.

The intended readers are scientists, engineers, lawyers, sociologists, politicians, and anyone interested in gaining an overview of risk analysis, wanting to become proficient in speaking the basic language of risk analysis, and understanding its applications in difficult risk management decisions. This book can be used as a textbook and reference for undergraduate, graduate, and other training courses in risk analysis. Also, the editor hopes that it will be used by legislators and their aides (local, state, and federal) to devise better laws to protect the public and to encourage responsible business development and profit increases, rather than using risk analysis to promote the status quo or reduce environmental safeguards. Several chapters demonstrate that application of the most enlightened environmental management increases profits (since pollution is equivalent to wasted resources). Thus, fiscal conservatism and emphasis on private property rights also mean increased environmental protection. ***Only in an unenlightened society are environmental safeguards mistakenly considered as being opposed to business interests and free markets. Better business with a cleaner environment is the paradigm for the 21st century.*** The old paradigm "business vs. environment" needs to be retired.

The book is divided into four sections. Section I, Theoretical Background of Risk Analysis consists of chapters demonstrating the scientific basis of risk analysis, types of risk analysis, and basic concepts. Chapters in this section discuss toxic chemicals risk analysis, epidemiological risk analysis, uncertainty and variability of risk analysis, Monte Carlo risk analysis modeling, probabilistic risk analysis of complex technological systems, ecological risk analysis, and the basic economics of risk analysis. Section II, Applications of Risk Analysis demonstrates applications of risk analysis to real-life situations. Examples come from agriculture (application of pesticides), indoors exposures, promoting pollution prevention, global climate change, etc. A chapter on computer software programs and use of the Internet in risk analysis is also added. Section III, Risk Perception, Law, Politics, and Risk Communication deals with differences between public perception of risks, scientific risk analysis and its legal applications, and how to communicate risks to those who may be affected. This section also has two chapters dealing with setting environmental priorities and comparative risk analysis and environmental justice. The insurability of risk deals with societal response to various risks of living. Section IV, Risk Management illustrates the use of risk analysis in devising better risk management in handling technologies (e.g., nuclear power plants) or general everyday environmental problems. Also, chapters deal with the management of natural risks such as earthquakes and floods and with the cleanup of radioactive hazardous waste sites on an Indian reservation. The final chapter integrates a worldview as seen by a risk analyst (Vlasta Molak, the editor).

The conclusion summarizes the topics elaborated in the chapters and suggests how the practice of risk analysis affects social management of environmental problems in view of the recent controversies in risk-benefit analysis applications in legislative proposals and regulations in the U.S.

3. HISTORICAL OVERVIEW OF RISK ANALYSIS

Historical perspective on risk analysis applications in society was given by Covello and Mumpower (1985).

Around 3200 B.C. in the Tigris-Euphrates valley, a group called Asipu served as risk analysis consultants for people making risky, uncertain, or difficult decisions. Greeks and Romans observed causal relationships between exposure and disease: Hippocrates (4th century B.C.) correlated occurrence of diseases with environmental exposures; Vitruvious (1st century B.C.) noticed lead toxicity; and Agricola (16th century A.D.) noticed the correlation between occupational exposure to mining and health.

Modern risk analysis has roots in probability theory and the development of scientific methods for identifying causal links between adverse health effects and different types of hazardous activities: Blaise Pascal introduced the probability theory in 1657; Edmond Halley proposed life-expectancy tables in 1693; and in 1792, Pierre Simon de LaPlace developed a true prototype of modern quantitative risk analysis with his calculations of the probability of death with and without smallpox vaccination. With the rise of capitalism, money use, and interest rates, there was an increased use of mathematical methods dealing with probabilities and risks. For example, the risk of dying was calculated for insurance purposes (life-expectancy tables). Physicians in the Middle Ages also observed a correlation between exposures to chemicals or agents and health: John Evelyn (1620–1706) noticed that smoke in London caused respiratory problems. He also noticed correlation of scrotal cancer with occupational exposures to soot in chimney sweeps.

4. RISK MANAGEMENT

Insurance, which started 3900 years ago in Mesopotamia, is one of the oldest strategies for dealing with risks. In 1950 B.C., the Code of Hamurabi formalized bottomry contracts containing a risk premium for the chance of loss of ships and cargo. By 750 B.C., Greeks also practiced bottomry. In 1583, the first life insurance policy was issued in England. In contemporary society, insurance has developed to deal with a wide variety of phenomena associated with adverse effects, from health insurance to mortgage insurance. Actuaries (people who calculate insurance premia, based on historical losses and estimates of the future income from premiums and losses) are probably the best risk assessors, since the failure in making accurate predictions about losses and premia income can result in the loss of the business. Companies with bad actuaries go bankrupt (see Chapter III.2).

Government interventions to deal with natural or manmade hazards are recorded in all great civilizations. In order to manage air pollution from burning coal in London, King Edward (1285) issued an order forbidding the use of soft coal in kilns, ***after an unsuccessful trial to voluntary decrease its use. Perhaps we can learn from this historical example that "voluntary" reduction in risks from pollution and technological risks in general are best achieved by designing and enforcing intelligent environmental and occupational laws. Carrots and sticks may be more effective in dealing with environmental and occupational risks (accidents or pollution) than either sticks or carrots alone!*** Thus, while we may choose to believe that industries and individuals sincerely have the public good in mind when dealing with industrial production, pollution, and waste management, it is helpful to have laws and regulations to insure responsible behavior in cases where promises are not kept because budgetary constraints have pushed environmental considerations out of the picture. ***The irony is that in most cases improvement in environmental management also improves the bottom line in the long run and often in the short run. Thus, budgetary***

constraints should encourage environmental protection and pollution prevention since they save money for the company and save on public health and litigation costs! However, as the great physicist Max Plank said, *"The new ideas do not win by the strength of their logic, but because their opponents eventually die!"* Hopefully, the idea of pollution prevention and safe environmental management, as one of the most obvious ways to improve profits, will prevail before all of its opponents die!

Water and garbage sanitation in the 19th and 20th centuries were extremely successful in decreasing the risk of mortality and morbidity, so were building and fire codes; boiler testing and inspection; and safety engineering on steamboats, railroads, and cars. A whole field of risk management was developed based on common sense risk analysis, which increased the longevity and generally improved the quality of life for most citizens in the developed world.

5. MODERN RISK ANALYSIS

Conceptual development of risk analysis in the United States and other industrially developed countries (referred to by the United Nations as "North") started from two directions: (1) with the development of nuclear power plants and concerns about their safety (this problem led to the development of the classical probabilistic risk analysis) and (2) with the establishment of the U.S. Environmental Protection Agency (EPA), Occupational Safety and Health Administration (OSHA), National Institute for Occupational Safety and Health (NIOSH), and equivalent governmental agencies in developed countries. These organizations developed in response to a rapid environmental degradation caused by indiscriminate use of pesticides, industrial pollution, and a public outcry, triggered by the publishing of Rachel Carson's book, *The Silent Spring*.

Modern industrial society underwent changes that must be factored into risk analysis and management associated with industrial development. However, one should keep in mind that in the underdeveloped countries (referred to as "South" by the United Nations) one still deals with infectious diseases, malnutrition, and other diseases of preindustrial society, in addition to environmental degradation due to either overpopulation or rapid, unregulated industrial development. In the North, the following applies for modern risks:

1. A shift in the nature of risks from infectious diseases to degenerative diseases
2. New risks such as from nuclear plant accidents, radioactive waste, pesticides and other chemicals releases, oil spills, chemical plant accidents, ozone depletion, acid rain generation, and global warming
3. Increased ability of scientists to measure contamination
4. Increased number of formal risk analysis procedures capable of predicting *a priori* risks
5. Increased role of governments in assessing and managing risks
6. Increased participation of special interest groups in societal risk management (industry, workers, environmentalists, and scientific organizations), which increases the necessity for public information
7. Increased citizen concern and demand for protection

Risk analysis can help manage technology in a more rational way and promote sustainability of desirable conditions for societies and eliminate conditions detrimental to the well-being of humans and ecosystems. However, in each particular case of risk assessment, the assumptions and uncertainties have to be clearly spelled out. All the models used in performing risk analysis have to indicate assumptions and uncertainties in conclusions.

Formal risk analysis can be organized into (Figure 1)

1. Noncancer chemicals risk analysis
2. Carcinogen risk analysis
3. Epidemiological risk analysis (which could include both cancer and noncancer chemicals or other nonchemical hazards, such as accidents, electromagnetic radiation, nutrition, etc.)
4. Probabilistic risk analysis associated with nuclear power plant safety and chemical plant safety
5. *A posteriori* risk analysis, which is applied in actuary science to predict future losses, either from natural phenomena, investments, or technology
6. Nonquantitative risk analysis, or "common sense" risk analysis, which can give only vague patterns of possible risks.

Chapters in Section I of this book will deal with these types of risk analyses and their limitations.

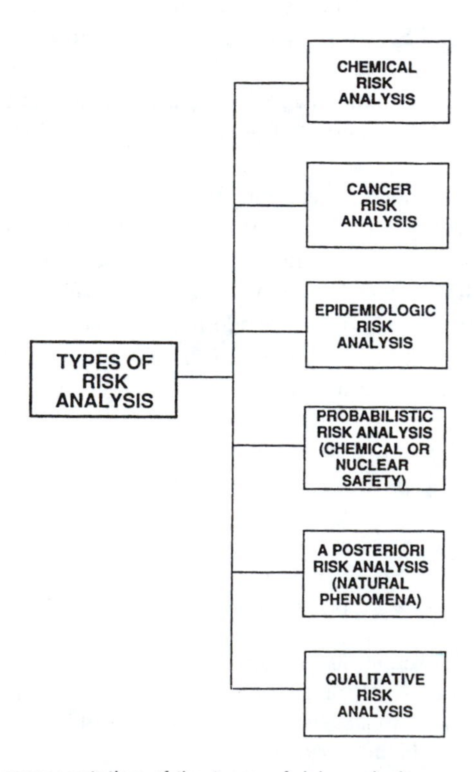

Figure 1 Schematic representation of the types of risk analysis.

For noncarcinogenic chemicals, it is assumed that an adverse effect occurs only if exposure to the chemical exceeds a threshold. Risk analysis is used both for establishing criteria and standards for chemicals in the environmental media and for evaluating risks in particular cases of exposures to toxic chemicals (such as contaminated water, soil, or air in the vicinity of a pollution source or evaluations of Superfund sites). It is assumed that there is no probability of harm if the exposure is below such a threshold. Criteria are based mostly on animal studies, and risk analysis methods deal with extrapolations from animal to human, from short-range to long-range exposures, and with similar scientific issues that require expert judgments and cannot be neatly put into a formula. Uncertainty in the derived criteria and standards is usually one to two orders of magnitude.

Risk analysis for establishing criteria for toxic substances is probabilistic only in the case of carcinogens. The probability of developing cancer or a cancer potency slope as a result of exposure to a particular level or concentration of a chemical is derived by modeling from animal data. Depending on the model applied, a variety of results may be obtained.

Probabilities of developing cancer or other diseases can also be obtained from epidemiological research correlating exposures to toxic substances with the development of cancer or other types of diseases. Epidemiological risk analysis deals with establishing correlations or causal relationships between exposures to chemicals or physical agents and diseases. Most frequently, retrospective, cohort, and mortality studies of occupational groups are used for assessing cancer risk. Standard morbidity or mortality ratios can be regarded as an increase in probability of a health risk with exposure. However, because of the large uncertainty in estimating exposure, the results of the epidemiological studies are combined with studies in animals, in order to confirm the causal relationship between exposures to an agent (carcinogen) and cancer.

Probabilistic risk analysis is applied to industrial process safety and nuclear plant safety (fault-tree and failure-tree analysis). The probability of an adverse outcome (failure of a component or a system) of a series of interconnected events is obtained by evaluating probabilities of failures of individual components. These probabilities are obtained either based on historical data or on assumptions of failure. Once a probability of failure of a chemical process is established, one can apply chemical risk analysis to establish the severity of consequences of a release of a particular toxic substance. This type of probabilistic risk analysis was the beginning of the modern discipline of risk analysis, when atomic energy promised a new way of tapping into an almost limitless energy resource. Until Chernobyl (see Chapter IV.1), the risk analysis numbers were very clear indicators of its safety. ***Chernobyl and the problems with disposing of radioactive waste from nuclear reactors demonstrated again that the technology that initially promised to be a panacea may not be all that was promised. Thus, it may be wise to be cautious when promoting technological fixes.***

Based on historical data, one can establish probabilities of adverse effects from natural phenomena (earthquakes, floods, etc.) or types of human activities (transportation accident rates, number of smokers with lung cancer, acute pesticide poisonings, etc.). This type of risk analysis is used extensively by insurance industries to establish insurance rates. Economic risk analysis also could be regarded as belonging to this category, because adverse economic effects are obtained from known prices of wasted chemicals and other costs associated with pollution (cost of cleanup of hazardous waste sites, legal costs, medical costs to society, etc.).

Some recent phenomena are not yet quantifiable. For example, risks from acid rain are not yet easily amenable to numerical analysis, neither are the risks from global warming. Therefore, one can only establish qualitative risks until more data is obtained to perform quantitative risk analysis. However, one should keep in mind that in the study of such complex phenomena we may never have sufficient data for accurate predictions and therefore we must base our risk management decisions on prudence.

5. LIMITATIONS OF RISK ANALYSIS

Each chapter in this book elaborates on topics in which the definition of risk analysis may vary, depending on the application. The reader will notice the wide diversity of definitions and controversy, which indicate that, unlike the physical sciences, there is much uncertainty associated with any risk analysis (assessment). While risk analysis may be a useful tool to evaluate relatively simple risks (such as health risks from toxic substances in a particular exposure scenario) and to compare them with alternative risks if different human actions were taken (e.g., replacement of particular chemicals or industrial processes and technology), it may be dangerous to apply it to more complex phenomena in order to derive definitive risk ranking or risk management plans. Thus, risk analysis should be applied with caution to the real-life problems, keeping in mind its limitations. The caution may be even more critical in risk-benefit analysis, where calculations of benefits may be even more uncertain and dependent on various underlying assumptions (see Chapter I.7).

A Nobel Laureate economist, Dr. Friedrich Hayek, expressed the dangers of applying science that dealt with "essentially complex phenomena" (such as risk analysis or economics) for sweeping policy decisions (Hayek 1991). His assessment of economics could be translated into a cautionary note on risk analysis:

> *There is as much reason to be apprehensive about long-run dangers created in a much wider field, by the uncritical acceptance of assertions which have the appearance of being scientific. There are definite limits to what we can expect science to achieve. This means that to entrust the science — or to deliberate control according to scientific principles — more than scientific method can achieve may have deplorable effects. This insight will be especially resisted by all who have hoped that our increasing power of prediction and control, generally regarded as the characteristic result of scientific advance, applied to the process of society, would soon enable us to mold it entirely to our liking. Yet the confidence in the unlimited power of science is only too often based on a false belief that the scientific method consists in the application of a ready-made technique, or in imitating the form rather than the substance of scientific procedure, as if one needed only to follow some cooking recipes to solve all social problems.*

The current controversy between industry, government, and environmentalists about the use of risk analysis follows the previous reasoning. Many environmentalists regard risk analysis as a devious tool used by the industry to maintain the status quo ("proving" that something is NOT dangerous) and totally deny its usefulness, while industry and governmental agencies in increasing numbers want to base all decisions on results of risk analysis. While it is true that risk analysis may be used by both sides in an issue to justify their actions, often based on some rather questionable numerical values, risk analysis could be useful to point out the dangers of pursuing one or another course of action. The most important thing is to always make risk assessment transparent to the public with all the assumptions and parameters clearly stated. ***The thought process that goes into evaluating a particular hazard is more important than the application of some sophisticated mathematical technique or formula, which often may be based on erroneous assumptions or models of the world.*** The controversy about the requirement for risk-benefit analysis **before** any law is enacted may lead the legislators into total regulatory deadlock, which may leave the public unprotected, even in obvious cases of environmental abuse.

Risk analysis can, under some circumstances, make general predictions about the outcome of our decisions; sometimes we can only obtain a very rough feeling about the possible outcomes. While in physical sciences the predictions are usually very accurate, in risk analysis our predictions could have a range of several orders of magnitude. If we were to build a bridge based on an assumption of the average value obtained for weight put on this bridge and, in reality, the weight may vary for one to two orders of magnitude, we would soon experience collapse if we did not allow ample space for uncertainty and caution.

The best we can hope in applying risk analysis to the complex problems that we face today (such as environmental exposure to chemicals and radiation, ozone hole, resource depletion, soil loss, global warming, etc.) is to ascertain patterns that could be useful for risk management. The numbers derived by risk analysis are at best crude and often misleading, if the uncertainty associated with them is not clearly spelled out. We could compare the risks of different cleaning methods at the hazardous waste sites or the risks of the use of different types of energy or transportation with more certainty than we could predict the global warming phenomena. Risk analysis can help us predict general economic, ecological, and human health impacts of certain decisions (e.g., either to use public transport modes or personal cars, nuclear energy, coal-powered plants, or conservation) (see Chapter IV.6), which then could help create more livable and equitable sustainable societies.

Compared with the accurate predictions we can get in the physical sciences, this sort of mere pattern prediction is not satisfying. ***However, to pretend that we possess the knowledge and power to shape the processes of society entirely to our liking, knowledge, which in the real world we do NOT possess, is likely to make us do a great deal of harm.*** As Dr. Hayek pointed out:

> *The recognition of the insuperable limits to his knowledge ought indeed to teach the student of society a lesson in humility which should guard him against becoming an*

accomplice in man's fatal striving to control society — a striving which makes him not only a tyrant over his fellows, but may make him destroy a civilization which no brain has designed but which has grown from the FREE efforts of millions of individuals.

REFERENCES

Carson, RL. 1987. *The Silent Spring.* Boston, MA. Houghton Mifflin. 1–448.

Covello VT and Mumpower J. 1985. Risk Analysis and Risk Management: A Historical Perspective, *Risk Analysis* 5(2):103–120.

Hayek FA. 1991. *Economic Freedom.* Cambridge, MA (Oxford, UK). Basil Blackwell Ltd. p. 287.

NAS. 1983. *Risk Assessment in the Federal Government: Managing the Process.* Washington, DC. National Academy Press.

Section I

Theoretical Background of Risk Analysis

CHAPTER I.1

Toxic Chemicals Noncancer Risk Analysis and U.S. Institutional Approaches to Risk Analysis

Vlasta Molak

SUMMARY

Most environmental problems that concern the public deal with exposures to toxic chemicals (by inhaling air, by ingestion of water or food, or by dermal exposure) originating from chemical or other industries, power plants, road vehicles, agriculture, etc. There are two types of noncancer chemical risk analysis uses: (1) to derive criteria and standards for various environmental media and (2) to characterize risks posed by a specific exposure scenario (e.g., at the Superfund site by drinking contaminated water; by consuming contaminated food; by performing some manufacturing operations; by accidental or deliberate spill or release of chemicals, etc.). Usually such exposure scenarios are complex and vary with each individual case, and, thus, methods in risk analysis must be modified to account for all possible exposures in a given situation.

Chemical risk analysis used for criteria development generally does not determine the probability of an adverse effect. Rather, it establishes concentrations of chemicals that could be tolerated by most people in our food, water, or air without experiencing adverse health effects either in short-term or long-term exposures (depending on the type of a derived criterion). These levels (either concentrations of chemicals in environmental media or total intake of a chemical by one or all routes of exposure) are derived by using point estimates of the average consumption of food and drink and body parameters such as weight, skin surface, metabolic rate, etc. Risk analysis is then applied to derive "criteria" for particular pollutants,

1-56670-130-9/97/$0.00+$.50

which are then modified by risk management considerations to derive standards. There are numerous criteria and standards established for various chemicals by the U.S. Environmental Protection Agency (EPA), the U.S. Food and Drug Administration (FDA), the National Institute for Occupational Safety and Health (NIOSH), and the Occupational Safety and Health Administration (OSHA). Since many of them were established before formal risk analysis techniques became available, they are undergoing revision, based on better risk analysis methods. For a particular pollution situation, one can measure or estimate exposures to a contaminant and compare them to the previously established criteria and/or standards. The likelihood of harm increases if the exposure levels exceed the derived "safe" levels. The exposure assessments could follow a deterministic model by assuming average parameter values (air, water, food consumptions, dermal intake, etc.) or could follow the Monte Carlo method, which uses real-world distribution data on various exposures, thus potentially giving more accurate and informative estimates of risk.

Key Words: toxic, chemicals, hazard, exposure, standard, criteria, dose response, acute, chronic, pollution

1. INTRODUCTION

Chemical risk analysis is generally divided into four parts (NAS 1983):

1. Hazard identification — identifying potentially toxic chemicals.
2. Dose–response relationships — determining toxic effects depending on amounts ingested, inhaled, or otherwise entering the human organism. These are usually determined from animal studies. Different "end points" of toxicity are observed, depending on the target organ of a chemical. Severity of a particular effect is a function of dose.
3. Exposure assessment — determining the fate of the chemical in the environment and its consumption by humans. Ideally, by performing environmental fate and transport of chemicals, and by evaluating food intakes, inhalation, and possible dermal contacts, one can asses total quantities of toxic chemicals in an exposed individual or population, which may cause adverse health effects. In criteria derivation, one uses either worse case exposure scenario or most probable exposure scenario and point values for various human parameters. Monte Carlo modeling uses real-world distribution data for those parameters.
4. Risk characterization consists of evaluating and combining data in Items 2 and 3. For establishing criteria and standards, assumptions are made about "average exposures," and the criteria are set at the concentration at which it is believed that no harm would occur. For example, reference dose (RfD) and health advisories (for 1-day, 10-day, and subchronic exposures) are derived for many chemicals with the use of safety (uncertainty) factors to protect most individuals. If an actual exposure to environmental pollutant (or pollutants) exceeds limits set by the criteria, efforts should be made to decrease the concentrations of pollutant. The magnitude of risk can be estimated by comparing the particular exposure to derived criteria or reference doses.

2. TOXICOLOGICAL BASES OF TOXIC SUBSTANCES RISK ANALYSIS

Over 110,000 chemicals are used in U.S. commerce. The Registry of Toxic Effects of Chemical Substance (RTECS) database, maintained by NIOSH, contains updated information on the toxicity of those chemicals (RTECS 1995). Since the number of chemicals potentially appearing in the environment is large, and the toxicological effects are very complex and differ depending on the chemical and conditions of exposure, it is sometimes difficult to determine how toxic is toxic. Risk analysis helps determine which chemicals are dangerous and under what circumstances. It can also help establish relative risks from various chemicals (ranking risks). If, for example, in a particular industrial setting the derived health risk from pollutant **A** is higher than from pollutant **B**, that may indicate that the action should first be taken to decrease the pollution by **A**. In order to be able to use information on such a large number of substances, the toxicologists have developed classification of chemicals by their acute, subacute, and chronic toxicity (Cassarett and Doull 1986).

2.1 Acute Toxicity

Acute toxicity is the most obvious and easiest to measure and is generally defined by the LD_{50} (lethal dose 50%). This is the dose expressed in milligrams per kilogram of body weight, which causes death within 24 hours in 50% of exposed individuals after a single treatment, either orally or dermally. LD_{50} is usually derived from animal studies (mice and rats). Measure of acute toxicity for gases is LC_{50} (lethal concentration of chemical in the air that causes death in 50% of animals if inhaled for a specified duration of time, usually 4 hours). Based on that definition, chemicals are divided into toxicity ratings of practically nontoxic, moderately toxic, very toxic, extremely toxic, and supertoxic (Table 1).

Table 1 Toxicity Ratings of Chemicals

Toxicity rating	Probable lethal oral dose for humans	Units/kg body weight	Example	
			Chemicals	LD_{50} (animals)
Practically nontoxic	>15	g/kg		
Slightly toxic	5–15	g/kg	Ethanol	10 g/kg
Moderately toxic	0.5–5	g/kg	Sodium chloride	4 g/kg
Very toxic	50–500	mg/kg	Phenobarbital	150 mg/kg
Extremely toxic	5–50	mg/kg	Picrotoxin	5 mg/kg
Supertoxic	<5	mg/kg	Dioxin	0.001 mg/kg

In the 16th century, the Swiss physician and alchemist Philippus Aureolus Paracelsus stated that "the dose makes the poison"; chemicals could be very useful at small doses and poisonous at high doses. For example, selenium, oxygen, and iron are nontoxic or not even useful at certain doses, but can be lethal at high doses. Generally, we are concerned with chemicals which are very toxic, extremely toxic, or supertoxic. Unless the chemical is a carcinogen or has some other chronic health or environmental effects (such as polychlorinated biphenyls [PCBs] or heavy metals), there is little concern for those chemicals in moderately toxic or less toxic groups.

2.2 Subchronic and Chronic Toxicity

In some instances, chemical substances can have very low acute toxicity, but can cause cancer (e.g., PCBs), birth defects (thalidomide), or ecological effects (DDT) (Cassarett and Doull 1986). Long-term exposures to relatively low concentrations of these chemicals can cause specific organ damage or cancer. Therefore, chemicals are also evaluated for their subchronic and chronic systemic toxicity, carcinogenicity potential, or reproductive and developmental toxicity. Data are usually obtained from animal studies and sometimes from epidemiological studies in humans.

2.3 Cancer Risk Assessment Models and Cancer Potency

Various cancer models can serve to determine cancer potency slope for a particular chemical (Johannsen 1990, Cassarett and Doull 1986). While for health effects other than cancer a threshold dose is assumed, for cancer it is assumed that any exposure may potentially cause cancer. However, the probability of getting cancer at low exposure concentrations may be so low as to be of no practical concern. The U.S. EPA defines negligible risk for cancer as that smaller than 1:1,000,000 (U.S. EPA 1980), and for OSHA a risk of less than 1:1000 is "acceptable" (OSHA 1989). This is a policy decision and has nothing to do with the science of risk analysis. The U.S. EPA has used a multistage linear model to establish potency slopes for approximately 140 cancer-causing chemicals, which can serve to establish the risks of pollutants in the air, water, and food (U.S. EPA 1988a). Since most of these potency slopes are derived from animal data, there is an uncertainty associated with their numerical values. An additional uncertainty is posed by high- to low-dose extrapolation, because animal studies are, for practical reasons, performed at relatively high doses in order to be able to observe effects.

3. DOSE–RESPONSE RELATIONSHIPS

For each chemical there are dose–response relationships for different types of toxicological effects (Figure 1). With an increasing dose, the percent of affected individuals with the same type of health effect increases. For noncarcinogens, a threshold dose is assumed which defines a no-observable-effect level (NOEL). It is assumed that exposure to a chemical that results in a dose smaller than a threshold is handled by the organism, and no adverse health effects occur. For carcinogens, however, it is assumed that no threshold exists and that even small number of molecules of carcinogen could potentially cause alterations in DNA, resulting in cancer (Upton 1988). The same curve could also be used for a dose–effect relationship, in which the severity of the effect in an individual increases with dose (Cassarett and Doull 1986, OSHA 1989).

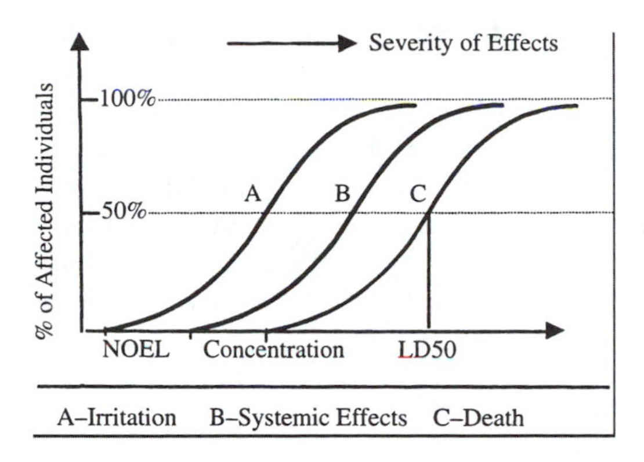

Figure 1 Dose–response relationships for different types of toxicological effects.

4. EXPOSURE ASSESSMENTS

Exposures are determined by measuring or estimating the concentration of the chemical in a particular environment and then establishing average amounts of a chemical consumed by an exposed person or population by ingestion of food and water, inhalation, or dermal contact during the studied time period.

In deriving criteria for a particular chemical, an average consumption of food and water is assumed, and a criterion is derived so that under normal conditions it does not result in a dose that would have adverse effects. For example, an average human weighs 70 kg, drinks 2 l of water, inhales 20 m^3 air per day, etc. (U.S. EPA 1989b). Based on an exposure assessment in a particular situation, one can derive total dose to an individual and compare it with existing criteria. Therefore, for chemicals with existing criteria, one only has to perform exposure assessments to establish possible adverse effects of a chemical by comparing it with the criterion.

Without exposure to a particular pollutant, there is no risk. Thus, the most important task is to establish or estimate true potential exposures and then estimate risk either for a maximally exposed individual, an average exposure, or use the Monte Carlo method to find distribution functions for various parameters of exposures. Frequently, such distributions are based on food surveys, census data, physiological data, etc. U.S. EPA Guidelines for Exposure Assessment (U.S. EPA 1986b, 1992b) are useful for deriving real-life exposures. If a company has reliable monitoring data on their pollutants, it should be relatively simple to estimate exposures to potentially exposed individuals. For performing proper exposure assessment, one needs to either measure the environmental concentrations and/or be able to realistically model the chemical fate and transport in the environment (bioaccumulation, degradation in the environment, chemical transformation, etc.). For each particular chemical or situation, different sets of parameters may apply. For better exposure assessment, it is also useful to know environmental pharmacokinetics. Substances that easily degrade and do not bioaccumulate are probably of less consequence than persistent compounds such as DDT, dioxins, and heavy metals.

5. EXAMPLES OF CHEMICAL RISK ANALYSIS

Most of the chemical risk analysis in the United States was developed by the U.S. EPA. NIOSH, OSHA, and the FDA have subsequently also started to use risk analysis for their evaluation of toxic substances (DHHS Committee to Coordinate Environmental and Related Programs 1985). The U.S. EPA has developed methods for dealing with toxic substances that contaminate the environment in general, and NIOSH, OSHA, and the FDA deal with occupational contaminants and food contaminants, respectively.

5.1 U.S. EPA Risk Analysis

The U.S. EPA has a long tradition of dealing with environmental pollutants and has developed criteria and standards for drinking water, ambient water, air, total intake reference dose (RfD), reportable quantities (RQs), and levels of concern (LOC) for many environmental pollutants from various lists of toxic chemicals. These lists, sometimes overlapping, contain over 1200 chemicals and/or chemical categories: Resource Conservation and Recovery Act (RCRA); Comprehensive Environmental Response, Compensation and Liability Act (CERCLA); and Superfund Amendments and Reauthorization Act (SARA), Title III (302 and 313) (U.S. EPA 1992a). Based on risk analysis for those chemicals, several types of criteria and standards for various media were derived using U.S. EPA-developed guidelines for carcinogen risk assessment, mutagenicity risk assessment, health risk assessment of chemical mixtures, suspect developmental toxicants, estimating exposures, and systemic toxicants risk assessment (U.S. EPA 1986a).

5.1.1 Criteria and Standard Derivation

Initially, risk analysis for chemicals at the U.S. EPA was developed in order to derive criteria and standards for chemicals that were polluting waters in the United States (U.S. EPA 1980). Gradually, risk analysis methods were expanded to all environmental media (U.S. EPA 1986a). Most of the criteria values are derived from extrapolation from animal studies using assumptions about inhalation, water consumption, food consumption, and weight of the average human. The details for criteria derivations and corresponding assumptions are available from the U.S. EPA (U.S. EPA 1986a,b). Generally, data are obtained from animal studies in which either NOAEL or lowest-observable adverse-effect level (LOAEL) is measured in acute, subchronic, or chronic studies. In order to extrapolate animal data to humans, an appropriate uncertainty factor (usually a multiple of 10) is applied in order to protect human populations and add an extra measure of caution. Criteria are derived using very simple arithmetic from experimental dose–response values and appropriate assumptions about weights and consumption patterns. When multiple animal studies exist, expert judgment is used to determine the most appropriate study. Usually, the most conservative studies and assumptions are used in order to provide a safety margin for error. In addition, since we are mostly exposed to multiple chemicals, which may have synergistic effects, it may be prudent to use conservative (protective) values with individual chemicals. Some of the criteria derived by the U.S. EPA are

1. **Ambient water quality criteria (AWQC)** were derived in 1980 for priority pollutants (U.S. EPA 1980). In derivation of these criteria, toxicity in fish and other aquatic organisms, as well as bioaccumulation, was considered.
2. **Health advisories (HA)** for drinking water indicate a "safe" concentration of particular chemicals in drinking water for 1-day, 10-day, and subchronic consumption. Usually, these are derived from short-term drinking water studies in rats and mice and application of a proper uncertainty factor (U.S. EPA 1988b).
3. **RfD (reference dose),** previously known as daily acceptable intake (ADI), is defined as the total daily dose of a chemical (in milligrams per kilogram of body weight) that would be unlikely to cause adverse health effects even after a lifetime exposure (Barnes and Dourson 1988). Or an RfD for a chemical is the estimation (with uncertainty spanning perhaps one order of magnitude) of a daily or continuous exposure to the human population (including sensitive subgroups) which is likely to be without an appreciable health risk. RfDs are established from all available toxicological data for several hundred chemicals, particularly those associated with Toxic Release Inventories (TRI). The RfDs and risk assessment methodologies used for their derivation are available from the on-line Integrated Risk Information System (IRIS 1995). The general formula for RfD derivation is

$$\mathrm{RfD} = \frac{\mathrm{LOAEL\ or\ NOAEL}}{\mathrm{UF} \times \mathrm{MF}}$$

   where UF is the "uncertainty factor" to account for the type of study used to determine NOAEL or LOAEL and MF is the modification factor (1 to 10), which depends on the quality of the toxicological database for a particular chemical. The establishment of MF is often rather subjective.
4. **The LOC (level of concern)** is defined as concentration of a toxic chemical in air that the general public could endure for up to 1 hour without suffering from irreversible health effects (U.S. EPA, FEMA, and DOT 1987). They were derived from IDLH (immediately dangerous to health and safety) values by dividing them with a factor of 10 or from LD_{50} by dividing them by 100. Since IDLH are derived using qualitative risk analysis (based mostly on expert judgment) for a healthy worker, there is a great uncertainly about their accuracy and protectiveness. Thus, the U.S. EPA used an additional uncertainty factor of ten.
5. **RQs (reportable quantities)** are derived for chemical spill reporting. The value of RQ is 1, 100, 500, 1000, and 5000 lb, and it depends on the acute toxicity, carcinogenicity, fate, and transport in the environment and reactivity (U.S. EPA 1987). The arithmetic is based on simple assumptions and toxicity of a chemical. These values are used for SARA, Title III and CERCLA reporting of chemical spills.
6. **Cancer potency (q*) slopes** are derived from animal studies using linear multistage analysis (U.S. EPA 1986c). The cancer potency slope is an indication of magnitude of a cancer threat; however, there is a great uncertainty about the accuracy of this number, because of various assumptions made in its derivation (U.S. EPA 1986c, 1988a). Chapter I.2 will address the issue in more detail.

7. **Reference concentrations (RfC)** for chronic inhalation from air were developed for some chemicals on Integrated Risk Information System (IRIS) (U.S. EPA 1989a). Although for many chemicals air criteria are established based on risk analysis, only six air standards exist (CO, SO_2, O_3, NO_x, lead, and particulates) (Cassarett and Doull 1986).

Standards for chemicals in air, water, or soil are derived with the consideration of criteria and other factors such as cost, policy issues, perception, etc. Generally, cost-benefit analysis is performed and alternative risks are considered. For example, although chlorination may cause cancer in a small number of individuals, chlorination removes the known risk of infectious diseases. An outbreak of cholera in Peru led to the death of more than 300 people because the officials decided that they did not want to expose the population to chlorine, which may cause cancer (Anderson 1991). However, in order to prevent a hypothetical risk of death of 1:1,000,000, the officials have introduced the far greater risk of cholera, a disease potentially deadly, that resulted in an actual death rate of 1:1000. This example illustrates that it is necessary to use common sense and comparative risk analysis when making decisions affecting a large number of people, rather than just mechanically apply risk analysis technique for a single chemical regardless of other possible risks.

5.1.2 Other Risk Analyses

The U.S. EPA derived risk analysis methods for a number of particular cases dealing with the adverse effects of chemicals on the environment. One of the most controversial and complicated analysis is the Risk Analysis for Superfund (U.S. EPA 1989b), which has been involved in numerous regulatory and societal gridlocks. The U.S. EPA manual essentially serves as a cookbook of procedures to follow in performing a risk assessment and feasibility study in a particular hazardous waste site. A student in scientific controversy may like to study this case.

With the passage of SARA, Title III law (or Community Right-to-Know Law), a method of hazard analysis was developed jointly by three agencies to assess the probability of accidental release of toxic chemicals in the environment of the U.S. (EPA/DOT/DOE 1987).

5.2 Risk Analysis by Other Institutions (NIOSH, OSHA, FDA, ATSDR)

For regulating chemicals in the workplace, OSHA uses permissible exposure limits (PELs) that are generally derived from threshold limit values (TLVs) developed by the Association of Governmental Industrial Hygienists. Although in 1989 (OSHA 1989) OSHA established PELs for over 600 substances, they were thrown out of court, and only old, less protective values are now in effect. NIOSH has similarly developed recommended exposure limits (RELs) for the same substances (NIOSH 1990). There was no formal risk assessment initially applied in the derivation of either TLVs (and PELs) or RELs, and the numbers were derived based on expert committees (qualitative and semiquantitative risk analysis). Frequently, such

TLVs were a compromise between technology and human health protection, not necessarily always protecting human health. The last several years have seen the development of epidemiologic risk assessment at NIOSH and cancer risk assessment at OSHA, similar to that at the U.S. EPA (Stayner 1992, OSHA 1989). The Agency for Toxic Substances and Disease Registry (ATSDR) has published *Toxicological Profiles*, which incorporates some of the EPA methods in evaluating risks to humans from exposures to toxic chemicals.

6. CONCLUSION

Risk analysis methods are always undergoing revisions, and, thus, appropriate organizations should be contacted for the latest applicable methodology for dealing with risks in a particular exposure scenario for a chemical. Hundreds of criteria documents, published or unpublished, are available from the U.S. EPA, NIOSH, ATSDR, and the FDA, containing risk analysis methods for a particular case. The information centers in those agencies can direct the reader to the most updated version of a document that contains method descriptions.

REFERENCES

Anderson C. 1991. Cholera Epidemic Traced to Risk Miscalculation. Peru Outbreak of Cholera as a Consequence of Faulty Risk Miscalculation. *Nature* 354(6351):255.

Barnes D.G., Dourson M. 1988. Reference Dose (RfD): Description and Use in Health Risk Assessments. *Regulatory Toxicology and Pharmacology* 8:471–486.

Doull, J. et al. 1980. *Casarett and Doull's Toxicology.* New York: MacMillan Publishing Company.

DHHS Committee to Coordinate Environmental and Related Programs. 1985. Risk Assessment and Risk Management of Toxic Substances. A Report to the Secretary. Department of Health and Human Services. April 1985.

IRIS. 1995. On-Line Integrated Risk Information System. User Support tel. 513/569-7254.

Johannsen F.R. 1990. Risk Assessment of Carcinogenic and Non-Carcinogenic Chemicals. *Critical Reviews in Toxicology* 20(5):341–366.

NAS. 1983. *Risk Assessment in the Federal Government: Managing the Process.* Washington, DC: National Academy Press.

NIOSH. 1990. NIOSH Pocket Guide to Chemical Hazards. U.S. Department of Health and Human Services.

OSHA. 1989. Air Contaminants; Final Rule (Codified at 29 CFR 1910). *Federal Register* 54:2332–2983.

RTECS. 1995. Registry of Toxic Effects of Chemical Substances-On-Line Toxicology Information Program. National Library of Medicine. tel. 301/496-113

Stayner L. 1992. Methodological Issues in Using Epidemiologic Studies for Quantitative Risk Assessment. Proceedings from Conference of Chemical Risk Assessment in the DOD: Science, Policy and Practice. Ed HJ Clewell, ACGIH, Cincinnati, Ohio. p. 43–51.

Upton A.C. 1988. Are There Thresholds for Carcinogenesis? The Thorny Problem of Low Level Exposure. *Ann. N.Y. Acad. Sci.* 534:863–884.

U.S. EPA. 1980. Water Quality Criteria Documents Availability. Appendix C. Guidelines and Methodology Used in Derivation of the Health Effect Assessment Chapter of the Consent Degree Water Criteria Document. *Federal Register* 45(231):79347–79379.

U.S. EPA. 1986a. The Risk Assessment Guidelines of 1986. EPA/600/8-87/045. August 1987.

U.S. EPA. 1986b. Guidelines for Carcinogen Risk Assessment. *Federal Register* 51:33992.

U.S. EPA. 1987. Health and Environmental Effects Profile for Hexachlorocyclohexanes. Environmental Criteria and Assessment Office. NTIS PB89126585XSP.

U.S. EPA, FEMA, DOT. 1987. *Technical Guidance for Hazard Analysis.* Washington DC: Government Printing Office.

U.S. EPA. 1988a. Evaluation of Potential Carcinogenicity of Acrylonitrile. Office of Health and Environmental Assessment. NTIS PB93181631XSP.

U.S. EPA. 1988b. Development of Maximum Contaminant Levels Under the Safe Drinking Water. U.S. EPA. Office for Cooperative Management. NTIS PB89225619XSP.

U.S. EPA. 1989a. Interim Methods for Development of Inhalation Reference Doses. EPA/600-8-88/066F. August 1989. Research Triangle Park, NC.

U.S. EPA. 1989b. Risk Assessment Guidelines for Superfund. Volume I — Human Health Evaluation Manual (Part A). Office of Emergency and Remedial Response. Washington DC.

U.S. EPA. 1992a. List of Lists. Consolidated List of Chemicals Subject to Reporting Under the Emergency Planning and Community Right-to-Know Act. NTIS PB92500792XSP.

U.S. EPA. 1992b. Guidelines for Exposures Assessment. *Federal Register* 57(104):22888–22937.

QUESTIONS

1. What is the general purpose of chemical risk analysis?
2. How does the U.S. EPA derive criteria for chemicals?
3. What standards are regulated by OSHA?
4. What is exposure assessment?
5. What is RfD?
6. What are uncertainty factors?
7. How does one calculate criteria?
8. RfD for chemical XYZ is 1 mg/kg/day. One-day health advisory (HA) for drinking water is 10 mg/l. Ten-day HA is 2 mg/l. It was found that neighboring groundwater and soil is contaminated by XYZ. The concentration measured in groundwater is 1 mg/l, and the concentration measured in soil around the community is 1 mg/kg/soil. What would be your recommendation about handling the possible public health problem based on this data?
9. The concentration of chemical Z in the Majestic River is given as 5 mg/l. Bioaccumulation factor for fish is 20. If the RfD for chemical Z is 2 mg/d, what would be your recommendation regarding the consumption of fish?

CHAPTER I.2

Epidemiology and Cancer Risk Assessment*

Herman J. Gibb

SUMMARY

A discussion of some of the leading documents on the use of human data in carcinogen risk assessment is provided. Types of epidemiologic studies and how such studies are evaluated are described. Also described are criteria for evaluation of the weight of human evidence. The variability of data from epidemiologic studies makes guidance with regard to dose–response assessment difficult. Risk assessors are cautioned to describe uncertainties and assumptions in the dose–response assessment. Future directions for human data and risk assessment include molecular epidemiology, examination of the variation in susceptibility to toxic substances, and increased international collaboration on epidemiologic research.

Key Words: epidemiology, biomarkers, dose–response assessment, susceptibility, risk, carcinogens, human data, carcinogenic risk assessment

1. BACKGROUND

In 1976, the U.S. Environmental Protection Agency (EPA) published the first guidelines for carcinogen risk assessment issued by a federal agency (U.S. EPA 1976). The U.S. EPA was a new agency at the time, having been created by an executive order in December 1970. The need for carcinogen risk assessment

* The views expressed in this chapter are those of the author and do not necessarily reflect the views of the U.S. Environmental Protection Agency.

guidelines had become apparent during deliberations on the restriction of certain pesticides.

The U.S. EPA's first guidelines on carcinogen risk assessment were quite brief in comparison to the guidance on the subject that was to follow, but it was an important first step. They were the first to recommend the separation of risk assessment from risk management. They were also the first to recommend the separation of qualitative from the quantitative assessment. On the subject of human data, the 1976 guidelines recognized the importance of human data in the identification of carcinogens and in deriving dose–response relationships, but provided little detail of how such data are used in an assessment.

In 1983, the National Research Council (NRC) published *Risk Assessment in the Federal Government: Managing the Process*, better known as the "Red Book" (NRC 1983). In that publication, the NRC said that risk assessment has four major elements, now commonly known as the risk assessment paradigm. The paradigm is central to an understanding of how both human and animal data are used in carcinogen risk assessment. The four elements are as follows:

- Hazard identification or "Does the agent cause the adverse effect?"
- Dose–response assessment or "What is the relationship between dose and incidence in humans?"
- Exposure assessment or "What exposures are currently experienced or anticipated under different conditions?"
- Risk characterization or "What is the estimated incidence of the adverse effect in a given population?"

The NRC noted that well-conducted epidemiologic studies that show a positive association between an agent and a disease are accepted as the most convincing evidence that a hazard exists. The NRC also indicated that human data are usually not available or are difficult to interpret and that both hazard identification and dose–response assessment are frequently forced to rely on animal data.

In 1985, the Office of Science and Technology Policy (OSTP) published a review of the science and principles of chemical carcinogens (OSTP 1985). They reviewed the strengths and limitations of epidemiology, the determination of causality from an epidemiologic study, types of epidemiologic studies, biochemical epidemiology, and implications of negative studies. It was the most complete discussion of the use of epidemiology in carcinogen risk assessment written at that time.

In 1986, the U.S. EPA issued new guidelines on carcinogen risk assessment (U.S. EPA 1986a). Similar to the OSTP, they provided details as to the kinds of epidemiologic data available and their evaluation. The 1986 guidelines also introduced a classification scheme with which to evaluate human and animal evidence of carcinogenicity for a suspected carcinogenic agent. They recommended that risk assessors classify both human and animal evidence on a particular substance as demonstrating "sufficient," "limited," "inadequate," "no data," or "no evidence." Risk assessors should then make an overall ranking of the potential of the substance to be a human carcinogen using the human and animal evaluations and any additional pertinent information such as short-term test findings and structure–activity relationship data. The International Agency for Research on Cancer (IARC) had used a similar but somewhat different classification scheme in its monograph series (IARC 1982).

In 1996, the EPA issued a review draft of "Proposed and Interim Guidelines for Carcinogen Risk Assessment" (U.S. EPA 1996). A major departure from the 1986 Guidelines is that the proposed guidelines recommend weighing the human, animal, and ancillary evidence for carcinogenicity in one step.

The discussion of human data in the 1995 draft reflected the results of an EPA workshop on human data in carcinogen risk assessment (U.S. EPA 1989). The 1995 proposed guidelines expand the discussion on human data from that in the 1986 guidelines. They describe the types of human studies, the adequacy of studies, and the determination of a causal relationship from human data. They provide recent references on biochemical epidemiology. A discussion of the combining of statistical evidence across studies is new in the proposed guidelines.

2. TYPES OF STUDIES

Epidemiologic studies are often described as either analytical or descriptive. Cohort and case-control studies are the two primary analytical types of studies. Correlation studies are generally considered descriptive.

In cohort studies, the epidemiologist studies the difference in disease occurrence between exposed persons and nonexposed persons. This may be done either prospectively or, using historical records, retrospectively. Because cancer usually involves a long latency period from exposure to disease, most of the cohort studies on cancer have been done retrospectively. In cancer case-control studies, the epidemiologist compares persons with the disease and persons without the disease for differences in exposure and other factors. Cancer correlation studies examine differences in cancer rates among groups in relation to factors, such as chemical exposure, to determine differences in disease occurrence.

The primary difference between the "analytical" and "descriptive" types of studies are that analytical studies consider individual exposure, while descriptive studies consider disease occurrence within a group. It is unknown if those who develop the disease in the group were the ones exposed. Individual exposure data are seemingly more attractive than group data, but group data can be very representative of the individuals within the group. Furthermore, the quality of individual exposure data is quite different across analytical studies. The assessor must employ expert judgment when evaluating the contribution of the different kinds of studies in the overall assessment.

Meta-analysis, which has received growing attention among epidemiologists and risk assessors over the last 10 years, is the comparing and synthesizing of studies dealing with similar health effects and risk factors. It can enhance the value of epidemiologic data in debates about environmental health risks. Meta-analysis may be particularly useful to formally examine sources of heterogeneity, to clarify the relationship between environmental exposures and health effects, and to generate information beyond that provided by individual studies or a narrative review. It may not be useful when the relationship between exposure and disease is obvious, when there are only a few studies of the key health outcomes, or when there is substantial confounding or other biases that cannot be adjusted for in the analysis (Blair et al. 1995).

The use of biomarkers in epidemiologic research, an approach also known as molecular epidemiology, has become more popular in the last 10 to 15 years. Biomarkers are cellular, biochemical, or molecular alterations measured in biologic media such as human tissues, cells, or fluids (Hulka 1990).

Biomarkers are classified by several different schemes, most of which are variations of the classification by Perera and Weinstein (1982). Included are biomarkers of (1) susceptibility (interindividual variation in response to a carcinogen), (2) internal dose (metabolism and tissue levels of carcinogens), (3) biologically effective dose (levels of covalent adducts formed between carcinogens and cellular macromolecules), and (4) early cellular response to carcinogen exposure (biological or biochemical changes in target cells or tissues that result from the action of the chemical and are thought to be a step in the pathologic process toward disease). Because of the long latency between exposure and cancer, the use of biomarkers in cancer epidemiology is somewhat problematic. The primary use of biomarkers in cancer epidemiology has been for screening (Hulka 1990), such as with sputum cytology (Tockman 1986) or micronuclei (Vine 1990).

Besides epidemiologic studies, human data include case reports. Case reports describe an effect in an individual or group without comparison to controls. An example would be a physician reporting to a medical journal that he/she has treated a case of a rare disease and that the case had been exposed to a particular substance. These reports may be very selective and, generally, are of limited use for hazard assessment. Such reports, nevertheless, are valuable because they raise the interest of the epidemiologic research community to do further study. Furthermore, case reports have been credited with identifying cancer hazards when there were unique features of the cases (e.g., vinyl chloride exposure and angiosarcoma of the liver, a very rare form of cancer).

3. EVALUATION OF EPIDEMIOLOGIC STUDIES

Because of ethical considerations, cancer epidemiology studies are observational, as opposed to the experimental conditions employed in animal studies. The necessity for observational study makes it more difficult to observe a carcinogenic response in humans than in animals and creates questions of interpretation not encountered with animal data. The risk assessor must conduct a critical analysis.

- For those epidemiologic studies that demonstrate no evidence of increased cancer risk, could the study have detected an increased risk of cancer? Were there enough people in a cohort study or enough cases and controls in a case-control study to be able to detect an effect? Was there sufficient exposure to the suspected carcinogen to have observed an effect? Were enough of the cohort study subjects followed

long enough to have detected an increased cancer risk, given that cancer has a relatively long latency period between exposure and effect?
- For those studies which demonstrate an effect, could the result be the effect of confounding? Confounding occurs when an increased risk of disease is attributed to a particular study variable when the true cause is another variable. For example, an increased risk of lung cancer observed among a group of industrial workers might be the result of an excess prevalence of smoking among the workers rather than the result of industrial exposure.
- What sorts of bias might be present? Differences in the way one obtains information on cases and controls or on exposed and nonexposed can create a bias in the results. Differences in the selection of the cases and controls or of the exposed and nonexposed can also prejudice the results. Many kinds of bias in epidemiologic studies have been identified, and it is not within the scope of the current discussion to describe these in detail. The reader is referred to any number of epidemiologic texts for further information (Kahn and Sempos 1989, Kelsey et al. 1986, Lilienfeld 1976, Rothman 1986, Checkoway 1989).
- Was there appropriate statistical evaluation of the data? Was there a description of the statistical methods provided? Did the authors articulate the assumptions and rationale for the use of such methods? Did the authors take appropriate steps to address confounding in the statistical analysis?

4. CRITERIA FOR CAUSALITY

No discussion is in the U.S. EPA 1986 guidelines on how the risk assessor evaluates the strength of the human evidence (U.S. EPA 1986a). The EPA workshop on the use of human data in risk assessment (U.S. EPA 1989) thought it important to identify criteria for causality. Bradford Hill had previously identified criteria for causality in the examination of cigarette smoking and lung cancer (Rothman 1986). The workshop adopted these criteria with some modification, and the criteria are in the U.S. EPA's 1996 proposed guidelines for carcinogen risk assessment. None of the criteria are considered conclusive by themselves, and the only criterion that is essential is the temporal relationship. The criteria are as follows:

- Temporal relationship: The development of cancer requires a latency period. Thus, the disease has to occur within a biologically reasonable time after initial exposure. This feature *must* be present if causality is to be considered.
- Consistency: The same result occurs in multiple studies.
- Magnitude of the association: A causal association is more credible when the risk is large and precise.
- Biological gradient: The risk is found to increase as the exposure increases.
- Specificity of the association: The likelihood of a causal interpretation is increased if a particular form of cancer is related to exposure in several studies (e.g., asbestos exposure and mesothelioma, cigarette smoking and lung cancer).
- Biological plausibility: The association makes biological sense with respect to metabolism, pharmacokinetics, etc.
- Coherence: The cause and effect are in logical agreement with everything known about the agent, exposure to the agent, and the disease.

5. DOSE–RESPONSE ASSESSMENT

One of the four major elements of the risk assessment paradigm is dose–response assessment. Dose–response assessment is the relationship of risk to dose and estimates of risk below the range of observation (ILSI 1995). With epidemiologic data, the relationship is termed exposure response since humans are not actually dosed.

The largest hurdle that the epidemiologist faces is usually the lack of information on exposure. The number of human studies that have sufficient exposure information with which to do exposure–response analysis is quite limited.

Even where exposure data exists, it is difficult to provide guidance to the risk assessor on exactly how to do a human exposure–response assessment. There is considerable variation in human studies with respect to design, quality of data available to the researcher, and varying presentations of results. Animal cancer bioassay studies are similar in their design and conduct so that guidance with respect to dose–response assessment with animal data is more straightforward.

Regardless of the type of human data available for exposure–response assessment, the risk assessor should always describe uncertainties in the data and sensitivity of the exposure assessment results to the potential variability. The assessor should also discuss the assumptions of the mathematical procedure(s) used to estimate the exposure response and other mathematical procedures that could reasonably be used.

The reader can best learn how human data is used in exposure–response assessment by examples. Suggested examples are the U.S. EPA's 1988 "Special Report on Ingested Arsenic" (1988), EPA's 1984 "Health Assessment Document on Chromium" (1984a), EPA's 1984 "Carcinogenicity Assessment Document on Coke Oven Emissions" (1984b), and EPA's 1986 "Health Assessment Document on Nickel" (1986b).

6. THE FUTURE OF HUMAN DATA IN RISK ASSESSMENT

While there is uncertainty associated with the results of every epidemiologic study, this uncertainty seems trivial compared to the extrapolation of risks from animals to humans (Smith 1995). As the National Research Council (1994) indicated, "laboratory animals are not human beings, and this obvious fact is one clear disadvantage of animal studies."

There is a current desire to expand the use of epidemiologic studies in risk assessment. It has been the subject of recent books (Gordis 1988, Graham 1995). The recent National Research Council report on risk assessment (NRC 1994) identified several areas for further epidemiologic research. A leading school of public health recently created a Center for Epidemiology and Policy to build a program of interdisciplinary research and training focused on the application of epidemiologic findings to public policy questions (Johns Hopkins School of Public Health 1995). The center is devoting its 1996 seminar series to the use of epidemiologic findings in regulatory actions, corporate decision making, legal decisions, media venues, and citizen action, among other policy arenas.

Guidelines on risk assessment have suggested that epidemiologic studies can detect only comparatively large increases in the relative risk of cancer (U.S. EPA 1986a, 1995). While this is true for the more traditional epidemiologic studies that use cancer morbidity and mortality statistics, molecular approaches should greatly increase the sensitivity of epidemiologic studies to detect increased risks of cancer. This will be a major direction in risk assessment over the next 5 to 10 years.

A second major focus in the coming years for epidemiology and carcinogen risk assessment will be the development of data on differences in individual susceptibility to carcinogens. As the NRC (1994) noted: "Human beings vary substantially in their inherent susceptibility to carcinogens, both in general and in response to any specific stimulus or biologic mechanism. No point estimate of the carcinogenic potency of a substance will apply to all individuals in the population." The NRC recommended that federal agencies undertake research to explore and elucidate the relationships between variability in each measurable factor (e.g., DNA adduct formation) and variability in susceptibility to carcinogenesis. They also recommended that guidance be provided on how to construct appropriate samples of the population for epidemiologic studies and risk extrapolation, given the influence of susceptibility variation.

A third major focus for epidemiology and risk assessment in the coming years will be increased international collaboration on epidemiologic research. As occupational and environmental controls have diminished exposure to carcinogens in developed countries, much attention has turned to epidemiologic study in the developing countries, including those in Eastern Europe. Collaboration on a variety of epidemiologic issues such as molecular methods and the collection and presentation of data on exposure response will maximize the data's utility for risk assessment.

REFERENCES

Blair, A., Burg, J., Foran, J., Gibb, H., Greenland, S., Morris, R., Raabe, G., Savitz, D., Teta, J., Wartenberg, D., Wong, O., Zimmerman, R., Guidelines for application of meta-analysis in environmental epidemiology, *Regulat Toxicol Pharmacol*, 22, 189, 1995.

Checkoway, H., Pearce, N., Crawford-Brown, D. J., *Research Methods in Occupational Epidemiology*, Oxford University Press, New York, 1989, Chap. 4.

Gordis, L., Ed., *Epidemiology and Health Risk Assessment*, Oxford University Press, New York, 1988.

Graham, J. D., Ed., *The Role of Epidemiology in Regulatory Risk Assessment*, Elsevier, Amsterdam, 1995.

Hulka, B., Overview of biological markers, in *Biological Markers in Epidemiology*, Hulka, B. S., Wilcosky T. C., Griffith, J. D., Eds., Oxford University Press, New York, 1990, Chap. 1.

International Agency for Research on Cancer, *IARC Monorgraphs on the Evaluation of the Carcinogenic Risk of Chemicals to Humans, Supplement 4*, International Agency for Research on Cancer, Lyon, France, 1982, Preamble.

International Life Science Institute, Low-dose extrapolation of cancer risks: current perspectives and future directions, ILSI Risk Science Institute, Washington, DC, 1995.

Johns Hopkins School of Public Health, *Public Health Newsletter*, Johns Hopkins School of Hygiene and Public Health Office of Public Affairs, December 1995, 7.

Kahn, H. A., Sempos, C. T., *Statistical Methods in Epidemiology*, Oxford University Press, New York, 1989, Chap. 10.

Kelsey, J. L., Thompson, W. D., Evans, A. S., *Methods in Observational Epidemiology*, Oxford University Press, New York, 1986, Chaps. 4 & 7.

Lilienfeld, A. M., *Foundations of Epidemiology*, Oxford University Press, New York, 1976, Chaps. 8 & 9.

National Research Council (NRC), *Risk Assessment in the Federal Government: Managing the Process*, National Academy Press, Washington, DC, 1983, Chap. 1.

National Research Council, *Science and Judgment in Risk Assessment*, National Academy Press, Washington, DC, 1994, 11.

National Research Council, *Science and Judgment in Risk Assessment*, National Academy Press, Washington, DC, 1994, 58.

National Research Council, *Science and Judgment in Risk Assessment*, National Academy Press, Washington, DC, 1994, 206.

Office of Science and Technology Policy, Chemical carcinogens: review of the science and its associated principles, 1985, *Federal Register* 50:10372–10442.

Perera, F. P., Weinstein, I. B., Molecular epidemiology and carcinogen-DNA adduct detection: new approaches to studies of human cancer causation, *J Chronic Dis*, 35, 581, 1982.

Rothman, K. J., *Modern Epidemiology*, Little, Brown and Company, Boston, 1986, Chap. 2.

Rothman, K. J., *Modern Epidemiology*, Little, Brown and Company, Boston, 1986, Chap. 7.

Smith, A. H., Bias, bias, everywhere? And not one drop of science?, in *The Role of Epidemiology in Regulatory Risk Assessment*, Graham, J.D., Ed., Elsevier, Amsterdam, 1995, 39.

Tockman, M. L., Levin, M. L., Frost, J. K., Ball, W. C., Jr., Cancer detection by exfoliative cytology, in *New and Sensitive Indicators of Health Impacts of Environmental Agents*, Underhill, D. W., Radford E. P., Eds., University of Pittsburgh, Pittsburgh, 1986, 115.

U.S. Environmental Protection Agency, Interim procedures and guidelines for health risk and economic impact assessments of suspected carcinogens, 1976, *Federal Register* 41:21402–21405.

U.S. Environmental Protection Agency, Health Assessment Document on Chromium, EPA 600/8-83/014F, Office of Health and Environmental Assessment, Research Triangle Park, 1984a.

U.S. Environmental Protection Agency, Carcinogenicity Risk Assessment Document on Coke Oven Emissions, EPA 600/8-82/003F, Office of Health and Environmental Assessment, Washington, DC, 1984b.

U.S. Environmental Protection Agency, The risk assessment guidelines of 1986a, *Federal Register* 51:33992–34005.

U.S. Environmental Protection Agency, Health Assessment Document on Nickel, EPA 600/8-83/012FF, Office of Health and Environmental Assessment, Research Triangle Park, NC, 1986b.

U.S. Environmental Protection Agency, Special Report on Ingested Inorganic Arsenic/Skin Cancer; Nutritional Essentiality, EPA/625/3-87/013, Risk Assessment Forum, Washington, DC, 1988.

U.S. Environmental Protection Agency, Workshop on EPA Guidelines for Human Risk Assessment: Case of Human Evidence, EPA/625/3-90/017, Risk Assessment Forum, Washington, DC, 1989.

U.S. Environmental Protection Agency, Proposed Guidelines for Carcinogen Risk Assessment, 1996, *Federal Register* 6(79):17960–18011.

Vine, M. F., Micronuclei, in *Biological Markers in Epidemiology*, Hulka, B. S., Wilcosky, T. C., Griffith, J. D., Eds., Oxford University Press, New York, 1990, Chap. 7.

QUESTIONS

1. What are some of the primary guidance documents on risk assessment?
2. What are the principal types of epidemiologic studies?
3. What is meta-analysis? What is its utility in the analysis of human studies?
4. How are biomarkers classified?
5. What sorts of questions should the risk assessor ask when reviewing cancer epidemiology studies?
6. Explain the criteria for determining causality between exposure and effect in human cancer studies.

CHAPTER I.3

Uncertainty and Variability in Risk Analysis

Richard Wilson and Alexander Shlyakhter

SUMMARY

The very word risk implies uncertainty. Conversely, if there exists an uncertainty whether a hazard exists, there remains a probability that it does and therefore a risk. However, some people argue that we should ignore "uncertain risks." This very phrase suggests a paradox and a contradiction. The meaning becomes clear in those cases where the existence of a hazard has not been convincingly proven. The risk caused by this hazard is then called an "uncertain risk." However, if we only attempt to reduce those risks with well-defined magnitude ("certain risks"), we will miss most of the opportunities to improve public health.

Common sense can guide us when scientific evidence is inconclusive. When sanitary engineers insisted on main drainage a century ago, they did so upon general principles, not upon the basis of reliable data showing that raw sewage or impure water caused bad health. The rule was to provide the best drainage and the purest water reasonably possible. There is now no question that this action was correct, even though the benefits at the time must have seemed very uncertain.

Quantitative analysis of uncertainty and variability is receiving growing acceptance in risk assessment. It is an important step forward from multiplying simple point estimates of the individual risk factors, as it provides a decision maker with more information about the reliability of the results.

In this chapter, we contrast several different types of uncertainty: stochastic uncertainties vs. uncertainties of fact and objective vs. subjective uncertainties. We also discuss the relationship between uncertainty and variability. Some of the discussion is from an earlier review (Wilson et al. 1985).

1-56670-130-9/97/$0.00+$.50

Key Words: uncertainty, risk, error, probability, stochastic, variability, common mode, mistake, weight-of-evidence

1. TYPES OF UNCERTAINTY

The uncertainty in a risk assessment changes with time as information develops. We may say that the lifetime risk of cancer is 25%, meaning that approximately 25% of all people develop cancer in their lifetimes. Once an individual develops cancer, we can no longer describe the situation by the term "risk." It is certain that he/she has cancer. Similarly, if a person lies dying after a car accident, the risk of his dying of cancer clearly drops to near zero. Thus, estimates of risks, insofar as they are expressions of uncertainty, will change as knowledge improves.

Different uncertainties appear in risk estimation in different ways. There is clearly a risk that an individual will be killed by a car if he/she walks blindfold across a crowded street. One part of this risk is stochastic; it depends upon whether the individual steps off the curb at the precise moment that a car arrives. Another part of the risk might be systematic; it will depend upon the nature of the fenders and other features of the car. Similarly, if two people are both heavy cigarette smokers, one may die of cancer and the other may not; we cannot tell in advance. However, there is a systematic difference in this respect between being a heavy smoker and a gluttonous eater of peanut butter with its aflatoxin content. Although aflatoxin is known to cause cancer (quite likely even in humans), the risk of eating peanut butter is much lower than that of smoking cigarettes. Exactly how much lower is uncertain, but it is possible to make estimates of how much lower and also to make estimates of how uncertain we are about the difference.

Some estimates of uncertainties are subjective, with differences of opinion arising because there is a disagreement among those assessing the risks. Suppose one wishes to assess the risk (to humans) of some new chemical being introduced into the environment or of a new technology. Without any further information, all we can say about any measure of the risk is that it lies between zero and unity. Extreme opinions might be voiced: one person might say that one should initially assume a risk of unity, because we do not know that the chemical or technology is safe; another might take the opposite extreme and argue that one should initially assume that there is zero risk, because nothing has been proven dangerous. Here and elsewhere, we argue that it is the task of the risk assessor to use whatever information is available to obtain a number between zero and one for a risk estimate, with as much precision as possible, together with an estimate of the imprecision. Within this context, the statement "we do not know" can be viewed only as procrastination and not a response to the request for a risk estimate (although this is not to condemn procrastination in all circumstances.)

The second extreme presented previously is surprisingly common, even in some government agencies which are supposed to take risks into account in their promulgation of regulations. This can arise whenever there is a propensity to ignore anything which is not a proven hazard. We claim that such an attitude is usually logically inconsistent and warn any users of risk assessments of this danger.

Fortunately, if risk assessors have been diligent in searching out hazards to assess, few hazards posing large risks will be missed in this way, so there may be minor direct danger to human health from a continuation of the attitude. This may lead to economic inefficiencies, however, and can easily lead to unnecessary anger between experts who disagree strongly.

Risk, and also uncertainty, has different qualitative meanings at different times. One may say that he/she has a risk of dying of cancer, meaning that it is uncertain whether or not he/she will develop cancer and die. If one should develop cancer, the risk would at once be changed. It is still not a certainty that one would die of cancer, since (some) cancers can be cured and there is the chance of spontaneous remission. But once all attempts at a cure have failed, the risk of death becomes a certainty.

The first type of uncertainty to consider is the stochastic uncertainty of certain processes. We consider, for example, the process of developing cancer. Some persons exposed to a large dose of carcinogens, for example, lifetime cigarette smoking, will develop lung cancer; others will not. Whether any particular smoker will develop lung cancer appears to be largely random: there is a stochastic component of uncertainty. Similarly, some persons crossing a crowded road blindfolded will be run down and killed, whereas others will not. Weather predictions are uncertain and probably stochastic; climate predictions cover a longer time and would not normally be considered stochastic.

It is easy to see that it does not really matter in this example whether the onset of cancer is actually a stochastic process or not. Scientists consider radioactive decay a stochastic process, but the Oxford English Dictionary goes further and traces "stochastic" from the Greek "to aim at a mark, guess". In aiming at a mark, we can specify a general distribution of hits, but not whether a particular point may be hit. Similarly, the details of why a cancer occurs in a particular individual at a particular time is unknown and, with our present and foreseeable knowledge, unknowable. Thus, arguments about whether the cancer is "really" started by a hit on an individual molecule are irrelevant.

We can list other examples of stochastic uncertainties. A typical risk assessment may estimate the probability that a particular person will be killed next year. For automobile accidents, this may be done on the basis of historical experience: of the U.S. population of 230 million persons, approximately 40,000 die in auto accidents each year, giving an average risk (population) of 180 per million per year. This estimate is fairly precise — it has been declining a few percent from year to year — so the probability of any one individual (randomly chosen from the population) being killed is also precise; but the individual cannot calculate his/her own time of death or whether he/she will in fact die from this cause. This uncertainty, inherent in the word "risk", is purely stochastic, provided the way we have analyzed the historical data is correct.

Sometimes, analysts try to make a sharp distinction between variability and uncertainty (Hoffman and Hammonds 1994, Hattis and Burmaster 1994). The distinction between the two cases can be blurred, and McKone (1994) calls these Type A and Type B uncertainties. By variability, we mean the measured and therefore known variation among members of a defined population potentially leading to differences in risks, whereas by uncertainty we mean the combination of all other effects that lead to variations in risks for the defined population. The distinction is different when viewed from different perspectives. One might ask what is the risk of exposure to a pollutant in a major city. Since the exposure obviously varies across the city, so also will the risk. If the person asking the question knows where he/she will live and the exposure at that location, he/she can calculate his/her risk at that location. This risk will vary across the city. But, if the person asking the question does not know where he/she will live and that this location will be "typical", the variability becomes an uncertainty to be folded in with other uncertainties of the risk calculation.

Similarly, in calculations of reactor safety, one must include a knowledge of how often crucial components are likely to fail. No one knows exactly whether a particular pump will fail, but an estimate of probability with its uncertainty can be gained from the historical record of the variability of pump failures.

Therefore, some of the arguments about whether a given parameter distribution is variability or whether it is an uncertainty is really a distinction between slightly different questions being addressed by the risk analysts.

People with different functions and responsibilities will see the uncertainties in different ways. A hospital administrator, whose responsibility is to provide emergency services, will only be interested in the total number of automobile deaths in his/her region in any one day. Although this will fluctuate around the mean, the uncertainty in his/her planning caused by this uncertainty will not be as great as that seen by an individual.

2. THE THEORY OF ERROR

The mathematical theory of measurement error is nearly two centuries old and is called "the theory of error." For example, the famous scientist Gauss, when describing his measurements of geographical locations of German mountains, invented the method of "least squares" and modestly suggested that no previous geographer had been as thorough. In his theory, each measurement is assumed to be statistically independent of every other. Therefore, errors of the measurements can be "added in quadrature" (the square of the combined error is the sum of the squares of the component errors).

The word "error" that is used in formal statistical theory has another connotation when used in discussions of public health and medicine and can mean "mistake" for which the perpetrator might be legally and economically liable.* Therefore, the words "uncertainty analysis" replace "theory of error". But, that does not mean that mistakes are not considered by risk analysts. In reactor safety analyses, for example, the postulated initiating event is often someone's mistake or error. An analysis of the frequency and distribution of these mistakes is an important input to any full probabilistic risk analysis (PRA).

* It has been said that when one mentions the word "error" to a physician, he telephones his lawyer at once.

3. COMBINING UNCERTAINTIES

For the risk of cancer caused by chemical carcinogens, the risk can be described by a formula with three factors (Crouch and Wilson 1981):

$$R = \beta \cdot K \cdot D \tag{1}$$

Beta is the carcinogenic potency in animals, K is an interspecies conversion factor, and D is the dose. It is self-evident that these factors are approximately independent of each other. Gauss' procedure for combining uncertainties is especially simple if each factor can be described by a lognormal distribution (the logarithm of each factor described by a normal distribution). Therefore, the risk itself is a lognormal distribution with variance equal to the sum of the variances of the individual distributions. Thus, we can understand the way in which uncertainties propagate by assuming that each term in the fundamental equation can be approximated, or bounded, by a lognormal distribution, and taking the logarithm of both sides,

$$LnR = Ln\beta + LnK + LnD \tag{2}$$

Each term in the modified equation is fitted by a normal distribution:

$$d(\ln D) \cdot dD + \frac{1}{\sqrt{2\pi} \cdot \sigma_D} e^{-\frac{(\ln D - \mu_D)^2}{2\sigma_D^2}} \tag{3}$$

If the process described by each term is independent of the others, then the distribution of ln$R$ in Equation 2 is also a normal distribution, (distribution of R is lognormal) with a standard deviation:

$$\sigma_R^2 = \sigma_\beta^2 + \sigma_K^2 + \sigma_D^2 \tag{4}$$

Of course, if the distributions are not lognormal, but are known, and if independence can still be assumed, the "risk distribution" (the distribution of the function R) can still be evaluated by a Monte Carlo program. But, it is useful to remember that if each distribution is smooth and can be approximated by a lognormal with the appropriate geometric mean and standard deviation, the distribution of the logarithm of R gets closer and closer to a normal distribution as the number of factors increases. With the advent of cheap computers, it is usual to do such calculations by computer (Thompson et al. 1992). But, we urge that in all cases the simple analytic calculations be done with approximate lognormal fits to ensure that no human errors are made.

Often overlooked is that while the median of the distribution of R is the product of the medians of the distributions for each factor, this does not apply to any other parameter of the distribution. In particular, the upper 98th percentile of R is *less than* the product of the upper 98th percentile of the individual distributions. A failure to realize this has resulted in many overestimates of risk.

4. UNCERTAINTIES IN ANALYSIS OF EXPOSURE AND DOSE

The three-parameter equation for risks of cancer from exposure to chemicals is commonly used. However, it assumes that the *dose* is known. In some situations that is true; the concentration in the blood of phenoxy chemicals or of lead can be, or is, measured. While early regulation explicitly used a pessimistic dose estimate (the Food and Drug Administration [FDA] discussed a "gluttonous consumer"), it is becoming more common to calculate that in a probabilistic fashion (Finley and Paustenbach 1994). This may enable regulation of chemical carcinogens to be less draconian. The calculation then depends on a large number of other factors, some of which are poorly known and others of which vary extensively over a population. Many of these factors are approximately independent, so we express

$$D = d_{1} \cdot d_2 \cdot d_3 \cdot \ldots \cdot d_n \quad \textbf{(5)}$$

A number of authors have written extensively on this topic and produced calculational procedures that, for example, distinguish the calculations of variability and uncertainty on the final answer (McKone and Ryan 1989, Green et al. 1993, McKone 1994, Bogen 1995). In this case, it is far from obvious that these variables are independent of each other. However Smith et al. (1992) argue that residual correlation is small and that assuming independence usually gives little error.

5. INDEPENDENCE, CORRELATION, AND COMMON MODE FAILURES

The importance of understanding whether or not two factors in a risk equation are statistically independent or not cannot be overstated. If they are independent, uncertainties combine in quadrature. For a technological system, oil refinery, space shuttle, or nuclear power plant, ensuring independence is a crucial part of system design. For a purely observational analysis, such as the analysis of chemical carcinogens or a study of global warming (see Chapter II.5 in this volume), the skill of the analyst is to choose those parts of the system which are approximately independent of each other, both for ease of calculation and for ease of understanding. In technical terms, this can be called "diagonalizing the error matrix."

For example, a reactor safety system is designed so that if a coolant pipe breaks (frequency P_1 in risk equation) an emergency core cooling system (ECCS) reinjects water into the system. The ECCS itself has a failure probability P_2. If that fails, the reactor containment should hold any radioactive fission products. But the containment might fail with probability P_3 for an overall probability of accident $P = P_1P_2P_3$. If each of the factors is small (1/100), P is much smaller (10^{-6}), which is often considered acceptable. A correlation between these terms is called in reactor accident analysis a "common mode failure." If there is a complete correlation between two factors, P immediately increases to 10^{-4}, which is usually considered unacceptable.

Typical examples of common mode failures in reactor safety calculations are fire that destroys all control cables that are (stupidly) installed in the same cable tray and intentional sabotage by someone who understands the weak points of the system. One must pay special attention to those cases where the factors are not statistically independent and add them in separately either "by hand" or by careful inclusion of correlations in the Monte Carlo program (Rasmussen et al. 1975).

6. THE INFLUENCE OF UNCERTAINTY IN RISK MANAGEMENT

It has become commonplace to insist that a manager must look at more than the number about a risk. The U.S. EPA, for example, attempts to be risk averse by regulating on the tail (upper 98th percentile) of a risk distribution or the upper end of the distribution of potency. But what should a manager do when the evidence for carcinogenicity is not statistically significant? Or alternatively stated, when the lower limit of the risk distribution encompasses zero risk? Should he/she still regulate on the upper limit? The U.S. EPA ignores the chemical if the evidence is not statistically significant, thereby setting the potency equal to zero with zero uncertainty! We suggest that a more logical and complete approach would be to follow a Bayesian procedure and assume that there is "prior" information on the carcinogenic potency that is updated by each experiment or new observation.

Zeise et al. (1984) noted that even when considering a chemical for which there have been no formal measurements a risk analyst starts with *some* information, however imprecise. First, one might assume that a new chemical is representative of the class of all chemicals that have been tested. The carcinogenic potency then spans seven decades with an uncertainty distribution and a geometric standard deviation of two or three decades. By measuring some parameter of acute toxicity, and using a toxicity to carcinogenicity correlation, this estimate can be updated and the uncertainty reduced to a geometric standard deviation of one decade or so. Further, measurement of the carcinogenicity in animals can reduce the uncertainty to one decade or less and the measurement in people (if exposure is well known) somewhat less.

Following the ideas in a classic paper by Schneiderman and Mantel (1973), Zeise et al. suggested that the prior information and way that the uncertainties diminish with research be explicitly recognized. Then, if regulation is always made on the upper 95th percentile of the distribution, it would become more lenient if the uncertainty reduces, but the central value does not change. Incentives for good research and honest reporting would then be automatically created without an onerous regulatory structure.

7. REGULATION ON RISK OR ON WEIGHT OF EVIDENCE

The procedure suggested earlier is very different from that used by The International Agency for Research on Cancer (IARC) (a division of the World Health Organization [WHO]). They describe the uncertainty by a categorization into group 1 carcinogens (those for which there is definite evidence on carcinogenicity in humans), group 2A carcinogens (those for which there is limited evidence of carcinogenicity in humans but sufficient evidence in animals), group 2B carcinogens (where there is less than sufficient evidence of carcinogenicity in animals), and group 3 carcinogens (where there the agent or exposure or circumstance is not classifiable) (IARC 1994). The U.S. EPA often uses a similar classification and implies that group 1 and group 2A carcinogens should be treated more severely than the others.

By contrast, if the weight of evidence is high, the uncertainty is presumably small. The uncertainty is presumably larger for group 2B or group 3 carcinogens for which the data are sparse, and regulation on the upper limit of a distribution would be very severe. The U.S. EPA procedure can be characterized and oversimplified by saying that it is "weight of evidence" for a chemical in group 2B or group 3, but "upper limit of risk" for those in group 1 or group 2A.

8. OBJECTIVE AND SUBJECTIVE UNCERTAINTIES

One may distinguish, somewhat arbitrarily, between objective and subjective risk estimates. The measure of uncertainty in animal to human extrapolations that we derive is based on animal cancer experiments. However, whether such an extrapolation is justified depends on knowledge of biological and chemical mechanisms and, as such, is very subjective. Evans et al. (1994a,b) have used expert judgements in an estimate of these uncertainties in several cases. Expert judgement is used in some aspects of reactor safety calculations. However, these expert judgements must be used with caution because experts are often overconfident (as discussed in the next section).

9. OVERCONFIDENCE IN RISK ESTIMATES

The reduction of uncertainty noted in the previous paragraph that occurs when research is done only applies to objective measures of uncertainty. Estimates of the nonstatistical errors are notoriously difficult to make. It is a natural trait to be confident in one opinion and therefore overconfident about the accuracy of one's estimate. It might be thought that "experts" are more careful than the public in such matters, but that is not the case.

Lawyers, in particular, have often noted that as research continues, the uncertainty increases. Of course, it is only the perception of uncertainty that increases — the actual uncertainty (whatever it is) stays the same or goes down. We illustrate this by two studies that have shown that this applies to the experts as well as to the public. Morgan et al. (1984) asked a group of experts about the risk that air pollution at present levels causes premature mortality. The spread of opinions was wide. After 5 years of research into air pollution, the same group was questioned again; the opinion spread was even wider! Morgan and Keith (1995) asked a group of experts their opinions on the temperature rise caused by increased greenhouse gas emissions. Then they asked the same group to develop a research program. Finally, their opinions about the temperature rise were elicited again. The stated uncertainty had increased!

This is part of a general phenomenon noted by Tversky that probability judgements are attached not to the events, but to the description of events (Tversky and Koehler 1994). As the description of events becomes more specific and detailed (the phenomenon is "unpacked"), the perceived probability of the sum of the descriptions becomes greater. Eventually, as further understanding is achieved, the situation should reverse.

Several authors have studied the overconfidence in estimates (Henrion and Fischoff 1986, Morgan and Henrion 1990, Cooke 1991). Analysis of historical trends in measurements and past projections allows quantification to the degree of overconfidence by fitting the empirical distribution and estimating the typical fraction of unsuspected errors for each type of data. Although the absolute error in measurements usually decreases with time, the estimated uncertainty is also smaller, so that the probability of "large" deviations relative to the estimated uncertainty is roughly the same (Shlyakhter and Kammen 1992, Shlyakhter et al. 1994).

There are two ways of contemplating and using this observation. The first is purely empirical: one can assume that if practitioners in a given field have understated the uncertainties in a particular way and by a particular amount, then other practitioners in the same field are likely to underestimate errors in the same way and by the same amount.

One may also approach the problem more theoretically. The standard uncertainty analysis can be supplemented with an analysis of the "uncertainty of uncertainty". If this second order uncertainty is itself assumed to have a Gaussian distribution, the resulting compound distribution of risk can be shown to have an exponential tail similar to that seen in the historical comparisons. The slope of the tail is characterized by one additional parameter, u, the relative uncertainty in the estimated standard error. For the cumulative probability of the actual error exceeding x times the estimated uncertainty, one can use the approximation $e^{-|x|/(0.7u+0.6)}$. Analysis of several historical datasets gives $u \approx 1$ for physical and environmental measurements and $u \approx 3$ for energy and population projections (Shlyakhter 1994a,b). One can use this approximation to account for unsuspected errors. For a Gaussian distribution, the 95% confidence interval (95% CI) is $\mu \pm 1.96\sigma$, where μ is the mean and σ is the standard error of the mean. If one takes into account the second order uncertainty, 95% CI is $\mu \pm Z\sigma$, where $Z = 3.9$ for $u = 1$, $Z = 6.0$ for $u = 2$, and $Z = 8.1$ for $u = 3$.

Standard uncertainty analysis provides an estimate of the width of the probability distribution around a simple point estimate. However, the commonly used 95% confidence intervals are determined by the tails of the distribution that are very sensitive to such underestimation of the uncertainties. In particular, the commonly used 95% upper bounds for normal and lognormal distributions are very sensitive to the underestimation of the true uncertainty. For example, the width of the 95%

CI is 1.19 times the width of the 90% CI ($Z_{0.95}/Z_{0.90}$ = 1.96/1.645 = 1.19) for a Gaussian (normal) distribution. This implies that an underestimation of the standard deviation (uncertainty) of such a distribution by 20% will underestimate by a factor of two the probability of the true value being outside the confidence interval (10% instead of the estimated 5%). This increase becomes even more important when it is realized that the underestimation of uncertainties particularly affects the tails of the distribution, as noted in the previous paragraphs.

10. CONCLUSIONS

Our discussion of the distinctions between different types of uncertainty, ways of looking at uncertainty, and ways of allowing for uncertainty may not be accepted by everyone. ***Risk assessors are sharply influenced by the various values of those using the risk assessments. However, we would like other risk analysts to consider seriously our procedure for discussing uncertainties, whether or not they agree with the values either for risk or for uncertainty that we have discussed. For the most important result of including uncertainties in a risk calculation, like the result of making the risk calculation itself, it is not the number, but the insight that the inclusion gives to the assessor or manager.***

ACKNOWLEDGMENTS

Our ideas, opinions, and views have been strongly influenced by many collaborators and other critics of our work. In particular, we would like to thank E. A. C. Crouch, D. Hattis, D. M. Kammen, F. Seiler, and L. Zeise for their interest over the many years we have discussed these matters.

REFERENCES

Bogen K.T. (1995) "Methods to Approximate Joint Uncertainty and Variability in Risk," *Risk Analysis*, 15:411–419.

Cooke R.M. (1991) *Experts in Uncertainty: Opinion and Subjective Probability in Science*, Oxford University Press.

Crouch E.A.C. and Wilson R. (1981) "The Regulation of Carcinogens," *Risk Analysis*, 1:47–57.

Evans J.S., Gray G.M., Sielken R.L., Jr., Smith A.E., Valdez-Flores C., and Graham J.D. (1994a) "Use of Probabilistic Expert Judgment in Uncertainty Analysis of Carcinogenic Potency," *Regulatory Toxicology and Pharmacology*, 20:15–36.

Evans J.S., Graham J.D., Gray G.M., and Sielken R.L., Jr. (1994b) "A Distributional Approach to Characterizing Low-Dose Cancer Risk," *Risk Analysis*, 14:25–34.

Finley B. and Paustenbach D. (1994) "The Benefits of Probabilistic Exposure Assessment: Three Case Studies Involving Contaminated Air, Water, and Soil," *Risk Analysis*, 14:53–73.

Green L.C, Armstrong S.R., Crouch E.A.C., Lash T.L., Luis S.J., and Perkins K.K. (1993) "Revised Protocol for a Multi-Pathway Risk Assessment for the WTI Facility in East Liverpool, Ohio," Cambridge Environmental Inc., Cambridge, Massachusetts.

Hattis D. and Burmaster D.E. (1994) "Assessment of Variability and Uncertainty Distributions for Practical Risk Analyses," *Risk Analysis*, 14:713–730.

Henrion M. and Fischoff B. (1986) "Assessing Uncertainty in Physical Constants," *American Journal of Physicians,* 54:791–797.

Hoffman F.O. and Hammonds J.S. (1994) "Propagation of Uncertainty in Risk Assessments: The Need to Distinguish Between Uncertainty Due to Lack of Knowledge and Uncertainty Due to Variability," *Risk Analysis*, 14:707–712.

IARC (1994) "IARC Monographs on the Evaluation of Carcinogenic Risks to Humans," *World Health Organization,* 61:30–32.

McKone T.E. and Ryan P.B. (1989) "Human Exposures to Chemicals Through Food Chains: An Uncertainty Analysis," *Environmental Science and Technology*, 23:1154–1163.

McKone T.E. (1994) "Uncertainty and Variability in Human Exposures to Soil Contaminants Through Home-Grown Food: A Monte Carlo Assessment," *Risk Analysis*, 14:449–464.

Morgan M.G. and Keith D.W. (1995) "Subjective Judgments by Climate Experts," *Environmental Science and Technology*, 29:468–476.

Morgan M.G. and Henrion M. (1990) *Uncertainty: A Guide to Dealing with Uncertainty in Quantitative Risk and Policy Analysis*, Cambridge University Press, New York.

Morgan G., Morris S.C., Henrion M., Amaral D.A.L., and Rish W.R. (1984) "Technical Uncertainty in Quantitative Policy Analysis — A Sulfur Air Pollution Example," *Risk Analysis,* 4:201.

Rasmussen N., et al. (1975) "Reactor Safety Study," U.S Atomic Energy Commission WASH 1400.

Schneiderman M.A. and Mantel N. (1973) "The Delaney Clause and a Scheme for Rewarding Good Experiments," *Preventive Medicine,* 2:165.

Shlyakhter A.I. and Kammen D.M. (1992) "Sea-Level Rise or Fall?" *Nature*, 357:25.

Shlyakhter A.I., Kammen D.M., Broido C.L., and Wilson R. (1994) "Quantifying the Credibility of Energy Projections from Trends in Past Data: The U.S. Energy Sector," *Energy Policy*, 22:119–130.

Shlyakhter A.I. (1994a) "Uncertainty Estimates in Scientific Models: Lessons from Trends in Physical Measurements, Population and Energy Projections," in *Uncertainty Modelling and Analysis: Theory and Applications*, editors: B.M. Ayyub and M.M. Gupta, pp. 477–496. Elsevier Science B.V.

Shlyakhter A.I. (1994b) "An Improved Framework for Uncertainty Analysis: Accounting for Unsuspected Errors," *Risk Analysis*, 14:441–447.

Smith A.E., Ryan P.B., and Evans J.S. (1992) "The Effect of Neglecting Correlations When Propagating Uncertainty and Estimating the Population Distribution of Risk," *Risk Analysis,* 12:467–474.

Thompson K.M., Burmaster D.E., and Crouch E.A.C. (1992) "Monte Carlo Techniques for Quantitative Uncertainty Analysis in Public Health Risk Assessments," *Risk Analysis,* 12:53–64.

Tversky A. and Koehler D.J. (1994) "Support Theory: A Nonextensional Representation of Subjective Probability," *Psychological Review*, 101:547–567.

Wilson R., Crouch E.A.C., and Zeise L. (1985) "Uncertainty in Risk Assessment," in "Risk Quantitation and Regulatory Policy," *Banbury Report* 10.

Zeise L., Wilson R., and Crouch E.A.C. (1984) "Use of Acute Toxicity to Estimate Carcinogenic Risk," *Risk Analysis,* 4:187–199.

QUESTIONS

1. Consider a situation where a chemical or a technology has not been tested. Should this be described as one of zero risk? Or high risk? Is there a bound?
2. If a risk is a product of four factors, and there is a correlation between two of them, show that the risk must lie between a low value, assuming that all four are independent, and a high value, assuming that the largest of two coupled risks and the other two independent risks are independent of each other.
3. Consider a risk that is numerically the product of two factors, which are independent of each other. If each factor can be represented by a lognormal distribution with logarithmic standard deviation σ, show that the resultant risk distribution is also lognormal and find the standard deviation.
4. If 400 chemicals are tested for carcinogenicity at a specific site, and each chemical is declared a carcinogen if the increase in tumors is significant at the level $p = 0.05$, how many will be found to be significant by chance? (False positives.) How is this modified if a search is made at five different sites at the same time?
5. For a lognormal distribution of risk, find the differences between the mean, the median, and the mode and find the upper 95th percentile bound in terms of the standard deviation σ of the distribution of the logarithm of the risk.

CHAPTER I.4

Monte Carlo Risk Analysis Modeling

David Vose

SUMMARY

Quantitative risk analysis plays an important part in the assessment of the risks and uncertainties surrounding a decision problem. Monte Carlo simulation is a very powerful and flexible way of performing such quantitative risk analyses. It allows the analyst to assign probability distributions to all uncertain components of a mathematical model of the problem and then, through random sampling of these distributions, determine the distribution of all potential outcomes that could occur under these uncertainties. This chapter offers a brief overview of the techniques involved in Monte Carlo risk analysis modeling including how to structure the model, assign distributions to the uncertain components within the model, model dependencies between uncertain components of the model, and finally how to present and interpret the results of the model.

Key Words: Monte Carlo, simulation, risk analysis, model

1. INTRODUCTION

Monte Carlo risk analysis modeling encompasses a range of techniques to mathematically describe the impact of risk and uncertainty on a problem. Each uncertain parameter within the model is represented by a probability distribution. The shape and size of these distributions defines the range of values that the parameters may take and their relative probabilities.

1-56670-130-9/97/$0.00+$.50

Quantities calculated by the model that are of interest are selected as outputs. Monte Carlo risk analysis software then randomly generates values from the probability distributions to calculate hundreds, or even thousands, of possible scenarios. The values that are calculated for these outputs are stored for each scenario (iteration). At the end of the simulation, these values are collected and analyzed to produce assessments of the uncertainty of the model's outputs.

The results of a Monte Carlo model are probability distributions of these outputs, the associated descriptive statistics, and measures of the relationship between the input and output uncertainties.

This chapter gives a brief overview of the various stages involved in producing a Monte Carlo risk analysis model, namely,

- Designing the structure of the risk analysis model
- Defining distributions that describe the uncertainty of the problem
- Modeling dependencies between model uncertainties
- Presenting and interpreting the risk analysis results

Monte Carlo simulations were once difficult to perform. However, with the rapid advance of PCs, software, and computer literacy, the technique has become widely accessible. Spreadsheets have become a very common modeling tool, and products like @RISK (Palisade Corporation, Newfield, New York) and Crystal Ball (Decisioneering Inc., Denver, Colorado) have extended the capabilities of spreadsheet programs to include Monte Carlo modeling. Software programs are also now available to add Monte Carlo capabilities to project planning tools, allowing uncertainty to be allocated to project task durations and costs.

It is perhaps an unfortunate side effect of people's faith in software that they believe that in purchasing risk analysis software they will be able to immediately perform credible risk analyses. A risk analysis model is considerably more complex than traditional single-point estimate (deterministic) models. This chapter considers the basic requirements of a Monte Carlo risk analysis and offers some introductory techniques to help ensure the accuracy of the reader's risk analysis modeling.

Monte Carlo risk analysis should not be thought of as the only technique for evaluating risk and uncertainty. There are a number of other numerical tools available for analyzing risks, and other, nonquantitative aspects will usually need to be considered also. The decision maker should use Monte Carlo risk analysis as one of several complementary inputs to the decision process.

2. DESIGNING THE STRUCTURE OF THE MODEL

The foremost consideration in designing a model is that it should provide the information that the decision maker needs, for example, the time and cost to complete a project, the profitability of a venture, or the probability of introducing a disease. The model must be set out and labeled to be easily understood by others. Furthermore, the model should be designed to be adaptable. Key inputs, even if deterministic (i.e., not uncertain), are best listed at the top of the spreadsheet, not hidden inside a formula somewhere. It will then be easier to make any necessary changes in light of new information or decisions. Uncertain parameters needed in more than one place in the model must only be input into one cell and that cell referenced wherever needed.

It is good practice to design a schematic diagram of the intended model and to circulate it to "stakeholders" for approval and comment. It may save having to make a lot of changes later on! It is also worth bearing in mind that the structure of some problems cannot be well defined, in which case a quantitative risk analysis simply may not be possible.

2.1 Disaggregation

Disaggregation means breaking the problem down into its smallest manageable components. In risk analysis, it is generally more convenient and more accurate to break down a problem into smaller components than one is used to doing in normal deterministic modeling.

Disaggregation generally makes the logic of the problem more apparent. For example, a marketing manager will be able to provide a much more detailed and accurate prediction of next year's sales volumes if he/she is allowed to break his/her sales down by product, region, etc. rather than being asked to produce a total sales estimate straight off. In this way, disaggregation may also reveal dependency relationships that were not immediately evident or simply had not been thought about.

However, disaggregation quickly becomes overcomplex and often unnecessarily detailed. The extra effort put into modeling a problem's finer details should be balanced against the benefits it provides.

It is often unnecessary to be very precise about most of the distributions of a well-disaggregated model. The final outcome of the model will usually be reasonably insensitive, for example, to whether most of the parameters are modeled by three-point estimate distributions like a BetaPERT or triangle, discussed later, or by more precisely determined distributions. This insensitivity is useful to know, as it may alleviate the need to spend further time or money on studying the variable. The degree to which the result of a model is affected by the distribution of one of its input variables is quite easy to determine. One can run two simulations of the model where the only difference between simulations is the type of distribution used to represent the variable in question, perhaps the BetaPERT and triangle. If there is no significant difference between the simulations' results, one can conclude that the model is insensitive to the exact shape of the distribution used and therefore can confidently select an approximate distribution like the BetaPERT.

3. DETERMINING THE INPUT UNCERTAINTY IN THE MODEL

This section looks at some methods for determining the distributions to model the uncertain variables within a risk analysis model. A number of commonly used distribution types are first discussed. Then we look at methods for modeling an expert's opinion of a variable's uncertainty. Finally, we look at how distributions can be determined by available data.

3.1 Examples of Typical Distributions Used in Modeling

Some of the types of distribution more commonly used in risk analysis modeling are described along with their applications.

3.1.1 Nonparametric and Parametric Distributions

Probability distribution functions can be placed into two categories: nonparametric and parametric distributions. A *parametric* distribution is based on a mathematical function which, combined with one or more distribution parameters, determines the distribution's shape and range. These parameters will often have little obvious or intuitive relationship to the distribution shapes they define. Examples of parametric distributions are lognormal, normal, beta, Weibull, Pareto, loglogistic, and hypergeometric — most distribution types, in fact.

Nonparametric distributions, on the other hand, have their shapes and range determined by their parameters directly in an obvious and intuitive way. Their distribution functions are simply mathematical descriptions of their shapes. Nonparametric distributions are uniform, general, triangle, cumulative, and discrete.

As a general rule, nonparametric distributions are far more reliable and flexible for modeling expert opinion. The questions that the analyst has to pose to the expert to define these distributions are intuitive and easy to respond to. Changes to these parameters will also produce an easily predicted change in the distribution's shape and range. There are three common exceptions to the preference for using nonparametric distributions to model expert opinion:

1. The BetaPERT distribution is frequently used to model an expert's opinion. Although it is, strictly speaking, a parametric distribution, it has been adapted so that the expert need only provide estimates of the minimum, most likely, and maximum values for the variable and the BetaPERT function finds a shape that fits. The BetaPERT distribution is explained more fully in Section 3.1.4.
2. The expert may occasionally be very familiar with using the parameters that define the particular distribution. For example, a toxicologist may regularly determine the mean and standard deviation chemical concentration in a set of samples. If a normal or lognormal distribution (which have mean and standard deviation as their parameters) were to be used to model the chemical concentration, it would be quite reasonable to ask the expert for estimates of the mean and standard deviation in this case.
3. The parameters of a parametric distribution *are* sometimes intuitive, and the analyst can therefore ask for their estimation directly. For example, a binomial distribution of the number of successes is defined by n, the number of trials that will be conducted, and p, the probability of success of each trial.

There are other problems associated with using parametric distributions that make their use undesirable when it is not completely necessary.

- The model that includes unnecessary parametric distributions is more difficult to review later because the parameters of the distribution have no intuitive feel, e.g., beta and lognormal.
- It is more difficult to persuade the decision maker of the validity of the model.
- It is more difficult to update the model in the light of new information.
- It is quite difficult to get the precise shape right when using parametric distributions to model expert opinion, as the effect of changes in the parameters are not usually obvious.

3.1.2 The Triangular (Triangle) Distribution

The triangle distribution is the most commonly used distribution for modeling expert opinion. It is defined by its minimum (a), most likely (b), and maximum (c) values. Figure 1a shows three triangle distributions: triangle(0, 10, 20), triangle(0, 20, 50), and Triangle(0, 50, 50), which are symmetric, right skewed, and left skewed, respectively. The triangle distribution has a very obvious appeal because it is so easy to think about the three defining parameters and to envisage the effect of any changes.

The mean and standard deviation of the triangle distribution are determined from its three parameters:

$$\text{mean} = (a + b + c) / 3$$

$$\text{standard deviation} = (a^2 + b^2 + c^2 - ab - ac - bc) / 18$$

From these formulas, it can be seen that the mean and standard deviation are equally sensitive to all three parameters. Many models involve the estimation of variables for which it is fairly easy to estimate the minimum and most likely values, but for which the maximum is essentially unbounded and could be enormous, for example, in estimations of cost and time to complete some task.

Central Limit Theorem, a fundamental principle of risk analysis modeling, tells us that, when adding up a large number of distributions (for example, adding costs or task durations), it is their means and standard deviations that are most important because *they* determine the mean and standard deviation of the risk analysis result. In situations where the maximum is so difficult to determine, the triangle distribution is not really appropriate, since the risk analysis result will depend a great deal on how the estimation of the maximum is approached. If the maximum is taken to be the largest *possible* value, the risk analysis output will have a far larger mean and standard deviation than if the maximum was assumed to be a practical maximum.

3.1.3 The Uniform Distribution

The uniform distribution is generally a very poor modeler of expert opinion, since all values within its range have equal probability density, but that density falls sharply to zero at the minimum and maximum in an unnatural way. It is rare indeed that the expert will be able to define the minimum and maximum, but have no

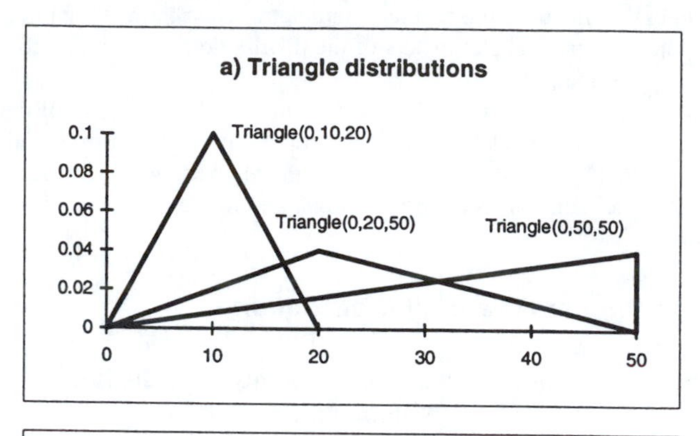

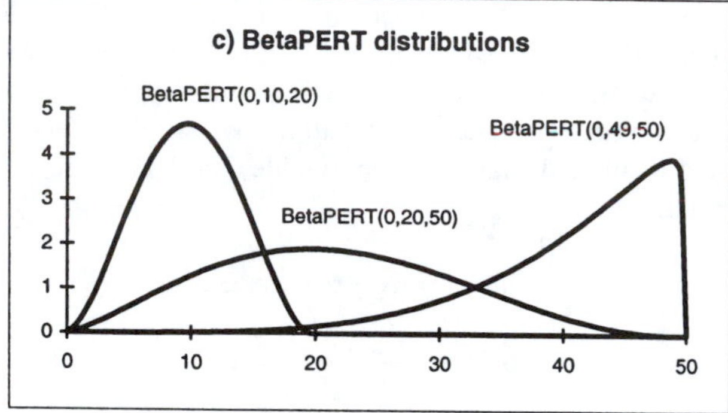

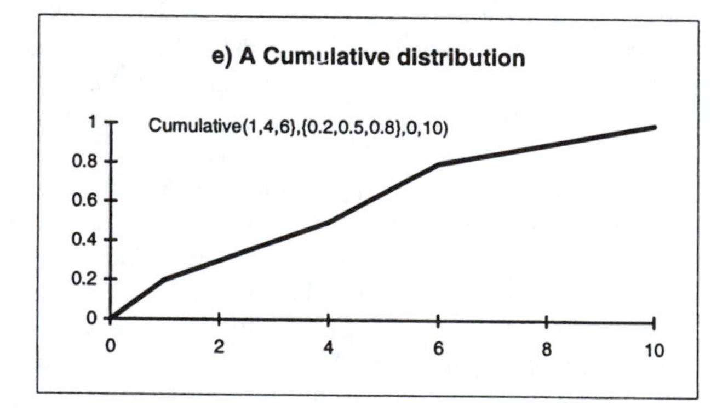

Figure 1 Examples of various types of distributions used in risk analysis modeling.

opinion to offer on central tendency. Figure 1b illustrates a couple of examples of the uniform distribution.

The uniform distribution does, however, have a couple of uses:

- To highlight or exaggerate the fact that little is known about the variable
- To model circular variables (like the direction of wind from 0 to 2π)

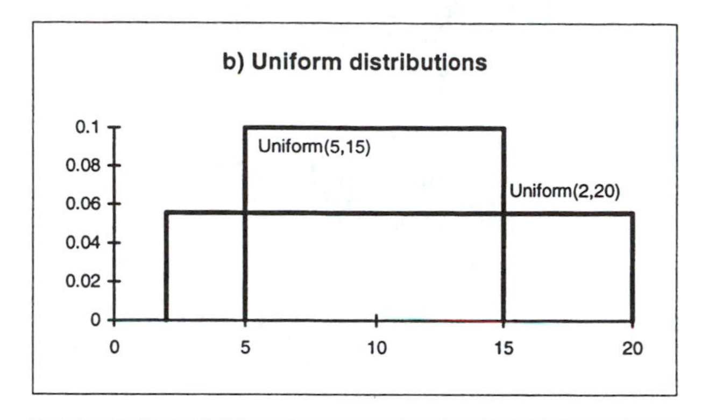

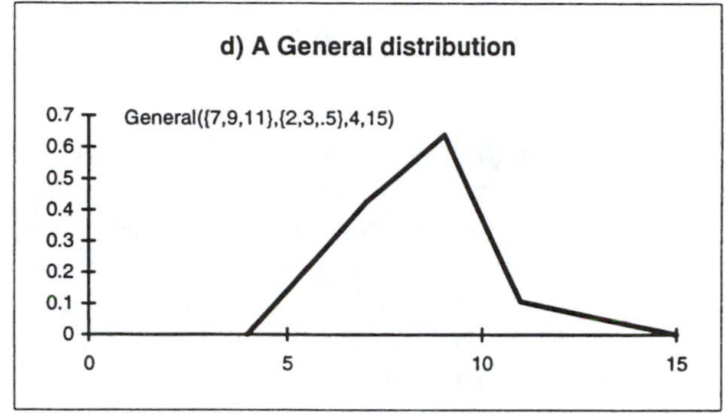

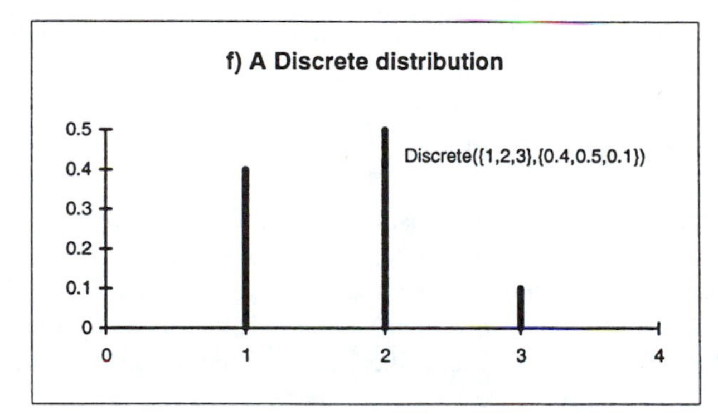

Figure 1 (Continued).

3.1.4 The BetaPERT Distribution

The BetaPERT distribution gets its name because of its use in PERT networks that estimate a project's duration and because it is a version of the beta distribution. It requires the same three parameters as the triangle distribution, namely, minimum (a), most likely (b), and maximum (c). The equation of a BetaPERT distribution is related to the beta distribution as follows:

$$\text{BetaPERT}\ (a, b, c) = \text{beta}(\alpha_1, \alpha_2)^*\ (c - a) + a$$

where

$$\alpha_1 = \frac{(\mu - a) * (2b - a - c)}{(b - \mu) * (c - a)}$$

$$\alpha_2 = \frac{\alpha_1 * (c - \mu)}{(\mu - a)}$$

$$\mu = \frac{a + 4 * b + c}{6}$$

The last equation for the mean (μ) is a restriction that is assumed in order to be able to determine values for α_1 and α_2. It also shows how the mean for the BetaPERT distribution is far more sensitive to the most likely value and correspondingly less sensitive to the minimum and maximum values than the mean of the triangle distribution. It therefore does not suffer to the same extent the potential systematic bias problems of the triangle distribution in producing too large a value for the mean of the risk analysis results. The standard deviation of a BetaPERT distribution is also less sensitive to the estimate of the extremes.

Figure 1c shows three BetaPERT distributions whose shape can be compared to the triangle distributions of Figure 1a.

3.1.5 The General Distribution

The general distribution is the most flexible of all of the distribution functions. It enables the analyst and expert to tailor the shape of the distribution to reflect, as closely as possible, the opinion of the expert. The general distribution has the form general($\{x_i\}$, $\{p_i\}$, minimum, maximum), where $\{x_i\}$ is an array of x-values with probability density weights $\{p_i\}$ and where the distribution falls between the minimum and maximum. Figure 1d shows a general({7, 9, 11}, {2, 3, 0.5}, 4 , 15) distribution. Use of the general distribution is discussed in more depth in Section 3.3.

3.1.6 The Cumulative Distribution

The cumulative distribution has the form cumulative($\{x_i\}$, $\{P_i\}$, minimum, maximum), where $\{x_i\}$ is an array of x-values with cumulative probabilities $\{P_i\}$ and where the distribution falls between the minimum and maximum. Figure 1e shows the distribution cumulative({1, 4, 6}, {0.2, 0.5, 0.8}, 0, 10) as it is defined in its cumulative form. The cumulative distribution is used in some texts to define expert opinion. However, I have found it largely unsatisfactory because of the insensitivity of its probability scale. Therefore, I usually prefer to use the general distribution instead. One circumstance where the cumulative distribution is very useful is in attempting to estimate a variable that may cover several orders of magnitude. For example, the number of bacteria in 1 kg of meat will increase exponentially with time. The meat may therefore easily contain 100 units of bacteria or 1 million.

3.1.7 The Discrete Distribution

The discrete distribution has the form discrete($\{x_i\}$, $\{p_i\}$), where $\{x_i\}$ are the possible values of the variable and $\{p_i\}$ are the relative likelihood of each x-value. The $\{p_i\}$ values do not have to add up to 1 as the software will normalize them automatically. It is actually often useful just to consider the ratio of likelihood of the different values and not to worry about the actual probability values. The discrete distribution has three distinct uses:

1. To model a discrete variable (i.e., a variable that may take one of two or more distinct values), e.g., the number of turbines that will be used in a power station. Figure 1f illustrates a discrete({1,2,3},{4,5,1}) distribution.
2. To model a variable that may be affected by a definable event, e.g., sales volume that will drop if a new competitor enters the market. This is known as conditional branching. Figure 2 illustrates an example.
3. To combine two or more dissimilar expert opinions (see Section 3.3).

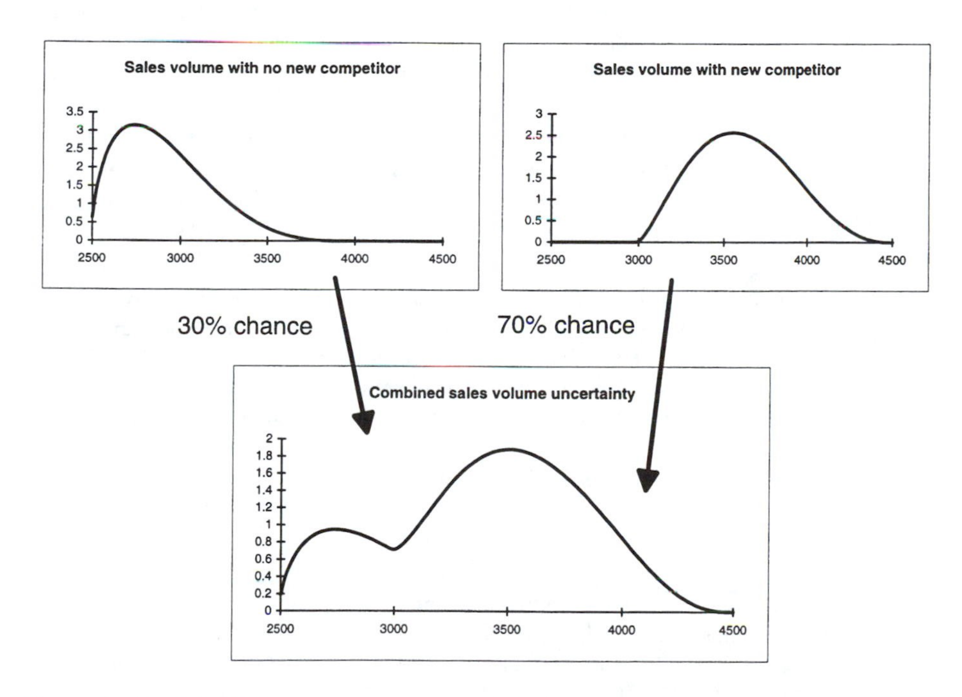

Figure 2 Using a discrete distribution to model conditional branching.

3.2 Developing Input Distributions from Data

Historic or test data are often available that relate to an uncertain variable within the risk analysis model. In such circumstances, it is clearly sensible to use that data to define the distribution of the variable's uncertainty. The following points should then be borne in mind.

- Is the distribution of uncertainty in the past data likely to be pertinent to the uncertainty of the future? Changes in the political, social, or commercial environment, for example, could make past data irrelevant.
- Is the sample size sufficiently large? The more data available for analysis, the more precisely the distribution can be defined. Can more data be collected easily and cheaply to improve the accuracy of the distribution?
- Is the parameter independent of all other factors in the model? For example, a bank's mortgage rate is strongly dependent on the base interest rate, both of which may appear in a model of the bank's revenue. Scatter diagrams of past data from two variables can be plotted together to assess the strength of any dependency.
- Is the data reliable? If there is some doubt, can it be cross-referenced against another source?

There may be a distribution whose theoretical derivation matches the variable being modeled. For example, a Weibull distribution is derived from a mathematical model of time until failure of a piece of equipment and is therefore likely to closely fit data of observed lifetimes of a device. Alternatively, there may be a distribution that is well known to empirically fit this type of variable. For example, the Rayleigh distribution is known to closely match the observed distributions of wave heights, even if the reason why is unclear.

If the observed data is continuous and reasonably extensive, it is usually sufficient to use a cumulative frequency plot of the data points to define its probability distribution. Figure 3 illustrates an example, fitting a distribution to 18 data points. The procedure is as follows:

- The minimum and maximum for the empirical distribution are subjectively determined based on the analyst's knowledge of the variable. For a continuous variable, these values will generally be outside of the observed range of the data. The minimum and maximum values selected here are 0 and 45.
- The data points are ranked in ascending order between the minimum and maximum values.
- The cumulative probability $P(x)$ for each x-value is calculated as follows:

$$P(x_i) = \mathrm{i} \,/\, (n + 1)$$

- The value $P(x_i) = i \,/\, (n + 1)$ is used because it places all of the x_is against the expected cumulative percentile that would be observed if the data points were randomly selected from a distribution. Therefore, it maximizes the chance of replicating the true distribution.

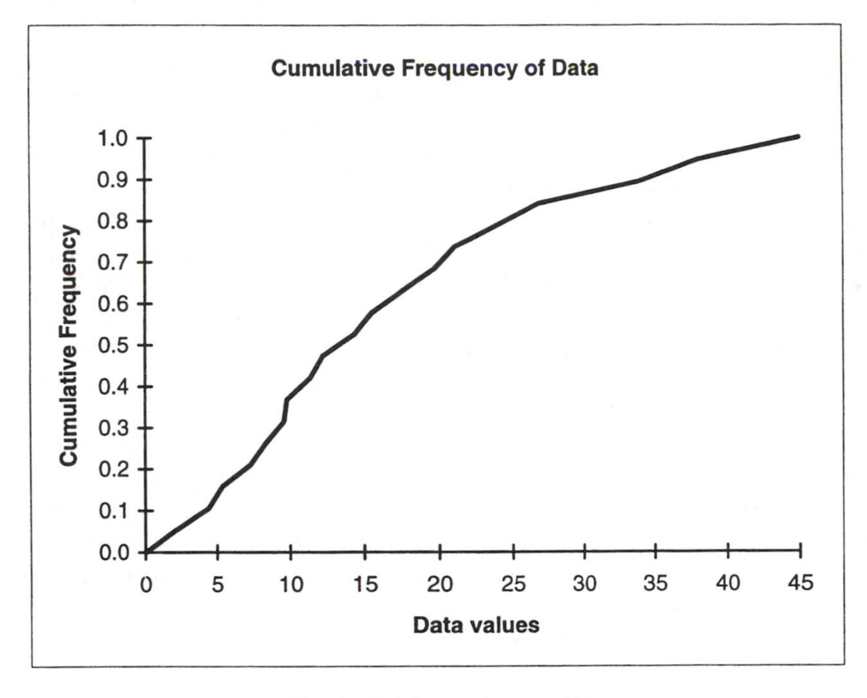

Number of data points n = 18

i	x	F(x) = i/19
	0.000	0.000
1	2.058	0.053
2	4.382	0.105
3	5.348	0.158
4	7.296	0.211
5	8.334	0.263
6	9.596	0.316
7	9.805	0.368
8	11.385	0.421
9	12.241	0.474

i	x	F(x) = i/19
10	14.392	0.526
11	15.566	0.579
12	17.615	0.632
13	19.745	0.684
14	21.077	0.737
15	23.894	0.789
16	26.804	0.842
17	33.879	0.895
18	38.018	0.947
	45.000	1.000

Figure 3 Fitting a cumulative distribution to data.

- These two arrays, $\{x_i\}$, $\{P(x_i)\}$, can then be used as direct inputs into a cumulative distribution.

3.3 Deriving Input Distributions from Expert Opinion

A great deal of risk analysis modeling relies to some extent on input from the opinion of experts. These experts are often inexperienced in providing their opinion of future outcomes in the form of probability distributions. It is the job of the risk analyst to try to guide them through the process as efficiently and accurately as possible.

First, there is often an objection by the expert — he/she thinks it is a waste of time with comments like "things are so uncertain at the moment, it is hard enough to give you a most likely figure, let alone a whole probability distribution." Since risk analysis is a relatively new field, the analyst may first have to persuade the expert of the value of the exercise to gain his or her cooperation.

It should be pointed out that, by giving the expert the opportunity to describe the whole range of possible values that he/she envisages a variable could take, rather than just giving a most likely value, risk analysis is relieving the expert from having to provide an estimate that he/she knows could never exactly occur.

Eliciting expert opinion may begin, where appropriate, by holding a brainstorming session with all the available people who are capable of providing an informed opinion of the problem being modeled. The analyst presents all of the relevant information to the group, and the group uses this time to discuss the structure of the problem and the sources of uncertainty within it. While this is expensive in terms of personnel time, the process has a lot to offer as a forum to focus on risks. It is probable that those present are not equally aware of the risks and may even be able to suggest tactics, etc. to reduce or eliminate them. In such a forum, the risk analyst acts as a catalyst for discussion, makes sure that the discussion is not dictated by the loudest voice, and tries to illustrate the structure of the problem as it develops.

After the brainstorming session, the analyst can interview each participating expert individually to elicit his/her opinion of distributions for uncertain variables within the model. Risk analysis texts offer several interviewing techniques, most of which I have found to be quite restrictive and difficult to explain to the expert. Therefore, I offer instead the following procedure that I have personally found to be very fruitful:

- Ensure that the expert has had access to all the relevant information, qualitative and quantitative, and has had time to digest it.
- Encourage discussion of the structure of the estimating problem and be open to modeling the parameter in a different way to the one you had visualized.
- If possible, break down the quantity to be estimated into manageable components.
- Through discussion of the problem, feel out the scenarios that would define the minimum and maximum values of the variable. Once the minimum and maximum have been determined, look at what value within that range would be the most likely to occur. These three figures can then be used to define a BetaPERT or triangle distribution. Leaving the estimate of the most likely value to the last also helps reduce a potential source of bias known as *anchoring* that has the effect of underestimating the degree of uncertainty.
- If the variable needs to be defined more precisely than a three-point estimate allows, draw out the range on graph paper and, through discussion, encourage the expert to sketch out a feel of the shape of the relative frequency distribution. Cross-reference this shape by comparing relative heights for some values within the range.

RISKview Pro (Palisade Corporation, Newfield, New York) is a unique software package that allows this to be done very neatly on the PC screen. Use decision-tree sketches if the parameter is dependent on discrete events.

- When a general shape has been agreed upon, get the expert to sign and date it. Offer the opportunity for revision at a later date, and give the expert a copy of the plot.
- Using a ruler and pen, convert the plot to a general distribution. This will go straight into the model as it is (RISKview Pro will do this automatically). Give a copy of the distribution to the expert.
- If the expert has drawn out the problem as a composite of several discrete but uncertain events, use the discrete distribution referencing the general distribution for each scenario with their associated probabilities.

The techniques described here avoid asking the expert for statistical parameters, like mean and standard deviation, to define the distribution. This is very important. People usually have little intuitive feel for the meaning of statistical parameters. For example, mode is often confused with mean and standard deviation is sometimes confused with range (maximum to minimum).

When trying to assign probabilities to a parameter, avoid using betting comparisons. The expert may be a real gambler, or quite the opposite, and inadvertently allow his/her attitude about betting to distort the representation of his/her opinion.

Avoid questions of differentiation at the extremes of the distribution, i.e., very low or high probabilities: e.g., What value do you think has a 95% chance of being exceeded? Now, what value . . . 97%? . . . 99%? Such questioning is of little use because people find it very difficult to distinguish between such small separations of probability. This degree of attempted exactness would generally have very little impact on the final result anyway.

3.3.1 Combining Two Dissimilar Expert Opinions

Experts will sometimes produce profoundly different probability distribution estimates of a variable. This is usually because the experts have estimated different things, made differing assumptions, or have different sets of information on which to base their opinion. However, occasionally two or more experts simply genuinely disagree. How should the analyst approach the problem? The first step is usually to confer with someone more senior and find out whether one expert is preferred over the other. If those more senior have some confidence in both opinions, a method is needed to combine these opinions in some way. I have used the following method for a number of years with good results.

Use a discrete($\{x\}$,$\{p\}$) distribution where the $\{x\}$ are the expert opinions and the $\{p\}$ are the weights given to each opinion according to the emphasis one wishes to place on them. Figure 4 illustrates an example where two experts have offered the differing opinions as shown. Expert A is considered by their management to have more experience than expert B, so their opinions are weighted 60:40.

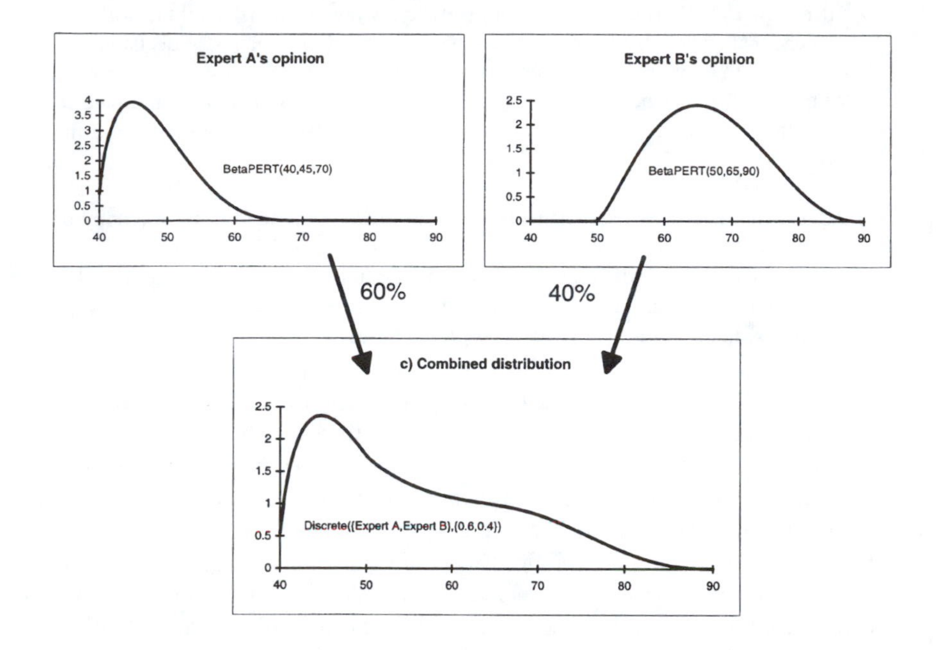

Figure 4 Combining two dissimilar expert opinions.

4. TECHNIQUES FOR MODELING DEPENDENCIES

A variable in a model is dependent on another if the value it can take is at least partially determined by the value of the other. For example, a bank's model of mortgage revenue might include distributions for the number of new mortgages sold and the mortgage rate. Past experience shows them that the higher the mortgage rate, the fewer new mortgages they will sell. Therefore, the sampling from these two distributions must be constrained so that the model will not produce impossible scenarios, like a very high mortgage rate at the same time as a very high number of new mortgages sold.

4.1 Using Rank Order Correlation to Model Dependencies

The easiest method for constraining the sampling of these two distributions is to correlate them using a rank order correlation, a feature common to most risk analysis software. Examples of the dependencies generated by rank correlation are shown in Figure 5. The closer the correlation coefficient (r) is to 1 or –1, the tighter the dependency. A positive correlation means that high values from both distributions will generally occur together, likewise with low values. A negative correlation means that a high value from one distribution is more likely to occur at the same time as a low value from the other distribution, and vice versa.

The limitation of this technique is that both distributions are being defined first and then attempts are made to manipulate the sampling from these distributions to reproduce the desired dependency. The more logical, but far more laborious, approach is to define the distribution of the independent variable and then develop a mathematical relationship that at least partially determines the value of the dependent variable.

In choosing which method to employ, it is practical to first estimate the impact on the model's results on the accuracy of your dependency relationship. If changing the degree of correlation produces little change in the model's output, the dependency is relatively unimportant, and the simpler rank order correlation approach will suffice.

4.2 Modeling Dependencies from Expert Opinion

This section explains a method for modeling an expert's opinion of a dependency relationship. First of all, it is presumed that the distribution for the independent variable has been defined and therefore its minimum and maximum known. The independent distribution may have come from expert opinion or from analysis of data. It is also presumed that the expert has had all the relevant past data possible made available to him/her to develop his/her opinion.

A graph is drawn with the independent's range on the x-axis and the estimated dependent's range on the y-axis (Figure 6a). We will use base interest rate and mortgage rate as an example here. Clearly, the interest rate is the independent as it affects the mortgage rate and not the other way round. There is also clearly a logical link between the two parameters. This should always be understood.

Looking at the minimum value for the interest rate, the expert is asked what minimum, most likely, and maximum value he/she would give for the mortgage rate at this value of the interest rate. This same question is asked at the maximum interest rate and a few values in between.

A least squares fit is then calculated through the minima, most likely, and maxima values, and these are then brought together into the one distribution.

In this example, we have used a triangular distribution triangle(minimum, most likely, maximum), so the final equation for the mortgage rate's (MR) dependency on the interest rate (IR) is

$$MR = \text{triangle}(1.4*IR-0.4, 1.6*IR+1.0, 1.9*IR+2.8)$$

Now, each time a value for the interest rate is generated, a corresponding related value for the mortgage rate is also generated. For example, if a value of 6 is generated for the interest rate,

$$1.4*IR-0.4, \rightarrow \text{minimum MR} = 8.0$$

$$1.6*IR+1.0, \rightarrow \text{most likely MR} = 10.6$$

$$1.9*IR + 2.8, \rightarrow \text{maximum MR} = 14.2$$

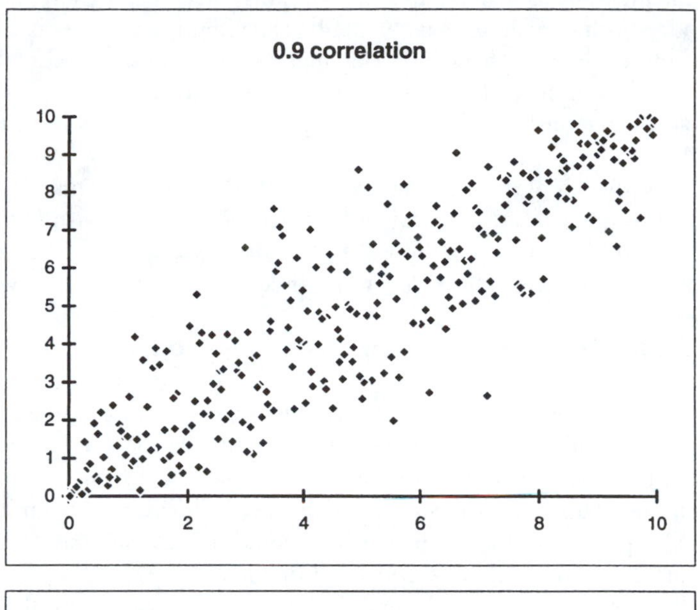

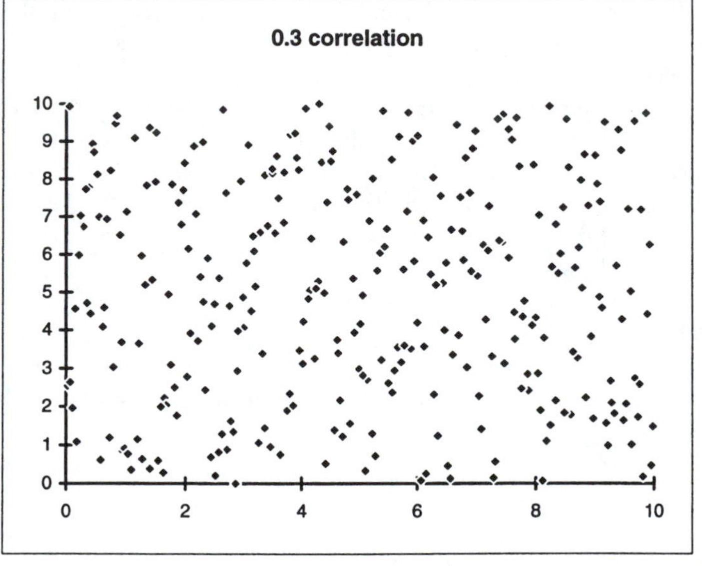

Figure 5 Examples of rank order correlations of two uniform(0,10) distributions.

So the mortgage rate will be generated from the triangle(8.0,10.6,14.2). Figure 6b illustrates how this formula behaves in a simulation of 300 iterations.

Although we have used a triangle distribution for this example, we could just have easily used a BetaPERT distribution. This is an excellent technique because it produces a dependency relationship that is logical to define and far more intuitive than rank order correlations.

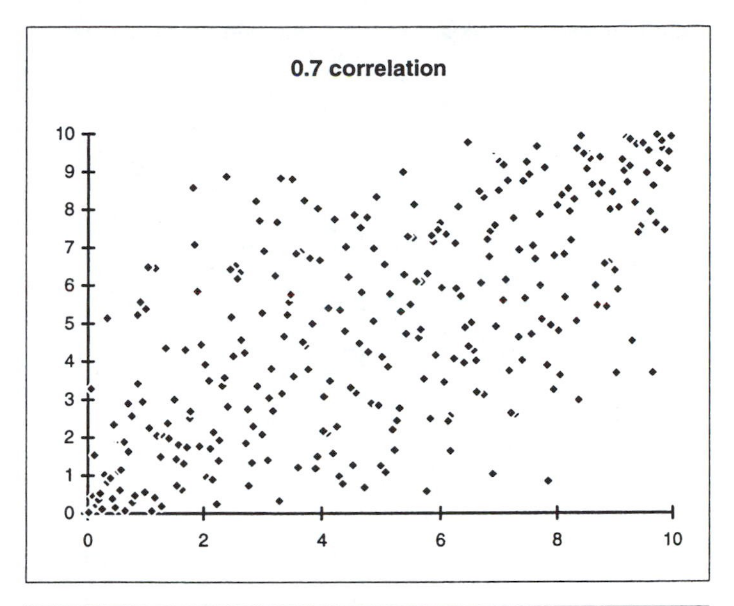

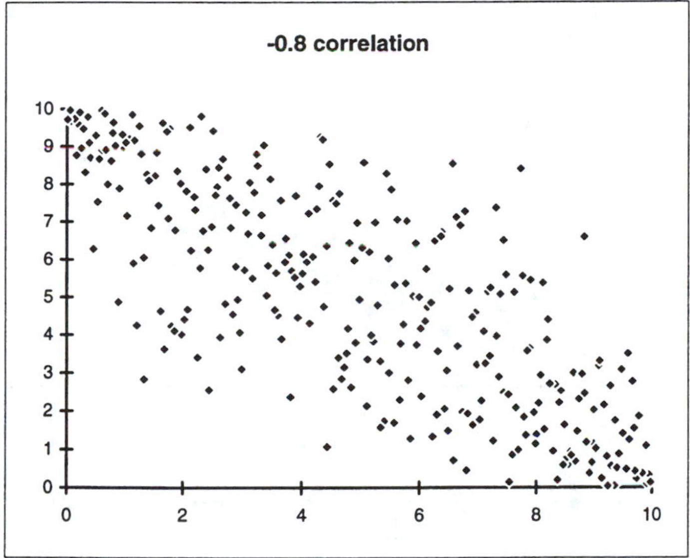

Figure 5(Continued).

5. PRESENTATION AND INTERPRETATION OF RISK ANALYSIS RESULTS

The risk analysis model and its results must be presented to the decision makers and other stakeholders in a manner that gives confidence in the validity of the model's structure and assumptions and presents the results of the risk analysis in a format

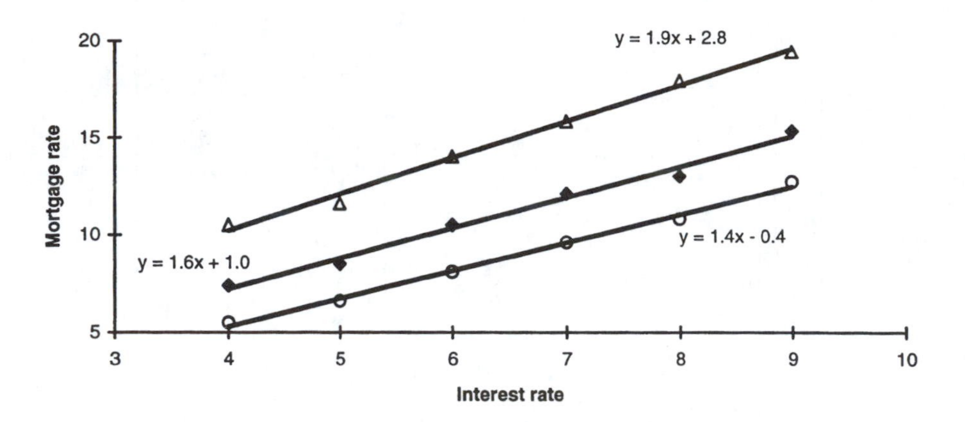

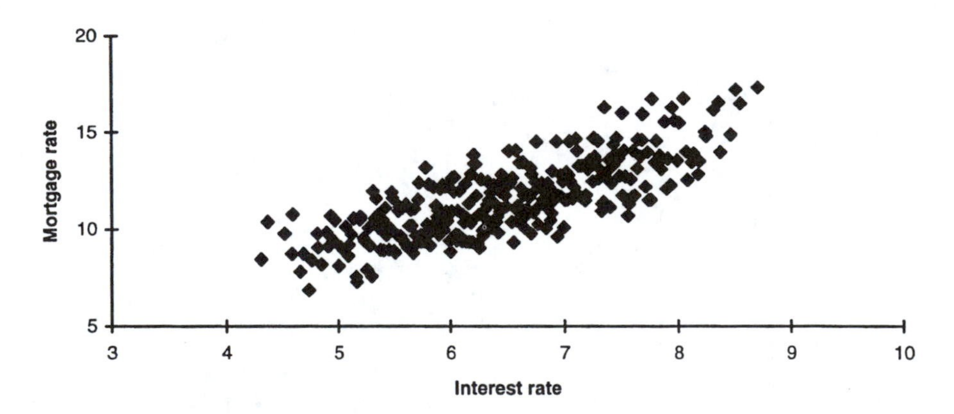

Figure 6 Modeling an expert's estimate of dependency between two variables.

that is easy to interpret and use. The remainder of this section considers these two aims in more detail.

5.1 Presenting the Model and Its Assumptions

Risk analysis models are inherently more difficult to verify than their deterministic counterparts. This means that the model's results may be viewed with suspicion by those who have to rely on them to make their decisions. Of course, unless the decision maker accepts the validity of the model, there will have been no point in going through the exercise in the first place.

Conversely, the difficulty in verifying a risk analysis model often leads to the blind acceptance of its results. Appropriate presentation of the model's structure and assumptions will help to identify any mistakes that have been made and to gain acceptance for a good model's results.

The structure of the model can be presented in any number of ways: from schematic diagrams, decision trees, and influence diagrams to printouts of spreadsheets and Gantt charts. The selection will obviously depend on the type of problem and its degree of complexity. A more complex model may be better illustrated by a set of diagrams rather than a single format.

The assumptions of the model can also be presented in a number of ways. Estimates of uncertain variables can be presented in tables if they have been simply defined (e.g., minimum, most likely, maximum). They may be easier to review as thumb-nail sketches if the modeling distribution is complex. Dependencies between distributions are best shown as scatter plots. A time series is usually most easily presented in the form of a summary chart. An example of an assumptions report showing these features is given in Figure 7.

5.2 Presenting Risk Analysis Results

The outputs from a risk analysis model should obviously be presented in a manner most appropriate to the questions the model is attempting to answer. Relative frequency plots (histograms) are very useful for giving a feel for the shape and degree of uncertainty (Figure 8). On the other hand, cumulative frequency plots enable the decision maker to read off relevant target values. For example, Figure 9 illustrates how the cumulative frequency plot can be used to determine a bid price, based on the expected cost to the contractor and the risk contingency they would add. One can read off statistics, such as the cost has an 85% chance of being below £148,000 and a 20% chance of being below £119,000 so there is also a 65% chance of being between these two values.

Scatter plots can be extremely useful for reviewing complex interrelationships between two components of a model. They are also often used as a check that the model is only generating realistic scenarios.

Very comprehensive statistical reports are a standard feature of Monte Carlo software. Most of the statistics calculated in these reports will have no relevance to, nor will they be easily interpreted by, the decision maker. It is therefore a shame that these reports are often relied upon as the primary vehicle for communicating the model's results. The statistics section of a risk analysis report should be pared down to provide only those statistics that are relevant to the problem. The mean, standard deviation, and a few selected cumulative percentiles are usually quite sufficient.

Tornado charts provide a further insight into the model by illustrating the relative contribution of the uncertain input variables to the uncertainty of the model's outputs (Figure 10). The length of the bars in the chart represent the rank order correlation between the input variables and an output. The higher the correlation, the greater the influence that the input variable has on the output. This is often very valuable information: it picks out the variables that would be worth more accurately defining,

KEY UNCERTAINTY ASSUMPTIONS

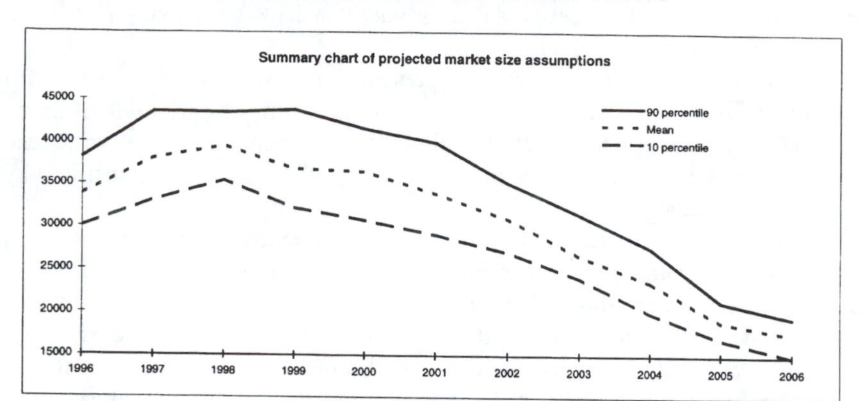

Year	1996	1997	1998	1999	2000	2001	2002	2003	2004	2005	2006
10 percentile	30000	33000	35300	32100	30700	29030	27000	24000	20000	17000	15000
Mean	33803	37949	39399	36690	36388	33848	30902	26600	23556	19020	17581
90 percentile	38131	43479	43355	43664	41429	39859	35101	31406	27547	21346	19485

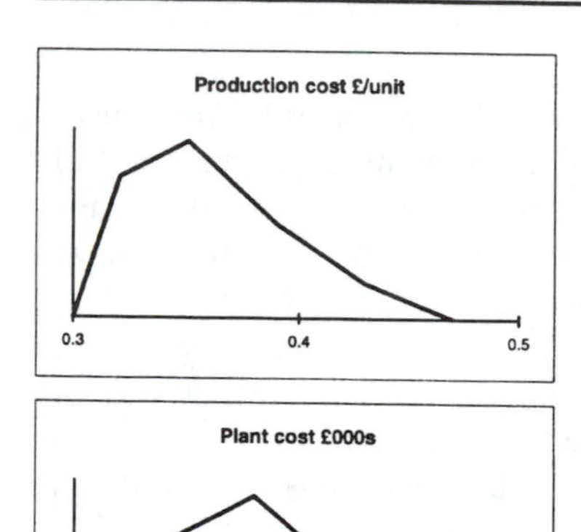

BetaPERT distribution parameters used for key quantities			
Variable	minimum	m. likely	maximum
Labour rate £/day	51	52	55.5
Advertising budget £k/year	17.2	20.3	24.1
Administration costs £k/year	173	176	181
Transient market share	0.13	0.17	0.19
Commission rate	0.14	0.145	0.18
Factory rental £k/year	172	174	181

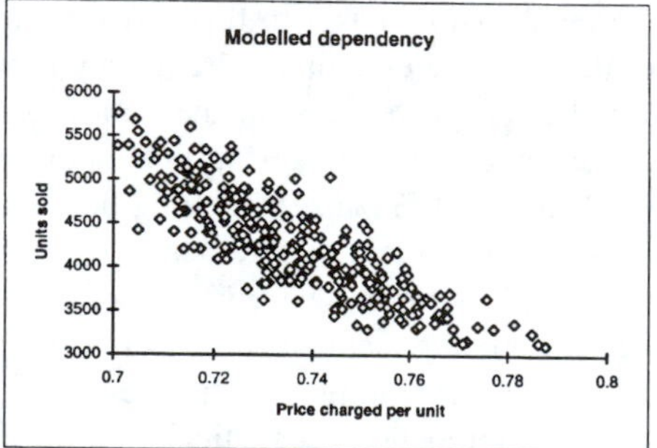

Figure 7 An example of the presentation of a model's assumptions.

thereby reducing the uncertainty of the problem; and it can be used to determine how many of the key uncertainties are under the management of the organization. The fewer key uncertainties that are under an organization's control, the more exposed it is to the problem's uncertainties.

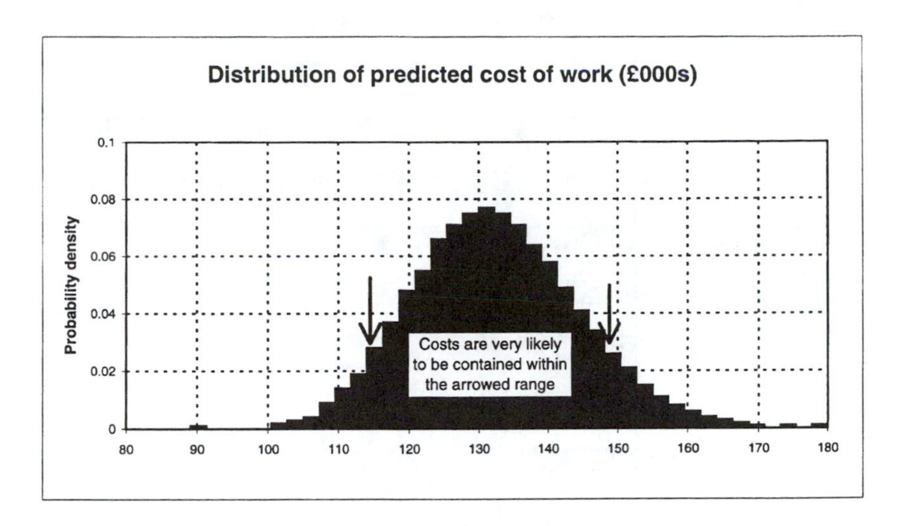

Figure 8 An example of a relative frequency plot for a model output.

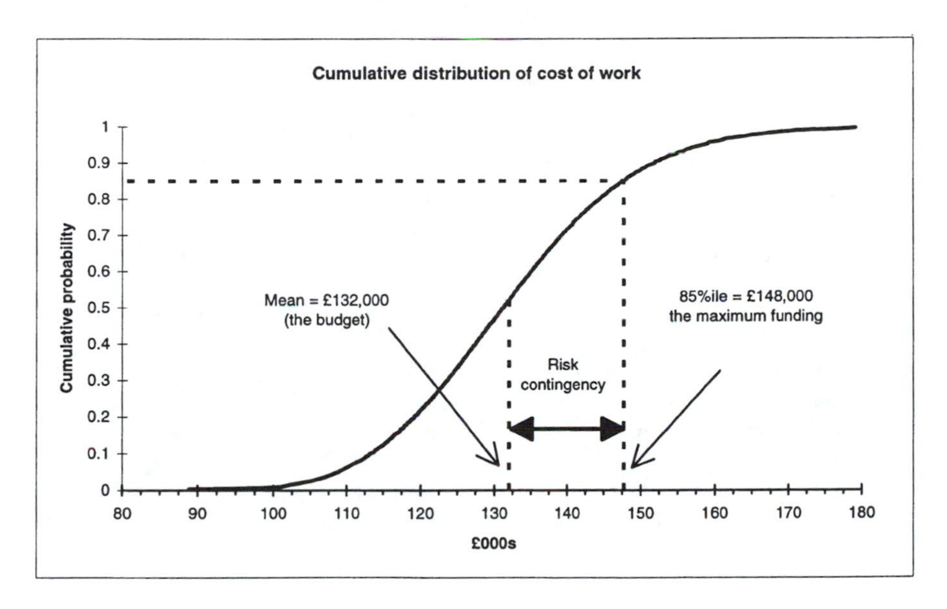

Figure 9 Using a cumulative frequency plot of the estimated total cost of a project to determine a bid price.

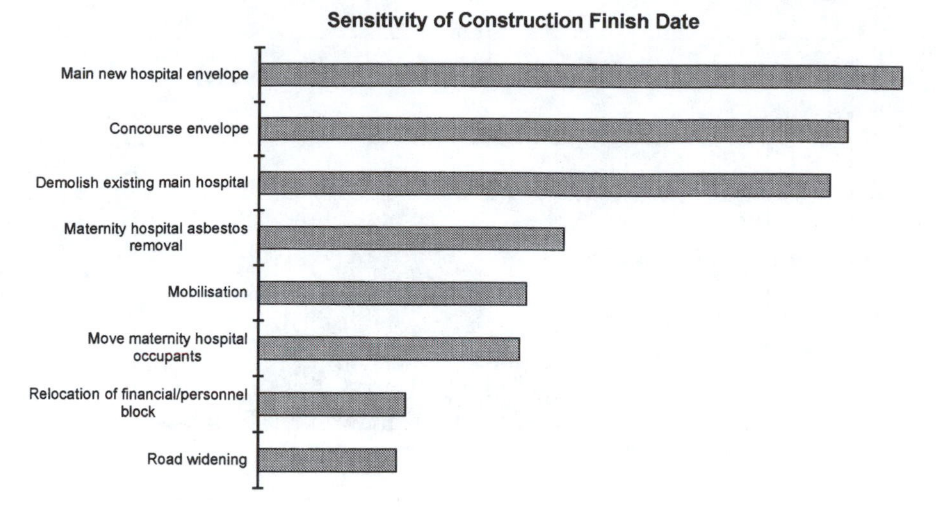

Figure 10 An example of a tornado chart.

QUESTIONS

1. How can the preparation of a risk analysis model benefit its participants?
2. Why is Monte Carlo modeling an improvement over single-point estimate models?
3. If a breed of pigs weight is normally distributed with mean of 100 kg and standard deviation of 10 kg,
   - What is the probability of any pig weighing between 80 and 110 kg?
   - What is the probability of a pig weighing exactly 100 kg?
   - What is the probability that 100 randomly selected pigs will weigh less than 10,200 kg?
4. What are the disadvantages of using bounded distributions in modeling expert opinion?
5. If BetaPERT distributions are used instead of triangle distributions, what would you expect to be the effect on the size of the risk contingency, as defined in Figure 9?
6. Name three examples in your area of work of variables that would need to be correlated.
7. If the result of a Monte Carlo model is the cost (£) of a project, what units do the following result statistics have — mean, mode, median, standard deviation, variance, skewness, kurtosis?

CHAPTER I.5

An Overview of Probabilistic Risk Analysis for Complex Engineered Systems

Vicki M. Bier

SUMMARY

Probabilistic risk assessment (PRA) was developed to facilitate the quantification of risks associated with complex engineered systems. It is particularly appropriate for analyzing the frequencies of extremely rare events, such as core melts in nuclear reactors, for which little if any accident data will be available. To the extent possible, PRA models are hierarchical in nature. This provides a way of structuring the vast quantities of information that go into a risk analysis. In particular, two reliability analysis techniques are commonly used for quantifying the likelihood of an accident: fault trees and event trees. Many PRAs use event trees to model major plant systems and fault trees to quantify the failure probabilities of the various systems.

A variety of different types of data are needed to support PRA quantification. This includes data on initiating event frequencies (i.e., the frequency of departures from normal operation), component failure rates, common cause failure rates (i.e., the frequency with which two or more components fail during a short period of time for the same reason), component maintenance frequencies and durations, component fragilities (i.e., component failure probabilities as a function of exogenous stresses such as earthquakes, fires, floods, or high temperatures), and human error rates. For most of these data needs, several different types of information may be available, including not only component-specific information (e.g., the number of observed failures of each component), but also expert opinion and data on the failure frequencies of similar components at other plants. Bayesian data analysis is often used to combine generic and component-specific information.

1-56670-130-9/97/$0.00+$.50

Once a plant-specific risk model has been developed and quantified, the model can then be used for risk management purposes, and in fact, the use of PRA has resulted in a number of examples of successful risk management involving relatively inexpensive but highly effective risk reduction options. In recent years, such risk management applications have increasingly been undertaken based on risk analyses performed by plant staff rather than consultants, reflecting the successful diffusion of risk analysis technology into the mainstream of engineering applications.

Key Words: probabilistic risk assessment, reliability analysis, fault trees, event trees, data analysis, risk management

1. WHY PROBABILISTIC RISK ANALYSIS?

Unfortunately, accidents happen, for a variety of reasons. Possible accident causes include equipment failures, natural disasters (such as earthquakes), human errors, and faulty operating procedures. As our technology becomes more powerful, the potential consequences of those accidents become more severe. Examples in recent years include the core melt accidents at Three Mile Island in Pennsylvania and Chernobyl in Russia, the loss of the U.S. Challenger space shuttle, and the toxic chemical release in Bhopal, India. Even accidents with little or no public health consequence, such as Three Mile Island, can be extremely costly, not only to the owner of the specific facility where the accident occurred, but also to an entire industry and to society as a whole. Therefore, while accidents can never be entirely prevented or eliminated, it is often important to reduce their frequency. A necessary first step in such risk reduction efforts is to be aware of the risks associated with the system in the first place.

Probabilistic risk assessment (PRA) was developed to facilitate the quantification of risks associated with complex engineered systems. It is particularly appropriate for analyzing the frequencies of extremely rare events, such as core melts in nuclear reactors or chemical plant accidents, for which little if any accident data will be available. Several features of PRA are worth noting.

1. PRAs are generally designed to model the response of a complex engineered system to *disturbances during operations*. Many systems, such as nuclear reactors or oil refineries, spend most of their time in steady state, and, as long as the system remains in steady state, it poses little if any risk. Even systems such as the space shuttle, which go through several different states over a relatively short period (e.g., launch, orbit, and landing), still pose little risk if they continue to operate as designed. Risk arises primarily when some event (e.g., an equipment failure, a natural disaster, or human error) disturbs the system from its intended mode of operation.
2. PRA provides an *integrated model of system response*. The ordinary engineering design process does a reasonably good job of designing the individual components or subsystems that make up a complex engineered system. However, to simplify the difficult and complicated task of system design, this is generally done by

specifying the boundary conditions under which each subsystem is expected to operate (e.g., sources of electric power, cooling water, etc.) and performing detailed engineering design of each subsystem individually. Thus, the dominant sources of risk usually arise either from *interactions between subsystems* (e.g., situations in which one subsystem fails and thus changes the environment faced by other subsystems) or from events such as *natural disasters* that take the system outside its usual operating envelope (so that the boundary conditions specified in the design process are no longer valid). PRA supplements the engineering design process by providing a holistic view.

3. Ideally, PRAs should identify the *types and levels of damage* that could result from different system responses. For example, there may be significant differences between accidents that result primarily in lost productivity and on-site repair costs and accidents that result in significant off-site damage or public health effects. Similarly, even accidents that jeopardize the health of the general public can differ according to their severity, the persistence of the hazard, the amount of lead time available for warnings or evacuation, etc. A comprehensive and well-designed PRA can help identify the plant conditions likely to result in different types of accidents and the different risk management strategies that might be appropriate for each one.
4. PRAs should provide not only qualitative assessments of system performance (e.g., safe or unsafe; high, medium, or low risk), but also *quantitative measures of risk*. Qualitative assessments provide little basis for evaluating the cost-effectiveness of alternative risk reduction actions or for determining the relative importance of different risk contributors. By contrast, quantitative estimates of accident frequencies or probabilities (even if only approximate or highly uncertain) provide a more rigorous basis for making such judgments.
5. A PRA should also include a *quantitative assessment of the uncertainty in the results*. Particularly when failure data are sparse (e.g., when components are either highly reliable or else relatively new and untested), large uncertainties may exist about component failure rates. The analyst may also be uncertain about issues such as the system success criteria (e.g., how many pumps must be operable in order for the system to perform its function) or the accuracy of the system model. While such uncertainties can be difficult to quantify, a point-estimate result that provides no indication at all of the extent of uncertainty is less valuable than even an approximate statement of uncertainty.
6. Finally, a PRA should provide not only an assessment of the current level of risk, but also information on *risk contributors and potential risk management actions*. Without such information, the only decisions available after the PRA is completed will be either to accept the status quo and continue operating or to shut the system down (typically at great cost). Instead, a well-designed PRA should be a tool to help facility owners and operators make good decisions about system design modifications, operations, and maintenance.

To summarize, a PRA assesses how well a plant or system responds to a variety of situations. As pointed out by Kaplan and Garrick (1981), the PRA should answer three basic questions:

1. What can go wrong?
2. How likely is it to go wrong?
3. What will be the consequences if it does?

The first question is answered by a structured list of possible accident scenarios. The second question is answered by quantifying the likelihood of each scenario (including the uncertainty about that likelihood). Finally, the consequences of an accident can be assessed in terms of a variety of damage indices. Examples include the state of the plant itself (e.g., the cost of repair or the operability of the remaining subsystems), the amount of material and/or energy released to the environment as a result of the accident, and the off-site consequences of the accident (e.g., property damage, public health effects).

Results can be presented graphically in a variety of formats. For example, probability distributions can be used to display uncertainty about scalar quantities, such as the frequency of an accident (see Figure 1). For damage types that involve different levels of severity (e.g., repair costs, numbers of fatalities), a complementary cumulative distribution function (CCDF) can be used to show the frequency, Φ, of exceeding any given damage level x (see Figure 2). However, such CCDFs still represent only point estimates of risk, since they do not display the uncertainty about the accident frequency (Kaplan et al. 1981). Uncertainty about such functions can be displayed by presenting a family of possible CCDFs $\Phi(x)$, possibly indexed by their probability, p, as shown in Figure 3.

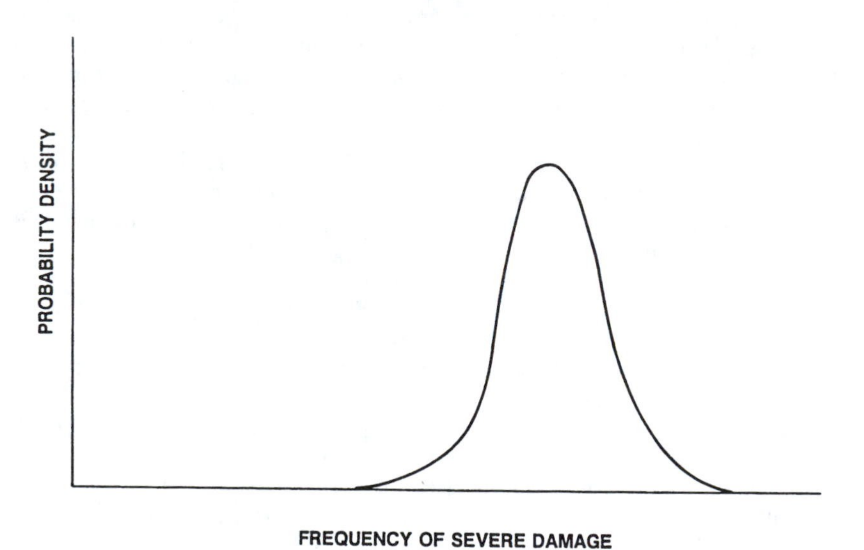

Figure 1 Probability distribution for an accident frequency. (From Kaplan et al., *Methodology for Probabilistic Risk Assessment of Nuclear Power Plants,* Pickard, Lowe and Garrick, Inc., Newport Beach, CA, 1991. With permission.)

2. STRUCTURE OF THE PRA MODEL

To the extent possible, PRA models are hierarchical in nature. This provides a way of structuring the vast quantities of information that go into a risk analysis. First, PRAs are categorized into three levels, according to scope (American Nuclear

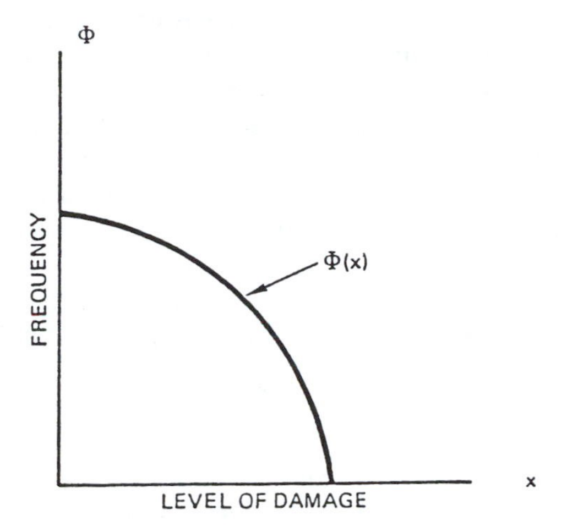

Figure 2 Complementary cumulative distribution function. (From Kaplan et al., *Methodology for Probabilistic Risk Assessment of Nuclear Power Plants,* Pickard, Lowe and Garrick, Inc., Newport Beach, CA, 1991. With permission.)

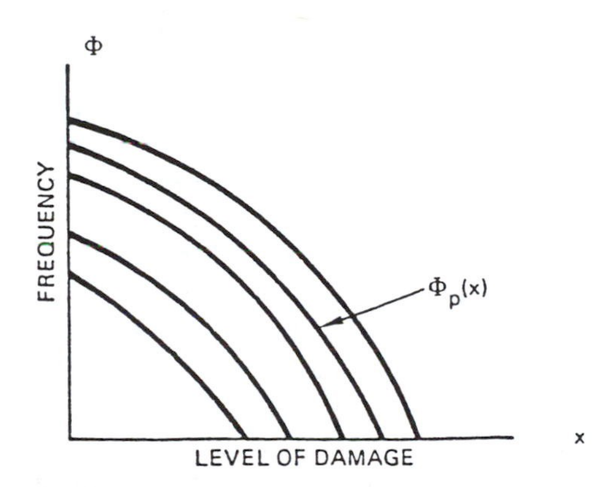

Figure 3 Family of possible complementary cumulative distribution functions.

Society and Institute of Electrical and Electronic Engineers 1983). A *level 1* PRA determines the likelihood of an accident and possibly also additional damage indices relating to the state of the plant itself (e.g., which postaccident safety systems are operational), but not the amount of material or energy released as a result of the accident.

A *level 2* PRA includes the postaccident phenomenology necessary to estimate the amount of material or energy released from the facility in question, but does not assess the off-site consequences that would result. For example, a level 2 PRA of a nuclear power plant would include analyses of potential postcore melt phenomena such as core/concrete interactions or hydrogen explosions in order to assess the structural response of the containment building and the amount of radioactivity released from the plant as a result of different types of accidents (corresponding to different plant damage states). Similarly, at a chemical facility, a level 2 PRA could include analyses of exothermic chemical reactions occurring due to unintended chemical mixtures. (Note that level 2 PRAs may not be meaningful for some types of facilities, such as holding tanks whose only failure mode is leakage.) A *level 3* PRA includes a model of how material released from the facility disperses in the environment, as well as models for public health impact and off-site property damage.

Throughout the remainder of this chapter, we will focus primarily on the elements of a level 1 PRA (or the level 1 portion of a larger-scope PRA). Within the level 1 model, two reliability analysis techniques are commonly used for quantifying the likelihood of an accident: *fault trees* and *event trees* (McCormick 1981). (*Reliability block diagrams*, which are logically equivalent to fault trees, are also sometimes used.) Fault trees and event trees are equivalent in the sense that it is possible to represent the same system or subsystem either way, and both techniques are useful. However, each technique has different strengths and weaknesses (Pate-Cornell 1984), and they therefore tend to be used in different contexts.

2.1 Fault Trees

Fault trees are constructed using *inductive or "backward" logic*. In other words, the process starts with a hypothesized system or subsystem failure (the so-called "top event") and works backward to identify which combinations of component failures could give rise to that top event. Figure 4 shows a highly simplified fault tree for a hypothesized auxiliary feedwater (AFW) system at a nuclear power plant. This fault tree implies that the top event (AFW failure) occurs if, and only if, either the AFW tank fails *OR* the pumps fail; similarly, the pumping capacity of the system is assumed to fail if, and only if, pump 1 fails *AND* pump 2 fails *AND* pump 3 fails (equivalent to a one-out-of-three success criterion). (The *AND* and *OR* logic is represented by the different-shaped gates in the fault tree.) Of course, a more realistic fault tree would include numerous components not shown in Figure 4, such as motor-operated valves and check valves, and it might span several pages.

Assuming that all components are independent, the failure probability of the AFW system shown in Figure 4 would be given by

$$P(AFW) = P(TANK) + \prod_{i=1}^{3} P(PUMP\ i) - P(TANK) \prod_{i=1}^{3} P(PUMP\ i)$$

(The cross-product of tank and pump failure is subtracted out of the equation in order to avoid double-counting events in which both the tank *and* the pumps fail.)

2.2 Event Trees

By contrast, event trees are constructed using *deductive or "forward" logic*. Rather than hypothesizing a system failure, the process starts by hypothesizing an

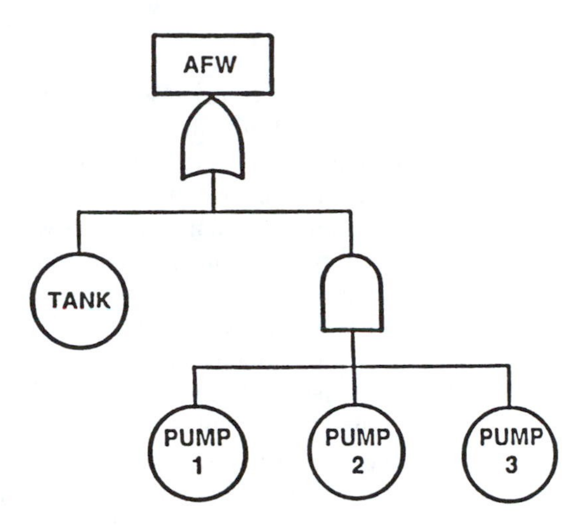

Figure 4 Simplified fault tree for a hypothesized auxiliary feedwater system.

initiating event (i.e., a departure from normal operations) and then works forward by identifying all possible combinations of subsequent events (i.e., successes or failures of particular components or subsystems) and determining which sequences of events could cause failure of the system as a whole. Figure 5 shows a simplified event tree representing an initiating event and the subsequent response of four subsystems (A, B, C, and D). For each subsystem, the upper branch represents success and the lower branch represents failure. Thus, the event sequence shown in bold consists of the initiating event I, followed by success of subsystem A, failure of subsystem B, success of subsystem C, and failure of subsystem D.

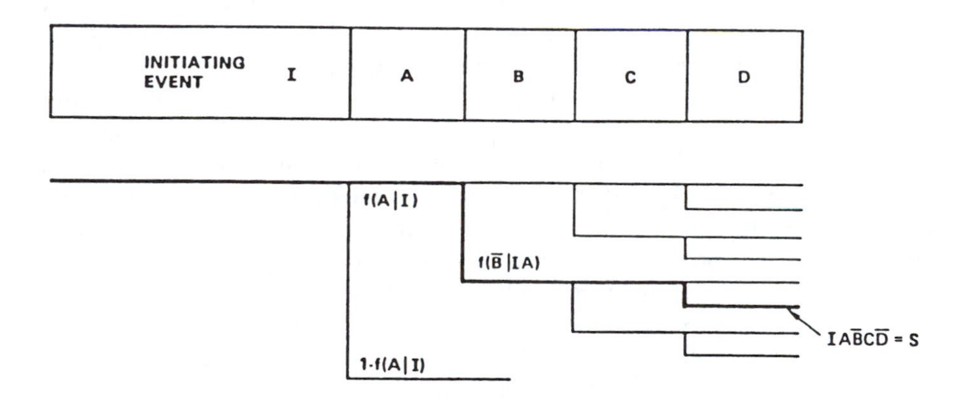

Figure 5 Simplified event tree for an initiating event and four subsystems.

The frequency of this scenario, S, can be quantified according to

$$\phi(S) = \phi(I)f(A \mid I)f(\overline{B} \mid IA)f(C \mid IA\overline{B})f(\overline{D} \mid IA\overline{B}C)$$

where

$\phi(S)$	=	the frequency of scenario S
$\phi(I)$	=	the frequency of initiating event I
$f(A \mid I)$	=	the conditional probability that subsystem A succeeds given that initiating event I has happened
$f(\overline{B} \mid IA)$	=	the conditional probability that subsystem B fails given that initiating event I has happened and subsystem A has succeeded
$f(C \mid IA\overline{B})$	=	the conditional probability that subsystem C succeeds given that initiating event I has happened, subsystem A has succeeded, and subsystem B has failed
$f(\overline{D} \mid IA\overline{B}C)$	=	the conditional probability that subsystem D fails given that initiating event I has happened, subsystem A has succeeded, subsystem B has failed, and subsystem C has succeeded

The conditional probabilities $f(\bullet)$ in the equation are sometimes referred to as *split fractions* (Kaplan et al. 1981), since they represent the long-run fraction of times that a developing scenario will follow a particular branch of the event tree. The conditional nature of these split fractions reflects the fact that subsystems may be dependent on each other. For example, if subsystem A provides electric power to subsystem B, then the failure probability of subsystem B will depend on whether subsystem A has succeeded or failed. This is also true for the initiating event I; for example, if the initiating event is an earthquake, this may affect the subsequent failure probabilities of other subsystems.

As mentioned previously, while both fault trees and event trees are useful, they have different strengths and weaknesses. In particular, event trees are well suited for displaying the *order of events* and also *dependencies between events* (e.g., the fact that the failure probability of subsystem B may depend on the status of subsystem A). Therefore, event trees are useful for facilitating communication about the assumptions made in the risk model, e.g., for presenting a risk model to plant staff for review and discussion. However, because combinations of subsystem successes and failures are explicitly shown, event-tree models can rapidly become extremely large, including literally billions of sequences. Fault trees, by contrast, provide a more compact way of representing *combinatorial numbers of events*, but can obscure dependencies and the chronological order of events.

Therefore, most PRAs use event trees to model major plant subsystems, e.g., frontline safety systems (such as AFWs), and sometimes also for electric power systems and other support systems (such as component cooling water or instrument air). Fault trees can then be used for the various split-fraction models, e.g., the conditional probability of AFW failure given a particular support system configuration. This approach is shown in Figure 6.

Differing schools of thought exist about how the system modeling effort should be allocated between event trees and fault trees; in particular, whether to use *large*

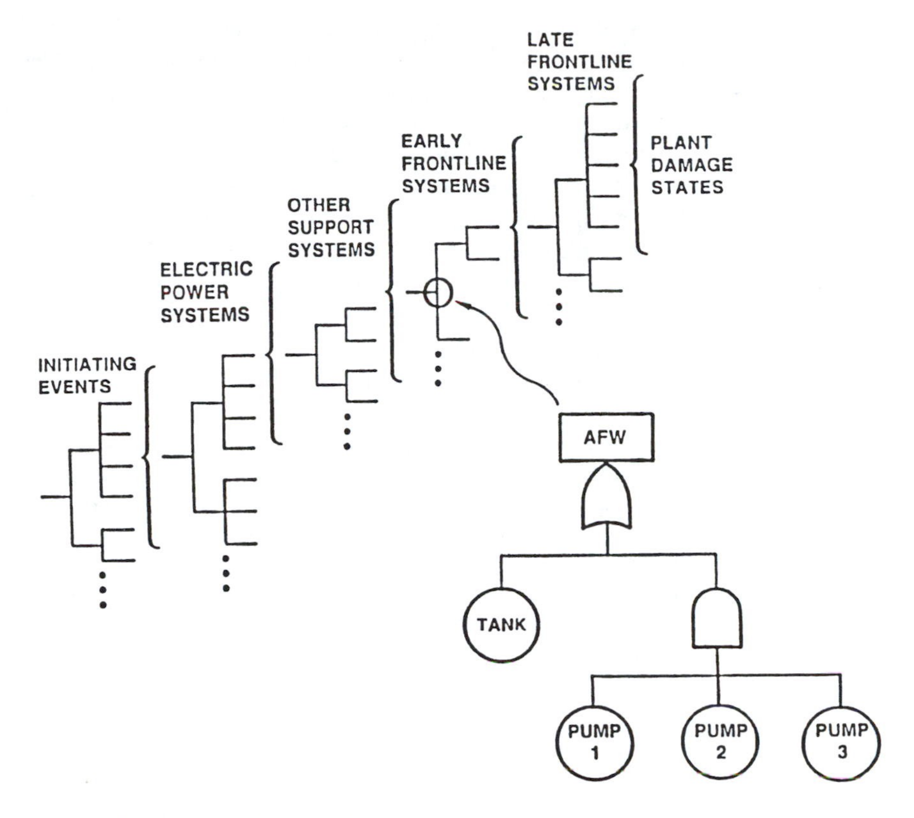

Figure 6 Relationship of split fraction models to event trees.

event trees (incorporating numerous top events, and hence permitting the use of smaller fault trees) or *large fault trees* (incorporating more detail into the subsystem models, and hence permitting the use of smaller event trees). In the large event-tree approach, the event trees will generally include support systems such as electric power, component cooling water, and instrument air, as shown in Figure 6; in fact, the large event-tree approach may even include separate top events for each train in a major system, if failure of different combinations of trains will have different effects on the plant as a whole. This makes it possible to explicitly show the dependence of safety systems on the status of the support systems; thus, for example, the split fraction model for the AFW system might vary depending on the availability of the specific buses that provide electric power to pumps 1, 2, and 3.

By contrast, in the large fault-tree approach, support systems such as electric power are generally included in the fault trees for the frontline safety systems that depend on them. Therefore, the same support system equipment may appear in fault trees for several different safety systems, necessitating *fault-tree linking*.

Both approaches have their advantages (e.g., see American Nuclear Society and Institute of Electrical and Electronic Engineers 1983). In particular, in the large fault-tree approach, the event trees will contain a relatively small number of sequences, and those sequences will involve relatively few top events. Hence, the sequences that are important to risk can be easily remembered and understood, and this approach can therefore facilitate communication. By contrast, in the large event-tree approach, the individual split fraction models are kept relatively simple by structuring the event trees to take advantage of independence and conditional independence relationships between the various top events. In this approach, the failure probability of a particular subsystem may depend on which other subsystems or top events have already failed (i.e., which branch of the event tree we are on), but will generally not depend on the specific *causes of failure* of those previous subsystems.

Several general guidelines for event-tree construction can be helpful in achieving conditional independence of the various top events.

1. Particularly if different combinations of failures do not have significantly different effects on the plant, it may be desirable to *group redundant components or trains* under a single top event. This is shown in Figure 6, where pumps 1, 2, and 3 are all grouped within the AFW top event. If the three trains of the AFW system were instead each assigned to a different top event, the corresponding top events would not be conditionally independent of each other, due to the potential for *common cause failure* (i.e., simultaneous failure of two or more AFW pumps due to the same cause). Modeling common cause failure between different top events would be complicated, since, for example, the failure probability of pump 2 might depend on whether pump 1 had failed due to common cause or not.
2. It is generally desirable to place *causally dependent events* (e.g., safety system failures) to the right of the events that influence them (e.g., support system failures). In principle, of course, a correct event tree could be drawn with the events in some other order, since *probabilistic* dependence is unrelated to *causal* dependence. However, this approach will introduce complex dependencies between the various top events. For example, if systems *A* and *B* both contain components that depend on electric power, and they appear to the left of the electric power system in the event tree, then the conditional failure probability for system *B* given that system *A* failed will need to take into account the probability that the failure of system *A* had been due to loss of electric power.
3. Similarly, it is often desirable to place *more severe events* toward the left side of the event tree. For example, in Figure 5, failure of subsystem A is assumed to result in guaranteed failure of subsystems B, C, and D; therefore, placing subsystem A on the left side of the event tree makes it possible to *prune* the event tree by eliminating the branches for subsystems B through D when subsystem A has already failed. Taking advantage of such functional and hardware relationships can greatly reduce the number of possible accident sequences that must be represented in an event tree, which would otherwise be on the order of 2^n, where n is the number of top events in the event tree (McCormick 1981).

When other concerns (such as those described earlier) do not dictate a particular order for the top events in an event tree, it can also be helpful to put the top events in *chronological order* (as shown for the early and late frontline systems in Figure 6). This often seems more natural and can therefore help to facilitate communication about the assumptions made in the risk model.

3. DATA REQUIREMENTS FOR PRA QUANTIFICATION

A variety of different types of data are needed to support PRA quantification (U.S. Nuclear Regulatory Commission 1984a). First, since PRAs focus on the consequences of departures from normal operations, estimates are needed for the frequencies of such *initiating events*. Possible initiating events include both events internal to the plant itself (e.g., equipment failures, loss of electric power, or human errors) and also external events (e.g., fires, floods, earthquakes, or severe winds).

In addition, data on *component failure rates* are needed to support quantification of fault trees and/or event trees. This includes not only the failure rates of individual components, but also data on *common cause failure rates*. In common cause failures, two or more components fail simultaneously (or during a relatively short period of time) for the same reason and therefore cannot reasonably be considered to fail independently. Common cause failures are often among the dominant contributors to system failure frequency for highly reliable systems. Among the reasons for common cause failure are common environmental conditions, such as corrosion, debris, or poor maintenance, and also common design defects. For example, debris can cause failure of multiple pumps or filters in the same subsystem. (Common cause failures generally do not include failures caused directly by failure of another component or subsystem, such as electric power, which can be explicitly modeled in a straightforward way in fault trees and/or event trees.)

Data on *component maintenance frequencies and durations* are also needed. While preventive maintenance is desirable in the sense that it can lead to lower component failure rates, the maintenance activities themselves result in components being temporarily out of service. And since accidents by definition happen at random times, there is no way to schedule preventive maintenance activities to ensure that needed components will be available in the event of an accident. So excessively high maintenance frequencies or needlessly long maintenance durations can increase accident frequencies, even if the intent of the preventive maintenance program is generally favorable.

A full-scope PRA (particularly one that includes external as well as internal initiating events) also requires data on *component fragilities* (i.e., component failure probabilities as a function of exogenous stresses). Examples of the types of stresses for which such fragility estimates may be needed include earthquakes, fires, floods, and high temperatures (e.g., in the event of air conditioning and/or ventilation failures). In some situations, vulnerability to shrapnel or missiles (e.g., due to turbine failures or boiler explosions) can also be important.

Finally, estimates of *human error rates* are also an important element of PRA quantification. Such errors include not only errors of omission (e.g., failure to perform a designated task or a particular step in a procedure), but also errors of commission (i.e., erroneous acts that are deliberately undertaken by operators). For example, errors of commission frequently stem from misdiagnoses of plant conditions, so that an operator may undertake an action that is appropriate for the apparent state of the plant, but not for the actual plant condition.

For most of the data requirements identified previously, several different types of information may be available. In addition to *component-specific information* (e.g., the number of observed failures of each component), generic information such as *failure frequencies for similar components* (i.e., at other plants or in other industries) can also be relevant. Such generic information can come either from raw data or from published sources (e.g., Nuclear Power Engineering Committee of the IEEE Power Engineering Society 1977, Idaho National Engineering Laboratory 1985, Pickard, Lowe and Garrick, Inc. 1991). Other useful generic information sources include *expert opinion* and *case histories*. Bayesian data analysis provides some particularly nice formalisms for combining generic and component-specific information (Martz and Waller 1982, Martz and Bryson 1983, Kaplan 1983), although other approaches can also be used.

Before beginning the actual data collection process, some preliminary steps must be performed. First, the analyst must specify the *level of detail* of the analysis, e.g., whether the analysis should be done at the level of individual resistors and capacitors or at the level of entire circuit boards. Second, the *components of interest* must be specified. For example, instruments such as temperature or pressure transducers are often not modeled in PRAs, unless failure of such instruments is considered to be a likely cause of human error. The analyst must also specify the *database study period*. For example, will all data be analyzed or only that collected after an initial design evolution or startup period?

For each component, the analyst must also determine the *failure modes* that are relevant to the PRA. For instance, valves may experience both failure to open on demand and failure to close on demand, but in some situations only one of these failure modes may be important to risk. Similarly, some valves and filters may also be at risk for leakage or plugging during system operation. Different failure modes must be distinguished if they have different effects on the system; thus, for example, a valve leak may have very different effects than a valve that is stuck in the wide open position. As a general rule of thumb, it is usually also a good idea to treat startup failures differently from run-time failures. This makes it possible to perform realistic analyses of accident scenarios with different durations and mission times. However, different *causes* of component failure (e.g., corrosion, fatigue, etc.) do not need to be distinguished in PRA, except possibly to aid in the prioritization of corrective actions.

In some cases, it may be desirable to pool information for similar components. This increases the total amount of information that is available for the analysis and therefore can decrease uncertainty. However, this is appropriate only if the components are sufficiently similar. Similarity depends on a number of factors, including not only *component type* (e.g., motor-driven vs. turbine-driven pumps, motor-operated vs. air-operated valves), but also *usage*, *environment*, and past *performance history*. For example, pumps that are in continuous operation are likely to have very different performance characteristics than pumps that spend most of their time in standby. Similarly, valves in rapid cycling operation may have different performance histories than valves that change position only occasionally. *Testing, monitoring, and refurbishment policies* also have a large effect on component performance. Therefore, if significant changes have been made in plant maintenance policies, it may be advisable to discard performance information collected prior to the change in policies.

Success or exposure data are also generally needed, in addition to failure histories. For example, for normally operating components or subsystems such as pumps, data will be needed on the number of hours of operation over which the observed failures occurred. Similarly, for passive components or subsystems such as tanks (which can fail even when the system as a whole is not in operation), data will be needed on the total number of hours of experience over which the observed failures occurred. Finally, data will be needed on the total number of demands experienced by standby components (such as emergency safeguard pumps) and also by cycling components (such as thermostatically controlled heating or cooling systems). To quantify the relevant exposure data, the analyst must identify the population of similar components for which data is being pooled (e.g., the number of similar valves vulnerable to a particular failure mode) and determine the appropriate units for each failure mode (e.g., hours vs. demands).

Once the risk assessment requirements have been identified, they can then be correlated with available data sources. Published and/or computerized data sources do not always provide sufficient detail for risk analysis purposes, and reporting to such sources may not always be consistent. Therefore, some creativity may be required in identifying suitable data sources. Among the data sources that may be available include maintenance requests; corrective action reports; significant event reports; anomaly reports; operational histories (e.g., plant logs or mission logs); and the results of surveillance, check-out, qualification, or acceptance tests.

Even after the raw data has been collected, data interpretation is still a significant task. Contrary to what one might wish, data do not generally speak for themselves, and a number of somewhat subjective decisions must be made by the data analyst. For example, analysts may need to revisit the *appropriate level of detail* for the analysis (e.g., should pump drivers be analyzed separately from pumps). Information may also be available on the performance of *noncritical components*, such as temperature or pressure transducers; the fact that information is available on the performance of such components does not necessarily mean that they should be included in the analysis. However, redundant equipment such as standby components should be included, even if only limited test data are available for those components, since the failure probabilities of such components can be an important aspect of system reliability.

The *applicability of test data* must also be assessed; this is particularly problematic for bench tests of uninstalled components under relatively benign laboratory conditions that may not be representative of normal use. Data also frequently include partial failures that may not be easily categorized as either failures or successes. For example, maintenance reports may indicate that a particular piece of equipment was making excessive noise and that inspection revealed a cracked turbine blade or a worn pump shaft; in the case of such *incipient failures*, the analyst must make a judgment about whether the equipment would likely have operated successfully for the mission time needed to prevent an accident. Similarly, failure reports may indicate that a pump was delivering only 85% of rated flow or that turbine speed was erratic, and the analyst must evaluate whether such *out-of-spec operation* would have been sufficient to prevent the component from performing its intended function.

Finally, many, if not most, event reports also include a discussion of *corrective actions*; in fact, some data reporting systems require that a corrective action be reported for each event. In this case, the analyst must decide whether the reported event is still relevant to predicting system performance once the corrective action has been undertaken. In the case of major system design changes, the answer will be relatively straightforward; for example, if the corrective action for a check valve failing to open is to remove the check valve from the system altogether, this failure mode will be effectively prevented (although other failure modes may be introduced). However, the effectiveness of other corrective actions, such as improved training and/or improved cleanliness procedures, may be more difficult to assess. In such cases, it is often more prudent to include the original failure in the database and assume that the effectiveness of the corrective action will be revealed over time as additional data are collected.

4. RISK MANAGEMENT USING PRA

Once an *integrated plant-specific risk model* has been developed, the model can then be used for risk management purposes. There are several steps to this process, as described by Garrick (1984a). First, the *contributors to risk* must be rank ordered so that the dominant contributors to risk can be identified. In cases where there are several damage indices (e.g., core melt, radioactive release, and public health effects), different damage indices may be associated with different dominant contributors to risk.

Next, the analyst must *decompose* the dominant contributors to risk into their specific identifiable elements. As an example, if the damage index of interest is public health effects, one might first identify which *type(s) of release* (of material and/or or energy) contributes the most to the risk of public health effects. It is important to note that the most important type of release from the point of view of any particular damage index may not be the most likely type of release overall. For instance, in nuclear power risk analysis, the most likely type of radioactive release is usually one in which the containment fails a number of hours after the initial core melt, but this is not usually the dominant contributor to acute health effects, since the time delay between the core melt and the release allows for significant deposition of radioactive material inside the containment. Next, the analyst can identify the *type of plant damage* leading to that type of release, e.g., which, if any, postaccident safety systems were operational to mitigate the accident consequences.

From there, the analyst can identify which *initiating event(s)* contributed the most to that particular type of plant damage. The event trees (or other plant model) can then be used to identify the dominant *event sequence(s)* resulting from those initiating events, as well as any major systems or subsystems whose failure contributes to that event sequence. For those major systems, the logic models for *system unavailability* can then be reviewed to determine the dominant failure modes of each system (e.g., independent failures, maintenance unavailability, human error, or common cause failures), as well as the specific *failure causes* (e.g., the specific component failures that cause system failure) and their frequencies and effects. Finally, the specific failure causes can be mapped back to the original *input data* (i.e., data on initiating events, components, maintenance, human error, common cause failures, environmental stresses, etc.) to determine which of the observed events in the database provided the basis for quantification of the various failure causes.

Once the elements involved in each of the dominant contributors to risk have been identified, the analyst can then *identify options* such as design and/or procedure changes for reducting the impact on risk of the dominant contributors. For each identified option, the analyst can then make the appropriate changes in the risk model and *recompute the risk* for each option. For each proposed design change (i.e., each proposed modification to an engineered system configuration) and/or procedure change, the analyst can then compute the *cost impacts* of the proposed modification relative to the base case. Present-value calculations may be involved in this computation, since some design changes may involve not only initial costs, but also annual operating or maintenance costs. Finally, the *costs, risks, and benefits* for each option can be presented to the decision maker.

The application of PRA has resulted in a number of examples of successful risk management. To some extent this is fortuitous. Before PRA had been widely applied, it may not have been immediately apparent that relatively inexpensive risk reduction options would generally be available. Certainly, if the most effective risk reduction options involved repouring a plant's foundation or building a new containment, there would be relatively little benefit to PRA, since these options would rarely, if ever, be justified. Fortunately, however, with PRA results and a little bit of engineering ingenuity, much less expensive risk-reduction options can frequently be identified; typical costs may range from a few thousand dollars for simple changes to several million dollars for more extensive plant modifications. A few examples of those are given, based on risk analyses performed by Pickard, Lowe and Garrick, Inc.; further detail and additional examples are given by Garrick (1984a, 1987).

For example, a PRA of one plant revealed that service water failure contributed approximately 20% to the total frequency of core damage. The failure mode of greatest concern involved transfer of the service water system from the cooling tower to a cooling pond to provide a source of makeup water for the AFW system. In particular, if the valves to the cooling tower were successfully closed, but the valves to the pond failed to open, then the service water system would cease functioning and leave the plant with no ultimate heat sink. An electronic interlock was proposed that would not allow the cooling tower valves to close unless the pond valves had successfully opened. This would ensure that service water flow was maintained while the operators attempted to identify a long-term source of makeup water to the AFW system. This rather inexpensive change reduced the unavailability of the service water system by about a factor of 10, significantly reducing its contribution to the core damage frequency.

Similarly, the Nuclear Regulatory Commission (NRC) mandated one utility to install an additional safety-grade AFW pump at each unit of the utility's two-unit plant. As a result of the PRA for this plant, it became clear that putting additional safety-grade pumps in the same rooms as the existing AFW pumps would not significantly enhance the reliability of the AFW system, since the existing pumps were installed in fire- and flood-protected rooms with tight doors, and therefore one of the primary reasons for failure of both existing AFW pumps was loss of cooling to the closed pump rooms. Therefore, rather than add an additional safety-grade pump at each unit, the utility received NRC approval to add a single additional nonsafety-grade pump in a remote location with natural ventilation. Since this pump was not dependent on the cooling system, it significantly reduced the overall contribution of the AFW system to the frequency of core damage, at less cost than the original NRC proposal.

In recent years, similar risk management applications have increasingly been undertaken based on risk analyses performed by utility staff rather than consultants. For example, Duke Power uses PRAs, along with data on accident "precursors" or near misses, as a basis for dialogue with plant maintenance staff about the importance of minimizing maintenance durations and returning key items of equipment to service as soon as possible (Busby et al. 1992). Similarly, based on an individual plant examination (IPE), Arizona Public Service rebalanced its electrical and ventilation loads, significantly reducing accident risk at only modest cost (Lindquist 1992).

Perhaps one of the most persuasive illustrations of the cost-effectiveness of PRA-based risk management was the use of PRA in the Atomic Safety and Licensing Board hearings on the safety of the Indian Point plants in New York. In this regard, Rowsome (then of the NRC) stated: "The mitigation conceptions, both those that were subject to contention and those that had been developed in the [NRC] staff action plan . . . , cost more than they were worth . . . . The prevention improvements volunteered by the licensees were worth roughly a quarter of a billion dollars in averted risk, a whopping big value . . . . In fact, if you count in the whole cost of the hearing and the inquiry and the PRAs . . . , the value accorded the fixes we found was about ten times the cost of the entire enterprise" (U.S. Nuclear Regulatory Commission 1984b).

5. CONCLUSION

A few key lessons learned from industry experience with PRA are summarized here; others have been discussed by Garrick (1984b). First, the PRA analyst cannot take short cuts. Every plant is significantly different, even nominally "identical" units on the same site. Therefore, each plant must be assessed *in depth.* In the nuclear power industry, this is particularly true for U.S. plants, with their lack of standardization (for example, virtually every U.S. plant has a different configuration for its service water system), and less true of overseas plants. However, even among plants with a high degree of initial standardization, the influence of operating and maintenance practices can far outweigh the inherent design reliability of the equipment, so that even plants that started out life as sister units, such as Millstone and Pilgrim, can have very different risk and reliability profiles.

Second, the failure of *support systems,* such as AC electric power or service water, are often the "Achilles heel" of highly reliable and redundant systems such as nuclear power plants. In fact, the more redundancy there is among other safety systems, the more important such support systems will be, since failure of support systems can effectively disable many redundant trains of equipment simultaneously. Similarly, plants are not always well designed against *severe acts of nature* such as earthquakes, fires, floods, or hurricanes. Once again, the importance of acts of nature becomes progressively greater as the inherent design reliability of the plant increases, since such acts of nature can again defeat the redundancy and reliability designed into the plant.

Finally, it is important to incorporate the knowledge of the plant's own *operations and maintenance staff* into the PRA. This enhances the study's quality by ensuring that the PRA captures the plant-specific details that may be relevant to risk. It also enhances the plant staff's understanding and acceptance of the study, thereby increasing the likelihood that risk management recommendations will actually be implemented in practice. Highly educated PRA analysts can sometimes fall prey to a tendency to view operations and maintenance personnel as glorified janitors, but this is far from the case. Such personnel have a wealth of plant-specific knowledge and experience at their fingertips and generally find the event sequences and other plant models of the PRA eminently understandable as long as the PRA analyst expends suitable effort to present these models in a clear and comprehensible fashion.

ACKNOWLEDGMENTS

Much of the material presented here is based on presentations and short courses prepared by my colleagues and myself while at Pickard, Lowe and Garrick, Inc. (now PLG, Inc.). In that regard, I would particularly like to acknowledge the leadership of Dr. B. John Garrick, as well as significant contributions by Dr. Dennis C. Bley, Mr. Karl N. Fleming, Dr. Michael V. Frank, Dr. Stan Kaplan, Dr. Ali Mosleh, and Mr. John W. Stetkar, among others.

REFERENCES

American Nuclear Society and Institute of Electrical and Electronic Engineers, PRA Procedures Guide: A Guide to the Performance of Probabilistic Risk Assessments for Nuclear Power Plants, Report NUREG/CR-2300, U.S. Nuclear Regulatory Commission, Washington, D.C., 1983.

Busby, B. E., Canady, K. S., and Abraham, P. M., Accident precursor program at Duke Power Company, *Transactions of the American Nuclear Society*, 65, 282, 1992.

Garrick, B. J., Experience and advancements in risk assessment, presented at Nuclear Power Reactor Safety Course, Massachusetts Institute of Technology, Cambridge, Massachusetts, July 16, 1984a.

Garrick, B. J., Recent case studies and advancements in probabilistic risk assessment, *Risk Analysis*, 4, 267, 1984b.

Garrick, B. J., Examining the realities of risk management, in *Uncertainty in Risk Assessment, Risk Management, and Decision Making*, Covello, V. T., Lave, L. B., Moghissi, A., and Uppuluri, V. R. R., Eds., Plenum Press, New York, 1987.

Idaho National Engineering Laboratory, Development of Transient Initiating Event Frequencies for Use in Probabilistic Risk Assessments, Report NUREG/CR-3862, U.S. Nuclear Regulatory Commission, Washington, D.C., 1985.

Kaplan, S., On a "two stage" Bayesian procedure for determining failure rates from experiential data, *IEEE Transactions on Power Apparatus and Systems*, PAS-102, 195, 1983.

Kaplan, S., and Garrick, B. J., On the quantitative definition of risk, *Risk Analysis*, 1, 11, 1981.

Kaplan, S., Apostolakis, G., Garrick, B. J., Bley, D. C., and Woodard, K., *Methodology for Probabilistic Risk Assessment of Nuclear Power Plants,* Report PLG-0209, Pickard, Lowe and Garrick, Inc., Newport Beach, CA, 1981.

Lindquist, R. C., Response to IPE and IPEEE results at Palo Verde NGS, *Transactions of the American Nuclear Society*, 65, 298, 1992.

Martz, H. F., and Bryson, M. C., On combining data for estimating the frequency of low-probability events with application to sodium valve failure rates, *Nuclear Science and Engineering*, 83, 267, 1983.

Martz, H. F., and Waller, R. A., *Bayesian Reliability Analysis*, Wiley Interscience, New York, 1982.

McCormick, N. J., *Reliability and Risk Analysis: Methods and Nuclear Power Applications*, Academic Press, New York, 1981.

Nuclear Power Engineering Committee of the IEEE Power Engineering Society, IEEE Guide to the Collections and Presentation of Electrical, Electronic and Sensing Component Reliability Data for Nuclear Power Generation Stations, Report IEEE STD-500, Institute of Electrical and Electronic Engineers, 1977.

Pate-Cornell, E., Fault trees vs. event trees in reliability analysis, *Risk Analysis*, 4, 177, 1984.

Pickard, Lowe and Garrick, Inc., Database for Probabilistic Risk Assessment of Light Water Nuclear Power Plants, Report PLG-0500, Newport Beach, CA, 1991.

U.S. Nuclear Regulatory Commission, Probabilistic Risk Assessment (PRA) Reference Document, Report NUREG-1050, Washington, D.C., 1984a.

U.S. Nuclear Regulatory Commission, Discussion of Indian Point Probabilistic Risk Assessment (open meeting), Washington, D.C., 1984b.

QUESTIONS

1. Why was probabilistic risk analysis developed, and what types of events are particularly well suited for it to analyze?
2. Why is it important for PRA to consider the interactions between multiple subsystems, rather than analyzing each system in isolation?
3. What is the primary advantage of providing quantitative measures of risk, rather than only qualitative assessments (e.g., safe or unsafe; high, medium, or low risk)?
4. Why is it important for PRA to provide information on risk contributors and not only an assessment of the total risk?
5. What are the three basic questions that must be answered in a PRA?
6. Why are PRAs structured in a hierarchical manner?
7. What are the relative strengths and weaknesses of fault trees and event trees?
8. What are the advantages and disadvantages of the large event-tree vs. the large fault-tree approach?
9. Summarize the guidelines for event-tree construction and explain why they are helpful.
10. What types of data are needed for PRA quantification, and where can they be obtained?

11. Discuss some of the issues that must be addressed as part of the data collection and interpretation process.
12. Why has PRA been so effective in risk management?
13. Why must PRA be performed on a plant-specific rather than a generic basis in order to be optimally effective?

CHAPTER I.6

Ecological Risk Analysis*

Robert T. Lackey

SUMMARY

Risk assessment has been suggested as a tool to help manage ecological problems. Ecological risk assessment is usually defined as the process that evaluates the likelihood that adverse ecological effects are occurring, or may occur, as a result of exposure to one or more stressors. The basic concept, while straightforward, is difficult to apply to any but the simplest ecological problems. Strong reactions, both positive and negative, are often evoked by proposals to use ecological risk assessment. Risk assessment applied to relatively simple ecological problems (chemical toxicity being the most common) is popular; there are many vigorous supporters, particularly among scientists, administrators, and politicians. Yet critics are equally vocal. The intellectual history of the risk assessment paradigm as applied to ecological problems does not follow a neat, linear evolution. A formidable problem in many risk assessments, and especially for complex questions such as addressing the challenge of ecological sustainability, is selecting *what* ecological component or system is to be considered at risk. This selection is entirely social and political, but estimating the actual risk is technical and scientific. The question of what is at risk must be answered within the political decision-making framework or the results of the risk assessment will be of limited utility. ***Performing credible risk assessments for complex ecological problems is difficult unless the boundaries of the assessment problem are highly constrained. However, narrowly defining ecological***

* This chapter is an abbreviated version of "Is Ecological Risk Assessment Useful for Resolving Complex Ecological Problems?" published in *Pacific Salmon and Their Ecosystems: Status and Future Options,* Deanna J. Stouder, Peter A. Bisson, and Robert J. Naiman, Eds., Chapman and Hall, Inc., New York, NY, 1996. This chapter does not necessarily represent the policy positions of the U.S. Environmental Protection Agency or any other organization.

problems produces risk assessments that are of limited relevance in resolving public policy questions.

Key Words: ecological risk assessment, risk assessment, risk management, environmental protection, decision analysis, expert opinion, conservation, ethics, modeling, multiple-use management, sustainability, bioassays, environmental impact assessment, ecological health, biological diversity

1. INTRODUCTION

Increasingly, there are calls for the use of risk assessment to help solve complex ecological problems (examples are Pacific salmon decline in the Pacific Northwest and the decrease in biological diversity). The basic concept underlying risk assessment is relatively straightforward. Risk is something that can be estimated (i.e., risk *assessment*). In turn, that estimate can be used to manage the risk (i.e., risk *management*). Ecological risk assessment is usually defined as "the process that evaluates the likelihood that adverse ecological effects are occurring, or may occur, as a result of exposure to one or more stressors" (U.S. EPA 1992). Analyses of the options and procedures for conducting risk assessment for human health issues are available in Chapters I.1, I.2, II.1, and II.2.

The basic concepts of risk assessment may be simple, but the jargon and details are not. Risk assessment (and similar analytical tools) is a concept that has evoked strong reactions whenever it has been used. At the extreme, some have even concluded that use of risk assessment in human health decision making is "premeditated murder" (Merrell and Van Strumm 1990). A number of philosophical and moral reasons for such strong reactions exist, but they are usually based on either (1) concerns that the analysis (risk assessment) and decisions (risk management) accept the premise that people will die to achieve the desired net benefits or (2) a belief that the process of risk assessment places too much power with technocrats.

Reaction to ecological risk assessment may be less harsh than reaction to risk assessment applied to human health problems, but even with ecological issues, both strong positive and negative responses occur. Several bills (e.g., Environmental Risk Reduction Act) have been introduced in the U.S. Congress mandating that federal agencies use risk assessment to set priorities and budgets. Several panels of prestigious scientists have made similar recommendations. Popular and influential publications argue for a risk assessment approach. On the other hand, some conclude that risk assessment is a disastrous approach, one that is "scientifically indefensible, ethically repugnant, and practically inefficient" (O'brien 1992).

Still, risk assessment has been used extensively to link environmental stressors and their ecological consequences. The risks associated with chemical exposure are the typical concern. Quantifying the risk of various chemicals to human health is a logical outgrowth of risk assessment as applied in the insurance industry and other fields. Over the past 20 years, a body of procedures and tools has been used for environmental risk assessment for human health. Risk assessment applied to ecological problems is more recent, but has also focused primarily on chemicals, with animals used as surrogates for "ecological health."

Adapting the risk paradigm from assessing insurance risks to assessing human health risks to assessing ecological risks has not been simple (Lackey 1994). Some view ecological risk assessment merely as using new labels for old ideas. It is still unclear whether ecological risk assessment will actually improve decision making and ultimately protect ecological resources.

2. RISK ASSESSMENT IN PRACTICE

In spite of the difficulties of defining problems in complex ecological policy questions, the use of risk assessment to help solve ecological problems is widely supported. Legislation recently debated in Congress would mandate the use of risk assessment by federal agencies for many problems. Clearly, many people think that risk assessment is a valuable tool and should be used extensively in solving ecological problems.

There is, however, a vocal group of critics of the use of risk assessment for ecological problems. They argue that risk assessment (and risk management) is essentially triage — deciding which ecological components will be "saved" and which will be "destroyed." The theme of "biospheric egalitarianism" is a mindset that makes risk assessment a real anathema. Many risk assessment critics have a strong sense of technophobia and often view mainstream environmental organizations as co-opted by industrial or technocratic interests.

Risk assessment is also challenged from a different, more utilitarian perspective. The assertion is that, while the concept of risk assessment is sound, the *process* of risk assessment is often controlled by scientists and others who have political agendas different from the majority. Critics contend that "risk assessors" use science to support their position under the guise of formal, value-free risk analysis. Risk assessment as thus viewed has the trappings of impartiality, but is really nothing more than thinly disguised environmentalism (or utilitarianism). The apparent lack of credibility and impartiality of the science (and risk assessment) underlying the policy debates over acid rain, stratospheric ozone depletion, global climate change, and loss of biological diversity are often offered as examples of how science has allegedly been misused by scientists and others to advocate political positions.

Risk assessment has historically been separated from management. Such separation requires that scientists play clearly defined roles as technical experts, not policy advocates; these distinctions are blurred when scientists advocate political positions. Further, some critics charge that scientists who use their positions to advocate personal views are abusing their public trust. The counterargument is that scientists, and all individuals for that matter, have a right to argue for their views and, as technical experts, should not be excluded simply because of their expertise. Others conclude that the execution of the scientific enterprise is value laden and therefore partially a political activity. Rather than attempting to be solely "scientifically objective," a scientist should also be an advocate. Either way, the role of the analyst must be clear to everyone using the results.

3. HISTORY OF THE PARADIGM

Neither risk assessment nor any other commonly used management tool is completely new, but draws on earlier tools and shares some of the core principles. For example, both assessment and management are based on the fundamental premise that all benefits are accruable to man. This is a utilitarian approach and a necessary assumption in all the models or paradigms that follow. "All benefits are accruable to man" encompasses the fact that society might choose to protect wilderness areas that few visit, preserve species that have no known value to man, or preserve natural resources for their scenic beauty. Benefits may be either tangible (fish yield, tree harvest, camping days, etc.) or intangible (pristine ecosystems, species preservation, visual beauty, etc.). It is easy to jump past the fundamental premise of a utilitarian assumption, but much of the political debate revolves around the issue of whether a person operates with a utilitarian worldview or ecocentric (or other, usually religiously based) worldview. It is not a trivial difference. In practice, however, the split between those with utilitarian and ecocentric (or other) worldviews is not complete; most of us manifest features of both (Herzog 1993).

The multiple-use model of managing natural resources has been the basic paradigm in North America during this century. Popularized by Gifford Pinchot and others, it has been used extensively and widely in fisheries, forestry, and wildlife. The idea is simple: there are many benefits that come from ecological resources (commodity yields, recreational fishing and hunting experiences, outdoor recreational activities, ecosystem services such as water purification, etc.) and that the mix of outputs needs to be managed to produce the greatest good for the greatest number over a sustained period of time (Callicott 1991). The concept is straightforward and works well if there is a high degree of shared values among the public.

A number of variations in the multiple-use model arose over the middle years of this century: maximum sustainable yield, maximum equilibrium yield, and optimum sustained yield. Widely used in teaching and management, these concepts have dominated mainstream professional thought and practice through current times. As with all natural resource management paradigms and goals, none of these evolved in a linear manner. The basic idea is that commodity yields (fish, trees, wildlife) could be produced annually from "surplus" production and could be continued in perpetuity with sound management. All these models suffered from the problem of a heterogeneous public with competing demands and with demands that change over time. Even today, there is still a struggle to control fishing, hunting, and logging levels in politically acceptable and managerially efficient ways.

Scientific management is a related management paradigm that includes operations research, management by objectives, optimization, linear programming, artificial intelligence, and other mathematical procedures (Lackey 1979). There are many outputs from ecosystems, both commodity and noncommodity, and these can (must) be measured, and the aggregate output optimized. The outputs are selected by experts, who then use mathematical tools to quantify and evaluate various combinations of outputs. ***Input from the public is not considered particularly important because there is a "correct" optimal set of decisions to maximize output. The***

natural resource "professional" is dominant in the process. The view that "if politicians and the public will just stay out of the process, we professionals will manage natural resources just fine" is characteristic of scientific management. There are many examples of the collapse of natural resources based on following this general management approach.

Ecosystem management, including variants such as watershed management, has become popular in the past decade. Both ecosystem and watershed management have ambiguous definitions, illustrated by the popular wall poster for ecosystem management: "Considering All Things." Usually other concepts, such as biological diversity, are embedded in ecosystem management, although biological diversity is an ill-defined concept in its own right. For example, in our quest to restore salmon stocks, should we eradicate squawfish (predators) and walleye (competitors), or do we restore ecosystems (habitat) to some desired state and let nature take its course? Does ecosystem management mean we "optimize" this mix of species? These and a myriad of others are policy questions and must be explicitly answered regardless of what management approach is used. They must be answered as policy questions, not scientific ones. Advocates of ecosystem management often see it as a fundamental shift in management and assessment thinking; skeptics see it as a "warmed-over" version of multiple-use management or, more pejoratively, as "policy by slogan."

A different approach is embodied in chaos theory and adaptive management; these approaches recognize the high degree of uncertainty in ecosystems. The basic idea is that ecosystems are unfathomably complex and react to unpredictable (chaotic) events; thus, it is pointless to develop sophisticated ecosystem models for decision making based on equilibrium conditions. There is also constant feedback between man's decisions and adjustments of the ecosystem to those decisions. Uncertainty is so great that it is not feasible to create useful predictive models. Also, for alternatives that preclude future options, adaptive environmental assessment and management will not work well (for example, construction of dams on the main stem of the Columbia River has had major ecological consequences for salmon, and each major project was an irrevocable decision). In general, the manager or analyst will make a series of "small" decisions, evaluate the results, and then make revised decisions. To make a "big" decision requires strong public support and acceptable ways to compensate the losers.

Total quality management (TQM) is a concept that became popular in business and government in the 1980s and 1990s. The widespread efforts to "reinvent government" have their basis in TQM. The core idea is that the customer comes first, and, in turn, management should be measured by what customers want. In natural resource management and environmental protection, the "customer" is often defined as the "public." Hence, TQM presupposes that an agency can find out what the public wants in terms of ecosystem management and protection and then deliver that "product." There are difficulties in defining the public, but TQM has been successful in some business applications. Its usefulness in managing and protecting public natural resources is open to question, however. In a pluralistic society, it is unlikely that there will be a common public goal for ecological resources that will allow the principles of TQM to be used effectively.

Risk assessment and management, the final management paradigm reviewed here, has been used as a tool in some of the previous paradigms or as a stand-alone approach. Strongly advocated by some, the approach has generally been used for assessing the role of chemicals in ecosystems or components of ecosystems. The basic idea of risk assessment and risk management is that there are many risks to the environment, ecosystems, and human health. These risks ought to be identified, quantified, and managed.

4. AVAILABLE TOOLS

There is a widely used set of tools and techniques to generate data for a risk assessment. Initially, the question of who assumes the "burden of proof" needs to be addressed. Do risk assessors assume that current ecological conditions are the norm and any proposed deviation from the status quo must be justified? Or do they assume some pristine ecological state as the norm? Or do they assume that the person or organization proposing the action must justify it? One of the reasons that the Endangered Species Act and Section 404 of the Clean Water Act are so potentially powerful is that they effectively shift the burden of proof to those who would change a defined condition (e.g., species must not go extinct or wetlands must not be altered unless there is explicit government approval). ***The practitioners of ecological risk assessment often overlook values, ethics, and burden of proof in defining the problem and operate instead on a purely technical level. To continue with the salmon example, why do we assume that the physical alterations of salmon rivers, such as the Columbia, are irrevocable? Is it not an option to demand that the organizations responsible for dams demonstrate that the dams are not adversely affecting salmon populations or alter their operations (including removal) so as not to adversely affect salmon? Why should the burden rest with those trying to protect or restore salmon?***

Bioassays are the most commonly used tools in producing the basic data for ecological risk assessment dealing with exposure to chemicals. There are many permutations of the basic bioassay, and the literature is extensive. Bioassays work well for certain types of ecological problems and especially for the "command and control" regulatory approach. Severe limitations, however, occur in assessing multiple, concurrent stresses, assessing effects on ecosystems or regions or assessing effects that are not chemically driven (e.g., land-use alterations). It is easy to lose sight of the fact that bioassays are simplifications of the ecosystems and regions with which risk assessors are concerned and are merely surrogates for the realistic tests or experiments that cannot be performed. On an administrative level, the use of bioassays has become institutionalized, and the public may now view such tests as more relevant to protecting the environment than is warranted.

Environmental impact analysis and monitoring are additional tools often used in risk assessment. Such analyses are relevant to real-world problems and are often targeted directly at public choice issues; there is an extensive literature on their many approaches and procedures. Because problems are "relevant," they are often complex scientifically, and, therefore, the resulting predictions lack the

scientific rigor typically seen in scientific journals. As a result, users often lack confidence in the reliability of the predictions. Moreover, the process of developing an environmental impact statement may be more important than the actual document produced.

Modeling and computer simulation are tools that have proved to be very popular in ecological risk assessment. These tools have many desirable features, such as the ability to deal with complex problems, the ability to evaluate alternative hypotheses quickly, and the ability to organize data and relationships into a defined whole. However, modelers often fall into the trap of substituting analytical rigor for intellectual rigor. Very simplistic (and incorrect) ideas can be masked by mathematical complexity. ***Even some of the most widely accepted and applied models in ecology illustrate the problem of developing and applying models to actual management issues. Further, the ease and beauty of tools such as computer-assisted geographic analysis can also cause the analyst to lose sight of intellectual rigor and common sense.***

Because most ecological risk assessment problems are complex and do not lend themselves entirely to laboratory experiments, field experiments, or modeling, the use of expert judgement and opinion is desirable and necessary. Expert opinion is useful, but is not without problems. For example, when experts have dramatically different opinions, how does a risk assessor handle this analytically? History is filled with examples of experts being completely in error. On the other hand, risk assessors trust that experts are less wrong on topics of their expertise than are nonexperts, and the use of experts and formal expert systems will continue to increase. There is the particularly insidious problem, when relying on the opinions of technical experts, of separating their personal and organizational values from their technical opinions.

Risk assessment, at least in the problem formulation stage, must include an explicit determination of what the customer wants. This is not as easy as it sounds. The customer is usually the public or a subset of the public (or an institutional surrogate such as a law or a court determination). Typical information from the public is that people want to "protect the environment," "protect endangered species," or "maintain a sustainable environment." The same people may also say that they want to "protect family-wage jobs," "maintain economic opportunities for our children," and "protect the sanctity of personal property." It is very difficult to move beyond such platitudes and obtain information that is really useful in risk assessment. On the other hand, individuals or elements of society with a direct and vested interest will have very specific preferences. Those less directly affected tend to have more general preferences. For example, studies show various elements of the public possessing at least nine different concepts of sustainability for forests, many of which are mutually exclusive (Gale and Cordray 1994).

5. APPLICATION

The first step in conducting any analysis of ecological risk is to clearly define the "problem." Unfortunately, this step is often overlooked or resolved simplistically. In many cases, agreeing on the problem is impossible because that is in itself the political impasse. There is also tension between analysts who want to simplify the problem so that it is technically tractable and politicians (who work in the real world) who must keep problem definition as realistic (which means technically complicated to analysts) as possible. Defining the problem is a political process requiring technical input, but it is based on values and priorities.

Considering the specific example of the Pacific salmon helps clarify the issues. An analyst must explicitly resolve whether the focus is on preserving some or all stocks (distinct populations) from extirpation or maintaining some or all stocks at "fishable" (high) levels. These are largely mutually exclusive alternatives. They are also not scientific decisions. Further, defining which species, communities, or ecosystems are to be evaluated in risk assessment is value based and not solely a scientific determination. Does the analyst consider the "baseline" condition to be 10,000 years ago; 200 years ago; or for the Columbia Basin, preimpoundment construction (basically before the Second World War)? Analysts should not decide these questions, society should. Depending on the baseline selected, the results of a risk assessment will differ.

Most practioners argue that, to be more useful, risk assessment (estimating risk) must be separated from risk management (making choices) both in practice and in appearance (Ruckelshaus 1985, Sutter 1993). There are counterarguments against separating assessment from management. Usually, the arguments recognize that it is impossible to separate a person's values from his/her technical (risk assessment) activities, and, therefore, the separation is illusionary. Separating the two activities (management and assessment) is not as easy as it might appear. Many scientists have strong personal opinions on public choice issues that concern ecological resources. It is difficult for anyone to separate purely technical opinions from personal value judgements. Even more difficult is convincing all elements of the public (all stakeholders) that the assessment is being conducted without a bias on the part of scientists.

The best scientists and most credible scientific information must be used in risk assessment. Besides being independent, the assessors must not advocate their organization's political position or their own personal agenda. If the risk assessment is not perceived to be independent, the results will be suspect. Further, the research and assessment function within an organization should be separated from the management and regulatory function. Credibility and impartiality are difficult to maintain, especially in the public eye.

Risk analysis will result in a number of options to "manage" the risk. These may range from drastic, expensive options to those that maintain the status quo, which may also be expensive. Options must be presented as clear alternatives with statements of ecological benefits and costs, and measures of uncertainty, for each. There is not a lot of rationality in most decision making, but there should be in decision analysis (Douglas and Wildavsky 1982). For example, risk analysts (and scientists) deal with estimates of ecological "change," while risk managers (and politicians) deal with ecological "degradation" and ecological "improvement." Such value-based statements move the scientist out of the scientific realm and into the political, value-driven realm. It may well be true that ecological conditions are better or worse from the *policy* perspective, but they are not better or worse from a *scientific* perspective.

My recommendations are (1) not to conduct a risk assessment unless there is a high likelihood that it will be used in decision making. If expectations are raised, and if no decision is made, the public senses that government institutions are not working. (2) Recognize that risk analysis of any significant ecological problem will result in options that create big winners and big losers. It serves no productive purpose to try to convince losers that they are really winners. ***If someone's property will be effectively expropriated for some larger societal good, that action should be clearly stated in the assessment. Conversely, if an owner is permitted to alter his property for short-term gain, but at huge expense to society at large or to future generations, that also should be clearly stated.***

6. SOME PROPOSED CHANGES

First, ecological risk assessment needs to be modified to create a paradigm of ecological *consequence* analysis. The concept of risk applied to natural resources will only work for a narrow set of problems where there is a clear public (and legal) consensus and on issues where there is an agreed-upon time frame of interest (are benefits and risks defined over 10 years or 10 centuries?). With all ecological "risks," a probability (of cause and effect or of ecological change) is neither good nor bad, it is only a probability. The resolution of many ecological problems is not limited by lack of scientific information or technical tools, but by conflict created by fundamentally different values and social priorities (e.g., for the salmon example, cheap food via irrigation water use vs. fishing; cheap power vs. free-flowing rivers; personal freedom vs. land-use zoning). If we are dealing with an ecological problem that is at an impasse because some of the stakeholders do not accept the utilitarian model, we should not be surprised when risk assessment and management do not resolve the issue. We need to do ecological consequence analysis, and let the political process select the desired option.

Second, the concept of ecological "health" needs to be better defined and understood by politicians and the public. The fundamental problem is not lack of technical information, but what is meant by health. Is a wilderness condition defined as the base, or preferred level, of ecological health? Is the degree of perturbation by human activity the measure of ecological health? The concept of ecological "degradation" is human value driven; the concept of ecological "alteration" is scientific. If the consequences of chaotic events in ecosystems are considered, what is "natural"? There are scientific answers for some of these questions, but political (social) answers to many others.

Third, risk assessors need better ways to use expert opinion. Most of the policy-relevant problems in ecology are too complex for easy scientific experimentation or analysis. An old rule in policy analysis is that if something can be measured, it is probably irrelevant to public choice. If problems are simplified to the point of making them scientifically tractable, then the result may lack policy relevance. Expert opinion must be used. Computer-generated maps and computer-assisted models may be elegant, but for really important decisions, the political process demands expert opinion.

Fourth, better ways need to be developed to evaluate and measure public preference and priorities in framing ecological issues. Public opinion polls always show that the public is very supportive of the environment, as it is with peace, freedom, and economic opportunity. The public is similarly supportive of preserving biological diversity, ecosystem management, and sustainable natural resource management. Unfortunately, this type of information is of limited use in helping make difficult environmental decisions. The public is not a monolith; it encompasses many divergent views, and individuals vary greatly in the intensity of their opinions. Individuals may argue forcefully for the industrial economic paradigm or for the natural economic paradigm, but practical political options are not framed in this context.

The fifth critical need is to develop better ways to present options and consequences to the public, to policy analysts, and to decision makers. Society is not well served by statements such as "it is a complicated problem and you need to have an advanced degree in ecology to understand it," or "you can select this option without significant cost to society" when there will be costs to some people. The main take-home message in risk assessment must be that there are no free lunches in environmental protection and that policy alternatives and the consequences of each must be explained in ways that the users of the assessments understand.

7. CONCLUSION

Biological and social science must be linked if public decision making is to be improved. Too often, forestry, fisheries, and wildlife problems are viewed as biological challenges. It is society that should define problems and set priorities, but the public speaks with not one, but many voices. Many of the stated public demands are mutually exclusive. Ecological "health," for example, is a social value defined in ecological terms. But, incorporating public input into risk assessment and management may be carried to the extreme (e.g., democratization).

Scientists must maintain a real and perceived position of providing credible ecological information — information that is not slanted by personal value judgements. Those involved in risk assessment cannot become advocates for any political position or choice, lest their credibility suffer. Such a position may be painful at times because no one can completely separate personal views from professional opinions. Risk assessors must be clear to the public (and political officials) on what scientific and technical information can and cannot do in resolving public choice issues.

We should not assume that complex ecological problems, such as the decline of the Pacific salmon, have only technological and rational solutions. Although tools such as risk assessment might help at the margins of the political process, they are not going to resolve the key policy questions. Nonrational ideas are extremely important in all significant public choice issues. Scientists and risk assessors should guard against technical hubris, a false sense of confidence in technology, technological solutions, and rational analysis, including risk assessment.

REFERENCES

Callicott, J. B., Conservation ethics and fishery management, *Fisheries,* 16, 22, 1991.

Douglas, M., and A. Wildavsky, *Risk and Culture,* University of California Press, Berkeley, California, 221 pp., 1982.

Gale, R. P., and S. M. Cordray, Making sense of sustainability: nine answers to 'what should be sustained?', *Rural Sociology,* 59, 311, 1994.

Herzog, H., Human morality and animal research: confessions and quandaries, *The American Scholar,* 62, 337, 1993.

Lackey, R. T., Ecological risk assessment, *Fisheries*, 19, 14, 1994.

Lackey, R. T., Options and limitations in fisheries management, *Environmental Management,* 3, 109, 1979.

Merrell, P., and C. Van Strum, Negligible risk: premeditated murder?, *Journal of Pesticide Reform,* 10, 20, 1990.

O'brien, M. H., A proposal to address, rather than rank, environmental problems. Presented at "Setting National Environmental Priorities: the EPA Risk-Based Paradigm and its Alternatives," Resources for the Future, Annapolis, Maryland, 15–17 November, 22 pp., 1992.

Ruckelshaus, W. D., Risk, science, and democracy, *Issues in Science and Technology,* 1, 19, 1985.

Suter, G. W. (editor), *Ecological Risk Assessment,* CRC Lewis Publishers, Boca Raton, Florida, 496 pp., 1993.

U.S. Environmental Protection Agency, Framework for ecological risk assessment, Risk Assessment Forum, Washington, DC, EPA/630/R-92/001, 41 pp., 1992.

QUESTIONS

1. What is the definition of ecological risk assessment? How does risk *assessment* differ from risk *management*?
2. Compare and contrast the application of risk assessment to ecological issues and human health issues.
3. What are the most important reasons offered for using risk assessment to help solve ecological problems? What are the major objections to the use of risk assessment to help solve ecological problems?
4. Compare and contrast the role of values, ethics, and science in formulating the "problem" in ecological risk assessment.
5. What are the commonly used alternatives to ecological risk assessment? What are their advantages and disadvantages?
6. Should the process of risk *management* be linked to risk *assessment*? What are the major benefits and dangers with the alternatives?
7. How is *adverse* determined in ecological risk assessment? Who decides what is adverse?
8. What role should scientists play in risk assessment and in risk management?

CHAPTER I.7

The Basic Economics of Risk Analysis

James A. Swaney

SUMMARY

Economics is about tradeoffs, and *opportunity cost* is the basic concept used to analyze tradeoffs and explore alternatives. Opportunity cost is the value of the next-best use of a resource, where value is measured as foregone alternative benefits. All costs are opportunity costs, and all opportunity costs are foregone benefits. Risk-reduction resources are being used wisely if it is impossible to reallocate those resources to produce more risk-reduction benefits. Many factors alter benefits over time, which introduces great uncertainty into analysis of long-lived projects. Whenever they can, economists use price data from markets to estimate benefits, not because the data are valid, but because they are cheap. ***However, markets miss and mismeasure benefits for many reasons, including "market failure"; buyer manipulation and confusion; and a variety of other technical, social, and economic factors that distort prices. Even without such distortions, market-generated benefit estimates are incomplete because they ignore "extra-market" concerns and values such as irreversibility, equity, and community need.*** Recognizing that weaknesses in benefit measurement and a tendency to ignore feasible alternatives often plague economic analysis, specific techniques such as risk–benefit, comparative risk, benefit–cost, and cost-effectiveness analyses are nevertheless capable of informing and improving risk management decisions. However, they should be applied with extreme caution, keeping in mind their limitations.

Key Words: risk analysis, risk–benefit analysis, comparative risk analysis, benefits, costs, benefit-cost analysis, discounting, cost-effectiveness analysis, opportunity cost, social cost, willingness to pay, cost shifting, discommodity, community need, market failure, externality, public good

1-56670-130-9/97/$0.00+$.50

1. INTRODUCTION

1.1 Three Basic Economic Questions for Risk Analysis

If politics is "all about compromise," then economics is "all about tradeoffs." The basic tradeoff in risk analysis is that resources used to assess and manage some risks are resources that otherwise could be used to manage other risks or, in some cases, to produce something else of value. A risk-reduction effort that fails to consider alternative use of risk-reduction resources risks wasting resources! If we want to avoid wasting resources, three economic questions should be considered. The first two questions deal with "the big picture."

First, could more benefits be produced if the resources were used elsewhere, possibly to reduce another risk? If the answer is "Yes," then resources have not been used economically. For example, in the United States, it is not unusual for workplace health and safety standards to require industries to spend several million dollars per life saved, yet we have been very slow to make simple highway improvements costing less than $100,000 per life saved (Rhoads 1985, pp. 17–20). Clearly, we have been wasting resources and lives by failing to improve our highways. Using economists's $4 million to $7 million value-of-life estimates (Viscusi 1993), so-called "excessive" regulations such as the coke-fume standards that cost the steel industry $4.5 million per worker saved appear reasonable, whereas a proposed OSHA (Occupational Safety and Health Administration) benzene exposure standard (one part per million averaged over an 8-hour working day) that would have cost *hundreds* of millions per life saved appear unreasonable (Rhoads 1985). If what we value is saving more lives, we are not using our resources wisely.

Second, could the same degree of risk reduction be accomplished at lower cost? For example, it is much cheaper to reduce infant mortality with prenatal care than with extended hospital stays for underweight infants. If what we value is healthy infants, prenatal care is a wise use of resources. However, the risk analyst may not be in a position to address these questions, since they are not usually trained in economics, nor are they typically empowered to shift resources between classes of risks. But, *someone* needs to address these questions if we are to extract the maximum value from our risk-reduction resources.

Once the risk analyst is confident that a particular risk-reduction strategy is a wise use of resources, a third, "microeconomic" question arises: at what point are the benefits of further risk reduction no longer worth the extra costs? Risk reduction often can proceed inexpensively at first, but after some point the cost of extra risk reduction typically rises rapidly, and some risks are simply too expensive to eliminate. The risk of death or serious injury in an automobile accident provides a simple example. This sizeable risk can be *eliminated* by simply avoiding automobile transportation, but this is not a viable option for most Americans. We do not want to give up the mobility and independence provided by the automobile, and most of us lack a feasible alternative mode of transportation. In economic terms, giving up our cars is simply too expensive. Risk *reduction* is another matter. As an individual you can avoid driving when you are likely to share the road with drunks, and you can ride in a more massive vehicle equipped with more advanced safety features operated by a better driver. As a citizen you can support a number of public (collective) efforts: remove drunks from the road, enforce slow and uniform speed limits, segregate traffic by vehicle mass, raise driver training/licensing standards, and advance the safety of both automobiles and highways. You might even support the use of gasoline tax revenue to produce alternative transportation, thereby lowering the cost of giving up your car. The fact that many people pursue some of these individual and social choices suggests that they find risk reduction to be worthwhile. Certain risk-reduction actions are inexpensive relative to the benefits of risk reduction, but few of us opt to totally eliminate automobile safety risks because the perceived cost is too high in comparison with perceived benefits.

1.2 The Economic Level of Risk Reduction Requires Accurate Benefit Estimation

In general, it is economical to continue to reduce risk so long as the marginal (extra or additional) benefit exceeds the marginal cost. At the point where marginal cost equals marginal benefit, the economical level of risk reduction has been achieved, because, beyond that point, any additional risk reduction will add more to costs than it produces in benefits. This approach is logically unassailable, but practically difficult because of problems in identifying and measuring benefits and costs. Benefit measurement is generally considered more challenging, but costs are really nothing more than the value of benefits foregone. Unfortunately, accurate identification and measurement of benefits would be a daunting task, even if technology, values, and property claims were unchanging. Since these and other environmental factors tend to change unpredictably, it is unwise to place too much confidence in estimates of the economic value of future consequences. In other words, there is a direct relationship between uncertainty and the "future distance" of benefit estimates. For example, projected benefits of automobile crumple zones and airbags would be substantially reduced by cost-effective "smart car" technology that prevented many types of automobile collisions. Health care provides another familiar example: life saving is a strong and persistent value, yet advances in medical science and technology are forcing us to reexamine this value in light of new possibilities and the costs they entail.

The conventional economic approach to measuring benefits is to estimate "willingness to pay" (WTP) and market prices. One major problem with using prices to measure WTP is that a wide array of "market failures" distort prices. Another key problem is that even where markets approximate the textbook ideal type of pure competition, prices are "produced" in the market by the interaction of demand and supply, where demand is the result of buyers' *willingness and ability to pay* and supply is the result of resource availability, technology, and other factors. This is not a problem if you believe that those with money should "call the economic shots" in direct proportion to the amount of money they have. In competitive markets, a dollar is like a vote in a one-citizen, one-vote democracy. Market "voice" is a direct function of wealth. Ross Perot, Warren Buffet, Steve Forbes, Ted Turner, and Michael Milken "speak" thousands of times louder than you or I, and people without money, including the world's poor and future generations, have no voice at all.* Where

* Of course, the numerous ecosystems and millions of other species have no voice either.

market prices are unavailable, economists typically utilize other methods to estimate WTP.*

People are being "good economists" when they take action to reduce risk so long as the *individual* benefits of the contemplated action outweigh the *individual* costs. They are being good citizens when they take action to reduce risk so long as the *social* benefits of the contemplated action outweigh the *social* costs. If everything of value conformed to the conditions of efficient property rights and all markets were purely competitive, social benefits and costs would equal the sum of individual benefits and costs.** Unfortunately, reality typically falls short of these conditions, leaving a wide disparity between private and social benefits and costs. Individuals face incentives to underinvest in collective (social) efforts, because the benefits of successful policy initiatives will be shared by everyone. Groups such as Mothers Against Drunk Driving (MADD) have been successful precisely because their members choose to be good citizens *instead of* good economists.

Economics is implicit in every aspect of risk analysis. Economics is less important for risk assessment than for risk management, but the fact that politics often determines which risks are addressed does not mean that economics is irrelevant. Since every risk assessment consumes resources that could have been used elsewhere, tradeoffs are inevitable. The limitations imposed by time, money, and the availability of technology and expertise suggest that economic decisions underlie every aspect of risk analysis.

Formal economic analysis "hits its stride" as risk analysis proceeds from risk assessment to risk management. However, the economists' tools are not all that precise, and there are often legitimate extraeconomic considerations that society uses to trump economic valuations.*** Yet the fact remains that risk management inevitably involves tradeoffs in the use of scarce resources, so risk managers should be familiar with basic economic tools *and* their limitations, and they should use those tools to build an economic perspective into their analysis. Even when society places other concerns above economic efficiency, good economic analysis informs the decision process and frequently uncovers lower cost alternatives.

There are several specific techniques for incorporating economics into risk analysis, including risk–benefit analysis (RBA), comparative risk analysis (CRA), and benefit–cost analysis (BCA). These and other risk analysis applications of economic reasoning are grounded in the concept of *opportunity cost* (described in the next section). Finally, the third section explores benefits and costs, beginning with a

* Common approaches to estimating WTP are travel cost, hedonic price, and contingent valuation. Any environmental economics text should explain these techniques (for example, see Goodstein 1995, pp. 94–99).

** For a clear and concise explanation of what is meant by efficient property rights, see Tietenberg 1994, pp. 32–34. The key property rights condition violated by "public goods" as discussed in this paragraph is exclusivity, which requires that all benefits and costs accrue only to the owner. Any introductory economics text and most environmental economics texts provide good explanations of the conditions of pure competition.

*** Not all economists agree with this statement. Many argue that if benefits and costs are defined so as to include all desirable and undesirable consequences, then economic analysis will be inclusive of all considerations, and there will be no reason for society to "trump" economics. This "all inclusive" approach requires that we somehow resolve ambiguities and conflicts in rights governing control of resources and that we put an accurate price tag on virtually everything of value.

discussion of the strengths and limitations of economists's tools and finishing with brief discussions of RBA, CRA, and BCA, the latter to include cost-effectiveness analysis (CEA), a less ambitious tool that is often more appropriate than "full blown" BCA, given data limitations and other uncertainties.

2. OPPORTUNITY COST

2.1 Opportunity Cost and Competitive Market Prices: Theory and Reality

Economists' key concept for evaluating tradeoffs is "opportunity cost." *Opportunity cost* is *the value of the most desirable foregone alternative* to a particular decision or course of action. If you are reading this chapter because you have nothing better to do and are simply "killing time," then your opportunity cost is essentially zero. But if you would otherwise earn \$100, your opportunity cost is \$100 (or what the \$100 is worth to you in terms of the satisfaction or pleasure derived from what you could purchase with it). Opportunity costs arise whenever resources have alternative uses and are scarce, which is most of the time. For example, the comic pages character Cathy buys a new car and exclaims that with her huge car payments she cannot afford a new fall wardrobe. Although she gives in to consumer impulse and purchases new clothes on credit, the fact remains that in buying the car she gives up all the things those hefty car payments otherwise could have purchased. (The credit card debt will not only "be there," but will continue to grow with interest.) The benefits she would have enjoyed from those consumer goods, which she can no longer afford, are the opportunity costs of the auto purchase. We all make such tradeoffs, whether consciously or not, when we make choices about how we use the valuable resources (such as our checkbook and our time) that we control.

Most people have less money than they would like, so it is instructive to think about opportunity costs in terms of money and prices. Money spent on one good is unavailable to spend on another good. For example, suppose a young couple with \$20 decides to spend it on either fast food (\$6) and a movie (\$14) or a more leisurely meal at a nicer restaurant (\$20). If these are the two highest value uses of \$20 in the eyes of this couple, whichever option they choose is presumed to provide the higher value, and the couple's opportunity cost is the value they would have obtained from the other option.

Since both options cost \$20, the market is "signaling" that the resources used to produce these two options are of equal value; that it costs society the same amount to produce {fast food + movie} as it does to produce a {casual restaurant sit-down meal}. Everything with a \$20 price tag should have an equal value of resources tied up in it, so the price is not only \$20 to the consumer, it is also \$20 to society. If every \$20 price tag indicates that society gave up exactly \$20 worth of other valuable stuff to produce and deliver that particular good or service, then the market price is providing an accurate measure of value. The only place where prices measure opportunity costs exactly is in economic theory, but *prices observed in properly functioning competitive markets provide good estimates of the value of the next-best* *(alternative) use of resources.** But what if our consumer couple is reacting to erroneous signals? What if the fast-food/movie option includes beef produced by Central American ranchers who displaced subsistence farmers, leaving them malnourished and homeless? Or what if that beef is produced by U.S. ranchers who graze their cattle on public lands at subsidized lease rates, using destructive range management practices? Such **shifted costs** are very real opportunity costs and should be included in the price of the fast-food burger. **Cost shifting** occurs when accounting costs are reduced, not by employing production or management methods that reduce opportunity costs, but by burdening workers, the larger community, the environment, or the future with those costs.** In short, the burger is "too cheap" because some opportunity costs are not included in the price. And if the burger's price is too low, then too many burgers are being sold, too many burgers are being produced, and too many resources are being allocated to making burgers. Many economists argue that the sit-down meal is similarly underpriced and conclude that *relative* prices are sending fairly accurate signals about the *relative* value of resources embodied in various products. Under these conditions, analysis based on market prices provides good guidance for choosing among similar projects or activities, but relative prices are generally an unreliable guide for choosing among activities that use different types of resources.

2.2 When Pure Competition Is Not Running Smoothly: Four Types of Market Failure

If all markets are *perfectly competitive*, all prices will be just high enough to pay all opportunity costs. That is,

$$\text{Price} = \text{Opportunity Cost}$$

Real-world markets never live up to the unworldly conditions of perfect competition, which requires perfect information, perfect foresight of market participants, and instantaneous adjustments. However, some real-world markets come close to perfect competition. These markets are called *purely competitive*. Pure competition is characterized by insignificant barriers to the entry or exit of buyers and sellers; an industry-wide standardized product; a large number of small, autonomous buyers and sellers; and extensive, accurate information. Prices are not equal to opportunity costs in pure competition, but *prices tend toward opportunity costs* so long as purely competitive conditions are maintained, with buyers receiving 100% of benefits and sellers bearing 100% of costs. When one or more of these conditions is not met, we have "market failure." When markets fail, the prices of resources and products do a poor job of measuring opportunity costs. Market failure is usually discussed in

* "Properly functioning" means that markets have few barriers to entry and good information, and few "shifted costs"; "good estimates" have to be taken in the context of the existing distribution of "market voice."

** The concept of "cost shifting" was, like nearly everything in economics, hinted by Adam Smith in his 1776 *The Wealth of Nations,* but credit for developing the idea belongs to the early 20th-century British economist, A. C. Pigou, and to K. William Kapp, an Austrian emigré who taught at Columbia University in the mid-20th century (see Kapp 1950, 1971, Swaney and Evers 1989).

terms of four general classes of problems: market structure, information, public goods, and externalities.

2.2.1 Market Structure: Imperfectly Competitive Markets

Market structure fails the test of pure competition wherever market participants are able to influence prices. That is, in purely competitive markets, all buyers and all sellers are such "small players" that virtually no one on either side of the market has any economic power. Everyone is what economists call a "price taker," which means that buyers and sellers alike can do no better than react to the market-dictated price. In other words, *pure competition is characterized by the total absence of price competition!* Examples of pure competition include the markets for major grains, such as wheat, corn, and rice, and certain financial instruments.* Beginning economics students often find this puzzling because their notion of competition is fast food, discount department stores, and other types of retail trade where advertising and price competition are intense. Fortunately, bagel shops and coffee bars are familiar retail markets that approximate pure competition. Information is good and entry is easy, so high profits by existing sellers attracted additional sellers, and the added supply drives prices back down toward opportunity costs (so long as there are no other sources of market failure). As barriers to entry become more formidable, market power increases. And market power means some discretion over price, indicating that price and opportunity cost will part company. Market power may also mean the ability to influence government policy, consumer preferences, or other markets, so it is very difficult to know if market power causes prices to over- or underestimate real opportunity costs. Neoclassical economists typically argue that firms with market power cut production in order to charge higher prices and bolster profits, but institutional economists argue that economic power begets political and marketing power, allowing powerful corporations to garner subsidies, shift costs, and keep consumer appetites whetted for the next "new and improved" product. These positive feedbacks loops allow many corporations to achieve low prices, sales growth, and enough profits to satisfy shareholders. To neoclassical economists, market power causes prices to be too high, overstating opportunity costs; to institutional economists, market power causes prices to be too low, understating opportunity costs.

2.2.2 Information Failure

Information failure occurs when buyers (or sellers) lack full knowledge of product characteristics. From automobiles and computers to insurance and investments, there are many markets where one side, usually the buyer, suffers from an information disadvantage. Markets beset by such "asymmetric information" are often regulated to reduce the incidence and severity of market failure, but "fleecing" of consumers and other forms of resource waste often persist. Even when sellers are

* Even these markets do not quite measure up because of government subsidies and regulations.

honest and law abiding, buyers often end up with less value than they expected, suggesting that market prices are inaccurate measures of value.

A related set of problems arises not from information imbalance per se, but from a combination of factors that produce "disappointed consumer syndrome" (DCS). DCS is a chronic and sometimes debilitating ailment where consumers are not only poorly informed, but their tastes and preferences are poorly defined and malleable. While this author knows of no definitive diagnosis, DCS symptoms are commonplace. Higher standards of living have not made people happier or more satisfied, and when more does not satiate the hunger, we keep eating anyway. Advertising and other forms of marketing contribute significantly to DCS, but the underlying causes probably have more to do with broader social processes wherein values, tastes, and preferences are defined.*

2.2.3 Public Goods

The third class of market failure is public goods. A "pure public good" is available to everyone (whether they pay or not) and can be enjoyed by additional people at no (opportunity) cost to anyone. A classic example of a pure public good is national defense. You benefit from it even if you are in prison and pay no taxes, and it does not cost any more to add "consumers," whether through births or immigration. Because you cannot be excluded from the benefits of the good, why bother to pay for it? (Economists refer to this as the "free-rider problem.") And if nonpayers benefit anyway, the producer of the goods will not earn sufficient revenue to cover costs. Another example of a public good is collective actions to improve automobile safety. The success of MADD in raising awareness and passing tougher penalties for drunk driving is a different type of collective good that will benefit everyone, whether they help in the effort or not. All automobile occupants, with the possible exception of drunks, gain when drunk driving is reduced, yet nothing forces drivers to share the cost of MADD's efforts. That is where the market failure comes in: The incentives of the marketplace are to underinvest in the "production" of such goods. "Let someone else pay, I'll still enjoy the benefits!" Most collective goods suffer this free-rider problem; hence, the economic rationale for taxes to pay for "necessary public works." Informal social rules such as common law conventions and community norms can also be viewed as public goods. They reduce uncertainty and, in so doing, provide an environment conducive to orderly economic activity. They also provide a stable environment for market transactions and keep enforcement and other policing costs low.

Another example of a public good is your local public radio or television station. Once the signal is produced, it is available to anyone with a radio receiver, whether or not that individual supports public radio.** Another characteristic of public goods is that the overall opportunity cost of additional users is essentially zero. When I receive the radio signal I do not leave any less of it for you. If an additional user adds nothing to overall opportunity cost, then from society's perspective the price should be zero because the opportunity cost is zero! So even if those who benefit

* These issues are revisited in Section 3.

** Commercial radio provides a rare example of private enterprise producing a public good.

can be made to pay, a price above zero will mean that the resource or good is underutilized. These are not conditions conducive to free enterprise: nonpayers cannot be excluded, and additional users should not be charged anyway!

Pure public goods are not all that common, but many goods have some "public goodness" to them. Familiar examples are uncrowded stadiums, streets, highways, trains, parks, classrooms, libraries, concert halls, and airplanes. Wherever there is unused capacity and additional users add little or no cost, the "opportunity cost price" should be low or zero. Since such pricing is no way to run a business, society ends up with the responsibility for funding the production of many of these "quasi-public goods" as well as most pure public goods. Even our private automobiles have some public goodness to them since it costs almost nothing to add passengers so long as you have empty seats. When priced so high that capacity remains unused, resources are wasted from society's perspective. When priced too low, however, congestion costs may arise. Once the excess capacity is utilized, the opportunity cost of an additional user is no longer zero and the "publicness" disappears. If access remains unrestricted, however, lots of resources will be wasted, as when cities suffer traffic gridlock. Raising the price of entry, as with carpool lanes, toll booths, or peak load fees is gradually gaining acceptance as an alternative to the conventional, "grow capacity" management strategy.

2.2.4 Externality (Social Cost)

The fourth class of market failure, externality, is probably the most relevant for risk analysis. Externalities occur whenever those who create foregone opportunities (costs) do not fully pay for them; that is, when actual costs are greater than accounting costs. If the Central American cattle rancher does not have to pay for all the lost opportunities he/she creates, his/her costs are lower than they should be, and so is the price of the fast food burger. Most environmental problems involve externalities, and, while some of them are accidental or unavoidable, most are the result of **cost shifting**. In the process of producing goods ("commodities"), nearly every economic activity also produces bads ("discommodities") (Coddington 1970). Everyone wants goods and no one wants bads, so it should come as no surprise that individuals looking out for their own interests, as they are supposed to do in capitalism, will devote more effort to "capturing" goods (benefits) than bads (costs).

Businesses succeed by occupying a niche: by providing a good or service that people want and are willing to pay for. Even where markets do not yet exist, if a market promises to capture enough benefits, someone will petition the government to establish (or clarify) property rights and rules for orderly transactions. Douglas North, a recent recipient of the Nobel Prize in Economics, asserts that the primary reason people set up institutions is "to create order and reduce uncertainty in exchange" (North 1991, p. 97). That is, markets will sprout up wherever the expected benefits from establishing markets exceeds the costs of setting them up. Those who petition the government for property rights to mine ore do not usually petition for restrictions on how they dispose of tailings. This "natural" incentive structure is asymmetrical: while it encourages markets to capture benefits, it fails to encourage markets to capture costs. The incentive to *capture* benefits of commodities becomes, for discommodities, the incentive to *escape* costs. Historically, this incentive has been "nourished" by abundant opportunities for "free" waste disposal. In addition to escaping costs by simply releasing them where the damage they do is not protected by established rights, costs can also be escaped by shifting them onto the unsuspecting or defenseless. After all, $100 of costs escaped has the identical effect on the bottom line as $100 of benefits captured.

Common practice has been to escape discommodities by discarding them into the atmosphere, land, or water, as convenient. Historically, when humans were few in number and weak in knowledge of how to manipulate their environments, people could soil their nest without spoiling it. "Natural garbage disposals" could usually break down discommodities and recycle their components. But the industrial revolution and burgeoning populations around the world changed things. A forward-looking community that drinks from a pristine river can petition the government to grant it the right to pristine river water. If the government has the means to enforce that property right, then upstream pollution can be averted (or a polluter would be required to pay the community for the right to pollute the river). But such market solutions to pollution are seldom practical, and despite the theoretical possibility of such solutions, advocates of "free market environmentalism" have searched far and wide and produced precious few examples. The absence of effective property rights to protect against discommodities arises in part from difficulties inherent in designing and enforcing efficient property rights for environmental media. The atmosphere, the oceans, rivers, and large lakes, by their "fugitive nature," do not lend themselves to private property solutions. If technological advances were to somehow overcome this and other obstacles, people could be granted the right to clean water and air (rights of refusal), and "reverse markets" could be established to force polluters to stop polluting or to compensate those who are injured. ***Such markets would have the effect of eliminating (internalizing) externalities, thereby improving efficiency. But politics would certainly intervene, because such markets could be set up only by taking away the long-established pollution privileges of politically entrenched and powerful "brown" interests.***

2.3 Measuring Opportunity Costs Where Markets Fail

Since contemporary markets are far removed from pure competition, market prices are very inaccurate measures of overall opportunity cost. Yet many professional economists contend that observed prices are "reasonably accurate" measures of opportunity cost! This argument has no theoretical support (Lipsey and Lancaster 1956–57), nor have economists produced any systematic empirical support, such as evidence that distortions tend to offset one another. Confronted with these facts, many economists admit to a very "liberal" definition of "reasonably accurate." Prices in capitalist economies are certainly better measures of opportunity cost than in socialist economies, and market prices are the *only* inexpensive source of information about opportunity costs. What economists really mean by "reasonably accurate" is that data is cheap and not uniformly bad.

While economists have developed tools for estimating opportunity costs when markets are absent, they generally use observed prices in parallel markets or other price data as their base. The general rule is "if available, use market prices to measure opportunity costs." As a general practice, the best that can be said for this approach is that it is expedient. It demonstrates acceptance of folklore and absence of critical judgment. There are, nonetheless, circumstances where relative price data produce valid results. For example, if alternative risk reduction projects use a similar resource mix, comparisons of inaccurate prices are likely to identify the more cost-effective project because similarly inaccurate prices will not affect relative values. But if the alternatives use different types of resources, with one project depending on resources whose prices grossly underestimate opportunity costs and the other project depending on more accurately priced or overpriced resources, cost analysis that relies on market prices will be unreliable.

Economists should probably devote more resources to adjusting market prices (mostly upward) to obtain more accurate opportunity costs measures. Lacking better opportunity cost data, price-based analysis may be no better than an educated guess. When alternatives use dissimilar resource mixes, risk analysts and managers should require their economic analysts to identify and discuss factors that cause resource prices to deviate from overall (social) opportunity costs and provide estimates of those deviations.

Few markets come close to meeting the conditions of pure competition, and few prices accurately measure opportunity costs. How inaccurate prices are, however, is a matter of speculation. While serious environmental risks are often associated with cost shifting, suggesting that many market prices are too low, no general conclusions can be drawn because other forms of market failure may produce prices that overstate opportunity costs.

Wise use of economic analysis techniques in risk management requires an understanding of the ideological and methodological underpinnings of economics as well as the practical shortcomings of specific economic measurements. A healthy skepticism regarding the validity of market data will often serve the risk analyst well, but for some risk management problems the entire enterprise of conventional economic analysis may obscure more than it reveals.

2.4 Other Shortcomings of the Market: Equity, Voice, Irreversibilities, Shifting Values

As mentioned previously, data from markets (or market proxies) allocate voice by willingness and ability to pay, in effect muting those without property or money, including the poor, future generations, and other species and ecosystems. Let us suppose for a moment that everyone considers this distribution fair and just, and also suppose that property rights for all resources are efficient and all resources are allocated through purely competitive markets. (Efficient property rights combined with pure competition for all resources is sufficient to prevent market failure.) Under these "all-inclusive market" conditions, prices would accurately measure opportunity costs for every conceivable resource allocation decision. Then, if all decisions are reversible (or if changes in system parameters, such as values, technology, and the natural environment are fully anticipated), the risk analyst could rely on economic analysis to determine optimal resource allocation for reducing risks. Irreversibilities present problems even for the all-inclusive market because something with no value today may become very valuable tomorrow. For example, the Pacific yew tree, an understory "junk" species of Pacific Northwest native forests that was typically bulldozed and burned when forests were clearcut, recently became prized for an accessible supply of a chemical in its bark, taxol, which doctors discovered to be an effective treatment for certain lethal cancers. Almost overnight, medical research turned a worthless species into a lifesaver.

Future technological advances may be able to reverse some apparent "irreversibilities," but it seems prudent to allocate more resources than economic analysis suggests when endeavoring to reduce the risk of an irreversible event. For example, scientists may be able to clone individuals of various extinct species from preserved DNA. However, this does not protect against loss of intraspecies diversity, nor does it protect the ecosystem in which this species played a role. Such procedures are also likely to be quite expensive relative to habitat protection.

Aside from scientific and technological advances that transform the "worthless" into the "priceless," market valuations are also affected by changes in individual preferences and social values. Until the mid-20th century, people in the United States used the term "wilderness" to refer to an unproductive area in need of human management and control. Today, as famous mountaineer Paul Petzoldt often remarked, the quickest way to destroy a wilderness is to declare it so. The demand for wilderness experience has grown even more rapidly than the supply of officially designated "Wilderness Areas," so when an area receives the "wilderness" label it tends to attract hoards of people who overrun it, degrading resources and sometimes threatening ecological systems. While human impacts are frequently reversible, resource recovery is often a very slow process. If we could take today's wilderness concept back to the year 1900 and measure the economic value of this country's substantial wilderness acreage, the value would be very, very low. The Sierra Club had been founded, but only a few scientists and spiritualists were interested in preserving wilderness. Shifts in values and advances in technology are nearly as unpredictable as they are certain, so even if we were an extremely risk-preferring and fatalistic society, we should be cautious about actions involving irreversibilities because, for all we know, our grandchildrens' society will share neither of these traits!

2.5 One More Caveat: Community Need

We have seen that even when everything of value is controlled and allocated by efficient property rights and purely competitive markets, markets fail to provide unambiguous guidance as to how to use resources wisely. We have also observed that market failure is commonplace, causing observed prices to diverge from actual opportunity costs. On top of these caveats we now add one more, the fact that social preferences frequently diverge from individual preferences. This problem is similar to public goods, but is "preference based." That is, public goods cause market failure because of their resource characteristics on the "supply side," while social preferences cause markets to misallocate resources on the "demand side."

There are two types of social preferences that in some cases should count more than individual preferences. One type of social preference arises from the individual's internal conflict between what he/she wants as a selfish individual and what he/she wants as a responsible member of the community. Market outlets for these community-oriented wants of individuals are usually absent, in part because of public good problems. The other type of social preference is community need (social value), typically "expressed" in capitalism through the political process. In a democracy, these two types of social preferences are connected insofar as the citizenry participates in developing public policy. Government-defined and enforced constraints on the behavior of market participants is, and always has been, necessary in capitalism because open, free markets require a peaceful, stable environment. The protection of property and persons, including everything from the definition of property to the traffic cop, are community needs and therefore government responsibilities. Just as the government provides the institutional environment for capitalism by defining and enforcing property rights, government also protects the community from adverse consequences of the market.

Many self-described "conservatives" want to define community need very narrowly to exclude all democratic deliberation and political action which interferes with their own particular notion of how property rights and responsibilities should be defined and distributed. But these are not questions that will go away because some people are content with the current arrangements. Community need would not disappear even if we all agreed on private ownership of all resources as well as a fair distribution of initial property claims. Rights of ownership, use, and disposition are seldom absolute because resources are seldom purely private; that is, the "exclusivity" condition of efficient private property seldom holds. As a result, extending property rights for one nearly always means reducing rights for others. Likewise, government restrictions on markets (laws, regulations, etc.) nearly always protect some by reducing the scope of choice enjoyed by others. Indeed, disposition of competing claims is, and always has been, a function of government, and modern economies are no exception. Some property rules define and protect vital community needs, while others support powerful narrow interests at great community expense. Economists would provide an invaluable public service if they could help the electorate identify even 5 or 10% on each end of this spectrum, so as to anchor and strengthen essential rules and legitimate claims on the one hand and undermine public subsidies on the other. Unfortunately, most economists fail to apply the opportunity cost concept to conflicting property claims, instead taking the existing structure as "given," "prior to analysis." Although implicit and passive, the effect is to condone and defend the existing distribution of wealth, power, and property claims.

Minimum wage and child labor laws illustrate how property rules are used to meet community need, and how such rules "distort" market prices. In pure competition, a $10 pizza has $10 worth of opportunity costs in it. Society has "invested" resources in this pizza that could otherwise have been used to produce exactly $10 worth of other benefits. A major "ingredient" of pizza is labor. When markets are purely competitive, each pizza worker is paid according to how much value is added by the last worker hired. If there is an overabundance of labor, competition among workers drives wages down, thereby reducing business costs and, eventually, prices. With a surplus of poor, really hungry children available to make pizzas, maybe the price would fall to $8, providing better value for consumers unaffected by falling wages. If wages fall to where workers cannot afford beans (let alone pizza), the market will "self-correct" as workers "voluntarily" drop out of the labor market!

In pure competition, there is no room for minimum wage or child labor laws or anything else that protects workers or the environment from the "discipline" of the market. In a truly free market, children would be "free to choose" to work longer hours for less pay! But we have minimum wage and child labor laws that interfere with the free market. The pizza that would cost $10 if produced in pure competition might cost us $11 or $12, maybe more. Society places various constraints on the marketplace that distort prices and interfere with the freedom of individuals to pursue their own preferences. Whether you prefer to think of these restrictions as "the Heavy Hand of Big Brother," "Government Pork and Other Special Interest Distortions," "Community Need," or some combination of the three, the fact remains that all modern economies have many such constraints. ***The message seems to be, "The preferences of individuals pursuing their self-interest in the marketplace are not the only preferences that count." Some regulations standardize practices and improve information and resource flows, leading to lower opportunity costs and lower prices, whereas other regulations, like child labor laws, push up business costs and prices. We do not know if prices are "too high" or "too low" as the result of the constraints society imposes, but we do know they are "distorted."***

Environmental, workplace health and safety, and product safety laws and regulations can be interpreted as society's efforts to correct for public good and externality market failures. But the market failure framework is bland and sterile, providing an incomplete and misleading view of the scope and purpose of society's interference in the market. The broad array of legal constraints imposed on markets suggests that societies are generally unwilling to allow people and nature to be "valued" by unregulated ("self-regulating") markets. Societies choose to insulate certain values (and valuables) from the "discipline of the market," in effect "trumping" the market's process for measuring and assigning value. If only we could agree on and write rules to (1) fulfill our social and moral obligations (for example, caring for the severely disabled, insuring opportunities for future generations, providing opportunities for others who "lack voice" in today's markets), (2) correct for market failures, and (3) eliminate the influence of narrow special interests, then prices would measure social opportunity costs because they would reflect private market values appropriately modified by social requirements!

3. BENEFITS AND COSTS

3.1 Accounting for Future Benefits and Costs

While most economists put far too much faith in market prices as measures of actual opportunity costs, the discipline has developed a number of sophisticated techniques to adjust existing prices for "market imperfections" and to measure "willingness to pay" for goods that have no prices. While these techniques are sometimes oversold by their practitioners, economists continue to make progress in measuring benefits and costs, *as they exist in the present.* But both individual and social preferences and values change over time (shifts in demand) and so do technologies and the availability of resources (shifts in supply). While forecasting tools can predict economic conditions in the near term, they cannot anticipate those major changes in priorities or advances in technology that come along every now and then, so longer time periods introduce greater uncertainty into estimates of benefits and costs.

When benefits and costs are spread over a number of years, economists mimic the market by discounting future benefits and costs. Discounting answers the question, "What are future values worth to us today?" (present values). Discounting is a simple financial technique that works like compound interest in reverse; in fact, discounting is the same thing as compounding. For example, if you put $100 in the bank today at 12% interest and allow the interest to compound, in 6 years you will have $200. Looking at the same process in reverse, a $200 payment 6 years from now is worth $100 today if the discount rate is 12%. So if an investment activity or project will produce $200 worth of benefits in 6 years, with a 12% discount rate the "present value" is $100. By using this technique, one can calculate the present value of all future benefits and costs and can subtract costs from benefits to arrive at the *present value of net benefits* (PVNB), which is the benchmark measure of benefit-cost analysis. For example, with a 10% discount rate, a $1 million cost 100 years from now has a present value of $75. The decision rule usually recommended is that PVNB should be positive; that is, the present value of a project's benefits should exceed the present value of its costs. This approach works well for projects of short duration, where the "time stream" of benefits and costs extends only a few years, but discounting is highly suspect when consequences are spread over many years; when the time distribution of benefits and costs is asymmetrical (e.g., benefits occur over 10 years, while costs are delayed for 100 years); when there is some probably, even if remote, that distant costs will be catastrophic; or when interests aside from today's resource owners and consumers are affected.

3.2 Benefits and Preferences

3.2.1 The Role, Validity, and Dynamics of Preferences

Recall that (opportunity) cost is the value of the next-best alternative use of a resource. That is, costs are really nothing more than measures of foregone benefits. So when economists look at benefits and costs, they are comparing the benefits of an activity or project with the benefits that could have been produced if the resources involved were instead used elsewhere. An activity is worthwhile if its benefits exceed the benefits that otherwise would be generated by other uses of resources required for the activity. As we have seen, economists' reliance on market measures of opportunity cost is based on expediency and doctrine, not on accuracy and reliability. But even if we obtained *accurate* measures of benefits, could we then conclude that these measures were also *valid*? Since the source of benefits is tastes, performances, and values, to ask if benefits are valid is to ask if tastes and preferences are any good.

Our preferences, both as individuals and as members of a community or society, are influenced by a variety of forces over time. As Thorstein Veblen argued at the turn of the (20th) century, our individual preferences are shaped by our desire to identify with and be accepted by others. That is, consumer wants and needs are primarily social. Frank H. Knight, one of the founders of the Chicago school of conservative economics, argued in the 1920s and 1930s that it was obvious that our wants are quite literally produced within the economy and that a crucial question was whether or not these wants are "any good." He concluded that they are not much good, but that any effort to improve them would involve what we might call "taste police" who would trample basic individual freedoms. Knight argued that the wants produced by capitalism are severely wanting, but his unhappy conclusion was that the alternative to the status quo was the slippery slope of socialism. The bulk of economists since Knight have ignored the issue, and the economists' tools reflect that neglect. All that matters are individual preferences as the economist finds them. Wherever individual preferences come from, they are presumed to be the product of free, self-reliant individuals. By ignoring the forces that shape markets in general and consumer demand in particular, conventional economic analysis of benefits and costs produces only a black and white snapshot of benefits and costs. Society may choose to add some color, but we still have only a snapshot. And when rapidly changing values or technologies are transforming the environments of the consumer or producer, economists' estimates of benefits and costs are likely to have a very short half-life. This is particularly true when positive feedbacks become established between social policy, technology, and consumer preferences. Automobile safety again provides an illustrative case.

In the 1950s, automobile "safety technology" was a padded dashboard; in the 1960s, it was a lap seatbelt. Progress was slow, and consumer demand was practically nonexistent. But Ralph Nader and other advocates convinced Congress to legislate safety standards for automobiles. Automobile manufacturers opposed legislative initiatives, accurately labeling them "paternalistic." Most economists agreed: if people wanted safer cars, they would demand safer cars, and the market would reward manufacturers who responded to that demand, thus shifting resources toward auto safety. "We live in a free country, and consumers should not be forced to pay for safety features they do not want." For decades producers resisted and consumers focused on other product characteristics. Yet by the early 1990s, consumer demand had shifted dramatically. Automobile salespeople in the 1990s tout safety features the way they touted horsepower in the 1960s and fuel economy in the 1970s, and manufacturers spend "megabucks" on advertising campaigns with safety themes. Safety technology is advancing far more rapidly because the market is being driven by both consumer demand and public policy. But where would we be today without "Crusader Nader" and congressional paternalism?

Consumer demand does indeed drive the market, but, in turn, a wide array of political, technical, social, and commercial forces drive consumer demand. Conventional economic analysis is literally designed to ignore these forces, and the risk analyst should be aware of the limitations that result. For example, in the early 1990s, the U.S. government required restaurants to make nutritional information available to their patrons, and the Center for Science in the Public Interest staged successful media events criticizing the nutritional composition of various restaurant menus. Some restaurants began to market reduced-fat menu items. Over time, the very simple and inexpensive production of nutritional information combined with a few doses of media-driven education will likely reduce the risk of diseases associated with diets high in saturated fat. Menus are likely to undergo a gradual change toward less saturated fat as consumer demands gradually shift. This suggests that inexpensive, unobtrusive policy changes combined with educational efforts often can be an effective risk-reduction strategy, especially when simple changes in consumer lifestyle promise significant risk-reduction dividends. More rapid change may be desirable, but many Americans react to sensible, scientifically based risk-reduction policy proposals as if the Bill of Rights was under assault, apparently giving the most trite lifestyle habit the significance of freedom of speech!

3.2.2 Individual Preferences and Social Consequences

Economists frequently argue that an individual will choose additional risk reduction whenever he/she expects benefits (for him/her) to exceed costs (to him/her). While true so long as moral obligation or duty does not intervene, this argument obscures two critical facts. First, an individual's expectations are often based on incorrect information, ignorance, or other distortions. Second, there are very few circumstances where the overall "social" benefits and costs are the same as private benefits and costs. Motorcycle helmet laws provide an example. Those who oppose laws requiring motorcyclists to wear helmets typically argue that the cyclist should be free to decide for him/herself. Many economists agree, assuming the cyclist is a well-informed rational decision maker, and assuming the benefits and costs of the choice are "purely private," enjoyed and incurred only by the individual making the decision. If these assumptions were true, helmet laws would be paternalistic and inefficient. Let us suppose that the first assumption is a reasonable approximation, and proceed to the second assumption. Motorcycle accidents entail many costs to society. State police and emergency medical crews and facilities are taxpayer funded. Some hospitals are private, but society ends up paying many medical bills. The opportunity costs of motorcycle accidents also include the social and economic losses to family and friends and to the larger community if the accident victim was a productive, contributing worker and citizen. If helmets substantially reduce these social costs, then paternalistic helmet laws may actually be economical. That is, the overall benefits of helmets (measured by avoided medical and morgue costs, avoided pain and suffering, and higher economic production) may outweigh the overall costs of helmets (measured by the value of resources used to produce helmets and the value of discomfort and reduced riding pleasure incurred in wearing them). Helmet laws may be (socially) efficient, even though most cyclists would choose to ride without a helmet. The point is that few risk-reduction decisions are purely private: individual choices typically have significant social consequences.

3.3 Applying Economic Principles to Risk: Three Techniques

Despite many economists' belief that risk reduction should be "left to the market," society has economic justification for interfering when a private decision maker's perceptions of benefits or costs differ markedly from social (overall) benefits and costs. Society may also decide to interfere on noneconomic grounds, as with child labor laws or environmental protection. The legal structure within which the market operates is (and always has been) society's construct. The rules for access to and control over resources, the definitions of property and the rights, privileges, and responsibilities attached thereto, and the rules of participation in markets and other resource allocation processes are all social constructs. Supply and demand interact and markets (fail to) clear, always within socially defined rules. Therefore, when society changes the rules, the conditions of the market change, and "efficient resource allocation" is *redefined.* ***Those who argue that society should not tamper with the rules and should act only to improve economic efficiency within those rules often claim to be practicing value-free, "positive economic science." The truth is that they are advocating the value-laden, normative premise that the current structure of rights and entitlements is good, fair, and just. Since the market operates within a socially constructed legal environment, there can be no a priori definition of "proper" boundaries for social constraints on the market. They are always subject to review.***

In the context of risk analysis, what is important is for society to make decisions based on good information, so that policy makers have at least thought about the likely consequences of their choices. Society has a long and well-established role in guiding the market, including education and standard-setting, both of which can produce cost-effective risk reduction over time. Economic analysis of risk-reduction alternatives can help policy makers in both the private and public sectors to separate risk-reduction efforts that make good economic sense from those that do not.

3.3.1 Risk–Benefit Analysis (RBA)

RBA compares the risks of an activity with its net benefits and asks the question, "Do the benefits of an activity justify the risks involved?" Generally, the greater the benefits, the greater the acceptable risk. For example, use of a drug that doubles the risk of stomach cancer may be an acceptable risk if it reduces the risk of heart failure, but an unacceptable risk for relief of minor headaches. The U.S. Food and Drug Administration (FDA) uses a risk–benefit approach when it expedites the approval of drugs for critical care circumstances. The risk-benefit approach is also found in environmental law; for example, the Toxic Substances Control Act (TSCA) requires the U.S. Environmental Protection Agency (EPA) to consider the benefits as well as the risks of chemical substances (Callan and Thomas 1996). Risk–benefit is a sensible risk management strategy because it seeks a balance between hazard reduction and net benefits, as opposed to a strategy that tries to reduce all risks, regardless of how many benefits will be lost in the process.* The objective of RBA, as with all economic analysis, is to weigh opportunity costs and choose the best (net-benefit-generating) alternative. The greater the benefits, the more risk should be tolerated, provided that those benefits are not available through a less risky activity.

When the decision to engage in a risky activity is purely private, with benefits and costs accruing only to decision makers, economists prefer to leave the decision to individuals. So long as individuals are well informed and the activity is voluntary, those individuals presumably expect the (total) benefits to exceed the (total) costs. For example, it is no secret that downhill skiing is hazardous, yet many people choose to ski. Apparently, people who ski expect the experience to deliver more pleasure than pain (including the "hurt" to their pocketbook), so skiing passes the risk-benefit test.

For risky activities where opportunity costs reach beyond the individual engaging in the activity, a risk-benefit management strategy should be supplemented by an evaluation of the distribution of benefits and risks. As with other costs, risks are sometimes shifted in the process of producing or delivering a good or service. If some are enjoying the benefits (e.g., consumers or businesses), while others are suffering the risks (e.g., households downwind from a factory or along an urban interstate highway corridor), society may choose to trump economic analysis for equity reasons. That is, society may veto the economist's judgment because the benefits go to one group, while the risks are imposed on another. In theory, if an activity passes the risk-benefit test, those receiving benefits should be able to remediate the risk and still come out ahead, either by reducing the risk or by compensating those facing the hazard. Such compensation rarely occurs, but, even where it is practical, society may judge such schemes unfair. For example, environmental hazards are often shifted onto the poor who are likely to be ill informed and who seldom have political clout. Because extra dollars typically have a higher value to poor people, they are likely to accept relatively small payments as compensation for exposure to a hazard. Like all economic analysis, a good RBA can inform the policy process, but should not preempt legitimate nonefficiency considerations.

A textbook example of RBA is gasoline combustion (Callan and Thomas 1996, p. 214). Burning gasoline produces many benefits that (to some degree) "balance" the environmental risks. But what of the alternatives? Lobbyists for the oil and automobile industries are not likely to point them out, but a good RBA will examine all the alternatives. Can the same transportation (net) benefits be delivered with reduced risk to human and ecosystem health? Risks can be reduced with cleaner burning gasoline and a wide array of substitutes, from alternative fuels (natural gas) to improved fuel efficiency to alternative transportation (bus, light rail, bicycle). Over time, we can redesign our lifestyles and develop new technologies to provide

* In the context of RBA, "benefits" refers to total benefits less total (accounting) costs, so that when we speak of "benefits greater than risks" or "risks balancing benefits," the opportunity costs of the risks are being compared with benefits net of other costs.

the benefits currently provided by burning gasoline. ***The key to a good RBA, as in all economic analysis, is to carefully account for all costs, social and environmental as well as private. Accounting for all opportunity costs requires consideration of all alternatives. When alternatives are left unexplored, economic analysis is incomplete and often wrong.***

3.3.2 Comparative Risk Analysis (CRA)

CRA compares many different types of risk, typically by ranking risks for the seriousness of the threat they pose to human health, ecosystems, or quality of life. CRA helps to identify "undermanaged" risks, where significant risk reduction is possible at relatively low cost. Some undermanaged risks are new or recently discovered, some are complex risks that are difficult to analyze, and some are risks that have simply been neglected. Merely bringing attention to neglected risks may lead to risk reduction through a process of media attention and education that produces behavioral or management changes. Initiated by the U.S. EPA's 1987 study, *Unfinished Business* (U.S. EPA 1987), CRA typically uses expert opinion to produce risk rankings. The idea behind *Unfinished Business* was to look at the big picture, both to identify relatively neglected risks with inexpensive risk-reduction opportunities and to involve and educate the public, the experts, and the environmental policy process itself.

"Comparative-risk-analysis-in-the-states" is an extension of this process, where information from both experts *and the public* is used to produce risk rankings. Several CRA-in-the-states projects have produced an "overall integrated risk ranking" that combines expert and public opinion. What is "high risk" to experts is often "low risk" on the public barometer, and vice versa. One reason for the discrepancy between expert and public opinion is that many of the risks that have received considerable media attention have also received considerable risk-reduction efforts. Hazardous waste, for example, is a top concern of the public, but not of the experts, in part because many hazardous substances are highly regulated and tightly controlled, having received years of research and regulatory attention. Another reason why experts and citizens disagree is that scientific information on chronic (rather than acute) risks is disseminated and assimilated rather slowly due to media inattention and public distrust of the experts. Finally, citizens are often concerned with risk dimensions such as catastrophic potential, voluntariness, and dread that the experts miss with their calculation of a risk's probability and severity (see Chapter III.1).

As always, opportunity cost is the economic concept "in play" in comparative risk analysis. Identifying economical risk assessment and management activities, where large quantities of risk reduction can be realized at low cost, is nothing more than a restatement of the idea that we should use our resources where they provide the greatest net benefits, always being careful to consider all reasonable alternatives. In light of the "mismatch" between scientific ranking and public perception, it may be useful to reconsider just how well-informed consumers are when they take their preferences and their pocketbooks to the marketplace, and, as a result, how accurately market prices measure opportunity costs. It is probably

safe to conclude that the opinion mismatch should not bolster our confidence in market prices as a guide to overall opportunity costs.

3.3.3 Benefit–Cost Analysis (BCA)

BCA,* a well-developed and highly refined technique used to evaluate many different types of projects and activities, attempts to produce a comprehensive "present value" measure of the net benefits of a project. As explained earlier (Section 3.1), present-value calculations use discounting to answer the question, "How much is some future benefit or cost worth in today's dollars?" Also, recall from the earlier discussion that uncertainty dominates estimates of future values as the time horizon is extended beyond a few years. A less difficult problem is that present values are sensitive to the discount rate used to calculate them. Since different discount rates produce vastly different present values whenever benefits or costs are spread over more than a few years; the common practice today is to calculate the present value of net benefits with several different discount rates, and present all of them with an explanation. This technique, called *sensitivity analysis*, allows the decision maker to view a range of net-benefit calculations to see how different discount rates impact the estimation of net benefits.

Another problem of BCA worth revisiting is that benefits are fluid (due to changing technology, resource claims, income distribution, and tastes and preferences), and difficult to measure. Here again, sensitivity analysis can be used to calculate and report benefits, giving the policy maker an idea of how different plausible assumptions affect benefit estimates. For a variety of reasons, from market failure to rights conflicts to community need to equity concerns, environmental law frequently mandates risk-reduction targets. With target in hand, **cost-effectiveness analysis** (CEA) can be employed to identify the least costly route to the specified goal. This makes the problem much more manageable because the range of possible alternatives has been narrowed significantly. But since costs are really nothing more than foregone benefits, the problems of benefit estimation remain. Noteworthy advocates of BCA in fact advocate CEA for risk management decisions. W. Kip Viscusi, whose specialty is estimating the value of life by studying what workers are willing to pay (in lower wages) to avoid workplace hazards, contends that "reasonable estimates" of the value of life are in the range of $3 million to $7 million, and concludes that these estimates "provide guidance as to whether risk reduction efforts that cost $50,000 per life saved or $50 million per life saved are warranted" (Viscusi 1993, p. 1943). Such advocacy of CEA by a leading proponent of BCA may suggest that even the loudest cheerleaders find it difficult to defend assumption-loaded and uncertainty-dominated BCA. Another BCA advocate who has used CEA examples to "prove" his case is Joseph Stiglitz, a prominent Stanford University and Clinton Administration economist. In hearings before the U.S. Senate, Stiglitz advocated increased use of BCA in government rule making, but the examples he gave were of CEA, where the goal is established by legislation and economics is employed to help identify the most cost-effective policies toward attaining that goal.

* Benefit–cost analysis and cost–benefit analysis are the same.

RBA, CRA, BCA, and CEA are capable of informing and improving risk management decisions so long as the opportunity cost concept is applied carefully, all feasible alternatives are considered, and the shortcomings of the techniques are effectively communicated.

This chapter has been successful if it has provided the noneconomist with an understanding of opportunity cost, a persistence in uncovering and analyzing alternatives, and an appreciation of the difficulties and limitations of economic analysis. Being able to ask good questions is probably the best protection against the tendency of economic practioners, like many others, to oversell their wares.

REFERENCES

Callan, S. J., and J. M. Thomas. 1996. *Environmental Economics and Management*, Chicago: Irwin.

Coddington, A. 1970. "The Economics of Ecology." *New Society*, April 9: 595–597.

Goodstein, E. S. 1995. *Economics and the Environment*, Englewood Cliffs, NJ: Prentice Hall.

Kapp, K. W. 1950; 1971. *The Social Costs of Private Enterprise*, New York: Schocken Books.

Lipsey, R. G., and K. Lancaster. 1956–57. "The General Theory of Second Best," *Review of Economic Studies*, 24: 11–32.

North, D. C. 1991. "Institutions," *Journal of Economic Perspectives*, 5 (Winter): 97–112.

Rhoads, S. E. 1985. *The Economist's View of the World,* New York: Cambridge University, pp. 17–20.

Swaney, J. A., and M. A. Evers. 1989. "The Social Cost Concepts of K. William Kapp and Karl Polanyi," *Journal of Economic Issues*, 23 (March): 7–33.

Tietenberg, T. 1994. *Environmental Economics and Policy*, New York: Harper Collins.

U.S. Environmental Protection Agency. 1987. *Unfinished Business: A Comparative Assessment of Environmental Problems,* Washington, D.C., Office of Policy Analysis.

Viscusi, W. K. 1993. "The Value of Risks to Life and Health," *Journal of Economic Literature*, 31 (December): 1912–1946.

QUESTIONS

1. What is the basic economic concept underlying the statement, "There's no such thing as a free lunch"?
2. When dry cleaners install vapor recovery equipment, they dramatically reduce their contribution to air pollution, *and* they increase their profits. Vapor recovery equipment is not cheap, but it has a 2- or 3-year payback because it reduces purchases of dry cleaning fluid. Aside from expecting lower prices on dry cleaning fluid, or technical improvements in vapor recovery equipment, can you think of any good reason for a business person to fail to invest in this equipment?
3. The dry cleaner example in the previous question is not unusual: A "prod" from the marketplace (higher prices) or government (regulations) often spurs technological or managerial innovation that improves productivity and cuts costs. Does the opportunity cost concept apply to such cases, or are "free lunches" sometimes available over time?

4. Aside from a few residents in Boston, San Francisco, and a few other cities, Americans use automobile transportation in spite of the sizeable risks. From an economic perspective, this is because the costs of giving up cars is too high. What types of policies or programs could make it more economical for individuals to choose to *eliminate* the risks of automobile transportation? (Hint: Improving the economics of alternative transportation involves *lowering* the price of alternatives and *raising* the price of automobiles.)
5. When people work for a cause that provides few benefits to them but ample external benefits (helping others or the community as a whole), they are being good citizens at the expense of being good economists. Can you think of three examples of such good citizenship in your community? In your workplace? In your profession? Over time, if people become better economists, will this help or hurt our economic prosperity?
6. According to Nobel Prize winner Milton Friedman, the sole responsibility of corporate managers is to maximize profits for shareholders. Is this argument more defensible in a world of purely competitive markets than in a world of cost-shifting, market failures, and easily influenced consumers?
7. If more people become "good economists" at the expense of being "good citizens," what effect will that have on cost shifting? Who will have fewer reservations about shifting or escaping costs, the good citizen or the good economist?
8. One point made repeatedly in this chapter is that observed market prices are very inaccurate measures of opportunity costs. If the government agencies that collect market data were able to produce more accurate opportunity cost data, what sorts of changes might occur over time? (Hint: What lessons can be learned from the 1986 federal law that requires facilities to report toxic releases?)
9. Economists, ethicists, and others have argued that all species and ecosystems hold "existence rights" because they are intrinsically valuable; that is, their value exists apart from how (or whether) we humans value them. Were we to accept this "intrinsic value" perspective, how would it effect our argument against actions that create irreversibilities?
10. In a democracy, how should community need be determined? Do we do a good job of this in the United States? With rising economic and ecological interdependence around the globe, how can we possibly (re)define community need democratically? (Hint: Good answers to this one should win some kind of very prestigious prize!)
11. List three private risk-reduction decisions or strategies (involving benefits and costs that affect only the decision maker). Now, try to think of consequences for the community. Can you come up with a nontrivial example of a purely private risk-reduction decision?
12. Within your area of expertise, list three important risks and associated risk-reduction strategies. Can you think of alternative strategies that would:
    a. accomplish somewhat less risk reduction, but at a far lower cost?
    b. accomplish considerably more risk reduction, at only a slightly higher cost?
13. Aside from economic fairness, can you think of other reasons why society might "trump" the most economical risk reduction strategies?

Section II

Applications of Risk Analysis

CHAPTER II.1

Assessment of Residential Exposures to Chemicals

Gary K. Whitmyre, Jeffrey H. Driver, and P. J. (Bert) Hakkinen

SUMMARY

Individuals in and around residences come in contact with a variety of chemicals from various potential sources, including outdoor sources that enter the residence, and from combustion sources and consumer products. Among the factors that determine the extent of exposure to a chemical are human exposure factors (e.g., body weight, types, frequencies and durations of various daily activities) and residential exposure factors (e.g., design and properties of a residence, including air exchanges per hour for the residence or the area of interest within the residence). The goal of this chapter is to provide readers with an overview of the assessment of residential exposures to chemicals. The chapter is organized as follows: Key Words, Introduction, Overview of General Issues, Lessons from the TEAM Studies, Assessment of Inhalation Exposures in the Residence, Assessment of Dermal Exposures in the Residence, Assessment of Ingestion Exposures in the Residence, Assessment of Exposures to Chemicals in Indoor Sources: Principles and Case Studies, Assessment of Exposures to Chemicals in Outdoor-Use Products: Principles and Case Studies, Data Sources for Residential Exposure Assessment, Discussion and Conclusions, References, Questions for Students to Answer.

Key Words: combustion appliances, consumer products, heating, ventilation, and air conditioning system (HVAC), human exposure factors, microenvironment, residential building factors, source characteristics, total exposure assessment methodology (TEAM), volatile organic compounds (VOCS)

1-56670-130-9/97/$0.00+$.50

1. INTRODUCTION

The general public is repeatedly in contact with time-varying amounts of environmental chemicals in air, water, food, and soil. On a daily basis, individuals are exposed in a variety of microenvironments that correspond to the daily activities that place persons in contact with environmental chemicals (e.g., soil contaminants during gardening, lawn chemicals during and following application, in-transit exposures to benzene from gasoline, environmental tobacco smoke [ETS] in residences and office buildings, volatile organic compounds [VOCs] from consumer products used in the residence). In response to the need to characterize multiple chemical exposures from multiple environmental media (e.g., soil, air, food, water), a number of ongoing efforts have been undertaken to develop methodologies to aid in quantifying these exposures (McKone 1991, Cal-EPA 1994).

In recent assessments of the human health impact of airborne pollutants, there has been increasing focus on the contribution of various microenvironments (e.g., indoors, outdoors, in transit) and sources (e.g., consumer products, combustion appliances, outdoor sources) to total human exposure to a given chemical. During the past 15 years, a number of studies, most notably the total exposure assessment methodology (TEAM) studies sponsored by the U.S. Environmental Protection Agency (EPA), have demonstrated that for a variety of contaminants, residential indoor air is often a more significant source of exposure than outdoor air (Thomas et al. 1993, Wallace 1993, Pellizzari et al. 1987). Some of the studies conducted in the past have found elevated indoor concentrations of certain pollutants, which raised questions concerning the types, sources, levels, and human health implications of indoor exposures (Spengler et al. 1983, Melia et al. 1978, Dockery and Spengler 1981). Assessment of potential consumer exposures has also been recognized by industry as a key part of the overall risk evaluation process for consumer products (Hakkinen et al. 1991). For example, several studies of potential indoor air exposures from use of consumer products have been conducted and published by industry and trade associations to support and confirm the safety of these particular products (Hendricks 1970, Wooley et al. 1990, Gibson et al. 1991).

2. OVERVIEW OF GENERAL ISSUES

Exposures to chemicals, in general, occur principally because humans engage in normal activities in various microenvironments that bring them into relatively close proximity with a number of chemical substances every day. These activities and concurrent sources of chemicals occur in outdoor air (i.e., via ambient levels of air pollutants such as nitrogen oxides, carbon monoxide, and particulates), in the work setting (e.g., exposure to industrial chemicals in factory jobs and exposure to carpet adhesive VOCs in office buildings), from pollutant exposures in vehicles while in transit or refueling (e.g., passenger-compartment benzene levels), and from chemical exposures in the residence. For the purpose of this chapter, the residential microenvironment is defined as indoor (i.e., inside the residence) as well as outdoor backyard areas.

There are a number of sources of residential exposures, including (1) consumer products such as cleaners, waxes, paints, pesticides, adhesives, paper products/printing ink, clothing/furnishings (e.g., which can off-gas VOCs); (2) building sources, which include combustive products from appliances and attached garages, building materials (e.g., which can release formaldehyde), and HVAC systems; (3) personal sources such as tobacco smoke and biological contaminants (e.g., allergens) of human, animal, and plant origin; and (4) outdoor sources of chemicals leading to infiltration of the residential environment. The latter include ambient combustive pollutants, contaminated soil particles that can infiltrate or be tracked into the home, drinking water (which can release volatile organics during showering or other use in the home), and contaminated subsurface water (e.g., infiltration of VOCs into basement areas).

The residential environment should be thought of in very dynamic terms. VOCs that enter the residential environment can be absorbed to surfaces, or "sinks," and then later be released as airborne levels that are depleted by various mechanisms, including air exchange with other rooms of the house and with outdoor air and with chemical/physical transformations in residential air. There is evidence that particulate contaminants, whether generated inside the residence or tracked in/infiltrated from the outdoor environment, are resuspended and recycled within the house by walking on floors and rugs, sweeping and dusting, and vacuuming (see Figure 1). Thus, the residence is the exposure unit.

There are a number of noninhalation exposure pathways that need to be addressed in characterizing and quantifying human residential exposures to chemicals. These include dermal exposure to dislodgeable residues on surfaces (such as pesticides on floors and carpeting and chemicals resulting from use of hard surface cleaners) and ingestion exposure to surface contaminants (such as that due to hand-to-mouth activity, particularly in infants and toddlers). There are several examples of studies and reviews that have addressed and provided examples of noninhalation residential exposures (Calvin 1992, CTFA 1983, ECETOC 1994, Turnbull and Rodricks 1989, Vermeire et al. 1993).

3. LESSONS FROM THE TEAM STUDIES

Since 1980, the U.S. EPA's Office of Research and Development has conducted a series of studies on human exposure to different classes of pollutants. These are commonly referred to as the total exposure assessment methodology (TEAM) studies. These studies have dealt with VOCs, carbon monoxide, pesticides, and particulates, often comparing indoor and outdoor exposures to these contaminants. When total personal exposures to VOCs (i.e., concentrations in the breathing zone) were measured via the presence of chemicals in exhaled breath, personal exposures most often exceeded outdoor air exposures. Median personal concentrations of VOCs were on the order of 2 to 5 times outdoor levels; maximum personal concentrations were roughly 5 to 70 times the highest outdoor levels (Wallace 1993). This observed variability in exposures indicates (1) the role of various human activities in bringing individuals into contact with chemicals indoors and (2) the importance of specific

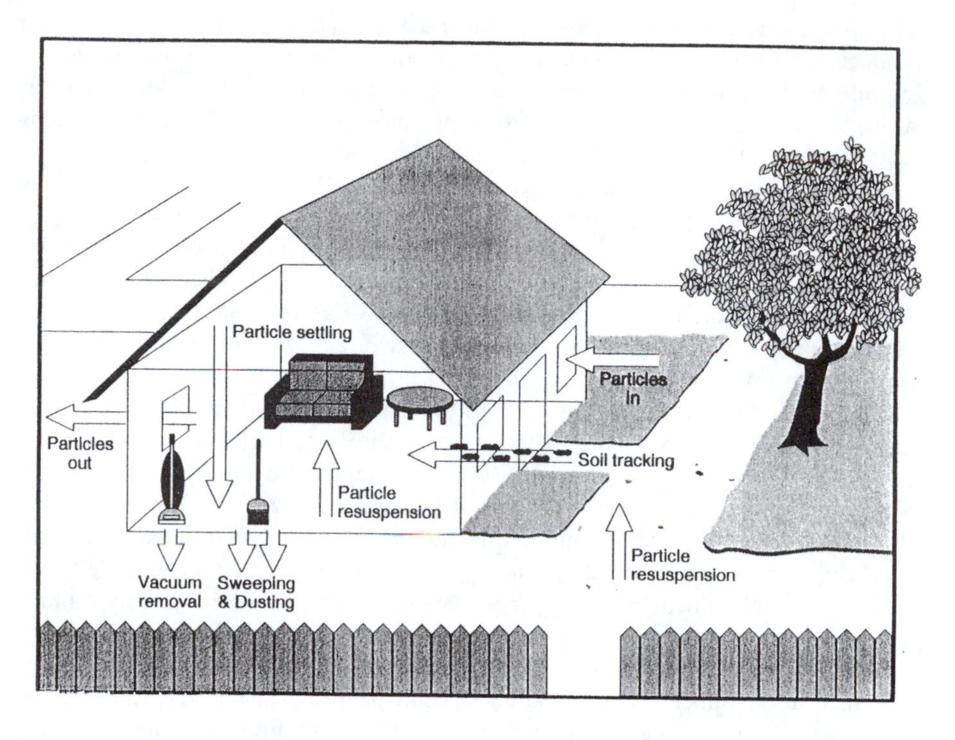

Figure 1 Potential pathways of human contact with contaminated soils. (Adapted from McKone, T.E. 1993. Understanding and Modeling Multipathway Exposures in the Home. Reference House Workshop II: Residential Exposure Assessment for the '90s. Society for Risk Analysis, 1993 Annual Conference, Savannah, Georgia.)

sources of exposures that may not be present in residential settings for all individuals. For example, (1) smokers had 6 to 10 times the personal benzene exposures of nonsmokers; (2) persons regularly wearing or storing freshly dry-cleaned clothes in the residence had significantly higher personal exposures to tetrachloroethylene; and (3) persons using mothballs and solid deodorizers in the residence were observed to have greatly elevated exposures to *p*-dichlorobenzene than nonusers (Wallace 1993).

The most recent study, known as PTEAM, focused on measuring personal exposures to inhalable particles (PM_{10}) of approximately 200 residents from Riverside, California, using specially designed indoor sampling devices. A major finding from this work is that personal exposures to particles in the daytime are 50% greater than either general indoor or outdoor concentrations. It has been hypothesized that these data suggest that individuals are exposed to a "personal cloud" of particles as they go about their daily activities, (Wallace 1993). Resuspension of household dust via walking in the residence, such as contaminated soil particles tracked into the home, and certain household activities such as vacuuming and cooking or sharing a home with a smoker, lead to significant particle exposures. The recent Valdez Air Health Study in Valdez, Alaska (Goldstein et al. 1993) generally supports the findings of the TEAM studies in terms of the importance of personal sources of exposure relative to outdoor sources. In the Valdez study, mean personal concentrations of benzene were roughly three to four times higher than outdoor levels, despite the presence of a significant outdoor source of benzene in the community (i.e., a petroleum storage and loading terminal).

4. ASSESSMENT OF INHALATION EXPOSURES IN THE RESIDENCE

An overview of factors that are commonly considered in assessing inhalation exposures to chemicals in the residence is provided in Figure 2. These factors include

- Source characteristics — Perhaps the most important factors determining the impact of chemical sources in the residence on inhalation exposures are the nature of the source (e.g., consumer product or residential construction material such as floor or wall surface), how it is released (fine respirable aerosols, nonrespirable coarse aerosols, vapor release [e.g., solid air freshener]), and the source strength (roughly proportional to the concentration of the chemical in the source or product).
- Human exposure factors — These include body weight, which varies between and within age and gender categories, and inhalation rates, which vary primarily by age, gender, and activity level.
- Physical-chemical properties — These include factors such as molecular weight and vapor pressure that determine the rate of evaporation into air of a chemical in an applied material (e.g., paint), or the release from aqueous solution (e.g., the role of the Henry's law constant in determining the release of volatile organics from tap water used in the home).
- Residential building factors — The basic characteristics of the room(s) and building in which residential exposures occur, as well as the ventilation configuration (i.e., number of windows and doors open, the rate of mechanical ventilation and air mixing, rate of infiltration of outside air), will determine the extent and rate of dilution of the chemical of interest in a specific indoor air setting.
- Exposure frequency and duration — The exposure frequency (i.e., the number of days per year, years per lifetime) and duration of exposure (i.e., minutes or hours of exposure to a chemical for a given day on which exposure occurs) are critical variables for estimating residential exposures to chemicals. These are a function of product-use patterns, human activities that bring individuals in contact with areas that may contain a chemical, and the nature of the population's mobility which limit the total number of years an individual may be exposed to a site-specific contaminated residence (e.g., radon).

As discussed in Whitmyre et al. (1992a,b), a number of these factors are associated with a wide range of variability across an affected population, resulting in a wide band of uncertainties; thus, the true distribution of exposures across the population would likely span several orders of magnitude.

A number of indoor air modeling tools are available for use in assessing inhalation exposures to a variety of contaminants from a variety of sources. Some are more oriented toward assessment of exposures to chemicals from consumer products when the specific emission term is not known, such as with the Screening-Level Consumer Inhalation Exposure Software (SCIES) developed by the Exposure

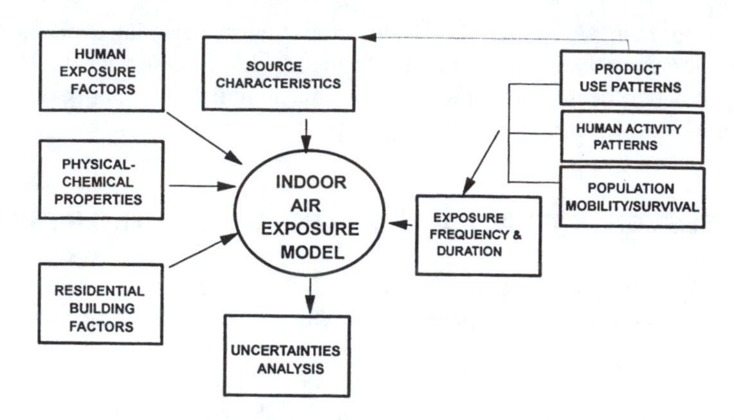

Figure 2 Components of indoor air residential exposure assessment.

Assessment Branch of the U.S. EPA's Office of Pollution Prevention and Toxics (U.S. EPA 1994). Another exemplary model is MAVRIQ, which can be used to estimate indoor inhalation exposures to organic chemicals due to volatilization from indoor uses of water (Wilkes and Small 1992).

A number of validated U.S. EPA modeling tools exist to address indoor airborne levels of chemicals from many types of emission sources. An example of an indoor air model that can be used when the emission term is known (e.g., aerosol product released at a rate of 1.5 g/sec for 3 min) is the Multi-Chamber Concentration and Exposure Model (MCCEM) developed for the Environmental Monitoring Systems Laboratory, U.S. EPA, Las Vegas (U.S. EPA 1991a). MCCEM is a user-friendly computer program that estimates indoor concentrations for, and inhalation exposures to, chemicals released from products or materials used indoors. Concentrations can be modeled in as many as four zones (e.g., rooms) in a building. The user provides values for emission rates, the zone where the source is located, the zone where exposure occurs, duration of exposure, air exchange rates, the nature of the building, and whether a short-term model (including average and maximum peak values) or long-term model is desired. The model contains room volume data and measured air flow rate data between different rooms for different building configurations and different geographic locations, or the user may build a hypothetical house or building, assigning the desired room (zone) volume and air exchange rates. Other examples of similar modeling tools include several U.S. EPA models, as well as the CONTAM model developed and updated regularly by the National Institute of Standards and Technology (NIST 1994).

A new database/model management tool developed by the University of Nevada at Las Vegas for the Environmental Monitoring Systems Laboratory, U.S. EPA, Las Vegas, is anticipated to revolutionize the modeling of indoor air exposures. This software tool is called the Total Human Exposure Risk Database and Advanced Simulation Environment (THERdbASE). This software integrates a number of indoor air models with distributional data on variables such as demographics, time activity, food consumption, and physiological parameter data that can be subset according to the needs of the assessment (Pandian et al. 1995). THERdbASE can also be used for estimating dermal and ingestion exposures and total human exposure via multiple agents and pathways, i.e., multiple agents present in more than one media and coming into contact with humans via multiple exposure pathways and routes. This software is now available for downloading via the Internet's World Wide Web at *http://eeyore.lv-hrc.nevada.edu* (ISEA 1995).

5. ASSESSMENT OF DERMAL EXPOSURES IN THE RESIDENCE

There are numerous opportunities for dermal exposure to chemicals in the residential environment. These include, but are not limited to, direct contact with cleaning/laundry products (e.g., cleanser, laundry detergent) during use, indirect contact with cleaning product residues (e.g., laundry detergent residues in washed clothing), contact with dislodgeable residues of a chemical after use (e.g., crawling infant contact with pesticide residues on rug); and direct contact with materials that are intentionally applied to the skin (e.g., soap, cosmetics).

There are basically two types of approaches to assessing dermal exposures: (1) the film-thickness approach and (2) dermal permeability-based approaches (U.S. EPA 1992). The film-thickness approach assumes that a uniform layer of a material (e.g., liquid consumer product) is present on a certain area of the skin and that all of the material in that layer is available for absorption. Default film-thickness data, in the absence of data on the actual product of interest, are available from the U.S. EPA (1987). Other variables that are unique to the film-thickness approach are the density of the product (grams per cubic centimeter, g/cm^3) and the percent dermal absorption anticipated during each event exposure period. Absorption can be assumed to be 100% for screening-level assessments, but severe overestimation of dermal exposure is likely to occur.

In contrast, dermal permeability-based methods recognize the fact that dermal absorption is a time-dependent process, and under controlled conditions, the dermal penetration can be expressed as a time-dependent parameter known as the dermal permeability coefficient (K_p). Measured and estimated dermal flux (micrograms per cubic centimeter per hour, $\mu g/cm^2/h$) and/or permeability coefficients (centimeters per hour, cm/h) have been published for various substances (U.S. EPA 1992, Driver et al. 1993). Additional discussion/information regarding dermal exposure assessment and percutaneous absorption kinetics can be found in U.S. EPA 1992, Kasting and Robinson 1993, and Wilschut et al. 1995.

Regardless of which general approach is taken, various additional factors must be taken into account to determine exposures.

- Human exposure factors — Besides body weight, which varies between and within age and gender categories, it is necessary to build an exposure scenario that specifies the amount of skin surface area exposed. One can use total surface area statistics and take a fraction representing the exposed area, or one can specify body parts that are exposed (e.g., both hands) and use body part surface area data (U.S. EPA 1989, AIHC 1995). Because skin surface area is closely correlated with body

weight, data on the ratio of surface area to body weight should ideally be used in calculating the dermal exposure (Phillips et al. 1993).
- Frequency and duration of exposure — The duration of exposure should represent the anticipated contact time with the skin prior to washing or removal.
- Concentration of the chemical on the skin — It is the estimation or measurement of vapor-phase or aqueous-phase concentration of a given agent in contact with the skin. For example, aqueous-phase exposures are usually expressed as micrograms (μg) of agent per cubic centimeter (cm^3) of aqueous solution.
- Surface area of skin exposed — The amount of surface area exposed is proportional to the amount of a given substance that may be percutaneously absorbed.

6. ASSESSMENT OF INCIDENTAL INGESTION EXPOSURES IN THE RESIDENCE

Ingestion of chemical residues can occur in the home beyond chemical residues (e.g., pesticides) consumed in food derived from nominally contaminated raw agricultural commodities (RACs) from spraying in the field. Primary examples of incidental residues include ingestion of cleaning agent and pesticide residues on plates and silverware following product use and ingestion of trace levels of organics (e.g., haloforms) in drinking water entering the home. Another important pathway for incidental ingestion exposure is hand-to-mouth behavior in infants and toddlers in particular; Vacarro (1992) has shown this to be actually the predominant exposure pathway (for this age group) for exposure to pesticide residues applied to carpets either directly or incidentally (e.g., through insecticide fogger use, such as a flea bomb), more so than inhalation or dermal contact through crawling on/touching contaminated surfaces. For food-related incidental contact, it will be necessary to consider the nature of the toxicological end point (e.g., short-term vs. long-term health effects) to determine which type of dietary consumption data is most appropriate (e.g., an upper bound on the amount eaten on 1 day in which the commodity is consumed or long-term averages which would include days on which the commodity is not consumed).

7. ASSESSMENT OF EXPOSURES TO CHEMICALS IN INDOOR SOURCES: PRINCIPLES AND CASE STUDIES

During the past 15 years, a number of studies, most notably the TEAM studies sponsored by the U.S. EPA, have demonstrated that residential air is often a more significant source of exposure to various chemicals (e.g., VOCs) than outdoor air. Many of the compounds of interest in residential air are present in consumer products that are used in and around the residence. Recent studies have investigated the relationship between use period/postuse period activities and exposures to a variety of chemicals in consumer products. While the resulting residential exposures are likely to be low in most cases, nonetheless, there is a need to characterize these exposures. For certain chemicals such as pesticides, postapplication exposures in particular may require characterization of various exposure pathways/routes and subpopulations to fully understand the magnitude of exposure associated with consumer uses of these chemicals. In performing such assessments, it is necessary to consider the range of approaches that can be taken, including use of body-burden modeling for intermittent exposures, use of indoor air modeling tools, incorporation of time-activity data, consideration of the form of the airborne concentration dissipation curve in determining postapplication exposures, and use and adjustment of emissions/concentration data for surrogate compounds to obtain an emission rate/airborne level for the compound of interest. The following case studies are provided to suggest the variety of possible exposure scenarios, sources of exposure, and chemical contaminants to which many individuals are exposed in the residence.

Case Study 1: Residential Exposure to Toluene During Use of Nail Polish. In one case study reported by Curry et al. (1994), inhalation exposures occurring during normal in-home use of nail lacquers were characterized. The study involved monitoring of personal, area, and background levels of toluene before, during, and after application of nail lacquer products. Based on the monitoring data, total personal exposures (during application plus postapplication) ranged from 1030 to 2820 μg per person per day. The dissipation kinetics for airborne toluene associated with this activity are shown in Figure 3 for a subject in a residence with poor ventilation (all outside doors and windows closed). Based on the log-linear regression curve, the estimated half-lives for toluene in the breathing zone of this subject and in the general area of the room of nail polish use (i.e., living room) were 67 and 89 min, respectively.

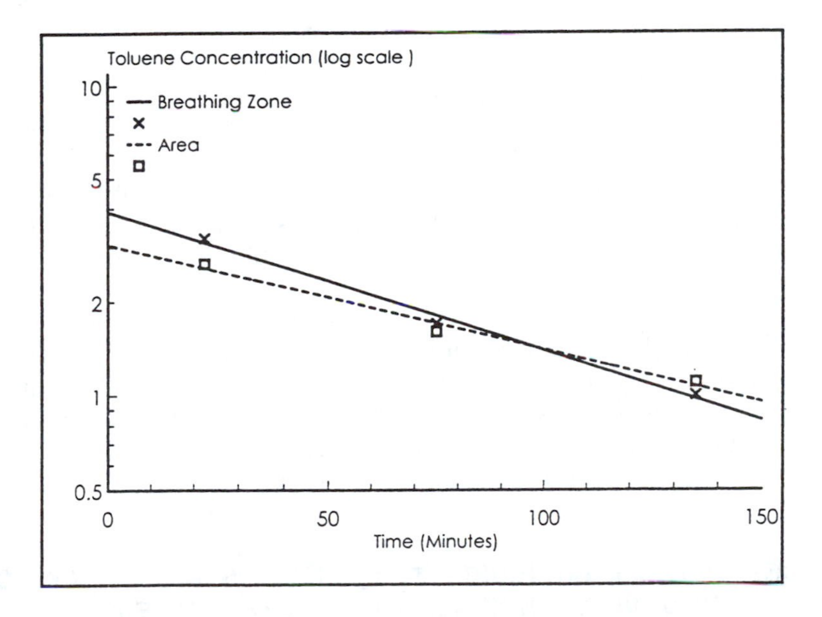

Figure 3 Log plot of area and breathing zone toluene concentrations (mg/m³) as a function of time during and following nail laquer application. (From Curry, K.K., et al. 1994. *Journal of Exposure Analysis and Environmental Epidemiology* 4 (4): 443–456. With permission of Princeton Scientific Publishing, NJ.)

Case Study 2: Para-Occupational Exposure to Perchloroethylene in the Home. The scientific literature contains numerous accounts of workers unintentionally transporting hazardous chemicals into their homes via clothing or personal body burdens that result in exposures to other individuals, such as family members, in the residence. Wallace et al. (1991) measured elevated levels of perchloroethylene (PERC) in the homes of dry cleaning workers. Thompson and Evans (1993) used a physiologically based pharmacokinetic model (PBPK) to verify that the workers' body burdens may be sufficient to explain elevated residential airborne levels (on the order of 100 $\mu g/m^3$), presumably attained by workers exhaling PERC into the home environment after work hours. The greater majority of the U.S. population is likely exposed to smaller, but detectable levels of PERC from various sources, including off-gassing from dry cleaning brought into the home.

Case Study 3: Exposures to Benzene from Attached Garages. Evaporative emissions of benzene from gasoline-fueled vehicles parked in residential garages have been measured and modeled. For garages that are an integral part of residences, the transfer of benzene-contaminated air to other parts of the residence may increase indoor concentrations of benzene, thus increasing the exposures of inhabitants to benzene (Furtaw et al. 1993). The rate of evaporation of benzene is dependent on the ambient temperature in the garage and the benzene content of the gasoline in the vehicle's tank (Furtaw et al. 1993). Monitoring and modeling studies have demonstrated that cars parked in garages that are an integral part of the residence, act as a considerable source of benzene to the residence. As part of the TEAM study in Bayonne and Elizabeth, New Jersey, mean benzene levels in four garages ranged from 10 to 100 $\mu g/m^3$; these were associated with mean benzene levels of 7.6 to 31 $\mu g/m^3$ measured for personal exposures inside the residence (Thomas et al. 1993). Temporal variations were noted in indoor and personal benzene levels over the six to ten monitoring periods at each home, although these changes were confounded by changes in outdoor benzene levels, that contributed to indoor and personal exposures (Thomas et al. 1993). Furtaw et al. (1993) reported similar results and concluded that from 4 to 50% of total benzene exposure for individuals in homes with attached garages may be attributable to evaporative emissions from parked vehicles.

Additional case studies can be found in the following publications: Calvin (1992), ECETOC (1994), Hakkinen et al. (1991), Hakkinen (1993), Turnbull and Rodricks (1989), and Vermeire et al. (1993). These publications provide exemplary exposure assessments to agents associated with consumer products, including gloves, hair spray, dish washing and laundry detergents, dentifrice, deodorants/antiperspirants, paint remover, baby pacifiers, teethers, and toys.

8. ASSESSMENT OF EXPOSURES TO CHEMICALS IN OUTDOOR-USE PRODUCTS: PRINCIPLES AND CASE STUDIES

A number of studies have been made of exposures to outdoor-use chemicals, most notably lawn chemicals, which include herbicides, insecticides, fertilizers, and other chemicals (e.g., lime). A number of opportunities exist for the general public to become exposed to lawn chemicals. Likely exposure pathways include dermal exposure to liquids during mixing/loading of formulation (e.g., in a hose-end spray unit), inhalation of aerosols and vapors (e.g., outdoor-use aerosol wasp spray), inhalation of dusts (e.g., dumping of granular formulation containing herbicide into mechanical spreader), and accidental/incidental spills (e.g., onto legs and feet). One would expect granular formulations to result in less exposure than liquid formulations because (1) the particle size for granular formulations is larger than the aerosols for liquids, limiting transport and exposure, and (2) the material incorporated into granules is likely to have a reduced bioavailability relative to the liquid formulation, particularly with regard to dermal exposures. Other factors that affect residential exposure include the use of protective equipment or additional layers of clothing, the frequency and duration of applications, and the use, rate, and percent of the active ingredient of the product used. Significant postapplication exposures may also occur from contact with dislodgeable residues of lawn chemicals during normal backyard activities.

Monitoring has been performed to collect compound-specific data with the intention of also being able to use such data as generic data to characterize exposures for specific application scenario/human-use patterns. Studies characterizing postapplication consumer exposures to lawn chemicals have used passive dosimeters (e.g., patch and partial/whole-body covers) and fluorescent tracers to characterize and quantify dermal exposures. These studies have often involved structured activities such as Jazzercise routines in order to standardize the exposures, such that inter-individual variability can be addressed. There are some significant method-related differences in measured exposures, in that the mean dermal exposures measured by dosimeter-based methods (e.g., fabric patches or whole-body covers) are about one order of magnitude higher than that quantified using fluorescent tracer techniques; thus, dosimeter-based methods may significantly overestimate dermal exposures to lawn chemicals (Eberhart 1994). In addition, attempts to remove and quantify dislodgeable residues from treated turf using methods such as polyurethane foam (PUF) rollers have allowed researchers to estimate transfer coefficients.

A residential exposure task force for turf chemicals known as the Outdoor Residential Exposure Task Force (ORETF) has been convened recently, comprised of approximately 30 member companies. It will focus on reviewing existing data, as well as conducting new studies that will provide the basis for development of a generic database for exposure assessment. This generic database will allow risk assessments to be conducted on both new and existing lawn care products.

Case Study 4: Residential Applicator Exposures to 2,4-D. Residential exposures to 2,4-D via use of herbicide formulations on lawns during application ($N = 22$) have been addressed by Harris et al. (1992). Normalized absorbed doses of 2,4-D (i.e., milligrams of exposure per pound of active ingredient handled) were estimated in the Harris et al. (1992) study based on postapplication urinary levels of 2,4-D. Under typical-use conditions, use of the granular formulation resulted in more than a tenfold lower exposure (mean of 0.0173 mg/lb a.i.; maximum 0.0639 mg/lb a.i.) compared to the liquid formulation (mean of 0.303 mg/lb a.i.; maximum 4.150 mg/lb a.i.) for normal clothing scenarios. The highest exposures occurred in those individuals not wearing protective clothing and were consistently associated with spills of liquid concentrate or excessive contact with the dilute mixture on the hands and forearms. Residues of 2,4-D were detected in 5 out of 76 air samples taken during applications by homeowners; however, inhalation exposures to lawn chemicals are generally lower in magnitude than dermal exposures.

Case Study 5: Residential Postapplication Exposures to 2,4-D. Harris and Solomon (1992) conducted a study that examined the exposures of ten individuals to 2,4-D from 1 hour of simulated activities on residential lawns starting at 1 and 24 hours after application. Wipe methods for the 30-m^3 test plots indicated that only 7.6% of the 2,4-D was dislodgeable (i.e., transferrable) from the lawn surface. This is consistent with the work of Thompson et al. (1984) that indicated that about 6% of 2,4-D applied to turf is dislodgeable shortly after application when applied at a rate of 0.89 lb a.i. per acre. The highest exposures were measured for those individuals who wore a minimum of clothing, i.e., shorts, short-sleeve shirt or no sleeves, and bare feet. The maximum exposure monitored during the study was 5.36 μg 2,4-D per kilogram of body weight.

Case Study 6: Reentry Exposures to Lawn Chemicals During Structured Activities. In one study (Eberhart 1994), dermal exposures and transfer coefficients were scaled from the adult subjects to children, based on relative surface area and time-activity data on duration of playtime, relative to subject monitoring time. The transfer factor (micrograms per square centimeter, $\mu g/cm^2$) has been suggested as the generic tie for estimating compound-specific dermal exposures, and it is the time-normalized dermal exposure (micrograms per hour, μg/h) divided by the transfer coefficient (square centimeters per hour, cm^2/h). Data from the Eberhart (1994) study showed approximately a loglinear or biphasic loglinear decline over time; the rate of decline for dislodegable residues is likely to be related to the vapor pressure and molecular weight of the chemical, chemical and biological degradation rates, and matrix effects (e.g., the extent to which turf may absorb and retain residues). Example transfer coefficients from this study were approximately 21,200 cm^2/h for adults, 12,400 cm^2/h for a 10-year-old child (extrapolated), and 9200 cm^2/h for a 5-year-old child (extrapolated).

9. DATA SOURCES FOR RESIDENTIAL EXPOSURE ASSESSMENT

A number of data sources exist for performing a residential exposure assessment. Human exposure factor data (e.g., distributions of body weights and skin surface areas, inhalation rates) can be obtained from the U.S. EPA's Exposure Factors Handbook (U.S. EPA 1989), which is currently being updated. Residential air exchange rate data have been summarized by Pandian et al. (1993) and refined by Murray and Burmaster (1995). Human time-activity data in the United States have been summarized by the U.S. EPA (1991b), compiled in the THERdbASE software (Pandian et al. 1995), and updated recently by John Robinson of the University of Maryland, College Park, MD. These data will be published as part of the U.S. EPA's upcoming revisions to the Exposure Factors Handbook. Dermal exposure assessment methods and dermal permeability coefficients for some organic chemicals are contained in the U.S. EPA's dermal exposure assessment guidance document (U.S. EPA

1992). Because skin surface area and body weight are closely correlated, total skin surface area to body weight ratios for use in residential exposure assessments are available from Phillips et al. (1993). Sources of food commodity consumption rate data for food-related incidental ingestion exposure analyses include software, such as DietRisk (Driver and Milask 1995) and the U.S. EPA's Dietary Risk Evaluation System (DRES), which is currently being updated and revised, the 1977–1978 and 1987–1988 United States Department of Agriculture (USDA) U.S. food consumption survey data, and specialty databases from various institutes and trade associations (e.g., National Institute on Alcoholism and Alcohol Abuse [NIAAA] database on wine consumption). An excellent source of data relevant to consumer product exposure assessments is ECETOC (1994).

10. DISCUSSION AND CONCLUSIONS

Given that most individuals spend more than 90% of their time in indoor environments, the need to develop methods for characterizing indoor exposures, in particular, has been recently evident. Jayjock and Hawkins (1993) have explored the complementary roles of indoor air modeling and research/data development in improving the level of confidence in estimations of inhalation exposures to indoor air contaminants. The use of real-world data to validate residential exposure models is critical to obtaining estimates that are more representative than the worse-case bounding estimates often obtained from unvalidated modeling approaches.

This chapter has focused exclusively on chemical agents in the residence and their implications for human exposures. While we have not addressed biological agents (e.g., allergens of biological origin) and physical agents (e.g., radon and electromagnetic fields), some of these additional agents encountered in the residential environment may be very important in terms of human health outcomes. These agents, and residential exposure assessment in general, will be discussed as a part of the Residential Exposure Assessment Project (REAP) being conducted by the Society for Risk Analysis, in cooperation with the International Society of Exposure Analysis (ISEA), and with funding from the U.S. EPA's Office of Research and Development and interested industries and trade associations. The objective of the REAP effort is to publish a textbook on residential exposure assessment by 1997.

REFERENCES

AIHC (American Industrial Health Council). 1995. *Exposure Factors Sourcebook.* Washington, DC.

Cal-EPA (California Environmental Protection Agency). 1994. CalTOX™, a multimedia total exposure model for hazardous-waste sites. Spreadsheet user's guide. Version 1.5. Sacramento, California: Office of Scientific Affairs, Department of Toxic Substances Control, Cal-EPA. NTIS Publication No. PB95-100467.

Calvin, G. 1992. Risk Management Case History — Detergents. In: *Risk Management of Chemicals.* M.L. Richards, Ed. The Royal Society of Chemistry, United Kingdom.

CTFA (Cosmetic, Toiletry and Fragrance Association, Inc.). 1983. Summary of the results of surveys of the amount and frequency of use of cosmetic products by women. Report prepared by ENVIRON Corporation.

Curry, K.K., D.J. Brookman, G.K. Whitmyre, J.H. Driver, R.J. Hackman, P.J. Hakkinen, and M.E. Ginevan. 1994. Personal exposures to toluene during use of nail lacquers in residences: description of the results of a preliminary study. *Journal of Exposure Analysis and Environmental Epidemiology* 4 (4): 443–456.

Dockery, D.W. and J.D. Spengler. 1981. Indoor-outdoor relationships of respirable sulfates and particles. *Atmospheric Environment* 15: 335–343.

Driver, J.H. and L. Milask. 1995. User's guide. DietRisk — chronic dietary exposure and risk analysis. Technology Sciences Group, Inc., Washington, DC.

Driver, J.H., R.G. Tardiff, L. Sedik, R.C. Wester, and H.I. Maibach. 1993. In vitro percutaneous absorption of [^{14}C] ethylene glycol. *Journal of Exposure Analysis and Environmental Epidemiology* 3 (3): 277–284.

Eberhart, D.C. 1994. Current activities in assessing human exposures to lawn chemicals. Presented at the Workshop on Residential Exposure Assessment, Annual Meeting of the International Society for Exposure Analysis and the International Society for Environmental Epidemiology, September 18, 1994, Research Triangle Park, North Carolina.

ECETOC (European Centre for Ecotoxicology and Toxicology of Chemicals). 1994. Assessment of non-occupational exposure to chemicals. Technical Report No. 58.

Furtaw, E.J., M.D. Pandian, and J.V. Behar. 1993. Human exposure in residences to benzene vapors from attached garages. Paper presented at and published in the *Proceedings of The International Conference: Indoor Air '93*, Helsinki, Finland, July 1993.

Gibson, W.S., F.R. Keller, D.J. Foltz, and G.J. Harvey. 1991. Diethylene glycol monobutyl ether concentrations in room air from application of cleaner formulations to hard surfaces. *Journal of Exposure Analysis and Environmental Epidemiology* 1 (3): 369–383.

Goldstein, B.D., R.G. Tardiff, S.R. Baker, G.F. Hoffnagle, D.R. Murray, P.A. Catizome, R.A. Kester, and D.G. Caniparoli. 1992. Valdez air health study. Anchorage, Alaska: Alyeska Pipeline Service Co. As cited in Wallace (1993).

Hakkinen, P.J., C.K. Kelling, and J.C. Callender. 1991. Exposure assessment of consumer products: human body weights and total body surface areas to use, and sources of data for specific products. *Veterinary and Human Toxicology* 33 (1): 61–65.

Hakkinen, P.J. 1993. Cleaning and laundry products: human exposure assessments. In: *Handbook of Hazardous Materials,* Academic Press, San Diego, CA, pp. 145–151.

Harris, S.A. and K.R. Solomon. 1992. Human exposure to 2,4-D following controlled activities on recently-sprayed turf. *Journal of Environmental Science and Health* B27 (1): 9–22.

Harris, S.A., K.R. Solomon, and G.R. Stephenson. 1992. Exposure of homeowners and bystanders to 2,4-dichlorophenoxyacetic acid (2,4-D). *Journal of Environmental Science and Health* B27 (1): 23–38.

Hendricks, M.G. 1970. Measurement of enzyme laundry product dust levels and characteristics in consumer use. *Journal of the American Oilers and Chemists Society* 47: 207–211.

ISEA (International Society of Exposure Analysis). 1995. ISEA Newsletter, Summer 1995 Issue. Argonne National Laboratory, Argonne, Illinois.

Jayjock, M.A. and N.C. Hawkins. 1993. A proposal for improving the role of exposure modeling in risk assessment. *American Industrial Hygiene Association Journal* 54 (12): 733–741.

Kasting, G.B. and P.J. Robinson. 1993. Can we assign an upper limit to skin permeability? *Pharmaceutical Research* 10: 930–931.

McKone, T.E. 1991. Human exposure to chemicals from multiple media and through multiple pathways: research overview and comments. *Risk Analysis* 11 (1): 5–10.

McKone, T.E. 1993. Understanding and Modeling Multipathway Exposures in the Home. Reference House Workshop II: Residential Exposure Assessment for the '90s. Society for Risk Analysis, 1993 Annual Conference, Savannah, Georgia.

Melia, R.J.W., C.duV. Florey, S.C. Darby, E.D. Palmes, and B.D. Goldstein. 1978. Differences in NO_2 levels in kitchens with gas or electric cookers. *Atmospheric Environment* 12: 1379–1381.

Murray, D.M. and D.E. Burmaster. 1995. Residential air exchange rates in the United States: empirical and estimated parametric distributions by season and climatic region. *Risk Analysis* 15 (4): 459–465.

NIST (National Institute of Standards and Technology). 1994. CONTAM93. User manual. Gaithersburg, Maryland: Building and Fire Research Laboratory, NIST, U.S. Department of Commerce.

Pandian, M.D., W.R. Ott, and J.V. Behar. 1993. Residential air exchange rates for use in indoor air and exposure modeling studies. *Journal of Exposure Analysis and Environmental Epidemiology* 3 (4): 407–416.

Pandian, M.D., E.J. Furtaw, et al. 1995. THERdbASE: Total Human Exposure Relational Database and Advanced Simulation Environment. Las Vegas, Nevada: Harry Reid Center for Environmental Studies, University of Nevada at Las Vegas. Developed under contract to the U.S. EPA, Office of Research and Development, Environmental Monitoring Systems Laboratory, Las Vegas, Nevada.

Pellizzari, E.D., T.D. Hartwell, R.L. Perritt, C.M. Sparacino, L.S. Sheldon, R.W. Whitmore, and L.A. Wallace. 1987. Comparison of indoor and outdoor residential levels of volatile organic chemicals in five U.S. geographic areas. *Environment International* 12: 619–623.

Phillips, L.J., R.J. Fares, and L.G. Schweer. 1993. Distributions of total skin surface area to body weight ratios for use in dermal exposure assessments. *Journal of Exposure Analysis and Environmental Epidemiology* 3 (3): 331–338.

Spengler, J.D., C.P. Duffy, R. Letz, T.W. Tibbets, and B.G. Ferris, Jr. 1983. Nitrogen dioxide inside and outside 137 homes and implications for ambient air quality standards and health effects research. *Environmental Science and Technology* 17 (3): 164–168.

Thomas, K.W., E.D. Pellizzari, C.A. Clayton, R.L. Perritt, R.N. Dietz, R.W. Goodrich, W.C. Nelson, and L.A. Wallace. 1993. Temporal variability of benzene exposures for residents in several New Jersey homes with attached garages or tobacco smoke. *Journal of Exposure Analysis and Environmental Epidemiology* 3 (1): 49–73.

Thompson, K.M. and J.S. Evans. 1993. Worker's breath as a source of perchloroethylene in the home. *Journal of Exposure Analysis and Environmental Epidemiology* 3 (4): 417–430.

Thompson, D.G., G.R. Stephenson, and M.K. Sears. 1984. Persistence, distribution, and dislodgeable residues of 2,4-D following its application to turfgrass. *Pesticide Science* 15: 353–360.

Turnbull, D. and J.V. Rodricks. 1989. A comprehensive risk assessment of DEHP as a component of baby pacifiers, teethers and toys. In: *The Risk Assessment of Environmental and Human Health Hazards: A Textbook of Case Studies,* Paustenbach, D.J., Ed. John Wiley & Sons, New York.

U.S. EPA (U.S. Environmental Protection Agency). 1987. Methods for assessing exposure to chemical substances. Volume 7. Methods for assessing consumer exposure to chemical substances. Washington, DC: Exposure Assessment Branch, Office of Pollution Prevention and Toxics. U.S. EPA Publication No. 560/5-85-007.

U.S. EPA (U.S. Environmental Protection Agency). 1989. Exposure factors handbook. Washington, DC: Exposure Assessment Group, Office of Health and Environmental Assessment, Office of Research and Development. U.S. EPA Publication No. 600/8-89-043.

U.S. EPA (U.S. Environmental Protection Agency). 1991a. MCCEM. Multi-chamber concentration and exposure model. User's guide. Version 2.3. Las Vegas, Nevada: Office of Research and Development, Environmental Monitoring Systems Laboratory.

U.S. EPA (U.S. Environmental Protection Agency). 1991b. Time spent in activities, locations, and microenvironments: a California-national comparison. Las Vegas, Nevada: Office of Research and Development, Environmental Monitoring Systems Laboratory. U.S. EPA Publication No. 600/4-91/006.

U.S. EPA (U.S. Environmental Protection Agency). 1992. Dermal exposure assessment: principles and applications. Washington, DC: Exposure Assessment Group, Office of Health and Environmental Assessment, Office of Research and Development. U.S. EPA Publication No. 600/8-91-011.

U.S. EPA (U.S. Environmental Protection Agency). 1994. Screening-level consumer inhalation exposure software (SCIES): description and user's manual. Version 3.0. Washington, DC: Exposure Assessment Branch, Office of Pollution Prevention and Toxics.

Vaccaro, J.R. 1992. Residential exposure to pesticides. Presentation at the Reference House Workshop, Society for Risk Analysis Annual Meeting, December 6, 1992, San Diego, California.

Vermiere, T.G., P. van der Poel, R.T.H. van de Laar, and H. Roelfzema. 1993. Estimation of consumer exposures to chemicals: applications of simple models. *Science and the Total Environment* 136:155–176.

Wallace, L. 1993. A decade of studies of human exposure: what have we learned? *Risk Analysis* 13: 135–139.

Wallace, L.A., E. Pellizzari, L. Sheldon, T. Hartwell, R. Perritt, and H. Zelon. 1991. Exposures of dry cleaning workers to tetrachloroethylene and other volatile organic compounds: Measurements in air, water, breath, blood, and urine. Presented at the Annual Meeting of the International Society for Exposure Analysis and Environmental Epidemiology, November 15–18, 1991, Atlanta, Georgia. As cited in Thompson and Evans (1993).

Whitmyre, G.K., J.H. Driver, M.E. Ginevan, R.G. Tardiff, and S.R. Baker. 1992a. Human exposure assessment I: understanding the uncertainties. *Toxicology and Industrial Health* 8 (5): 297–320.

Whitmyre, G.K., J.H. Driver, M.E. Ginevan, R.G. Tardiff, and S.R. Baker. 1992b. Human exposure assessment II: quantifying and reducing the uncertainties. *Toxicology and Industrial Health* 8 (5): 321–342.

Wilkes, C.R. and M.J. Small. 1992. Inhalation exposure model for volatile chemicals from indoor uses of water. *Atmospheric Environment* 26A: 2227–2236.

Wilschut, A., W.F. ten Barge, P.J. Robinson, and T.E. McKone. 1995. Estimating skin permeation: the validation of five mathematical skin permeation models. *Chemosphere* 30: 1275–1296.

Wooley, J., W.W. Nazaroff, and A.T. Hodgson. 1990. Release of ethanol to the atmosphere during use of consumer cleaning products. *Journal of the Air & Waste Management Association* 40: 1114–1120.

QUESTIONS

1. Identify an example of each of the following types of residential exposures, categorized by source:
   A. consumer product
   B. building related
   C. ingestion
   D. ambient air, water, or soil
2. Discuss how indoor exposure to outdoor contaminated soil might occur, i.e., by what mechanisms of entry into the residence, by what mechanisms of distribution within a residence, and by what potential routes of exposure.
3. Provide some examples of how people vary in their "human exposure factors," and the impact this has on their exposures to chemicals within a residence.
4. Provide some examples of how residences vary in their "residential exposure factors," and the impact that this has on the exposures that occupants may have to a chemical within a residence.

CHAPTER **II.2**

Pesticide Regulation and Human Health: The Role of Risk Assessment*

Jeffrey H. Driver and Gary K. Whitmyre

SUMMARY

Pesticides are an integral part of modern agricultural and urban and rural pest control programs. They contribute significantly to the abundance and quality of food, clothing, and forest products and to the prevention of disease. Pesticides are developed specifically for their ability to interact and interfere with a variety of biological targets in the pests at which they are directed. Because of the fundamental similarities of organisms at the subcellular level, human and environmental health hazards must be evaluated. The role of risk assessment in characterizing the potential health effects associated with dietary, occupational, and residential exposures to pesticides continues to provide an important mechanism for the use of sound science in the risk management decision making for these chemicals. The manufacture, distribution, and use of pesticides in the United States are strictly regulated under the Federal Insecticide, Fungicide and Rodenticide Act (FIFRA). This statute, which is administered by the U.S. Environmental Protection Agency (EPA), requires that any pesticide registered in the United States must perform its intended function without causing "unreasonable adverse effects on the environment." Thus, implementation of the statutory requirements of FIFRA includes consideration of the economic, social, and environmental costs and benefits of the use of a given pesticide. This chapter is intended to provide an overview of how potential human health risks are assessed under FIFRA with regard to the agricultural, occupational, and residential

* Adapted in part from Driver, J. and C. Wilkinson. 1995. Pesticide and human health: Science, regulation, and public perception, In: *Risk Assessment and Management Handbook for Environmental, Health & Safety Professionals.* Eds. Kolluru, R., S. Bartell, R. Pitblado, and S. Stricoff. New York: McGraw-Hill.

1-56670-130-9/97/$0.00+$.50

uses of pesticides. The chapter is organized as follows: Introduction, Balancing Benefits Against Risks, Pesticides and Food Safety, Evaluation of Occupational Exposures to Pesticides, Evaluation of Residential Exposures to Pesticides, Questions for Students to Answer, and References.

Key Words: pesticides, U.S. Federal Insecticide, Fungicide and Rodenticide Act (FIFRA), risk benefit, risk assessment, dietary, occupational and residential exposure, uncertainty analysis

1. INTRODUCTION

A pesticide is defined under the U.S. Federal Insecticide, Fungicide, and Rodenticide Act (FIFRA) as "any substance or mixture of substances intended for preventing, destroying, repelling, or mitigating any insects, rodents, nematodes, fungi, or weeds, or any other forms of life declared to be pests, and any substance or mixture of substances intended for use as a plant regulator, defoliant, or desiccant." In the United States, pesticide use is regulated under FIFRA (1947 and as amended in 1972, 1975, 1978, 1980, 1988, and 1990) on the basis of a risk-benefit standard. This balancing process considers "the economic, social and environmental costs, as well as the potential benefits of the use of any pesticide" [7 U.S.C., §136(a) (1978)].

Under FIFRA, pesticide use is controlled through a registration process that is administered by the U.S. EPA. A given pesticide may have many different uses, each of which must be individually approved. U.S. EPA registration of a pesticide for a given use and approval of a label describing the legally binding instructions for that use are required before a pesticide can be distributed and sold. For a pesticide to be registered, manufacturers must develop and submit to the U.S. EPA extensive data in support of the product to take account of a broad range of potential environmental and human risks as part of the regulatory evaluation of a pesticide. These data include product chemistry; efficacy; inherent toxicity to mammals (as surrogates for humans), wildlife and plants; environmental fate; and occupational and residential exposure data, where relevant. These requirements have been applied not only to new pesticides, but also to older pesticides through an ongoing reregistration program. A comprehensive discussion of the FIFRA registration process can be found in Conner et al. (1993).

The role of risk assessment in pesticide regulation has evolved dramatically since the late 1960s. Under the 1947 FIFRA, primary concern was given to the effectiveness of the product and proper labeling regarding use and protection of users from acute hazards. Some long-term data were required by the U.S. Food and Drug Administration (FDA) in establishing tolerances for pesticides used on food. However, the early 1960s saw the publication of Rachel Carson's *Silent Spring* (Carson 1962), which stimulated public concerns over the potential adverse effects of pesticides then in wide-scale use and the scientific concerns over long-term impacts of many pesticides on human health reported by the HEW secretary's "Commission on Pesticides and their Relationship to Environmental Health" (the so-called Mrak Commission Report) (HEW 1969). These events triggered a major change in the regulatory process, which led to greater emphasis on the potential long-term hazards to humans and the environment and to the banning of many commonly used pesticides such as dichlorodiphenyltrichloroethane (DDT), chlordane, heptachlor, aldrin, dieldrin, and 2,4,5-T.

The increased emphasis on risk also led to the 1972 amendments to FIFRA and the shift of pesticide regulation from three separate agencies, the U.S. Departments of Agriculture and Interior and the U.S. Food and Drug Administration, to the then newly formed U.S. EPA. The 1972 amendments to FIFRA completely revamped the regulatory framework from essentially a consumer protection and labeling law into a comprehensive regulatory framework extending into all aspects of pesticide sales, distribution, use, and disposal. At the heart of this new conceptual framework was the introduction in the statute of an explicit requirement to balance the risks of a pesticide against its benefits as the fundamental test of whether a pesticide should be allowed on the market.

The resource requirements placed first upon industry to conduct the expanded test regimens in response to comprehensive regulatory requirements and second upon government regulators to review and evaluate these data are resulting in greater stimulus for international harmonization of data requirements, test protocols, standards for interpretation, and methods of risk assessment and risk management (U.S. EPA 1994a). Major efforts are underway with Canada and Mexico under the North American Free Trade Agreement (NAFTA) umbrella and through the Organization for Economic Cooperation and Development (OECD) and World Health Organization (WHO). One major impediment to this harmonization process, however, is the different approach taken to cancer risk assessment in the United States compared to Europe and international organizations such as WHO, which place more emphasis on whether the pesticide is or is not genotoxic in assessing its potential cancer risk. However, the U.S. EPA has recently issued revisions to the agency's 1986 Cancer Guidelines. Proposed changes include greater qualitative consideration of the relevance of animal tumors to potential human oncogenicity, increased consideration of mechanisms of action, and more flexibility to incorporate new scientific developments.

2. BALANCING BENEFITS AGAINST RISKS

As noted in the introduction to this chapter, the manufacture, distribution, and use of agricultural chemicals in the United States are strictly regulated under FIFRA, which is administered by the U.S. EPA. FIFRA requires that any pesticide registered in the United States must perform its intended function without causing "unreasonable adverse effects on the environment." The latter phrase is defined as meaning "any unreasonable risk to man or the environment taking into account the economic, social, and environmental costs and benefits of the use of the chemical." It is important to recognize that FIFRA is a risk-benefit statute. While use of the term "unreasonable risk" implies that some risks will be tolerated under FIFRA, it is clearly expected that the anticipated benefits will outweigh the potential risks when the pesticide is used according to commonly recognized, good agricultural practices.

Risk can be defined as the probability that some adverse effect will occur. In the case of a pesticide, risk is a function of the intrinsic capacity of the material to cause a given adverse effect (e.g., neurotoxicity, cancer, developmental, or immunotoxicological effects) and of the level of exposure. Since pesticides are developed specifically for their biological activity or toxicity to some form of life and because, at the subcellular level, organisms have many similarities with one another, most pesticides are associated with some degree of toxicity. The degree of risk, however, will vary, depending on the nature of the inherent toxicity of the pesticide and the intensity, frequency, and duration of exposure which, in turn, relate to the circumstances under which exposure occurs. The potential health risks to a pesticide applicator or farm worker exposed to pesticides occupationally, for example, are likely to be greater than either the risks to residential users of pesticides (i.e., homeowners) or the risks to individuals in the general population who are exposed to traces of pesticides in food and/or water.

Methods for characterizing exposures to pesticides includes (1) collection of monitoring data (i.e., airborne concentrations, dermal or surface dislodgeable residues) for the specific pesticide and use scenario of interest, (2) use of monitoring data on surrogate chemicals for the same use scenario, (3) determination and use of body burden/tissue levels of pesticides, and (4) use of mathematical models to estimate exposures associated with pesticide application or postapplication periods (e.g., as in a residence).

The burden of providing the data to demonstrate that a given pesticide meets these registration requirements rests with the manufacturer. Current registration requirements include, as an example, a comprehensive battery of tests to evaluate potential acute, subchronic, and chronic mammalian toxicity (see Table 1) and environmental transport, fate, and impact on nontarget species. Information on product composition, stability, and analytical methodology and, in some cases, data on residue levels (e.g., in food crops, dislodgeable residues on surfaces and foliage) are also required. A separate registration must be approved by the U.S. EPA for each use pattern (e.g., crop, consumer product). This information, along with the approved conditions of use and any special restrictions or hazard warnings, must be incorporated into the product's label.

3. PESTICIDES AND FOOD SAFETY

One of the key scientific issues in evaluating food safety is the confidence (based on the estimated level of uncertainty) associated with quantitative estimates of dietary exposure to pesticides and the associated health risk(s). As noted previously, pesticides that are to be registered for use on food crops must be granted a tolerance by the U.S. EPA. Tolerances constitute the primary means by which the U.S. EPA limits levels of pesticide residues in or on foods. A tolerance is defined under the Federal Food, Drug and Cosmetic Act (FFDCA, 1954) as the maximum quantity of a pesticide residue allowed in/on a raw agricultural commodity (RAC) and in processed food when the pesticide has concentrated during processing (FFDCA, §409).

Table 1 Toxicity Data Requirements[a] Proposed by the U.S. EPA under FIFRA for Food[b] and Nonfood[c] Uses of Pesticides

Acute Testing	Developmental Testing
Acute oral toxicity—rat	Developmental toxicity —two species, rat and rabbit
Acute dermal toxicity—rabbit, rat, or guinea pig	Reproduction—rat
Acute inhalation toxicity—rat	Postnatal developmental toxicity—rat and/or rabbit
Primary eye irritation—rabbit	
Primary dermal irritation—rabbit	Mutagenicity Testing
Dermal sensitization—guinea pig	*Salmonella typhimurium* (reverse mutation assay)
Delayed neurotoxicity—hen	Mammalian cells in culture
Acute neurotoxicity—rat	*In vivo* cytogenetics
Subchronic Testing	General Metabolism—rat
90-d oral—two species, rodent and nonrodent	
21-d dermal—rat, rabbit, or guinea pig	Special Testing
90-d dermal—rat, rabbit, or guinea pig	Domestic animal safety
90-d inhalation—rat	Dermal penetration
28-d delayed neurotoxicity—hen	Visual systems studies
90-d neurotoxicity—rat	
Chronic Testing	
Chronic feeding—two species, rodent and nonrodent	
Carcinogenicity—two species, rat and mouse	

[a] Different testing requirements exist for food vs. nonfood uses, for the manufacturing- or end-use product vs. the technical grade of the active ingredient, and for experimental use permits. For a complete discussion of data requirements, specific conditions, qualifications or exceptions see NRC (1993; Chapter 4, *Methods for Toxicity Testing*).
[b] Food uses include terrestrial food and feed, aquatic food, greenhouse food, and indoor food.
[c] Nonfood uses include terrestrial nonfood, aquatic nonfood outdoor, aquatic nonfood industrial, aquatic nonfood residential, greenhouse nonfood, forestry, residential outdoor, indoor nonfood, indoor medical, and indoor residential.

Adapted from NRC, 1993 and 40 CFR, Part 158.

Tolerance concentrations on RACs are based on the results of field trials conducted by pesticide manufacturers and are designed to reflect maximum residues likely under good agricultural practices.

Section 408 of FFDCA requires that the U.S. EPA should consider "the necessity for the production of an adequate, wholesome and economical food supply" in setting tolerances. Under this statute and the risk-benefit balancing requirements of FIFRA, it has not been unusual for the U.S. EPA to register and set food tolerances for pesticides considered to be potential carcinogens. Section 409 of FFDCA, however, concerns tolerances of materials classified as food additives. This applies to pesticide residues only when the residue occurs as a result of pesticide use during processing or when a residue present in a RAC is concentrated during processing. The problem with Section 409 is that it contains the Delaney Clause, which specifically prohibits the presence of residues of materials found "to induce cancer in man or animal." This creates a regulatory paradox that while residues of "carcinogenic" pesticides are allowed in RACs under Section 408 of FFDCA, they are not allowed under Section 409. In practice, the U.S. EPA has historically used a "negligible risk" standard for the regulation of some potentially carcinogenic pesticides. The legal inconsistency created by the Delaney Clause has been the subject of legislative and regulatory debate (NRC 1987).

Human dietary exposure to agricultural chemicals in food is a function of food consumption patterns (i.e., grams of a commodity consumed per day within a relevant population strata), the residue levels of a particular chemical on (or in) food, and body weight. Thus, in general, dietary exposure (milligrams per kilogram per day, mg/kg/d) can simply be expressed as a function of consumption and chemical concentration:

$$\textit{Dietary exposure} = f\,(\textit{consumption, chemical concentration, body weight})$$

In reality, however, estimation of dietary exposure (and risks) to chemicals such as pesticides is a very complex endeavor. The complexity can be attributed to factors such as the occurrence of a particular pesticide in more than one food item; variation in pesticide concentrations; person-to-person variation in the consumption of various food commodities; changing dietary profiles across age, gender, ethnic groups, and geographic regions; the percentage of crop treated with a given pesticide; the potential effects on pesticide concentrations due to "aging," i.e., during transport and storage, and during food processing or preparation; and distribution of the raw commodity or processed product throughout regional areas or the entire United States. Thus, both food consumption and pesticide concentration data are characterized not by a single value, but rather, by broad distributions reflecting high, low, and average values. The inherent variability and uncertainty in food consumption and pesticide concentration data should be reflected in dietary exposure estimates of pesticides. Therefore, it is now common to describe pesticide exposures as a distribution of exposures for individuals in a particular population subgroup, e.g., hispanic, female children, ages 1 to 2 years. The distribution of dietary exposures (and thus, risk) is determined by combining or convoluting the distribution of food consumption levels and the distribution of pesticide concentrations in food.

An example of a unique U.S. food consumption distribution is shown in Figure 1. This multimodal lognormal distribution is presented as the cumulative frequency of daily grape juice consumption (on days that grape juice is consumed) for females 18 to 40 years old (ordered data, i.e., smallest to largest, in log scale are plotted against their expected normal scores), based on the results of the USDA's 1987–1988 Food Consumption Survey (USDA 1983, 1993). This illustrates the importance of not assuming that any single food commodity consumption rate across a population can be described by a single "representative" value or an inferred distribution form (i.e., an estimated distribution, rather than the actual underlying empirical data distribution).

Because both commodity consumption rates and residue levels are represented as a distribution of values across a population, dietary exposure estimates (as with assessments of other exposure pathways) are associated with uncertainties that relate to the inherent variability of the values for the input variables (Whitmyre et al. 1992). Thus, great benefit can be derived from conducting stochastic analyses of exposure based on the distributional data, in that quantitative measures of the uncertainties can be derived and reported (e.g., 10th, 50th, 90th percentiles). Given adequate data

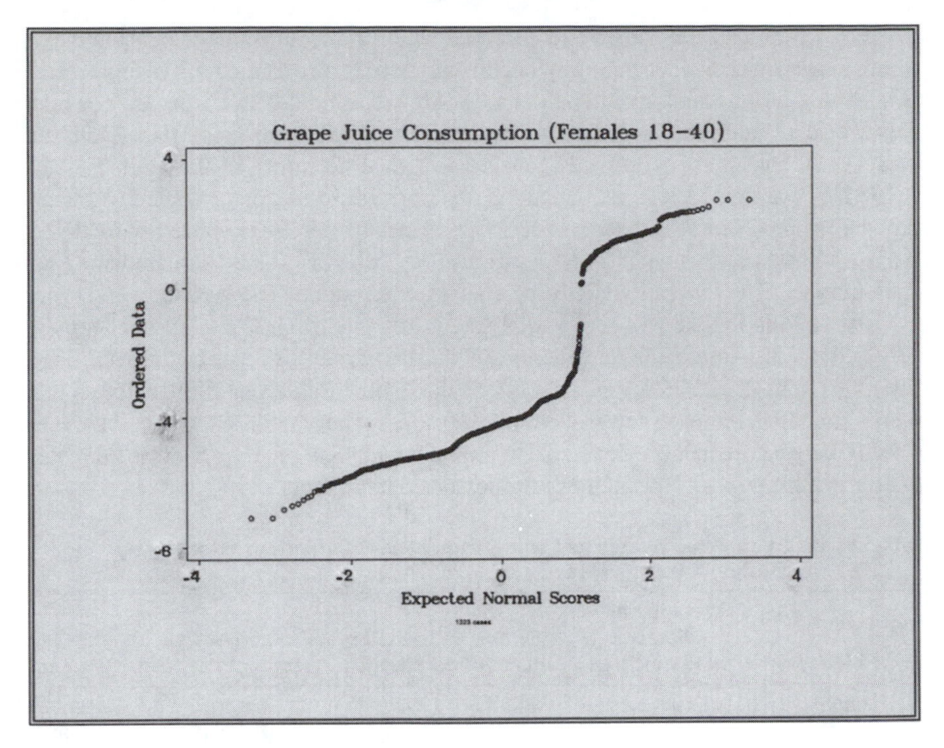

Figure 1 Daily grape juice consumption distribution for U.S. females (18 to 40 years old): ordered data (min to max; log scale) vs. expected normal scores.

on food consumption and pesticide concentrations, the National Academy of Sciences (NAS) has recently recommended the use of distributions rather than single-point data to characterize dietary exposures (and risks) associated with pesticides or food additives in/on food (NRC 1993).

In addition to uncertainty analysis, recent scientific, regulatory, and political attention has been focused on the potential exposures and health risks associated with pesticide residues in the diets of infants and children (Vogt 1992a,b, NRC 1993). The scientific and medical community has recognized for many years that significant quantitative (e.g., differential absorption, metabolism, detoxification, and excretion) and qualitative differences (e.g., differential susceptibility) may exist in infants and children vs. adults for a given chemical (e.g., pesticide, pharmaceutical) (Guzelian et al. 1992). However, the NAS (NRC 1993) found that quantitative differences in toxicity between children and adults are usually less than a factor of approximately 10. Further, differences in diet and, therefore, dietary exposure to pesticide residues account for most of the differences in pesticide-associated health risks (infants and children have distinctly different food consumption patterns and consume more calories of food per unit of body weight than do adults). Thus, differences in dietary exposure were generally a more important source of differences in risk than were age-related differences in physiological sensitivity to the effects of the chemical (NRC 1993).

To place dietary exposures and potential health risks into proper perspective, it is informative to consider naturally occurring sources of carcinogens (using cancer as an exemplary toxicological end point) vs. those pesticides that are carcinogenic. Based on Doll and Peto's dietary risk proportion to total cancer burden (Doll and Peto 1981), and the proportion of cancer-associated mortality in the United States in 1984, Scheuplein (1992) estimated the risk of death from cancer related to dietary exposure is 0.077 or 7.7 lifetime excess cancer deaths per 100 exposed individuals. Using estimates of dietary intake for various food categories and estimated amounts of carcinogens (and their associated potencies), Scheuplein (1992) developed cancer risk attributions for the respective food categories (see Table 2). While Scheuplein (1992) did not quantify the uncertainty (including variability) associated with these "point estimates," the author justifiably concludes that this illustration suggests that even a modest attempt to lower the dietary risk associated with natural carcinogens would likely be much more beneficial to public health than regulatory efforts devoted to eliminating traces of pesticides residues or contaminants.

Table 2 Risk Estimates Associated with Various Food Categories[a] Containing Carcinogenic Substances

Food category	Average daily dietary intake of food	Average daily dietary intake of carcinogen	Cancer risk estimate of total	Percent
Traditional food	1000 g	× 0.1% = 1000 mg	7.6×10^{-2}	98.82
Spices and flavors	1 g	× 1% = 10 mg	7.6×10^{-4}	0.98
Indirect additives	20 mg	× 10% = 2 mg	1.5×10^{-4}	0.2
Pesticides and contaminants	200 µg	× 50% = 0.1 mg	7.6×10^{-6}	0.01
Animal drugs	1 mg	× 10% = 0.1 mg	7.6×10^{-6}	0.01
Food preparation (charred protein only)	1 g	× 0.01% = 0.1 mg	7.6×10^{-6}	0.01
Mycotoxins	10 µg	× 10% = 0.001 mg	7.6×10^{-8}	0.0001

[a] Examples of traditional foods include grains, fruits, vegetables, meat, poultry, etc.; examples of spices and flavors include mustard, pepper, cinnamon, poppy seed, vanilla, etc.; examples of indirect additives include packaging migrants, contact and surface residues, lubricants, etc.; examples of pesticides and contaminants include insecticides, herbicides, fungicides, etc.; examples of animal drugs include antibiotics, sulfonamides, anthelminthics, growth promotants, etc.

Adapted from Scheuplein, R.J. 1992.

Another important dietary exposure issue involves exposure to pesticide residues in water. The discovery, in the late 1970s, that groundwater in some parts of the country was contaminated with pesticides created legitimate public concern. Approximately 53% of the U.S. population (more than 97% in rural areas) obtains its drinking water from groundwater sources (USGS 1988); groundwater also supplies 40.1% of the water in public water systems (USGS 1990). Surface water represents 59.9% of the water in public systems (USGS 1990) and is the major source of drinking water for approximately 47% of the U.S. population (USGS 1988). These potential sources of pesticide residues have been the subject of several studies (Hallberg 1989, Baker and Richards 1989, U.S. EPA 1990, Holden and Graham 1990). The U.S. EPA conducted a 5-year National Survey of Pesticides in Drinking Water Wells, the first survey undertaken to estimate the frequency and occurrence with which pesticides and their degradation products occur in drinking water wells (U.S. EPA 1990). The study surveyed 1349 drinking water wells for 126 pesticides and products as a statistical representation of the more than 10.5 million rural domestic wells and 94,600 wells operated by the 38,300 community water systems that use groundwater. The study reported that 10.4% (6.8 to 14.1%, 95% confidence interval [CI]) of the community water system wells and 4.2% (2.3 to 6.2%, 95% CI) of the rural domestic wells contain more than one pesticide (U.S. EPA 1990, NRC 1993). In the majority of cases, the levels that were found are below the U.S. EPA health advisories and, while still of concern, are not considered to constitute a significant threat to human health.

4. EVALUATION OF OCCUPATIONAL EXPOSURES TO PESTICIDES

Human exposure to pesticides may occur occupationally, usually involving dermal and inhalation exposure routes. Occupationally exposed populations include workers at pesticide manufacturing facilities, plant growers and harvesters (e.g., greenhouses; vegetable, vine, and tree crops), farmers, professional grounds applicators (e.g., farms, parks, roadsides, etc.), lawn care professionals, structural applicators (e.g., factories, food processing plants, hotels, hospitals, offices, homes, etc.), agricultural mixers/loaders and applicators, and field workers (e.g., harvesters). Significant pesticide exposures may occur for workers who mix, load, and/or apply pesticides and for workers involved in postapplication activities such as harvesting (U.S. EPA 1984, Maddy et al. 1990).

Numerous exposure studies of widely varying quality have been conducted in a variety of occupational settings. Occupational exposure data on inhalation and dermal exposures to specific pesticides can be found in scientific publications, registration standards and special review documents published by the U.S. EPA, state regulatory agencies, and regulatory agencies in other countries (Honeycutt et al. 1985, Plimmer 1982, Wang et al. 1989, U.S. EPA 1994b). The first exposure monitoring of pesticide handlers occurred in the early 1950s, following an episode of poisoning among applicators (Griffiths et al. 1951). This study assessed inhalation exposures by trapping airborne parathion using respirator filters. Direct exposure monitoring in the period between 1951 and the mid-1970s provided critical data for evaluating and improving workplace hygiene practices, such as protective clothing, based on direct dermal monitoring with gauze patches (Durham and Wolfe 1962, Durham et al. 1972).

Additional engineering controls to mitigate occupational exposures to pesticides were subsequently developed, including enclosed cabs (some with filtered air), closed transfer/mixing systems for pesticides, improved hose fittings and couplings, personal protective clothing and equipment, and lower-exposure formulations and packaging (e.g., water-soluble packets) (Krieger et al. 1992). More recently, efforts in measuring and evaluating potential occupational exposures and health risks have focused on (1) the development of exposure- and risk-based reentry intervals for harvesters; (2) more rigorous guidance on direct measurement of dislodgeable foliar residues, transfer factors (see Table 3), and potential inhalation or dermal exposures; and (3) reliable measurements of body burden (i.e., biomonitoring) (Krieger et al. 1992, U.S. EPA 1994b).

Table 3 Harvest Activities and Corresponding Transfer Factor Estimates

Work task	Potential transfer factor range (cm^2/h)	Primary dermal contact	Example
Sort/select	50–800	Hand	Mechanical (e.g., garlic)
Reach/pick	500–8,000	Arm/hand	Tomato, strawberry
Search/reach/pick	4,000–30,000	Upper body/hand	Tree fruit
Expose/search/reach/pick	20,000–150,000	Whole body/hand	Raisin and wine grapes

$$\text{Absorbed daily dose } (\mu g/kg) = \text{Dislodgeable foliar residue } (\mu g/cm^2) \times \text{Transfer factor } (cm^2/h) \times \text{Time (h)} \times \text{Clothing penetration } (\%) \times \text{Dermal absorption } (\%) \times \text{Body weight } (kg^{-1})$$

Adapted from Krieger et al. 1992.

Under FIFRA, safe occupational (e.g., professional applicators, harvesters) exposure levels must be demonstrated based on data provided by the registrant in accordance with U.S. EPA's guidelines as described under Subdivisions U and K. Agricultural worker (i.e., mixer/loader, applicator, harvester) exposure studies are often required for pesticide registrations. Sometimes data on surrogate compounds or use of generic exposure factors (e.g., normalized exposures [micrograms of exposure per pound of active ingredient applied] for a given application method and transfer coefficients for dislodgeable residues [see Table 3] from foliage to determine reentry exposures) are accepted by the U.S. EPA as part of a registration package. Surrogate worker exposure data for mixer/loaders and applicators are contained in the recently updated Pesticide Handlers Exposure Database (PHED) released by the U.S. EPA (1995a). PHED was developed by the U.S. EPA in collaboration with Health Canada (previously Health and Welfare Canada) and the American Crop Protection Association (ACPA) (previously the National Agricultural Chemistry Association [NACA]). PHED provides a very useful tool for modeling/predicting potential pesticide exposures based on consideration of numerous factors such as active ingredient application rate, formulation type, mixing and application methods, and protective clothing.

PHED is being used by registrants and government agencies to supplement or replace field exposure studies and as an evaluation tool for analysis of field exposure data. PHED contains over 1700 records of data on measured dermal and inhalation exposures and on various parameters that may affect the magnitude of exposures. Each data record represents one replicate for a single worker involved in 1 day or less of a given activity. The basic premise of PHED is that the exposures are more a function of the application equipment, formulation type, level of protective clothing, and individual work practices than the specific chemical nature of the active ingredient. Guidelines have been developed for proper use and reporting of PHED data (U.S. EPA 1995b,c). Table 4 shows the summary dermal exposure statistics from PHED for open mixing/loading of liquid formulations using standard work clothing (long pants, long-sleeve shirt, protective gloves), when the PHED data are subsetted for only those worker records associated with adequate quality assurance per U.S. EPA guidance (U.S. EPA 1995b,c). The high degree of variability in worker exposures to pesticides is reflected in the wide confidence interval for dermal exposure noted in the example PHED output report in Table 4.

Table 4 PHED Summary Statistics for Dermal Exposures: Open Mixing/Loading of Liquid Formulations

Scenario: Long pants, long sleeves, gloves						
		Micrograms per lb AI mixed				
Patch location	Distrib. type	Median	Mean	Coeff. of var.	Geo. mean	Obs.
Head (all)	Other	2.99	128.9568	493.8357	4.0992	121
Neck, front	Lognormal	1.695	23.2318	360.9199	1.74	103
Neck, back	Lognormal	0.341	15.7106	381.706	0.5427	109
Upper arms	Other	0.582	157.6735	903.2036	1.4925	90
Chest	Other	3.905	19.2219	262.7404	3.4337	89
Back	Other	0.8875	11.009	221.7177	1.8891	88
Forearms	Other	0.6655	4.4266	211.9821	0.8927	84
Thighs	Lognormal	3.82	16.8134	196.8466	4.0237	71
Lower legs	Other	0.952	38.271	819.5203	1.1162	81
Feet	Lognormal	5.371	346.998	180.1404	19.5296	25
Hands	Lognormal	3.5883	34.7596	316.3227	3.5782	80
Total derm:		39.3962	24.7973	797.0722	42.3376	

Note: 95% C.I. on Mean: Dermal: [–12060.5932, 13654.7376]; Number of Records: 137; Data File: MIXER/LOADER; Subset Name: OPENMIX.LIQ.DERM.MLOD.

As part of an ongoing effort to improve and harmonize existing guidelines, the U.S. EPA is also currently revising Subdivision K of the Pesticide Assessment Guidelines under FIFRA (U.S. EPA 1994b). Listings of existing guidelines and proposed guidelines are provided in Table 5. The new guidelines, which will be referred to as Series 875 — Occupational and Residential Exposure Test Guidelines Group B: Post-Application Monitoring Test Guidelines, provide guidance to persons required to submit postapplication exposure data under 40 CFR 158.390. Generally, these data are required under FIFRA when certain toxicity and/or exposure criteria have been met for a given pesticide.

Table 5 Existing and Proposed U.S. EPA Pesticide Guidelines

Existing U.S. EPA Pesticide Assessment Guidelines	
Subdivision D	Product Chemistry
Subdivision E	Wildlife and Aquatic Organisms
Subdivision F	Hazard Evaluation: Human and Domestic Animals
Subdivision G	Product Performance
Subdivision I	Experimental Use Permits
Subdivision J	Hazard Evaluation: Nontarget Plants
Subdivision K	Reentry Exposure
Subdivision L	Hazard Evaluation: Nontarget Insects
Subdivision M	Microbial and Biochemical Pest Control Agents
Subdivision N	Chemistry: Environmental Fate
Subdivision O	Residue Chemistry
Subdivision R	Spray Drift
Subdivision U	Applicator Exposure Monitoring
Proposed Harmonized U.S. EPA OPPTS Test Guidelines	
Series 810	Product Performance Test Guidelines
Series 830	Product Properties Test Guidelines
Series 835	Fate, Transport, and Transformation Test Guidelines
Series 840	Fate and Transport Field Studies Test Guidelines
Series 850	Ecological Effects Test Guidelines
Series 860	Residue Chemistry Test Guidelines
Series 870	Health Effects Test Guidelines
Series 875	Occupational and Residential Exposure Test Guidelines
	(Group A — Applicator Exposure Monitoring Test Guidelines)
	(Group B — Post-Application Exposure Monitoring Guidelines)
Series 880	Biochemicals Test Guidelines
Series 885	Microbial Pesticide Test Guidelines

When the Subdivision K guidelines were first published in 1984 (U.S. EPA 1984), they were designed to establish an acceptable scientific approach to the postapplication/reentry data requirements for typical agricultural exposure scenarios. Since 1984, there have been significant changes in the U.S. EPA's data needs and requirements resulting from (1) the reregistration process for pesticides which often has required postapplication studies; (2) an emphasis at the U.S. EPA on the evaluation of residential exposures in response to the expanding usage of pesticides in this environment; (3) revisions to the Good Laboratory Practice (GLP) Practice Standards in 1989 which focused more attention on quality assurance and quality control (QA/QC) (the U.S. EPA's recent data rejection rate analysis [U.S. EPA 1993] indicated that the most common cause for rejection of studies was inadequate or lack of QA/QC); and (4) the need to harmonize data requirements within the U.S. EPA (e.g., with requirements under the Toxic Substance Control Act [TSCA] for industrial chemicals, inerts, and consumer products) and with international organizations (e.g., North Atlantic Treaty Organization [NATO] and the Organization for Economic Cooperation and Development [OECD]). A description of some of the key types of studies that are likely to be required under the proposed Series 875, Group B guidelines are presented in Table 6.

Table 6 U.S. EPA/OPPTS Series 875, Group B: Description of Required Studies

Dissipation Studies
Foliar Dislodgeable Residue (FDR) Dissipation Study
Soil Residue Dissipation (SDR) Study
Indoor Surface Residue (ISR) Dissipation Study
Measurements of Human Exposure
Dermal exposure (passive dosimetry)
Inhalation exposure
Biological monitoring
Other Relevant Data
Human activity data
Toxicity data
Detailed use information

Adapted from U.S. EPA, 1994b.

5. EVALUATION OF RESIDENTIAL EXPOSURES TO PESTICIDES

As noted earlier, the proposed Series 875, Group B guidelines will include a new emphasis on nonoccupational, residential exposures to pesticides. Health risks associated with residual pesticides in air and on surfaces have only recently been examined from a public health perspective. However, the limited monitoring data that are available indicate that nonoccupational pesticide exposures in the general population are likely to be low relative to occupational exposures. As expected, the major source of pesticide exposure for the general population appears to result from the intermittent use of pesticides in and around the home (U.S. EPA 1994b, Whitmore et al. 1994), including both application and postapplication exposures.

Pesticide use in residential environments by professional pesticide applicators and consumers can be grouped into several general categories, including (1) indoor uses (e.g., broadcast floor sprays for fleas) vs. outdoor uses (e.g., treatment of pest activity areas such as wasp nests and ant mounds, use of antimicrobial products in swimming pools), (2) turf uses (e.g., granular applications for control of soil-dwelling insect pests, preemergent and postemergent herbicide sprays) and ornamental uses (e.g., foliar sprays for shrubs), (3) home garden uses (e.g., fungicide dusts for tomatoes), and (4) structural pest control uses (e.g., termiticides). Other sources of indoor exposure to pesticides for the general population may be from ambient air, food, water, ambient particles, and indoor house dust (Whitmore et al. 1994, Wallace 1993, Wallace 1991, 1993, Pellizzari et al. 1993, Jenkins et al. 1992, Vaccaro et al. 1991, Vaccaro et al. 1993).

The U.S. EPA has collected monitoring data on residential pesticide exposures (Whitmore et al. 1994). The Non-Occupational Pesticide Exposure Study (NOPES) was designed to assess total human exposure to 32 pesticides and pesticide degradation products in the residential environment. The NOPES program included 24-h indoor air, personal air, and outdoor air measurements. Integrated body burden/tissue data on pesticides are also available from the National Human Adipose Tissue Survey (NHATS), which was conducted by the U.S. EPA to determine levels of a variety of toxic substances in human fat tissue. The NHATS survey, however, was largely restricted to highly lipophilic persistent compounds such as chlorinated hydrocarbons and may be of limited value for hydrophilic pesticides that are more rapidly metabolized and excreted.

Other residential pesticide monitoring studies have included general surveys of multiple pesticides and measurements of air and surface concentrations of pesticides following specific applications of products such as termiticides, pest strips and crack and crevice or baseboard treatments, total release aerosols or foggers, broadcast applications, and hand-held sprays (Fenske et al. 1990, Racke and Leslie 1993, Whitmore et al. 1994). These studies generally demonstrate that measurable, but relatively low levels of pesticide residues exist in homes and that indoor and personal (e.g., breathing zone) exposures are higher than outdoor exposures. In most cases, negligible human health risks are associated with these exposures (Whitmore et al. 1994). However, residential exposures to infants and children associated with adverse health effects have raised concern and prompted further investigation (Zweiner and Ginsburg 1988, Berteua et al. 1989).

Additional research activities related to residential exposure assessment that are currently being sponsored by the U.S. EPA include the ongoing update of the U.S. EPA's "Exposure Factors Handbook" (U.S. EPA 1989). In addition, the Office of Research and Development (ORD) has initiated a cooperative agreement (referred to as the Residential Exposure Assessment Project or REAP) with the Society for Risk Analysis (SRA) and the International Society of Exposure Analysis (ISEA) to develop an authoritative series of reference documents describing relevant methodologies, data sources, and research needs for residential exposure assessment. The REAP will compliment other U.S. EPA initiatives, such as the development of the Series 875 guidelines, and will facilitate an exchange of information and talent between the U.S. EPA, other federal and state agencies, industry, academia, and other interested parties.

REFERENCES

Baker, D.B. and R.P. Richards. 1989. Herbicide concentration patterns in rivers draining intensively cultivated farmlands of northwestern Ohio. In: *Pesticides in Terrestrial and Aquatic Ecosystems.* ed., Weigmann, D. Blacksburg, VA: Virginia Water Resources Research Center, Virginia Polytechnic Institute and State University.

Berteau, P.E. et al. 1989. Insecticide absorption from indoor surfaces. In: *Biological Monitoring for Pesticide Exposure.* ACS Symposium Series 382. eds., Wang, R.G. C.A. Franklin, R.C. Honeycutt and J.C. Reinert. pp. 315–326. Washington, D.C.: American Chemical Society (ACS).

Carson, R. 1962. *Silent Spring.* Boston: Houghton Mifflin.

Conner, J.D., L.S. Ebner, S.W. Landfair, C.A. O'Connor et al. 1993. *Pesticide Regulation Handbook.* Fourth Edition. New York: Executive Enterprises Publications Co., Inc.

Doll, R. and R. Peto. 1981. The causes of cancer: quantitative estimates of avoidable risks of cancer in the United States today. *J. Natl. Cancer Inst.* 66:1193–1981.

Durham, W.F. and H.R. Wolfe. 1962. Measurement of the exposure of workers to pesticides. *Bull. World Hlth. Org.* 26:75–91.

Durham, W.F., H.R. Wolfe and J.W. Elliot. 1972. Absorption and excretion of parathion by spraymen. *Arch. Environ. Hlth.* 24:381–387.

Fenske, R.A., K.G. Black, K.P. Elkner, C. Lee, M.M. Methner and R. Soto. 1990. Potential exposure and health risks of infants following indoor residential pesticide applications. *AJPH.* 80:689–693.

Griffiths, J.T., C.R. Stearns and W.L. Thompson. 1951. Parathion hazards encountered in spraying citrus in Florida. *J. Econ. Entomol.* 44:160–163.

Guzelian, P.S., C.J. Henry and S.S. Olin (eds). 1992. *Similarities and Differences Between Children and Adults: Implications for Risk Assessment.* International Life Sciences Institute (ILSI). Washington, D.C.: ILSI Press.

Hallberg, G.R. 1989. Pesticide pollution of groundwater in the humid United States. *Agric. Ecosyst. Environ.* 26:299–368.

HEW (Department of Health, Education and Welfare). 1969. Report of the Secretary's Commission on Pesticides and Their Relationship to Environmental Health. Washington, D.C.: DHEW.

Holden, L.R. and J.A. Graham. 1990. Project Summary for the National Alachlor Well Water Survey. Monsanto Final Report. MSL-9629. St. Louis, MO.

Honeycutt, R.C., G. Zweig and N.N. Ragsdale (eds). 1985. *Dermal Exposure Related to Pesticide Use.* ACS Symposium Series 273. Washington, D.C.: American Chemical Society.

Jenkins, P.L., T.H. Phillips, E.J. Mulberg and S.P. Hui. 1992. Activity patterns of Californians: use of and proximity to indoor pollutant sources. *Atmos. Environ.* 26:2141–2148.

Krieger, R.I., J.H. Ross and T. Thongsinthusak. 1992. Assessing human exposures to pesticides. *Rev. Environ. Contam. Toxicol.* 128:1–15.

Maddy, K.T., S. Edmiston and D. Richmond. 1990. Illnesses, injuries and deaths from pesticide exposures in California, 1949–1988. *Rev. Environ. Contam. Toxicol.* 114:57–123.

NRC (National Research Council). 1987. *Regulating Pesticides in Food: The Delaney Paradox.* Washington, D.C.: National Academy Press.

NRC (National Research Council). 1993. *Pesticides in the Diets of Infants and Children.* Committee on Pesticides in the Diets of Infants and Children. Board on Agriculture and Board on Environmental Studies and Toxicology. Commission on Life Sciences. Washington, D.C.: National Academy Press.

Pellizzari, E.D., K.W. Thomas, C.A. Clayton, R.W. Whitmore, R.C. Shores, H.S. Zelon and R.L. Peritt. 1993. Particle Total Exposure Assessment Methodology (PTEAM): Riverside, California Pilot Study — Volume I. EPA/600/SR-93/050. Research Triangle Park, NC: U.S. EPA.

Plimmer, J. 1982. *Pesticide Residues and Exposure.* ACS Symposium Series 182. Washington, D.C.: American Chemical Society.

Racke, K.D. and A.R. Leslie (eds). 1993. *Pesticides in Urban Environments: Fate and Significance.* Washington, D.C.: American Chemical Society (ACS).

Scheuplein, R.J. 1992. Perspectives on toxicological risk — an example: foodborne carcinogenic risk. *Crit. Rev. Food Sci. Nutr.* 32:105–121.

USDA (U.S. Department of Agriculture). 1983. Nationwide Food Consumption Survey 1977–78. Human Nutrition Information Service, Hyattsville, MD.

USDA (U.S. Department of Agriculture). 1993. Nationwide Food Consumption Survey 1987–88. Human Nutrition Information Service, Hyattsville, MD.

U.S. EPA (Environmental Protection Agency). 1984. Pesticide Assessment Guidelines. Subdivision K, Exposure: Reentry Protection. Washington, D.C.: U.S. EPA.

U.S. EPA (Environmental Protection Agency). 1988. Pesticides in Drinking Water Data Base 1988 Interim Report. Washington, D.C.: U.S. EPA, Office of Pesticide Programs.

U.S. EPA (Environmental Protection Agency). 1989. Exposure Factors Handbook. Washington, D.C.: U.S. EPA.

U.S. EPA (Environmental Protection Agency). 1990. National Survey of Pesticides in Drinking Water Wells. Phase I. NTIS Doc. No. PB-91-125765. Springfield, VA.

U.S. EPA (Environmental Protection Agency). 1993. Pesticide Reregistration Rejection Rate Analysis: Occupational and Residential Exposure. EPA/738-R-93-008. Washington, D.C.: U.S. EPA.

U.S. EPA (Environmental Protection Agency). 1994a. Pesticides Industry Sales and Usage: 1992 and 1993 Market Estimates. 733-K-94-001. Washington, D.C.: U.S. EPA, Office of Prevention, Pesticides and Toxic Substances.

U.S. EPA (Environmental Protection Agency). 1994b. Series 875 — Occupational and Residential Exposure Test Guidelines, Group B — Post Application Exposure Monitoring Test Guidelines. Version 3.2. Washington, D.C.: U.S. EPA.

U.S. EPA (Environmental Protection Agency). 1995a. Pesticide Handlers Exposure Database (PHED) User's Guide. Version 1.1. Washington, D.C.: U.S. EPA, Office of Pesticide Programs (in conjunction with Health Canada and the American Crop Protection Association).

U.S. EPA (Environmental Protection Agency). 1995b. PHED: Pesticide Handlers Exposure Database Reference Manual. Version 1.1. Washington, D.C.: U.S. EPA, Office of Pesticide Programs (in conjunction with Health Canada and the American Crop Protection Association).

U.S. EPA (Environmental Protection Agency). 1995c. Guidance for Reporting PHED Exposure Evaluations. PHED V1.1. Washington, D.C.: U.S. EPA, Office of Pesticide Programs (in conjunction with Health Canada).

USGS (U.S. Geological Survey). 1988. National Water Summary 1986: Hydrologic Events and Ground-Water Quality. U.S. Geological Survey Water Supply Paper 2325. Denver, CO: U.S. Government Printing Office.

USGS (U.S. Geological Survey). 1990. National Water Summary 1987: Hydrologic Events and Ground-Water Quality. U.S. Geological Survey Water Supply Paper 2350. Denver, CO: U.S. Government Printing Office.

Vaccaro, J.R. et al. 1991. Evaluation of dislodgeable residue and absorbed doses of chlorpyriphos to crawling infants following broadcast application of a chlorpyriphos-based emulsifiable concentrate. The Dow Chemical Company. Indianapolis, IN.

Vaccaro, J.R. et al. 1993. Chloropyriphos: exposure to adults and children upon re-entry to domestic lawns following treatment with a chlorpyriphos-based mixture. The Dow Chemical Company. Indianapolis, IN.

Vogt, D.U. 1992a. Proposed Changes to Policies Governing Pesticide Residues in Foods. CRS Issue Brief. Congressional Research Service. Library of Congress. Washington, D.C.

Vogt, D.U. 1992b. Food Safety: Issues and Concerns Facing Congress. CRS Issue Brief. Congressional Research Service. Library of Congress. Washington, D.C.

Wallace, L.A. 1991. Comparison of risks from outdoor and indoor exposure to toxic chemicals. *Environ. Hlth. Perspect.* 95:7–13.

Wallace, L. 1993. A decade of studies of human exposure: what have we learned? *Risk Analysis.* 13:135–138.

Wang, R.G.M., C.A. Franklin, R.C. Honeycutt and J. Reinert (Eds). 1989. Biological Monitoring for Pesticide Exposure. ACS Symposium Series 382. Washington, D.C.: American Chemical Society.

Whitmore, R.W., F.W. Immerman, D.E. Camann, A.E. Bond, R.G. Lewis and J.L. Schaum. 1994. Non-occupational exposures to pesticides for residents of two U.S. cities. *Arch. Environ. Contam. Toxicol.* 26:47–59.

Whitmyre, G.K., J.H. Driver, M.E. Ginevan, R.G. Tardiff, and S.R. Baker. 1992. Human exposure assessment I: understanding the uncertainties. *Toxicol. Industrial Hlth.* 8 (5):297–320.

Zweiner, R.J. and C.M. Ginsburg. 1988. Organophosphate and carbamate poisoning in infants and children. *Pediatrics.* 81:121–126.

ADDITIONAL REFERENCES AND RESOURCES

ACSH (American Council on Science and Health). 1989a. *Pesticides and Food Safety.* New York: ACSH.

ACSH (American Council on Science and Health). 1992. *ALAR: Three Years Later — Science Unmasks a Hypothetical Health Scare.* New York: ACSH.

ACSH (American Council on Science and Health). 1989b. *Pesticides: Helpful or Harmful?* New York: ACSH.

ACSH (American Council on Science and Health). 1990. *ALAR: One Year Later — A Media Analysis of a Hypothetical Health Risk.* New York: ACSH.

Agriculture Canada. 1992. Pesticide Information Backgrounder B92–01. Pesticides Directorate.

Ames, B. N. 1983. Dietary carcinogens and anticarcinogens. *Science.* 221:1256–1263.

Ames, B. N., R. Magaw and L. S. Gold. 1987. Ranking possible carcinogenic hazards. *Science.* 236:271–280.

Ames, B. N. and L. S. Gold. 1989. Pesticides, risk and applesauce. *Science.* 244: 755–757.

Ames, B.N., M. Profet and L.S. Gold. 1990. Dietary pesticides (99.99% all natural). *Proc. Natl. Acad. Sci. U.S.A.* 87:7777–7781.

Archer, D. L. and J. E. Kvenberg. 1985. Incidence and cost of foodborne diarrheal disease in the U.S. *J. Food Prot.* 48:887–894.

Archibald, S.O. and C.K. Winter. 1990. In: *Chemicals in the Human Food Chain.* eds., Winter, C.K., J.N. Seiber and C.F. Nuckton. New York: Van Nostrand Reinhold.

ATSDR (Agency for Toxic Substances and Disease Registry). 1988. National Exposure Registry. Atlanta, GA: U.S. Department of Health and Human Services, Centers for Disease Control, Agency for Toxic Substances and Disease Registry.

Australia. 1989, September. *Requirements For Clearance of Agricultural and Veterinary Chemical Products.* 1st Edition. Australian Agricultural and Veterinary Chemicals Council.

Baker, S.R. and C.F. Wilkinson (eds). 1990. *The Effects of Pesticides on Human Health. Advances in Modern Environmental Toxicology.* Volume XVIII. Princeton, NJ: Princeton Scientific Publishing Co., Inc.

Beier, R.C. 1990. Natural pesticides and bioactive components in foods. *Rev. Environ. Contam. Toxicol.* 113.

Belgium. 1992. *Directive Homologation des Produits Phytopharmaceutiques.* Institut D'Hygiene et d'Epidemiologie.

BNA (Bureau of National Affairs). 1992. Working paper for considering draft revisions to the U.S. EPA Guidelines for Cancer Risk Assessment. *Daily Environ. News.* 232:E1–E32.

BNA (Bureau of National Affairs). 1994. Draft revisions to Guidelines for Carcinogen Risk Assessment — Review draft. *Daily Environ. News.* 154–155:E1–E45.

Canada. 1990. Biological Safety Factors in Toxicological Risk Assessment. Department of National Health and Welfare.

Canada. 1990, January. Risk Management in the Health Protection Branch. Health Protection Branch, Health and Welfare Canada.

Canada. 1991. Carcinogen Assessment. Health Protection Branch, Health and Welfare Canada.

CAST (Council for Agricultural Science and Technology). 1994. *Foodborne Pathogens: Risks and Consequences.* Ames, IA: CAST.

China. Undated. Data Requirements for Pesticide Registration. Institute for the Control of Agrochemicals, Ministry of Agriculture.

Denmark. 1992. Risk Management and Risk Assessment in Different Sectors in Denmark. Danish Academy of Technical Sciences.

Denmark. 1989, May. *Embryo-Foetal Damage and Chemical Substances.* Ministry of Health, The National Food Agency of Denmark.

Doll, R. and R. Peto. 1981. The causes of cancer: quantitative estimates of avoidable risks of cancer in the United States today. *J. Natl. Cancer Inst.* 66:1193–1981.

DowElanco. 1993. Could You Eat 340 Oranges Every Day? Form No. 533-00-013. The Dow Chemical Company. Indianapolis, IN.

Dragula, C. and G. Burin. 1994. International harmonization for the risk assessment of pesticides: results of an IPCS survey. *Reg. Toxicol. Pharmacol.* 20:337–353.

FAO (Food and Agriculture Organization). 1985. *FAO Guidelines for the Registration and Control of Pesticides.* Rome: FAO.

France. Undated. *Estimation des Risques Cancerogenes des Pesticide. Note II.*

Government Accounting Office (GAO). 1993. Pesticides: A Comparative Study of Industrialized Nations' Regulatory Systems. Report to the Chairman of the Committee on Agriculture, Nutrition, and Forestry. Washington, D.C.: U.S. Senate.

HEW (Department of Health, Education and Welfare). 1969. Report of the Secretary's Commission on Pesticides and Their Relationship to Environmental Health. Washington, D.C.: DHEW.

HHS (U. S. Department of Health and Human Services). 1988. Let's Improve Our Health. National Health and Nutrition Examination Survey III, Fact Sheet. Washington, D.C.: U. S. Department of Health and Human Services.

IFT (Institute of Food Technologists). 1989. *Assessing the Optimal System for Ensuring Food Safety: A Scientific Consensus.* Chicago: Institute of Food Technologists.

IPCS (International Programme on Chemical Safety). 1993. Report of an IPCS Planning Meeting on Harmonization of Approaches to the Assessment of Risk from Exposure to Chemicals. Unpublished paper available from IPCS. IPCS/PAC/93.11.

Japan. 1984. *Agricultural Chemicals Laws and Regulations.* Society of Agricultural Chemical Industry.

Johnson, E.L. 1982. Risk assessment in an administrative agency. *Am. Statistician.* 36:232–239.

Johnson, E.L. 1989. Pesticide residues. In: *International Food Regulation Handbook: Policy, Science, Law,* Eds., Middlekauff, R.D. and P. Shubik. pp. 253–281.

Kimbrough, R. D. 1987. Early biological indicators of chemical exposure and their significance for disease. In: *Mechanisms of Cell Injury: Implications for Human Health,* ed., Fowler, B. A., pp. 291–301. New York: John Wiley & Sons Ltd.

Kohn, G. K. 1987. Agriculture, pesticides and the American chemical industry. In *Silent Spring Revisited,* Eds., Marco, G.J., R. M. Hollingworth and W. Durham, pp. 159–174. Washington, D.C.: American Chemical Society.

Marco, R. M. Hollingworth, and W. Durham (Eds). 1987. *Silent Spring Revisited.* Washington, D.C.: American Chemical Society.

McEwen, F. L., and G. R. Stephenson. 1979. *The Use and Significance of Pesticides in the Environment*. 538 pp. New York: John Wiley & Sons.

Moolenaar, R.J. Carcinogen risk assessment: international comparison. Submitted for publication in *Regul. Toxicol. Pharmacol.*

NACA (National Agricultural Chemicals Association). 1993. From Lab to Label: The Research Testing and Registration of Agricultural Chemicals. Washington, D.C.: NACA.

NACA (National Agricultural Chemicals Association). 1994. Annual Industry Survey. Washington, D.C.: NACA (as cited in EPA, 1994a).

NFPA (National Food Processors Association). 1992a. Fact Sheet: What Benefits are Realized from Pesticide Use? Washington, D.C.: NEPA.

NFPA (National Food Processors Association). 1992b. The Health Significance of Pesticide Residues in Food. Washington, D.C.: NFPA.

Nilsson, R., M. Tasheva, and B. Jaeger. 1993. Why different regulatory decisions when the scientific information base is similar? — Human risk assessment. *Regul. Toxicol. Pharmacol.* 17:292–332.

Nordic Countries. 1985. *Potensgradering av Kreftfremkallende Stoffer — Rapport fra en nordisk arbeidsgruppe*. Nordisk Ministerrad. Miljo-Rapport.

NRC (National Research Council). 1984. *Toxicity Testing: Strategies to Determine Needs and Priorities*. 382 pp. Washington, D.C.: National Academy Press.

NRC (National Research Council). 1980. *Committee on Urban Pest Management*. Washington, D.C.: National Academy Press.

NRC (National Research Council). 1973. *Toxicants Occurring Naturally in Foods*. Washington, D.C.: National Academy Press.

NSC (National Safety Council). 1985. *Safety Facts*. Chicago: National Safety Council.

Office of Technology Assessment (OTA). 1993. Appendix A: international risk assessment. In: *Researching Health Risks*. OTA-BBS-570. pp. 187–210. Washington, D.C.: U.S. Government Printing Office.

TSG (Technology Sciences Group). 1993. Summary of the chemical regulations of the European Community. In: *International Environmental Monitoring Library*, TSG, Washington, D.C.

U.S. EPA (Environmental Protection Agency). 1984. Standard Evaluation Procedures — Oncogenic Potential: Guidance for Analysis and Evaluation of Long Term Rodent Studies. Washington, D.C.: U.S. EPA.

U.S. EPA (Environmental Protection Agency). 1992. April 20, Draft. Standard Evaluation Procedures — Developmental Toxicity Studies. Washington, D.C.: U.S. EPA, Office of Pesticide Programs.

U.S. EPA (Environmental Protection Agency). 1993, March, Draft. Standard Evaluation Procedures — Reproductive Toxicity Studies. Washington, D.C.: U.S. EPA, Office of Pesticide Programs.

U.S. EPA (Environmental Protection Agency). 1987. The Risk Assessment Guidelines of 1986. Washington, D.C.: U.S. EPA, Office of Health and Environmental Assessment. EPA/600/8-87/045.

U.S. EPA (Environmental Protection Agency). 1987. Pesticide Assessment Guidelines Subdivision U — Applicator Exposure Monitoring. Washington, D.C.: U.S. EPA, Office of Pesticide Programs.

United Kingdom. 1989. Guidelines for the Testing of Chemicals for Mutagenicity. Department of Health.

United Kingdom. 1991. 42 Guidelines for the Evaluation of Chemicals for Carcinogenicity. Departement of Health, Committee on Carcinogenicity of Chemicals in Food, Consumer Products and the Environment.

WHO (World Health Organization). 1993. Classification Systems on Carcinogens in OECD Countries — Similarities and Differences, prepared by Norway and The Netherlands.

WHO (World Health Organization). 1990. *Environmental Health Criteria: Principles for the Toxicological Assessment of Pesticide Residues in Food.* Volume 104. International Programme on Chemical Safety. Geneva: WHO

WHO (World Health Organization). 1979. Environmental Health Criteria 9: DDT. *Arch. Environ. Hlth.* 33:169.

Whysner, J. and G.M. Williams. 1992. International cancer risk assessment: the impact of biologic mechanisms. *Regul. Toxicol. Pharmacol.* 15:41–50.

QUESTIONS

1. Explain the concept of risk benefit as it is implemented in FIFRA.
2. Identify and briefly discuss two scientific issues associated with the evaluation of potential dietary exposures and human health risks associated with pesticides residues in food.
3. Identify two examples of how occupational exposures to pesticides can be measured.
4. Describe three exposure pathways that may be relevant to potential human exposures inside a home after using a flea and tick fogger product.
5. Briefly describe two benefits that result from the international harmonization of testing guidelines and protocols for studies related to pesticide registration (e.g., acute, subchronic, and chronic mammalian toxicity testing).

CHAPTER **II.3**

Ionizing Radiation Risk Assessment

Joseph L. Alvarez

SUMMARY

There is little difference between the analysis of risk from ionizing radiations and that from chemicals. Modes and pathways of exposure must be considered as well as concentrations and length of exposure. The differences from chemical exposures are particularly important, not the least being regulatory methods of limiting risk.

An important difference is that dose is dependent on the type of radiation. Types of radiation are limited but the differences in how the types of radiation interact with tissue are important in determining the dose and risk. Regardless of the type of radiation or radionuclide that is the source of radiation, dose or effective dose is applied in risk analysis in a simple linear equation.

The next important difference is that although risk is simply determined from dose, dose is the quantity used in regulation of radiation. Dose is the regulatory limit partly because effects were historically compared to measured dose and acceptable dose was considered in relation to the dose from natural radiation in the environment.

Several important distinctions in how dose is received determine both limits and methods of calculation. These distinctions are external dose, internal dose, effective dose, and population dose.

External dose is dose received from radiation outside the body.

Internal dose is dose received from radionuclides deposited in the body.

Effective dose is a means of equating external and internal dose to the same measurement quantity. Different types of radiation may have different effective doses even when the same amount of energy is deposited in the body. Even the same type of radiation may deliver different effective doses for the same amount of energy deposited if the energies of the radiations differ.

1-56670-130-9/97/$0.00+$.50

Population dose is individual dose summed over a population. In general, the regulatory limit for population dose is risk rather than dose.

Methods for calculating dose follow conventions developed by national and international committees on radiation protection. Many of these conventions follow idealized schemes for which modifying methods have been developed depending upon the radiation type, energy, and method of receiving the dose.

Key Words: dose, dose-rate, background, X-ray, gamma ray, beta ray, alpha ray, energy deposited, effective dose, whole-body dose, external dose, internal dose, radioactivity, radionuclides

1. INTRODUCTION

Ionizing radiation is radiation that has sufficient energy to ionize atoms or molecules. Ionizing radiation risk assessment follows the usual definition of risk:

$$R = (P,S,C) \tag{1}$$

where the risk R is identified as the probability P of a consequence C for a given situation S. In this definition, ionizing radiation risk is no different from any other type of risk. (For the remainder of this chapter, radiation will mean ionizing radiation.) Several differences from chemical risk are found in regulations governing radiation risk which tend to cause differences in how radiation risk is perceived and regulated. The most important risks for regulation are carcinogenic and mutagenic risks, which are considered probabilistic, but there are also injury risks and injury leading to death risks which are considered deterministic.

Probabilistic risks are considered to have a probability of occurrence down to zero dose, while deterministic risks have a threshold dose, below which no effect is expected. Probabilistic or stochastic risks have a probability of initiation. Once initiated and established, there is progression to a single end result regardless of dose. Deterministic risks require a minimum dose for expression of the effect. The magnitude of the deterministic effect is determined by the dose above the threshold.

Within these two definitions of type of risk, there are several methods for calculating risk based on the type of radiation, either alpha, beta, gamma, X-ray, or neutron. The calculation is further dependent upon how the radiation enters the body.

Much of the methods for calculating risk are dependent upon the history of the development of knowledge of the risk and the consequent regulations developed for different sources of radiation.

1.1 History

Knowledge of the existence of ionizing radiation dates to only shortly before 1900 (Roentgen 1895). Knowledge that it could pose a risk to living organisms dates to nearly the time of its discovery (Bergonie and Tribondeau 1906, Furth and Lorenz

1954). The ability to characterize and quantify the risk continues to evolve and to evolve in controversy. It was initially known that large doses resulted in burns or death (Furth and Lorenz 1954). The ability to protect against these immediate effects was quickly learned (Bergonie and Tribondeau 1906). Later, the discovery of genetic effects was thought to be the limiting factor for dose (Muller 1927). Finally, the carcinogenic effects of radiation were discovered (McCombs and McCombs 1930, Folley et al. 1952). It is the carcinogenicity of ionizing radiation that is the current basis for establishing standards for protection against ionizing radiation (ICRP 1976).

1.2 Background Radiation

Living organisms have evolved in a background of ionizing radiation (Eisenbud 1987). This background is both internal and external to the organism. The background radiation is also highly variable with time and place (Eisenbud 1987). The average background dose to an organism is, nevertheless, fairly constant worldwide, with only a few percent coefficient of variation. Even when the difference from the mean background is greater than one order of magnitude or more, no observable adverse effects are seen (Luckey 1991, Kondo 1993, Jaworowski 1995). The assumption has long been made that since living organisms have evolved in this natural background and no effects are observed for large departures from the mean, there are no effects at background doses or, if there are effects, the effects are exceedingly small, natural, and acceptable. It should then follow that small, artificial doses above the background mean will result in no effects or exceedingly small, negligible and acceptable effects (ICRP 1976). This logic has been abandoned in current regulatory and standard-setting theory for the more conservative assumption that all ionizing radiation, including background, causes adverse effects (ICRP 1991). Nevertheless, the concept of acceptable dose persists with the result that regulations concerning radiation and risks are dose based rather than risk based. The understanding persists that, since it is impossible to legislate background, it is difficult to legislate small differences from an uncertain background.

1.3 Types of Radiation

There are several types of ionizing radiation, each of which deposits energy in a different manner in living organisms. These radiations are X-ray, gamma, beta, alpha, and neutron. The first two are electromagnetic radiation, the second two are charged particles, and the last is a neutral particle. Dose is principally the sum of the deposited energy, but, in some cases, especially that of the neutron, there is an energy-dependent factor that is necessary to convert energy deposited to effective dose. Effective dose is an estimate of the risk, while energy deposited is the quantity measured (ICRP 1991, BEIR V 1990).

Each of the types of radiation has energy-dependent patterns for depositing energy that determine the manner of assigning dose and where in the organism dose is deposited.

1.4 National and International Standards

Current national and international standards are based on the conservative theory that the stochastic or carcinogenic effects of ionizing radiation act linearly with dose and with no threshold. This theory is espoused on the basis that there is no ready means of disproving the linear, no-threshold assertion and that following such a precept will clearly be conservative. These linear based standards involve a standard risk calculation, resulting in a risk per unit dose. The linear calculational method obtains a probability of dying of cancer for an individual exposed to a given amount of ionizing radiation or the projected total cancer deaths resulting in a population for a given population dose (ICRP 1991, BEIR V 1990). The probability distribution in Equation 1 is a simple coefficient of dose. A simple coefficient presumes that risk is a purely random quantity, equally distributed to each individual. The current, suggested form of Equation 1 is

$$R = aD \tag{2}$$

where $a = 0.1\ Sv^{-1}$ and D = dose in Sv (ICRP 1991). Regulatory agencies may use a different value for a (U.S. EPA 1992, 1993). The International Commission on Radiation Protection (ICRP) further recommends that when the dose is small or the dose is delivered at a low dose rate, coefficient a be divided by 2 (ICRP 1991). No guidance was given to regulatory agencies as to the magnitude of a small dose or a low dose rate. It can be presumed that the factor of 2 is obtained for doses and dose rates on the order of background.

Risk, as calculated in Equation 2, is not the usual standard or regulatory method for setting limits for ionizing radiation dose. The usual method is to specify dose as the limit rather than risk as a limit. There are two reasons for this standard-setting methodology. The first is that it is the traditional method which is based on actual measurement of dose and standards developed for deterministic effects. The second reason is that natural background is unavoidable and perforce is acceptable. Small increases or even multiples of the background dose must also be acceptable.

The dose standards are developed in two forms: acceptable doses for workers and acceptable doses for the public. In general, allowable public doses are a factor of 10 lower than worker doses for individuals (ICRP 1991, U.S. NRC 1993). In addition, distinctions are made in the standards for past and future practices. This distinction recognizes that it is easier, and therefore less hazardous, to include dose-reduction measures in a future activity than to modify an existing situation for purposes of reducing doses. The distinction further recognizes that there is uncertain risk at low doses and therefore uncertain benefit in lowering doses. It is therefore possible that more actual harm may result from lowering a hypothetical and uncertain low-dose risk.

The dose standards further modify the concept of acceptable dose on a risk-benefit basis. This risk-benefit basis assumes that the dose is only acceptable if the benefit derived exceeds the risk from the dose. This risk-benefit basis and the concept of acceptable dose translates to the further practice that, although the dose is acceptable, it does connote risk, so if it is reasonable to reduce even acceptable dose, it should be done. This concept results in regulations that define different levels of acceptable dose for different activities (U.S. EPA 1993). The dose limit is that which is technically feasible and reasonable.

Risk is determined by assigning dose. The essential task in radiation risk assessment is dose assessment, since risk is determined by comparison to dose standards.

2. EXTERNAL DOSE

External dose is dose received from a source outside the body. The source may be fixed, mobile, or intermittent. It is only important for radiations that can penetrate to living or vital cells. The distance from the source is important, depending upon the source shape and intensity. The dose received is dependent upon the time exposed. The dose may be reduced or eliminated by shielding and or increasing the distance from the source. The dose further depends on the time spent near the source. The dose to organs, parts of the body, or the whole body are important, depending upon the type of radiation or the part of the body being irradiated.

The most general type of dose is external, whole-body dose. Most risk is determined in relation to external, whole-body dose. Risk from dose to an organ or part of the body is determined as a fraction or multiple of the whole-body risk coefficient (ICRP 1991). The risk so determined is converted to effective dose for purposes of regulation. As an example, the hands and forearms do not contain blood forming bone marrow, so neither damage to the hemopoetic system nor leukemia is likely if a similar dose were received by bone marrow in the trunk of the body; therefore, the allowable dose to the hands and forearms is higher than the whole-body dose (U.S. NRC 1993).

Dose received externally from X-ray or gamma radiation is often assigned based on simple measurement of dose rate, dose to a dosimetry device, or calculation based on source strength. When precision is required, simple techniques are not adequate, and attention must be paid to dose distribution in tissue.

Dose distribution in the body or organs is dependent upon the source shape, intensity, and energy; the orientation of the source; and the recipient of the dose (source to subject geometry).

Photons passing through matter deposit energy with depth in the matter by the relationship

$$E = E_0 e^{-\mu x} \tag{3}$$

where E is the rate of energy deposition, E_0 is the energy deposition rate of the beam at the surface of the matter, μ is a coefficient of attenuation, and x is the depth in the matter (Cember 1969). The coefficient μ is dependent upon the photon energy. The attenuation increases nonlinearly with decreasing energy of the photon. Despite this attenuation relationship, dose at 1 cm depth in tissue (the depth that whole-body dose is nominally assessed) follows a different relationship that is dependent upon

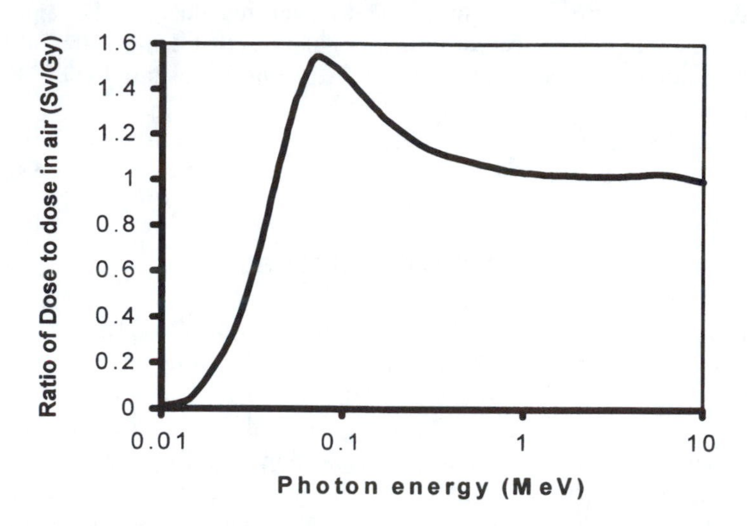

Figure 1 The ratio by photon energy of H_{10} in Sv to dose in air in Gy.

an additional factor of scatter, which is further dependent upon the size and direction of the photon beam (ICRU 1985). For a narrow, parallel beam normally incident upon tissue, the dose to energy relationship is shown in Figure 1. Separate relationships must be determined for other geometries.

When the incident radiation is energetic electrons or beta particles, yet another relationship is required. Electrons deposit energy much more rapidly than photons. More complicated equations are necessary to describe the energy loss than Equation 2 (Bethe and Askin 1953). Nevertheless, the energy loss can be predicted reasonably well with the important result that virtually all the energy is deposited within 1 cm of the surface of tissue for electrons up to several megaelectronvolts (MeV). For this reason, external dose for energetic electrons concerns only near-surface organs such as the skin and lens of the eye (U.S. NRC 1993). Beta particles are emitted in a spectrum of energy from those radionuclides that are beta emitters (Bethe and Askin 1953). The spectrum of energy further complicates the prediction of energy deposition. The actual spectrum will depend upon any intervening material between the emitting radionuclide and the tissue surface. Only few radionuclides are pure beta emitters, the majority also emit gamma photons. The gamma dose must also be considered when present, as well as photons produced by bremsstrahlung of the beta particles from containing or intervening materials (Alvarez 1983). External beta dosimetry must include considerations of the beta energy spectrum and the source to tissue geometry, including intervening material.

Neutron dosimetry is further different from either photon or electron dosimetry (Auxier 1966). Neutrons interact with tissue by absorption and nuclear reactions in atomic nuclei and ionization by displacement by collision of atoms from molecules. These interactions are energy dependent over a wide range of neutron energies and require specialized knowledge to perform the dosimetry. Neutron doses and the need for dosimetry is nearly confined to the weapons industry and nuclear reactors.

3. INTERNAL DOSE

Internal dose is dose received from radionuclides within the body. These radionuclides may enter the body by inhalation, ingestion, absorption, or through wounds. Dose or energy deposited is the result of the amount of radionuclides, the residence time in the body, the organ or organs where the radionuclide locates, the disintegration rate of the radionuclide, the type of radiation emitted, and the energy per disintegration. Determining internal dose requires knowledge of the metabolism of the radionuclide in the body. Dose is assigned using models of the metabolism (Cember 1969, ICRP 1979). As with external dose, internal risk is calculated using fractions or multiples of the whole-body risk coefficient.

3.1 Conventions for Assigning Internal Dose

Assigning internal dose is not confined to determining the energy deposited and applying the proper equivalent factor for the type of radiation. Organs differ in their radiosensitivity. The same energy per gram of tissue in one organ may have a completely different carcinogenic end point from that of another organ. Table 1 lists tissue weighting factors for various organs. Organ doses are assigned as multiples of permissible organ dose, permissible organ concentrations, or an equivalent whole-body dose by multiplying by an appropriate fraction. This dose assignment practice is a furtherance of the acceptable dose concept.

Table 1 Tissue Weighting Factors

Organ	Weighting factors
Gonads	0.20
Bone marrow (red)	0.12
Colon	0.12
Lung	0.12
Stomach	0.12
Bladder	0.05
Breast	0.05
Esophagus	0.05
Liver	0.05
Thyroid	0.05
Skin	0.01
Bone surface	0.01
Remainder	0.05

Adapted from ICRP 1991.

Internal dose results from the accumulation of radionuclides in organs of the body. These radionuclides may be accumulated from one exposure incident, several exposure incidents, or continuous exposure over some time interval. (Natural radionuclides in food are a source of lifetime exposure that leads to internal dose.) The radionuclides that accumulate in an organ are removed from the organ by natural metabolic processes and by radioactive decay. The dose received from radionuclides must be summed from the exposures and calculated based on the energy deposited in the organ over time (ICRP 1979). The dose is determined from the radioactivity in the organ (or simply activity) at a given time or the accumulated activity over time. The activity at a given time after a single deposition can be determined from

$$A = A_0 e^{-\lambda_E t} \tag{4}$$

where A is the activity at some time t after the initial deposition, A_0 is the initially deposited activity, and λ_E is the effective removal coefficient which is a combination of the metabolic removal and the radiological decay. Since dose is the energy deposited in the organ, the total energy deposited integrated over time is the required quantity. The total energy is determined from the energy deposited in the organ by each disintegration of the radionuclide. The dose is determined from

$$D = \int_0^T E_{ave} A_0 e^{-\lambda_E t}\, dt \tag{5}$$

where E_{ave} is the average energy per disintegration and T is the time interval of the integration. The concept of committed dose is used for each exposure such that $T = 50$ for adults and $T = 70$ for children (ICRP 1991).

Determining the activity of a radionuclide in an organ can be accomplished in some instances by performing radiological measurements of the body. These measurements may be of photons emitted from the body or of radionuclides in excreta from the body or blood samples. In other cases, the activity can be estimated from exposure to measured concentrations of radionuclides in air breathed or food and water consumed. In all cases, the metabolic behavior of the radionuclide in the chemical and physical form of the exposure must be known for the body and organs. For example, an ingested radionuclide in a given chemical form will be partially or totally absorbed into or eliminated from the body. The amount absorbed must be known, as well as the mode of adsorption. The absorbed material will then partition to various organs and systems of the body. The amount of retention of the radionuclide in the organ and its eventual elimination determine the energy deposited in the body and individual organs.

4. EFFECTIVE DOSE, DOSE AS RISK

A new dose quantity was defined by ICRP in 1991 as the "effective dose," ***E***. The effective dose considers the risk to the organ irradiated and the type or quality of the radiation (ICRP 1991). The effective dose is based on risk factors determined for separate organs based on their respective radiosensitivity. This is a little different from the risk factor determined for whole-body irradiation (uniform radiation throughout the body), since the risk for whole-body irradiation should be obtainable from the sum of the risk to all the organs when the whole body is irradiated. The difference is in how dose is determined and assigned. Whole-body dose is a term that applies to external radiation and is the usually assumed equivalent when the term dose is used. This equivalent to the whole-body dose is the effective dose.

For internal dose to a specific organ where the radiation source is deposited only in that organ and all the dose is received by that organ, the dose calculation involves simple multiplication by an effective dose factor. For deposition in several organs and/or irradiation of another organ from the organ of deposition, effective dose factors are required for the amount of radiation received by the several organs. The total effective dose is the sum of the effective dose to the several organs.

For external dose, effective dose considers particular differences from the whole-body dose. It is normally assumed that when dose is estimated from measurements in air using instruments or dose is measured using dosimetry devices that the dose is whole-body dose. The whole-body dose is further assumed to be what is defined as the $\boldsymbol{H}_{10}$ dose (ICRU 1985). This is the dose that would be measured by the instrument or dosimeter at 1 cm depth in a tissue equivalent sphere 30 cm in diameter. This definition further requires that the radiation be uniform, parallel, and incident at 90° upon the plane of measurement. This very useful definition clearly has practical limitations. The first is that few instruments and devices can mimic the $\boldsymbol{H}_{10}$ definition for all photon energy ranges. Figure 1 shows the relationship of $\boldsymbol{H}_{10}$ in Sv to dose in air in Gy for an ideal instrument. Most instruments and dosimeters perform near the ideal for much of the range of energies, usually above about 0.3 MeV, and are acceptable approximations for $\boldsymbol{H}_{10}$ (Alvarez 1983). It is necessary to know the response of the instrument over the energy range of interest when performing measurements to be used for dose.

Instruments and dosimeters are usually calibrated to approximate the $\boldsymbol{H}_{10}$. These conditions require uniform, parallel, and normally incident photon radiation. Effective dose considers the irradiation of separate organs, most of which are well removed from $\boldsymbol{H}_{10}$ conditions. Effective dose requires that each organ be considered separately, unless uniform, whole-body irradiation can be demonstrated. Effective dose requires that the source to body geometry be known or well approximated (Zankel et al. 1992). As an example, the case of a body standing and centered on a plane that is a uniformly distributed source is considered. Figure 2 shows the effective dose in Sv compared to the dose as measured in air in Gy (Zankel et al. 1992). Important differences can be seen from the $\boldsymbol{H}_{10}$ energy relationship in Figure 1.

4.1 Combining Internal and External Dose

The method for assigning dose as a whole-body effective dose is particularly useful to combining internal and external dose. Dose is assessed as the total of internal and external dose. It is important to have a convention for combining dose to arrive at a single risk value. The single risk value is whole-body effective dose, whether the dose is internal, external, partial body, or some combination of the various types of dose.

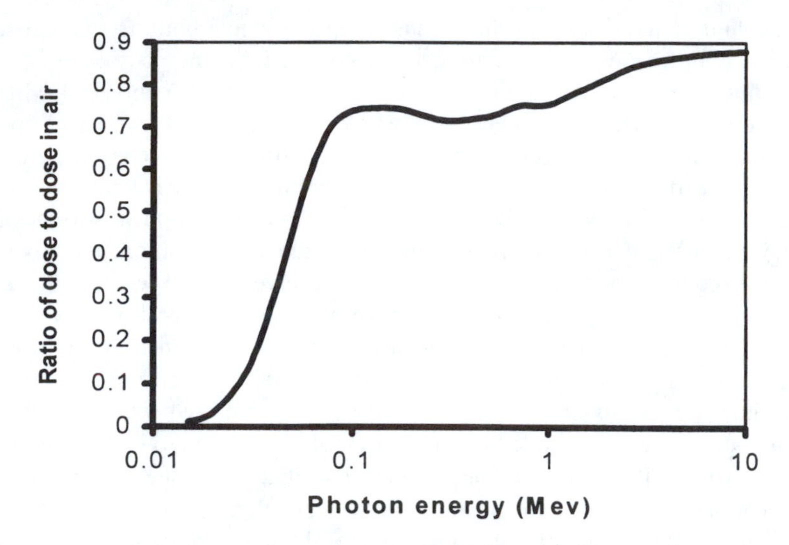

Figure 2 The ratio by photon energy of the equivalent dose (***E***) in Sv to the dose in air in Gy for a body standing on an infinite planar source.

4.2 Additive and Cumulative Dose

The concept of risk may be determined as the dose from a single incident, as the total dose for a given time period, or as the total lifetime dose. This is possible using effective dose since radiation dose is considered additive and cumulative by convention. It is well known that dose given in separate increments separated by time is less effective than dose given in a single exposure. Nevertheless, no discounting of dose received in separate exposures is taken in summing dose for conservatively estimating the risk. This is the usual regulatory approach for estimating the risk. Recent national and international standards have recommended the use of dose rate factors in discounting risk, but no guidance has been given as to when or at what dose rate discounting is recommended (ICRP 1991).

4.3 Population and Individual Dose

A particular consequence of Equation 2 is that mathematically risk or dose calculated for populations or individuals is identical. If individual A receives a dose D_A, and individual B receives a dose D_B, then the risk of either A or B having the health effect is

$$R = a(D_A + D_B) \tag{6}$$

In this case, we have calculated both the individual risks to A or B and the population risk of the population A and B. The validity of the operation in Equation 6 is in scientific dispute, but is generally accepted in the radiation protection community as a conservative measure and usually employed by regulatory agencies.

Population risk is used to justify actions when individual risk is acceptable. An individual risk of 10^{-4}, if deemed acceptable to one individual, may not be acceptable to 100 individuals, since the population risk could be considered to be 10^{-2}. This type of calculation presents both mathematical and philosophical problems. The mathematical problem concerns the uncertainty in the operation. The calculation of an individual risk of 10^{-4} is statistically very uncertain and contains inherent conservatism. The operation in Equation 6 introduces further statistical uncertainty, and the validity of the operation itself has not been established. The philosophical problem concerns how to justify action based on population risk when the risk to individuals is acceptable and the action will lower no individual's risk appreciably.

4.4 Calculating Dose

External dose is calculated simply using an $\boldsymbol{H}_{10}$ dose rate and the time spent at that dose rate,

$$D = \boldsymbol{H}_{10}T \tag{7}$$

In general, dose is accumulated from a variety of locations or in several positions within an area, with each location or position having a different source to subject geometry. The dose is then calculated from an effective dose rate $\boldsymbol{E}$ for different times $\boldsymbol{T}$,

$$D = \sum_i E_i T_i \tag{8}$$

Internal dose usually involves a source receptor exposure pathway. This pathway may be very simple, as airborne radionuclides in a room that the receptor enters for only a short time. Far more complicated source receptor exposure pathways may be encountered, involving, for example, airborne radionuclides deposited on plants which are then consumed by animals which are in turn consumed by humans. These more involved pathways require modeling and measurement to complete the assessment of exposure which then must be subjected to a time integral as in Equation 5. Important aberrations of the pathways may occur that are not readily predictable by the suspected chemistry of the pathway. These aberrations are caused by the atomic rarity of the radionuclide. Radionuclides are found in radiologically important concentrations far below the limits of normal chemical analysis and therefore chemical understanding. Predictable chemical behavior occurs only when carrier atoms are present in far higher concentrations than the radionuclide, i.e., nonradioactive isotopes of the same element in the same chemical form. Lack of sufficient carrier atoms may necessitate measurement of the radionuclide at every stage of the pathway.

5. CONCLUSIONS

Risk is determined for ionizing radiation from dose. In many cases, it is not necessary to make a translation from dose to risk. The calculation of or determination of dose requires differing levels of skill, depending upon the type of radiation and how it is delivered. Internal dose may be the most difficult determination, but arriving at equivalent dose can be difficult for any situation. The amount of effort expended in determining dose may depend upon the accuracy required. Many simple dose determinations involve varying degrees of conservatism. It is impractical to make decisions concerning risk if the degree of conservatism is not known or controlled. Accurately determining dose is a means of controlling conservatism.

REFERENCES

Alvarez, J. L. Beta measurements in a photon field. *Proc. of the International Beta Dosimetry Symp.*, NUREG/CP-0050. Washington DC: U.S. Nuclear Regulatory Commission; 1983.

Auxier, J. A.; Jones, T. D.; Snyder, W. S. Neutron interactions and penetration in tissue, *Radiation Dosimetry*, F. H. Attic and W. C. Risk, Ed. New York: Academic Press; 275, 1966.

BEIR V, Committee on the Biological Effects of Ionizing Radiation. *Health Effects of Exposure to Low Levels of Ionizing Radiation*, Washington, DC: National Academy Press; 1990.

Bergonie, J.; Tribondeau, L. De quelques resultats de la radiotherapie et essai de fixation d'une technique rationelle. *C.R. Seances Acad. Sci.* 143, 983, 1906.

Bethe, H. A.; Askin, J. Passage of radiations through matter, *Experimental Nuclear Physics*, Vol. 1, Segre, E., Ed. New York: Wiley; 1953.

Cember, H. *Introduction to Health Physics*, New York: Pergamon Press; 1969.

Eisenbud, M. *Environmental Radioactivity*, New York: Academic Press; 1987.

Folley, J. H.; Borges, W.; Yamasaki, T. Incidence of leukemia in survivors of the atomic bomb in Hiroshima and Nagasaki. *Am. J. Med.* 13, 311, 1952.

Furth, J.; Lorenz, E. Carcinogenis by ionizing radiation, *Radiation Biology*, Vol. I, Part II, Hollaender, A., Ed. New York: McGraw-Hill; 1145, 1954.

ICRP. *Recommendations of the International Commission on Radiological Protection.* Oxford: Pergamon Press; ICRP Publication 26; 1976.

ICRP. *Recommendations of the International Commission on Radiological Protection.* Oxford: Pergamon Press; ICRP Publication 39; 1979.

ICRP. *Recommendations of the International Commission on Radiological Protection.* Oxford: Pergamon Press; ICRP Publication 60; 1991.

ICRU. International Commission on Radiation Units and Measurements. *Determination of Dose Equivalents Resulting from External Radiation Sources*. Bethesda, MD: ECRU; ICRU Report 39; 1985.

Jaworowski, Z. Hormesis: The beneficial effects of radiation. *Nukleonika* 41, 22, 1995.

Kondo, S. *Health Effects of Low-Level Radiation*. Madison, WI: Medical Physics Publishing; 1993.

Luckey, T. D. *Radiation Hormesis*. Boca Raton, FL: CRC Press; 1991.

McCombs, R. S.; McCombs, R. P. A hypothesis on the causation of cancer. *Science* 72, 423, 1930.

Muller, H. J. Artificial transmutation of the gene. *Science* 66, 84, 1927.

Roentgen, W. C. Ueber eine neue Art von Strahlen. *Vorlaeufige Mitteilung. Sitzungsber. Physik.-Med. Ges.* Wuerzburg; 137, 1895.

U.S. EPA (United States Environmental Protection Agency). High-level waste standards. *Federal Register* 58, 66398, 1993.

U.S. EPA (United States Environmental Protection Agency). *Health Effects Assessment Summary Tables.* OHEA ECAO-CIN-821; 1992.

U.S. NRC (United States Nuclear Regulatory Commission). Standards for protection against ionizing radiation. *CFR* 10, 20, 1993.

Zankel, M.; Petoussi, N.; Drexler, G. Effective dose and effective dose equivalent — the impact of the ICRP definition for external photon irradiation. *Health Phys.* 62, 395, 1992.

QUESTIONS

1. What is the quantity for limiting radiation risk?
2. What international agency sets standards for radiation risk?
3. What U.S. agency sets acceptable limits for the radiation industry?
4. Why are these limits different for different industries?
5. What is the meaning of whole-body dose?
6. How does effective dose differ from whole-body dose?
7. How does energy deposited relate to dose?
8. An individual receives doses of .05 and .03 Sv at age 25, .12 Sv at age 28, a 50-year committed dose of .23 Sv from an internal exposure at age 32, and a dose of .05 Sv at age 45. If she retires at age 55 what is her occupational radiation dose?
9. The dose rate in air from a 0.1 MeV photon is .02 Gy/h. What is the effective dose to an individual who is exposed for 2 h?
10. An individual receives 2 Sv to the lung from a 30-year exposure to radon. What is the effective dose?

CHAPTER **II.4**

Use of Risk Analysis in Pollution Prevention

Vlasta Molak

SUMMARY

Industrial pollution can be defined as the presence of toxic substances in air, water, or soil, often resulting from inefficiencies in production processes. The presence of these substances can present a health risk to humans or ecological systems. These risks can be estimated and compared using risk analysis methods. Therefore, risk analysis can serve to establish a priority of pollution problems based on the magnitude of risk that they pose either to human health or ecological systems.

Pollution can also be regarded as resources distributed in the wrong places, and pollution prevention at the source can be regarded as saving on resources and decreasing costs of production. Since economic risk analysis can indicate economic losses resulting from pollution, it can be used to encourage pollution prevention at the source as a means of improving the bottom line. Risk communication explains the data derived from toxicological risk analysis and economic risk analysis to decision makers in the most compelling manner to encourage pollution prevention at the source. This chapter will demonstrate some of the applications of risk analysis in dealing with industrial pollution.

Key Words: pollution prevention, source reduction, industrial, system analysis

1-56670-130-9/97/$0.00+$.50

1. MODERN RISKS ASSOCIATED WITH POLLUTION

The following are examples of the changes in a modern industrial society that must be factored into risk analysis and risk management associated with industrial pollution.

1. Increased new risks such as those from nuclear plant accidents, radioactive waste, pesticides and other chemicals releases, oil spills, chemical plant accidents, ozone depletion, acid rain generation, and global warming.
2. Increased ability of scientists and equipment to measure contamination.
3. Increased number of formal risk analysis procedures capable of predicting *a priori* risks.
4. Increased role of governments in assessing and managing risks.
5. Increased participation of special interest groups in societal risk management (industry, workers, environmentalists, and scientific organizations), which increases the necessity for public information.
6. Increased citizen concern and demand for protection.
7. Realization that pollution is resources distributed in the wrong places; thus, by implementing clean technology, the companies have a chance to increase efficiency and profits.

Risk analysis can help manage technology in a more rational way to promote sustainability of desirable conditions of societies and eliminate those conditions that are detrimental to the well-being of humans (and ecosystems). Figure 1 gives a schematic presentation of technological activities that can result in environmental pollution, which in turn results in risks to human and ecological health. Since many of these human and ecological health risks are associated with industrial pollution (air, soil, and water contamination), risk analysis can play a role in industrial pollution prevention by establishing the magnitude of risk associated with each pollution case and indicating which pollution prevention option will result in the highest reduction of risk (Shorthouse 1990/1991).

2. APPLICATION OF RISK ANALYSIS TO INDUSTRIAL POLLUTION PROBLEMS

In an ideal world, with unlimited resources, we could technically eliminate all industrial pollution and develop closed loop production systems with no waste generated. However, since our resources are limited, we should concentrate on the pollution problems causing most human health problems and environmental damage. Application of risk analysis to industrial pollution provides an opportunity to derive neutral sets of data and compare the magnitudes of problems in terms of human and environmental health and costs of alternatives. Until recently, most decisions regarding pollution management were made based on public perception. For example, a heavy smoker might tolerate a risk of 1:2 of emphysema, cancer, or heart disease, but would not tolerate a risk of 1:10,000 imposed on him by a factory releasing pollutants. Generally, the public accepts much greater voluntary risks than smaller

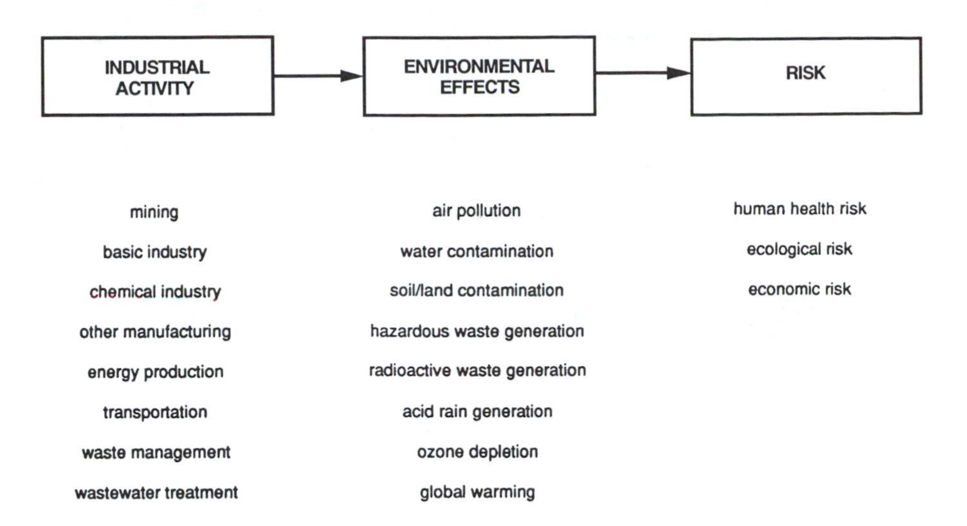

Figure 1 Technological activities and their effects.

imposed (involuntary) risks (Slovic 1987). For example, dealing with hazardous waste sites is using up most of the U.S. Environmental Protection Agency's (EPA) budget (Superfund), even though the health risks to population may be minimal. In last few years, the U.S. EPA and other federal agencies have emphasized risk analysis as a basic tool to determine the priority of environmental problems, rather than being pushed into action by the perception of risks (U.S. EPA 1990). Chapters on comparative risk analysis and setting environmental priorities address this problem in more detail (see Chapters III.3 and III.4).

Figure 2 depicts how a particular industrial plant can apply risk analysis at various points of production. Risks due to transport of raw materials, their input into process, production process, toxic releases, and products leaving the production process and their transport can be evaluated by using historical data, and measures could be taken to decrease the risks of accidents and resulting pollution. Within the plant, one can evaluate process safety (and probability for spill and/or accident that would result in pollution) by applying probabilistic risk analysis (see Chapters I.4 and I.5). Based on this analysis, additional levels of safety may be introduced, and/or a change of equipment and process may be implemented. Based on probable magnitude of release resulting in air, water, and/or soil/land contamination, one can perform risk assessment of the effects of pollutants on human and ecological health. Similarly, "routine" releases can be evaluated using chemical risk analysis or carcinogen risk analysis.

With the passage of SARA, Title III law, a new opportunity appeared for an environmentally conscious design in manufacturing. Most manufacturing facilities have to submit a yearly inventory of pollution: routine toxic releases (as expressed in Toxic Release Inventories) and accidental releases in all media (air, water, soil, POTW, underground injection, off-site waste treatment facilities, etc.). Also, they have to reveal their pollution prevention, control, and waste management practices. Each industry filing a Toxic Release Inventory (TRI) report has an opportunity to

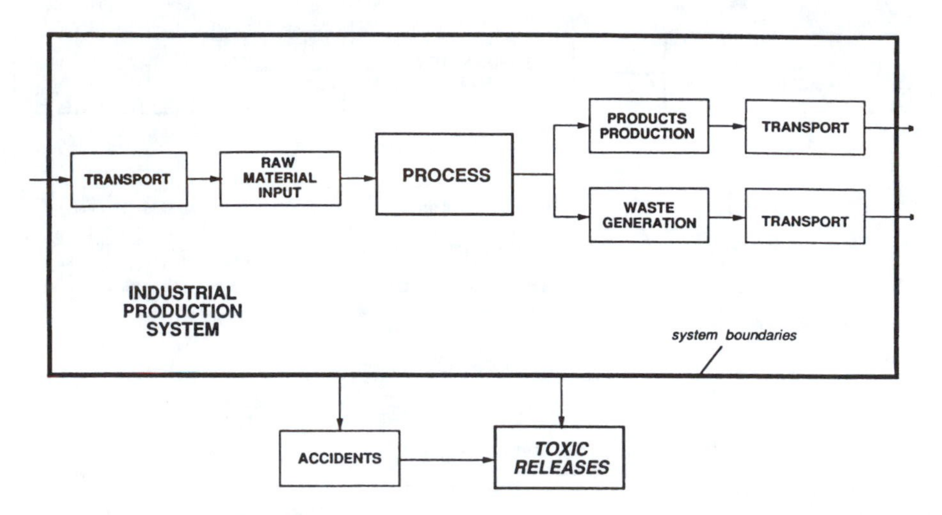

Figure 2 Schematic presentation depicting various points of production in an industrial plant when risk analysis can be applied.

evaluate its practices using risk analysis. From measured concentration of pollutants in air, water, and/or soil, one can perform comparative risk assessments for each particular chemical or process and rank them according to the magnitude of risk (either probability of cancer or magnitude of exceeding reference doses [RfDs], air, or water criteria). Although the company may not violate any permits, it is conceivable that the potential exposures to toxic chemicals released from the plant may exceed safe levels. Each company can perform their own risk analysis. Numerous computer programs exist to simulate and evaluate health risks from chemicals (RISKWARE 1994; see Chapter II.6 on computer software for risk analysis). A good starting point for performing chemical risk analysis may be TRI reports, since they represent an inventory of pollution (even if incomplete).

Chemical risk analysis and economic risk analysis can be applied to evaluate various waste disposal practices. The impacts of alternatives could be evaluated using economic risk analysis, ecological risk analysis, and human health risk analysis, which could derive an optimal solution for a given problem. Based on such analyses, one can set priorities in pollution prevention. Description of risk assessments that could be performed by industries is described next.

2.1 Chemical Industry

Since the chemical industry deals with the largest number of toxic chemicals and processes, the application of risk analysis is most appropriate.

1. One could study the exposure of workers (from the existing data or from modeling routine releases) to raw materials, intermediates, releases, and final products (occupational safety standards). Based on these studies, one can devise reduction of releases of pollutants into the environment, with priority given to those posing the

greatest risk. Probabilistic risk analysis of the process safety (fault-tree analysis) could be used to pinpoint week points in the system and thus decrease unintentional catastrophic releases. Attempts should be made to minimize the loss of chemicals (releases) during the process and establish safety measures to minimize the risk of other types of injuries (explosions, falls, etc.).

2. Hazardous wastes and/or toxic releases to air, water, and/or soil are other targets for risk analysis because of potential exposure of surrounding populations and ecosystems as the whole. TRI reports are an ideal starting point for such an analysis. Also, an economic analysis of various disposal options, recycling, and pollution prevention could point out optimal solutions.
3. Products themselves maybe highly toxic and, thus, may cause problems to the users that may outweigh benefits. However, it may be difficult to persuade a producer to stop producing a chemical that may be harmful and not really necessary (although the market has gotten used to it for various reasons), but brings financial rewards to the producer.

2.2 Other Manufacturing

One can perform risk assessments for any industry involving chemical pollution in a same manner as for chemical industry. Additional analysis may be performed on packaging material from a waste reduction point of view or economic risk analysis.

2.3 Hazardous Materials Handling and Pollution Control

Another argument for industrial pollution prevention could be by performing risk analysis of various pollution control options, which in many cases have led to environmental problems of great magnitude (soil contamination at Love Canal, dioxin contaminated soil in Missouri). Risk analyses of numerous hazardous waste sites indicates the magnitude of the problem, which was caused by the choice of inappropriate industrial pollution control option. Superfund risk assessments are part of standardized procedure in evaluating sites (U.S. EPA 1992b). Potential health and ecological risks associated with pollution control methods are

1. Exposure of workers and public to toxic chemicals (groundwater and surface water contamination):
    a. Superfund sites (remedial investigation/feasibility study RI/FS) — Numerous such studies were performed both by the U.S. EPA and by industry to determine necessity and priority of cleanup of the site (U.S. EPA 1992a,b).
    b. Other sites — Industry was often storing the waste on the premises or the plant, either in drums or partially degrading toxic chemicals and/or evaporating them by using lagoons. Such treatment of hazardous waste often results in site contamination and groundwater and surface water contamination.
2. Potential for generation of other, more potent, toxic chemicals by municipal solid waste incinerators. One of the byproducts in incineration performed at temperatures under 900°C is dioxin, a potent human toxicant and carcinogen (Mukerjee and Cleverly 1987). Also, incineration of other chlorinated organics results in chlorine generation, which in itself is highly toxic.

3. Contamination of waste water systems and waterways. Frequently, the largest number of toxic chemicals from the TRI list are released into sewers, eventually reaching POTW (publicly owned waste water treatment works) (Molak 1989). If such releases are done in a short period of time, they may cause die-off of microorganisms and subsequent pollution of waterways. Comprehensive ecological and economic risk analysis may point out the necessity for pollution prevention. For example, pollution of surface waters upstream increases cost of drinking water purification downstream in the communities depending on the river for drinking water. Such hidden costs could make a stronger argument from a business point of view for pollution prevention.

When pollution is evaluated from a system point of view, one can find that previously used options of treatment of hazardous waste are more risky (from a human health point of view and often from economic point of view) than pollution prevention. Thus, performing a risk analysis may be useful to make a case for pollution prevention.

2.4 Energy Production

Energy production is unfortunately often connected with industrial pollution. An increase of efficiency in energy consumption by industry can be regarded as pollution prevention. Different types of energy production, however, introduce different risks that could be evaluated and quantified. Chapter IV.1 indicates some of the ways to improve safety in the nuclear industry. However, the Chernobyl accident may have altered the assumptions that have been made in conventional nuclear plants risk analysis, and some other energy sources may be preferable.

Although, in most cases, manufacturing plants do not have an option about which energy source to use, in some instances, they could opt for oil or gas vs. coal, based both on chemical risk assessment (lowest pollution option) or economic risk analysis.

2.5 Transportation

Transport of raw materials, finished products, and, very often, waste by industry also contributes to industrial pollution.

1. Air pollution. 60 to 70% of total air pollutants in any large city are generated by transport, both by industrial and individual vehicles. Major pollutants are CO, NOx, and organics which result in ozone development on the ground.
2. Routine and catastrophic spills of toxic chemicals during transport from production to consumers. The most frequent risks are posed by overturned trucks carrying gasoline or other oil products and toxic chemicals.
3. Routine and catastrophic oil spills (water and land). For example, the Exxon Valdez has alerted the public about ecological consequences of oil transportation. However, routine spills are even more hideous because total oil released by routine spills is twice that of the catastrophic spills.

Because of the pollution potential of transportation, more efficient flow of raw materials, products, and waste can be designed to minimize the amounts of materials carried and the types of carrier used. For example, per ton of material used, the least polluting and least expensive alternatives to trucks may be trains and/or boats (where applicable). From a system point of view, developing local markets for the products may be another way of decreasing pollution from transporting the products long distances.

3. APPLICATION OF RISK ANALYSIS FOR A PARTICULAR POLLUTION SOURCE

For each industrial pollution source, one can perform an environmental exposure assessment for the pollutants coming from that source. By comparing the particular exposure with the previously established criteria for that chemical (or derive the criteria using the risk assessment guidelines), one can establish possible human health and ecological risks from that industrial source. However, even in cases where the pollutant concentration in air, water, and/or soil do not exceed criteria, one could make a case for pollution prevention based on economic risk analysis. The scheme presented in Figure 2 can be applied to a chemical plant that produces organic chemicals. For example, if plant X was found to use five processes that may result in catastrophic releases, a fault tree can be constructed for each process and for the overall plant (Shorthouse 1990/1991). Probabilistic risk analysis can evaluate the likelihood of an accident. Also, it can indicate amounts of toxic chemicals potentially released from such an accident. Chemical risk analysis can then evaluate possible adverse effects of such a spill on humans and ecological systems.

The human and ecological risks from routine releases can be evaluated separately based on known data of exposures and properties of the chemical. One can study both human health risk and economic risk. In many cases, it would be difficult to do a complete health risk analysis (air modeling, exposure modeling, etc.) for chemicals that companies have to report as TRI, since TRI provides only total yearly releases. However, in cases where patterns of releases are known, one can perform a valid chemical risk analysis and set priorities for pollution prevention.

Economic risk analysis based on TRI can be used to decrease pollution around companies, because TRI enables us to see the inefficiency of the process (forces company to evaluate its material flow). Data from TRI have used this approach to demonstrate economic losses in Hamilton County, Ohio, due to pollution (Molak 1989, 1991).

Based on a premise that pollution is resources distributed in the wrong places, economic losses of wasted resources in Hamilton county were calculated from total TRIs to amount to $25 million (Molak 1989). Since TRIs report only 5 to 10% of released toxic chemicals, the estimated real value of loss is $250 to $500 million or $250 to $500 per person (Molak 1991). A rough estimate based on TRI in the United States is $60 to $120 billion. These numbers indicate that the cost analysis argument for waste reduction and pollution prevention may be compelling (Molak 1990).

A major objection to that argument is that it would take a lot more money to recapture pollutants from the smokestacks, waste water, or soil. This statement is valid only in the old paradigm of pollution control, where pollution is controlled at the end of the tailpipe. It is too late to capture pollutants at the end of the pipe, since at that point the entropy has already increased and large amounts of energy are needed to concentrate chemicals. However, prevention of losses of chemicals in the process itself or recycling of materials can bring about savings.

Often, it is difficult to prove that exposure to a given concentration of a particular pollutant causes harmful health effects, but since it cannot do any good, and companies are losing money on raw materials, it is prudent to prevent the pollution at the source.

4. USE OF CHEMICAL INFORMATICS IN POLLUTION PREVENTION

In a complex industrial production plant, one way to understand the system under observation is to define its boundaries, as well as input and output material flows (Olbina 1991). A flow-chart diagram helps establish points of potential human and ecological health risks due to the release of toxic chemicals. Numerous computer risk assessment programs are available, which incorporate the previously or similarly described methods of risk analysis (RISKWARE 1994). Chapter II.6 on software available for risk analysis can give the reader an update on the field and help the reader choose the most appropriate software for company needs. Based on the flow-chart diagrams and performed risk assessments, it is possible to find those processes where changes can be made to decrease risk and subsequently pollution. Pollution prevention strategies can be based on priorities obtained by carefully executed comprehensive risk analysis. All material and energy inputs and outputs can be accounted for and efficiency in production can be maximized. Reduction in material loss through releases to air, water, and/or soil can then be regarded as improvement in efficiency or process optimization. Comprehensive cost analysis could indicate where the profits could be most improved if pollution prevention were implemented.

5. CONCLUSION

Application of risk analysis to industrial pollution has demonstrated that

1. Risk analysis can help establish priorities for pollution prevention in a particular production system by first dealing with toxic chemicals and processes that result in the highest human and ecological health risks.
2. Pollution can be regarded as resources distributed in the wrong places. The old paradigm of pollution control is economically less desirable than pollution prevention, as can be demonstrated by application of economic risk analysis.
3. Both economic and often public health cases for introduction of waste minimization and pollution prevention can be made by using risk analysis.

REFERENCES

CERCLA 1993. Studies Under CERCLA. Office of Emergency and Remedial Response. NTIS PB93126787XSP.

Freeman H. et al. 1992. Industrial Pollution Prevention: A Critical Review. *J. Air Waste Management Association* 42(5): 618–656.

Molak V. 1989. Waste Minimization and Community-Right-to-Know Law. Proceedings of 1st International Conference on Waste Minimization and Clean Technology (Geneva, May 29–June 1 1989). ISWA, pp. 404–410.

Molak V. 1990. Pollution as a Terminal Disease. *Risk Analysis* 10(4): 605–607.

Molak V. 1991. Over $100 Billion/Year Wasted by Industry into the Air, Down the Drain, and into the Countryside of the United States. *J. Clean Technology and Environmental Sciences* 1(2): 155–157.

Mukerjee D. and Cleverly D.H. 1987. Strategies for Assessing Risk from Exposures to Polychlorinated Dibenzo-*p*-Dioxins and Dibenzofuranes Emitted from Municipal Incinerators. Waste Management and Research 5: 269–278.

NIOSH 1990. NIOSH Pocket Guide to Chemical Hazards. U.S. Department of Health and Human Services

Olbina R. 1991. Computer Supported Modelling of Hazardous Waste Management. Dissertation. University of Ljubljana, Faculty of Science and Technology, pp. 271.

OSHA 1989. Air Contaminants; Final Rule (Codified at 29 CFR 1910). *Federal Register* 54: 2332–2983.

RISKWARE 1994. SOFTSTRACTS. Society for Risk Analysis. 1994 Annual Meeting (tel. 703/790-1745).

RTECS 1995. Registry of Toxic Effects of Chemical Substances—Online Toxicology Information Program. National Library of Medicine.

Shorthouse B.O. 1990/1991. Using Risk Analysis to Set Priorities for Pollution Prevention. *Pollution Prevention Review* 1(1): 41–53.

Slovic P. 1987. Perception of Risk. *Science* 236: 280–285.

U.S. EPA, FEMA and DOT 1987. *Technical Guidance for Hazard Analysis*. Washington DC. Government Printing Office.

U.S. EPA 1990. Reducing Risk: Setting Priorities and Strategies for Environmental Protection. SAB-EC-90-02.

U.S. EPA 1992a. List of Lists. Consolidated List of Chemicals Subject to Reporting Under the Emergency Planning and Community Right-to-Know Act. NTIS PB92500792XSP.

U.S. EPA 1992b. Guide for Conducting Treatability Studies Under CERCLA. Office of Emergency and Remedial Response. NTIS PB93126787XSP.

QUESTIONS

1. Define industrial pollution.
2. How can industrial pollution be viewed in terms of resource?
3. What is pollution prevention?
4. What is waste minimization?
5. What are the modern risks associated with industrial pollution?
6. List major industrial activities.
7. List major environmental effects of modern industrial activities.
8. How can risk analysis be used in pollution prevention?

CHAPTER **II.5**

Integrated Risk Analysis of Global Climate Change*

Alexander Shlyakhter and Richard Wilson

SUMMARY

This chapter discusses several factors that should be considered in integrated risk analyses of global climate change. We begin by describing how the problem of global climate change can be subdivided into largely independent parts that can be linked together in an analytically tractable fashion. Uncertainty plays a central role in integrated risk analyses of global climate change. Accordingly, we consider various aspects of uncertainty as they relate to the climate change problem. We also consider the impacts of these uncertainties on various risk management issues, such as sequential decision strategies, value of information, and problems of interregional and intergenerational equity.

Key Words: climate change, model uncertainty, probability distributions, tails, truncation, equity, value of information surprise

1. ANALYZING THE RISKS OF GLOBAL CLIMATE CHANGE

Integrated risk analyses of global climate change seek to arrive at answers to the following basic question: "What are the likely impacts of global warming upon the world, and can the possible adverse impacts be eliminated or reduced?"

For the assessment of the potential risks of global warming, we use a general layout of the progression of the physical processes involved stimulated by ideas

* Based on Shlyakhter, A.I., Valverde, L.J., and Wilson, R. 1995. *Chemosphere,* 30: 1585.

originally put forth by Kates et al. (1985). We divide the processes leading to global warming into a sequence of roughly independent steps. Our diagram is simplified by considering only CO_2 as a greenhouse gas. This simplification is made because CO_2 is the most important greenhouse gas that humans can alter. A more complete diagram would show other entries, such as methane, nitrogen oxide, and chlorinated fluorocarbons. The difficult scientific question of the role of water vapor, the most important greenhouse gas, is discussed later in this chapter.

For clarity, we enumerate the steps of the main sequence running from top to bottom in the center of Figure 1 by the numbers 1 to 6 referred to in the text. Equation 1 represents the final environmental outcome of interest as the product of six factors corresponding to these steps. The first factor is the world population; the second factor is energy production *per capita*; the third factor is the total CO_2 emissions per unit of energy production; the fourth factor is the increase of atmospheric concentration of CO_2 per unit emission; the fifth factor is the temperature rise per unit of CO_2 concentration; and the sixth factor is the environmental outcome of interest (e.g., sea level rise) per unit temperature rise.

Factor 1 2 3 4 5 6

$$\Delta h = Popul \cdot \left(\frac{energy}{person}\right) \cdot \left(\frac{CO_{2_{emit}}}{energy}\right) \cdot \left(\frac{CO_{2_{atmos}}}{CO_{2_{emit}}}\right) \cdot \left(\frac{\Delta T}{CO_{2_{atmos}}}\right) \cdot \left(\frac{\Delta h}{\Delta T}\right) \qquad \textbf{(1)}$$

The relationship of Equation 1 to Figure 1 is explained as follows: the product of the first two factors is the world energy use; the product of the factors 1, 2, and 3 is the total of world CO_2 emissions; the product of factors 1 to 4 is CO_2 concentration; and so on.

In writing Equation 1, we assume that each factor is independent of all the others. Indeed, the factors are chosen so that this is approximately true. This is a simplifying assumption that enables us to make a first approximation to the environmental outcome(s) of interest. Further refinements would be to identify, for example, those environmental outcomes that arise from correlations that exist between the factors, such as the combined effects of CO_2 and temperature on plant growth.

Further, it is a *static* representation of the problem, whereas, in reality, the physical situation evolves with time. This means that when the CO_2 concentrations have reached double preindustrial levels, the temperature rise will not have reached the value given by the static calculation. Missing in this static representation are the effects of large heat sinks. These simplifications notwithstanding, Equation 1 is a useful preliminary framework to discuss the uncertainties that arise from an incomplete understanding of the physical processes.

At the top of Figure 1 is a line suggesting that we can modify world population (e.g., upward by reducing war, famine, and pestilence or downward by birth control) by societal action. The next line suggests that humans may modify the energy use per capita (e.g., upward by increasing the global standard of living or downward by increased efficiency of energy use). The third line suggests that we can modify CO_2

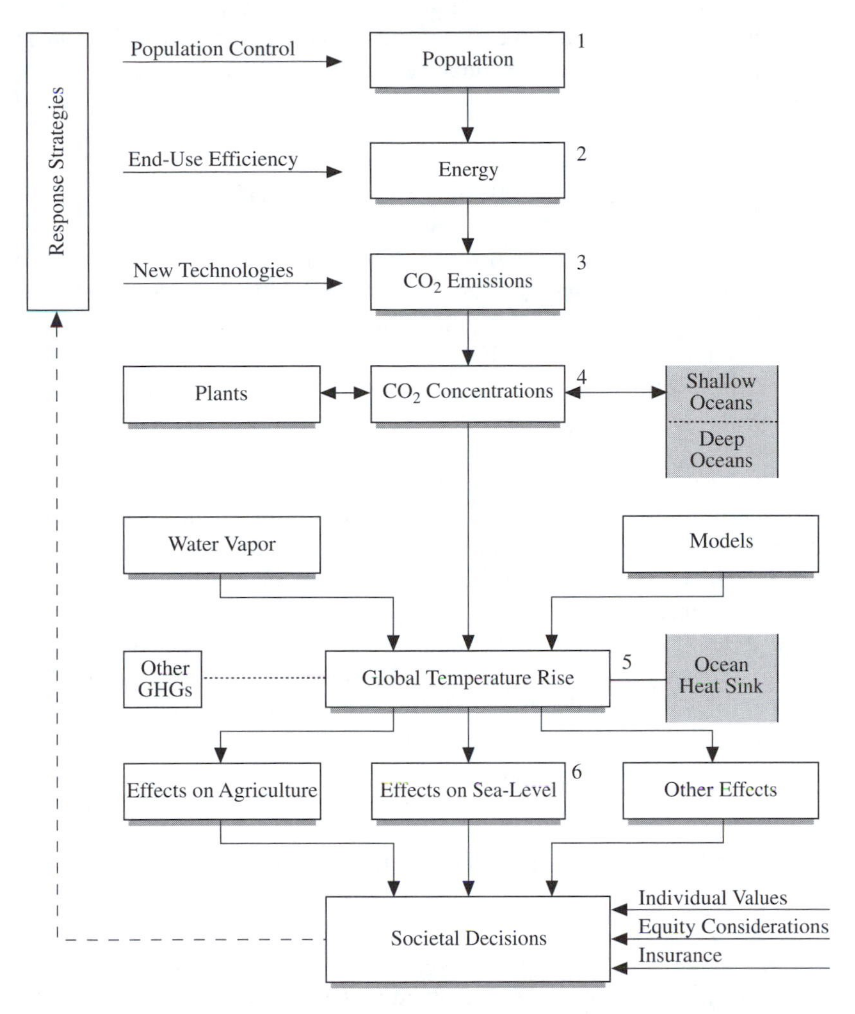

Figure 1 The proposed causal framework for global climate change consists of three parts: climate change assessment, impact assessment, and risk management. Population and energy policy studies, together with the models of climate system, serve as inputs to climate change assessments. Impact assessment is concerned with the effects of climate change. World population, energy consumption, and CO_2 emissions appear as end points in risk management decisions about climate change.

emissions per unit of energy (e.g., upward by abandoning nuclear energy or downward by replacing fossil fuels — especially coal — by alternative fuels such as nuclear, hydro, and solar).

Although it is intuitively attractive to create a sink for CO_2, we do not draw a line in Figure 1 to modify the ratio of concentrations to emissions, because, at this time, the scientific consensus seems to be that this is not possible on the necessary scale. Nor do we draw a line suggesting a modification of the ratio of temperature rise to CO_2 concentration, in that we know of no suggestion that this can be done. We do, however, draw a line suggesting a possible mitigation of the outcome given a temperature rise. If the outcome is defined generally (e.g., the effect on gross national product, GNP), humans can modify this factor by adaption, such as moving north as the temperature goes up.

2. ADDRESSING UNCERTAINTY

The uncertainties that characterize the global climate change problem are both factual and stochastic in character. Stochastic uncertainty arises at the end of the causal chain. Without detailed information concerning regional impacts, it appears to be almost random how changes in CO_2 would affect a particular community.

Further, it is clear that the uncertainties in the first few factors are different still. They are largely uncertainties concerning what society will do in response to the global climate change problem. Although these uncertainties can often be addressed by examining past experience, people have a habit of surprising analysts. Indeed, one of the purposes of analyzing the risks of global warming is to encourage people to behave in productive ways that are not predictable from past behavior.

In the United States, the approach that most regulatory bodies take toward uncertainty is very conservative and does not always take into account the best analytical methods that may be available. The U.S. Environmental Protection Agency's (EPA) approach to uncertainty propagation, for example, takes a conservative upper limit for each risk factor. The upper limits are then multiplied to arrive at a total risk level for regulation. If this approach were taken for global warming, there would be a very broad distribution.

3. DISCUSSION OF THE INDIVIDUAL FACTORS

3.1 Factor 1: Population

The study of population is a fairly mature science, and predictions of world population over the next few decades have smaller relative uncertainty than estimates of some of the other factors in Equation 1, such as sea-level rise per unit warming.

3.2 Factor 2: Energy per Person

Energy use *per capita* has been discussed by Hafele et al. (1981) and by Goldemberg et al. (1988). The most effective way to reduce energy consumption is through improving end-use efficiency (Goldemberg et al. 1988). Hafele et al. point out that as countries develop, the energy use per capita increases sharply, and even the energy use per unit of GNP rises. Such effects are often associated with the migration of population from the countryside to the towns. But later in the development process, energy use per unit of GNP falls. This decrease comes about for various reasons. After a country passes a threshold of energy use, and a technological foundation is established for a new technology, further increases of GNP are inherently less energy intensive. Naturally, this observation raises the question of how efforts to improve energy efficiency in developing countries can accelerate this historical process. Economists argue — with considerable historical justification — that the most effective way of encouraging energy efficiency is by increasing the price of energy. The use of taxes or charges to reduce energy use per capita has been discussed by Nordhaus and Yohe (1983), Nordhaus (1991), and Jorgenson and Wilcoxen (1991).

3.3 Factor 3: CO_2 Emissions per Unit of Energy

The amount of CO_2 emitted per unit of energy produced is not constant and can be changed by societal action. Combustion of natural gas produces half the amount of CO_2 produced by the combustion of coal. In addition, natural gas is easier to use, so that 52% thermodynamic efficiency has been achieved in a combined cycle turbine vs. 42% for the best coal burners.

3.4 Factor 4: Fraction of the Emitted CO_2 That Stays in the Atmosphere

Keeling et al. (1994) have measured CO_2 concentrations over many years. If one naively assumed that all of the CO_2 emitted from fossil fuel burning stays in the atmosphere, then the CO_2 concentrations would be increasing at twice the rate that has been observed. This leads to a discussion of the *carbon cycle* or carbon budget (Revelle and Suess 1957, Revelle and Munk 1977, Bacastow and Keeling 1981, Moore and Bolin 1986, Keeling et al. 1994, Siegenthaler and Joos 1992, Moore and Braswell 1994). Thus, a critical scientific uncertainty is the environmental sinks for CO_2. Terrestrial plants and soils are both potential sinks, and the oceans are another. Although the deep oceans are effectively unlimited in the amount of carbon they can absorb, the rate of absorption is limited by chemical partitioning rates, as well as the transfer rates between the surface water layers and the deep oceans. Sophisticated models use a number of time constants to describe this process and use a number of terms to describe the most important aspects. The crucial question remains, "If we cease CO_2 emissions now, how long will it be before the atmosphere returns to equilibrium concentration?"

3.5 Factor 5: Global Temperature Rise per Unit Increase in Atmospheric CO_2

The central issue of the scientific debate on global warming is the temperature rise resulting from an increase in the atmospheric concentration of greenhouse gases. These greenhouse gases include — in addition to carbon dioxide (CO_2) — methane (CH_4), nitrous oxide (N_2O), Freon, and, most importantly, water vapor. Factors that determine concentrations of these greenhouse gases (except water vapor) from known emissions are moderately well understood. Global temperature rise does not directly affect the atmospheric concentrations of these gases. But, as anyone can see by observing the earth's clouds, the concentration of water vapor varies rapidly in space and time, and this variation arises from feedback mechanisms that are less well understood.

If there were no change in the concentration of water vapor (such as would be the case if the earth were completely dry), the global mean surface temperature would increase by $\Delta T_d = 1.2°C$, given a static doubling of CO_2. But, concentration of water vapor is expected to increase with increasing temperature as water evaporates, and, since water vapor is the most important greenhouse gas, this could amplify warming. Water can introduce interactive feedbacks in numerous ways, such as water vapor, clouds (especially cirrus clouds), and snow-ice albedo. These feedbacks introduce considerable uncertainties into the estimates of the mean surface temperature rise, ΔT_s.

The value of ΔT_s is roughly related to ΔT_d by the formula $\Delta T_s = \Delta T_d / (1 - f)$, where f denotes the sum of all feedbacks. The water vapor feedback is relatively simple: warmer atmosphere contains more water vapor, which is itself a greenhouse gas. This process gives rise to a positive feedback, in that an increase in one greenhouse gas, CO_2, induces an increase in another greenhouse gas, namely, water vapor. The net effect of cloud feedback is determined by the amount of clouds, their altitude, and their water content. The values of ΔT_s from different models vary from $\Delta T_s = 1.9°C$ to $\Delta T_s = 5.2°C$ (Cubasch and Cess 1990). This wide-range change is brought about by including one large contribution from water vapor. Typical values for the equation relating ΔT_s to ΔT_d are $\Delta T_d = 1.2°C$ and $f = 0.7$, so that $\Delta T_s = 4°C$. It is important to recognize that some feedbacks of water vapor may not have been identified yet. The outputs of the models used by the International Panel on Climate Change (IPCC) (Houghton et al. 1990) and the National Academy of Sciences (NAS) (1991) were given as "bounds" on the global temperature rise ΔT_s. We assume that the parameter f has a normal (or a lognormal) distribution and adjust the parameters so that the "limits" of IPCC or NAS correspond to upper and lower 95th percentiles of the distribution of f. This procedure gives rise to a long tail, particularly to the upper portion of the distribution. It is this tail, and this simple approach, that leads to statements, such as those made by Dickinson (1986), that there is a 1% chance of ΔT being above 5°C for a static CO_2 doubling. We discuss the possible meaning of this claim below. We also note that general uncertainty analysis suggests that even the magnitude of this tail could be underestimated.

3.6 Factor 6: Sea Level Rise per Unit Rise in Global Temperature

The present best estimate for sea level rise in the IPCC "Business-as-Usual" scenario is 66 cm by the year 2100, and the estimate is based upon the work of Oerlemans (1989), who calculated sea level rise $\Delta h/\Delta T$ using a simple fit to the temperature rise predicated by a specific global emissions scenario (Houghton et al. 1990). This model assumes a simple extrapolation from past behavior for emission of CO_2. Letting $\Delta T = \alpha(t - 1850)^3$, where t is time, we have $\alpha = 27 \times 10^{-8}$ K yr^{-3}, and the uncertainty Δ is 35% of the mean for each variable.

Although we assume independence for each of the factors in Equation 1, the physical stresses that global climate change places on the environment have the potential to compound synergistically. It has been suggested that if surface temperatures increase, but temperatures in the troposphere do not, then the strength of storms would (contrary to model predictions) increase (Emanuel 1987). In addition, the reach and severity of storms may be increased by a rising sea level. As alluded to earlier, ecosystems are also at risk. For example, Bazzaz (1990) and Bazzaz and Fajer (1992) have studied the combined effects of rising concentrations of CO_2, rising temperatures, and increased ultraviolet radiation on plants and ecosystems. Their findings suggest that these factors can give some species distinct advantages over others. For instance, most weeds are more resilient to stresses than cultivated plants. One possible remedy would be to increase the production of pesticides. Such action, however, would likely lead to increased energy use and potential health risks.

These synergistic effects of temperature rise and CO_2 concentration increase do not invalidate the concept of calculations derived using the assumption of independence, but they do form an exception that must be evaluated separately. This situation is analogous to the deviations from independence in reactor safety calculations due to common mode failures. Rasmussen set up a procedure for analyzing nuclear reactor accidents by constructing an *event tree* that follows the progression of a nuclear power accident from the initiating event to the ultimate consequence (Atomic Energy Commission 1975). The probability of failure was calculated at each step, and it was assumed that each step was independent of the previous one. However, sometimes several events occur simultaneously or several pieces of equipment fail simultaneously. The overall usefulness of the procedure (now called Probabilistic Risk Assessment or PRA) is not invalidated by the existence of *common mode failures*; on the contrary, the procedure has proven to be an excellent means by which to uncover these common modes.

4. CONCEPT OF ACCEPTABLE RISK

Differences in lay vs. expert perceptions of risk can often be illuminated by comparisons. For example, Dickinson (1986), as noted earlier, used a lognormal fit of various estimates of global warming to estimate that there is a 5% in a lifetime chance that an increase in greenhouse concentrations would, by the year 2100, lead to a temperature rise of 10°C. If this 5% chance were obtained during that time period, such an event would almost surely come as a "surprise" to many people, the reason being that it would give rise to adverse consequences that were largely unanticipated by them. It is interesting to note, however, that public opinion polls suggest that many people are unconcerned about a 5% (calculated by an "expert") chance of a climate-related catastrophe within their lifetime. We are pressed to attribute this view to ignorance, because climate change has now entered the political consciousness with the election of Al Gore to the vice presidency of the United States; he is the author of a popular book on this subject (Gore 1992).

These observations notwithstanding, it is worth noting that the public *is* concerned about a 1% chance of a nuclear accident (also calculated by an expert) in the same time period. We also note that an airliner with a calculated chance of failure far lower than 5% in its 30-year life would not be allowed to fly in commercial service. Does that mean that the public trust the experts on climate change more than the experts on nuclear power? Most students of risk assessment would assign the difference to an *outrage factor* associated with involuntary, insidious, or unfair practices of the nuclear and airline industries. That could be either considered as a reason for less trust or as an important qualitative difference that cannot be altered.

There are no simple answers for the particular reasons why people differ in their perceptions of, and reactions to, risk. Nevertheless, if the nature of the uncertainties that underlie problems such as global climate change is not clearly articulated and understood, then confusion may arise even among the best experts. For example, Clark (1989), referring to Dickinson's analysis, notes that the chance that the world of 2100 will have witnessed a single nuclear power catastrophe is anywhere from 10 to 100 times less than the chance that everyone in the world will be living in the Mesozoic greenhouse. He concludes that "this assessment jars common sense, which is exactly why we need to reexamine the assessment methods and philosophies that produced it as an urgent task of understanding global environmental change."

For chemical carcinogens, it is common to discuss a risk to an individual of 10^{-6} in a lifetime of 70 years. This is a far smaller number than the probabilities of a huge temperature rise and catastrophic effect in the next 70 years. Following Clark (1989), we ask whether it means that the U.S. EPA is too conservative in taking this small number for chemical carcinogens, too optimistic about global warming, or whether the comparison is altogether invalid? As discussed earlier, in order to avoid the confusion shown in Clark (1989), we have to make a clear distinction between stochastic uncertainties and uncertainties of fact.

The distribution may be truncated by bringing in other information not considered by the GCMs. In particular, we can examine historical global climate trends to determine whether they are consistent with the extremes of such a distribution. This procedure is used in Figure 2 to compare the observed global mean temperature changes during the last century with predicted values (Wigley and Barnett 1990). The global temperature rise attributable to CO_2 doubling can be estimated from a visual inspection of such curves. Figure 2 clearly illustrates that the main rise occurred before 1940, and we see the rise in temperature since 1980, which brought the subject to public attention, although it is smaller than the model predictions.

The models before 1992 did not include the effects of fine particulates, including sulfates, that have spread over the Northern Hemisphere from fossil fuel burning. These particulates have a cooling effect (Hansen and Lacis 1990), as does the ozone depletion in the upper atmosphere that has been observed in recent years. That there might be cooling due to aerosols has been known for a long time, but has only recently been included. It is probably masking the effect of the rise in the CO_2 concentrations (Charleson et al. 1992, Taylor and Penrose 1994). There is pressure to reduce aerosols from those who believe that they contribute to adverse effects on public health. To the extent that these efforts are successful, global warming may become worse and therefore be identifiable.

Michaels et al. (in press) have carefully compared the predictions of the models with recent temperature records. They note that if the sulfate explanation is correct, then the models should work best in regions where sulfate concentrations are lowest, namely, in the Southern Hemisphere and high latitudes. However, this expectation does not seem to hold. Michaels et al. (in press) also find no correspondence between observed and predicted temperature trends during the times of longest day (in the summer) and longest night (in the winter) when the models predict the largest effects.

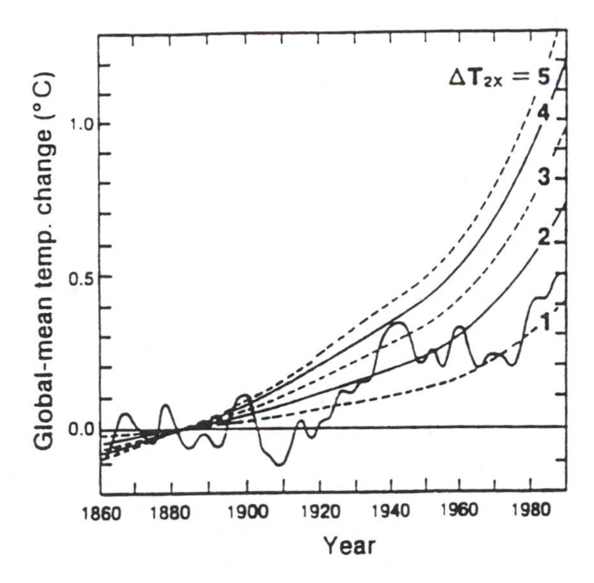

Figure 2 Observed global mean temperature changes (smoothed to show the decadal and longer time scales trends more clearly) compared with predicted values for several values of climate sensitivity to doubling CO_2 concentration, ΔT_{2X}, (shown on the curves). (Modified from Wigley, T.M.L. and Barnett, T.P. 1990 "Detection of the greenhouse effect in the observations," *Nature,* 369:239–255.)

This result shows that the models *in their present form* do not seem to describe what is occurring in as much detail as we would like. The observations, however, do seem consistent with the idea of a negative feedback effect canceling the warming trend.

It is crucial to emphasize, however, the fact that a comparison of the historical record to the models does not unequivocally show that an effect of increased greenhouse gases *does not* prove the inverse, i.e., that no effect is occurring. This point was emphasized by Duffy (1994). It seems that there are at least three effects of approximately equal magnitude: CO_2-forced warming (described by the models), aerosol-forced cooling (only recently described by the models), and natural variability with causes of source unknown. In evaluating public policy options, we are therefore left with the general arguments referred to in the Introduction, namely, that society is making large changes in an important climate parameter — CO_2 concentrations — on a short global time scale. Given this state of affairs, should society wait until an adverse effect has been definitively proven, or, alternatively, should society try to reduce the changes in CO_2 concentrations until our models show definitively that nothing adverse is occurring? Nevertheless, it does seem that a temperature rise, ΔT_s for static CO_2 doubling larger than 6°C, although possible, is unlikely (Wigley and Barnett 1990, Wigley and Roper 1991). Kondratyev and Galindo (1994) emphasize that the argument that global warming may be more remote than the IPCC assumes is best addressed by a more careful look at climate change and ecodynamics, and they summarized the recent progress to this end.

We have discussed the issues presented in this section with many scientists more expert on climate change than ourselves. As the analysis suggests, the distribution of opinion about the predicted temperature rise is wider than that suggested by the IPCC or NAS; some say that $\Delta T/CO_{2_{atmos}}$ (Factor 5) is close to zero; another who disbelieves the models, because of their inability to describe the historical record, nonetheless puts it at the lower bound of the distributions; still others might put it at three times the IPCC value. This observed variability is formally being addressed by Morgan and Keith (1993).

5. SEQUENTIAL DECISION-MAKING STRATEGIES

As discussed previously, the simplified model of Section 2 is a static model. The risk is estimated, with its uncertainty, at one period of time, and the decision on what, if anything, to do is made essentially simultaneously with the assessment. Of course, real life is not that simple. In an important subject such as global climate change, decisions on measures to avert or mitigate the effects of climate change will be made frequently over time. The assessment must then be an *iterative* process.

Any reasonable analysis of the potential risks of global climate change must address the general concern shared by many scientists that once the effects of global warming appear above some agreed upon "noise" level, the CO_2 atmospheric concentration will be so far advanced that the effects of warming will take centuries to reverse, if at all possible. It is here that the time constant for coupling to the deep oceans is critical. If this time constant is large, then it seems reasonable to suppose that action to prevent increased CO_2 concentration should be taken *before* the effect of this increase is conclusively verified.

The crucial question that must be continuously addressed therefore becomes "What actions, if any, should be taken about global warming when the effects have not yet demonstrated themselves unequivocally?"

The first decisions might be to take those actions where the cost is not high. In this regard, a much publicized article addressed the minimal agreement of three distinguished scientists with widely different views (Singer et al. 1991).

1. Conserve energy by discouraging wasteful use globally.
2. Improve efficiency in energy use.
3. Use nonfossil fuel energy sources wherever this makes economic sense.

This minimal set of actions seems quite reasonable, yet none of these actions are currently being undertaken in the United States. It seems that, in spite of much political rhetoric, the United States and other countries are taking very little action.

As noted earlier, actions 1 and 2 are most easily achieved by an increase in the cost of the relevant fuels; yet, in 1993, an increase was rejected by the U.S. Congress. In the past, Revelle et al. (1991) have advocated expansion of nuclear power. Needless to say, the present U.S. administration, while emphasizing global warming, is not following this particular course of action. Naturally, one cannot reasonably expect that the administration will accept all of the recommendations that scientists put before them. In particular, any emphasis on nuclear power might be considered by some members of the public an overreaction to an uncertain threat, and introducing what, in their view, is a worse threat. Many people prefer to wait until society can try a combination of biomass, wind, and photovoltaic electricity. In the framework of risk analysis, this preference can be construed as a willingness to accept a delay in reducing global warming and a willingness to accept the uncertainty of whether the hoped-for result (economically attractive nonfossil and nonnuclear energy) can be achieved.

Consensus on a more draconian set of actions might be achieved if the uncertainties in the assessment of outcomes were reduced and, moreover, if it could be demonstrated that CO_2 — forced global warming is actually occurring. However, a decade-long time scale is anticipated for narrowing the uncertainties in predictions of the rate of climatic change through improved coupled atmosphere-ocean models (McBean and McCarthy 1990).

These possible research programs are intended to provide information for making better decisions in the future. In general, there are two ways to approach the problem of valuing information (Hammit 1994). First, new information can refine the fundamentally correct, but imprecise, information characterized by a prior distribution. Second, the new information can reveal that a prior distribution reflects overconfidence or fundamental misunderstanding. The conventionally defined value of information measures the first type of information value, but, because it is fundamentally dependent on the prior distribution, it cannot capture the second type. Stated another way, if the gain in information is measured by the expected decrease in the variance of a parameter, then an overly narrow prior distribution produces an underestimate of the gain, since the gain cannot be larger than the prior variance. For this reason, the probability of surprise should be explicitly taken into account when discussing the value of future research. The probability of surprise can be quantified by a statistical analysis of the frequency and magnitude of past errors (Shlyakhter 1994).

6. INTERREGIONAL AND INTERGENERATIONAL EQUITY

One of the most obvious public policy concerns about global climate change is that of interregional equity. Each person who emits, or allows someone else to emit, one of the greenhouse gases gains (or perceives that he gains) some benefit. However, the increase in global warming happens all over the world, and the risk initially caused by the gaseous emission may be incurred by completely different people. Clearly, intergenerational equity is important also in that future global warming depends, in part, on past CO_2 emissions.

Society has developed a variety of tools for coping with interregional inequities. The most obvious is a transfer of payment, by taxation or otherwise, from those gaining the benefit to those incurring the risk. It is easier, and perhaps fairer, to make a risk-related decision when the risks are borne by the same person or group to whom the benefits accrue. If the risk of an action exceeds the benefit perceived by that person, then the action will not proceed. However, if the person who bears the risk is different from the person to whom the benefit accrues, and if the risk bearer is willing to value risk lower than the benefactor values the benefit, then it may be possible to achieve a net excess of benefit over risk for each party; this might be achieved by some charge of payment, whereby the party who benefits compensates the party who bears the risk. Although by such a monetary transfer the risk/benefit decision for each party becomes favorable, there is the complication of deciding upon the exact payment; one party may benefit (overall) more than the other, and negotiation(s) may be time consuming. In fact, the time and effort needed to make the negotiated transfers themselves become an additional cost.

Even when there are several groups, this procedure might be generalized to ensure that the risk-benefit balance is positive for each affected group of importance. A similar concept applied to technological risk has been advanced by Fischhoff (1994). Clearly, the manner and degree to which *existing* institutions are able to effectively apply this conception of acceptable risk is a matter of considerable debate. Nevertheless, it is useful to consider how simple conceptions of interregional equity can be applied to the problem of global climate change.

The Earth Summit at Rio de Janeiro in June 1992 illustrated the importance of interregional equity. In particular, Third World countries with little industrial development argued that it was not fair for them to reduce emissions of greenhouse gases, or, for that matter, to encourage their absorption by retention of rain forests, unless appreciable transfer payments came from those countries who are producing the major emissions.

This important fact suggests a possible change of policy. Instead of spending money in the United States to reduce emissions of greenhouse gases, why not consider paying another government to reduce them? While initial applications of this idea are for financial contributions of the United States government to an overseas government, this might be extended to payments for specific tasks, such as helping otherwise unprofitable hydroelectric, nuclear, or solar power plants. Such fiscal measures as tax incentives, based upon the number of CO_2 molecules reduced, might bring the private sector into a productive role. Interesting studies are currently underway on whether paying the Chinese to consider nonfossil fuels for their energy expansion will be a cost-effective way for the United States to spend its CO_2 reduction dollars.

This line of reasoning suggests that economic adjustments can be considered for intergenerational equity, as well as for interregional equity. A person (or society) can, and perhaps should, make appropriate investments to pay for the cost of possible future consequences. Raiffa et al. (1977) argue that future "lives" in the risk-benefit equation should be discounted at the same rate as money. Their argument is that money can be invested now, at the monetary discount rate, and when the hazard arrives, the money has been appropriately increased by the interest accumulated. If it is proper to discuss a relationship between the amount one is willing to pay to reduce a hazard and the benefits one gets from the reduction, then it is also appropriate to discount that amount with the usual monetary discount rate. The money could be set aside for "balancing" the risk over future generations, as well as for finding a way to avoid the risk. Money might be invested in avoiding some other comparable risk, such as cancer, which in the future would otherwise add to the risk of global warming.

7. CONCLUSION

By contemplating the process of integrated risk analysis of global climate change, we have shown how the calculation of risk proceeds by considering a number of nearly independent steps. Serious problems remain, however. Most of the GCMs that have been designed to describe the effect of increasing CO_2 concentrations fail to describe important regional details; few seem to get the changes correct over a 10-year time span, and all may be completely wrong over time scales spanning many thousand years.

An important feature of risk management related to climate change seems to be the feeling among both scientists and the lay public that we should take an insurance policy, which simply amounts to considering the upper limit of a probability distribution of impacts (Schelling 1991). This upper limit is, in many respects, ill defined, but can be quantified using a *de minimis* concept of risk. In this context, however, it must be defined for nonstochastic uncertainty. This conception of risk distinguishes between scenarios that are believed possible and those that are rejected as improbable. ***Empirical evidence suggests that overconfidence in predictions of future development results in long tails of the distribution and, therefore, in unexpectedly high probabilities of surprise. These tails can, in principle, be truncated by using additional information, such as model-independent restrictions on climate change from paleoclimatic data or from volcanic eruptions. To this end, efforts must now focus on learning how to translate the large body of contradictory information about past and present climate into defensible upper limits on the probability of surprise. We believe that integrated risk analysis of global warming and its impacts should become a working tool for decision makers and risk managers in illuminating the important scientific issues and uncertainties that frame the global climate change debate.***

REFERENCES

Atomic Energy Commission (1975) "An Assessment of Accident Risks in US Commercial Nuclear Power Plants," U.S. Atomic Energy Commission, WASH-1400.

Bacastow, R.B. and Keeling, C.D. (1981) "Atmospheric carbon dioxide concentration and the observed airborne fraction," in *Carbon Cycle Modelling, SCOPE 16,* Bolin, B. (Ed.), International Council of Scientific Unions, John Wiley & Sons, New York, pp. 103–112.

Bazzaz, F.A. (1990) "Response of natural ecosystems to the rising global CO_2 levels," *Annual Review of Ecology and Systematics*, 21:167–196.

Bazzaz, F.A. and Fajer, E. (1992) "Plant life in CO_2-rich world," *Scientific American*, January, pp. 66–72.

Charleson, R.J. et al. (1992) "Climate forcing by anthropogenic aerosols," *Science*, 255:1623.

Clark, W. (1989) "Towards useful assessments of global environmental risks," in *Understanding Global Environmental Change: The Contributions of Risk Analysis and Management,* A Report on the International Workshop, October 11–13, Kasperson, R., Dow, K., Golding, D. and Kasperson, J. (Eds.), Earth Transformed Program, Clark University, Worcester, Massachusetts.

Cubasch, U. and Cess, R. D. (1990) "Processes and modelling," in *Climate Change: The Intergovernmental Panel on Climate Change Scientific Assessment,* Report Prepared for the IPCC by Working Group I, Houghton, J. T., Genkins, G. J., and Ephraums, J. J. (Eds.), Cambridge University Press, NY, pp. 75–91.

Dickinson, R.E. (1986) "Impacts of human activities on climate — a framework," in *Sustainable Development of the Biosphere*, Clark, W.C. and Munn, R.E. (Eds.), International Institute for Applied Systems Analysis, Laxenburg, Austria, Cambridge University Press, NY, pp. 252–291.

Emanuel, K.A. (1987) "The dependence of hurricane intensity on climate," *Nature*, 326:483–485.

Duffy, P.B. (1994) "Comments upon 'Global Warming: a reduced threat'," *Bull. Amer. Meteorological Soc.*

Fischhoff, B. (1994) "Acceptable risk: a conceptual framework," *Risk: Health, Safety & Environment*, Winter 1994, pp. 1–28.

Goldemberg, J. et al. (1988) *Energy for a Sustainable World*, John Wiley & Sons, New York.

Gore, A. (1992) *Earth in the Balance: Healing the Global Environment,* Houghton-Mifflin, Boston, MA.

Hafele, W. et al. (1981) *Energy in a Finite World: Paths to a Sustainable Future,* International Institute for Applied Systems Analysis, Ballinger Press, Cambridge, MA.

Hammit, J. (1994) "Can more information increase uncertainty?," *Chance*, 8(3):15–18.

Hansen, J.D. et al. (1981) "Climate impacts of increasing carbon dioxide," *Science*, 213:957–966.

Hansen, J.D. and Lacis. (1990) "Sun and Dust versus Greenhouse gases: an assessment of their Relative Roles in Global Climate Change," *Nature,* 346:713–718.

Houghton, J.T., Jenkins, G.J., and Ephraums, J.J. (Eds.) (1990) *Climate Change,* Working Group I, Intergovernmental Panel on Climate Change Scientific Assessment, Cambridge University Press, New York.

Jorgenson, D.W. and Wilcoxen, P.J. (1991) "Reducing U.S. Carbon Dioxide Emissions: the Cost of Different Goals," John F. Kennedy School of Government, Center for Science and International Affairs, Discussion paper number 91–9.

Kates, R.W., Hohenemser, C., and Kasperson, J.X. (eds.) (1985) *Perilous Progress: Managing the Hazards of Technology*, Westview Press, Inc., Boulder, Colorado.

Keeling, C.D., Bacastow, R.B., Carter, A.F., Piper, S.C., Whorf, T.P., Heimann, M., Kondratyev, K.Y., and Galindo, I. (1994) "Global climate change in the context of global ecodynamics," *Geophysica International,* 33(3):487–496.

McBean, G. and McCarthy, J. (1990) "Narrowing the uncertainties: a scientific action plan for improved predictions of global climate change," in *IPCC,* 1990, Chap. 11.

Michaels, P., Knappenberger, P.C., and Day, D.C. "Sulfate aerosol and global warming," *Theoretical and Applied Climatology,* in press.

Michaels, P., Knappenberger, P.C., and Day, D.C. "Predicted and observed long day and night temperature trends," *Atmospheric Research,* in press.

Mook, W.G. and Roeloffzen, H. (1989) "A three-dimensional model of atmospheric CO_2 transport based on observed winds: 1. Analysis of observational data," in *Aspects of Climate Variability in the Pacific and the Western Americas*, Geophysical Monograph 55, American Geophysical Union, Washington, D.C.

Moore, B. and Bolin, B. (1986) "The oceans, carbon dioxide and global climate change," *Oceans*, 29(4):9–15.

Moore, B. and Braswell, B.H. (1994) "The lifetime of excess atmospheric carbon dioxide," Global Biogeochemical Cycles, 8(1):23–38.

Morgan, M.G. and Keith, D.W. (1995) "Subjective judgments by climate experts," *Envir. Sci. Tech.* (Oct. 1995) 29:468–477.

National Academy of Sciences (1991) *Policy Implications of Greenhouse Warming,* Synthesis Panel, Committee on Science, Engineering, and Public Policy, National Academy Press, Washington, D.C.

Nordhaus, W.D. (1991) "The cost of slowing climate change: a survey," *Energy Journal,* 12(1):37–64.

Nordhaus, W.D. and Yohe, G.W. (1983) "Future paths of energy and carbon dioxide emissions," in *Changing Climate*, Report of the Carbon Dioxide Assessment Committee, National Academy of Sciences, Washington, D.C.

Oerlemans, J. (1989) "A projection of future sea level," *Climatic Change*, 15:151.

Raiffa, H., Schwartz, W.B., and Weinstein, M.C. (1977) "Evaluating health effects of societal decisions and programs," in *Decision Making and Environmental Protection Agency,* Selected Working Papers, National Academy of Sciences, Washington D.C., vol. 2B.

Revelle, R. and Suess, H.E. (1957) "Carbon dioxide exchange between atmosphere and ocean and the question of an increase of atmospheric CO_2 during past decades," *Tellus*, 9:18–27.

Revelle, R. and Munk, W. (1977) "The carbon dioxide cycle and the biosphere," in *Energy and Climate,* Studies in Geophysics, National Academy of Sciences, Washington, D.C., pp. 140–158.

Schelling, T. (1991) Talk presented at the John F. Kennedy School of Government, Harvard University, Cambridge, Massachusetts.

Shlyakhter, A.I., Valverde, L.J., and Wilson, R. (1995) "Integrated risk analysis of global climate change," *Chemosphere*, 30:1585–1618.

Shlyakhter, A.I. (1994) "An improved framework for uncertainty analysis: accounting for unsuspected errors," *Risk Analysis*, 14:441–447.

Siegenthaler, U. and Joos, F. (1992) "Use of a simple model for studying oceanic tracer distributions and the global carbon cycle," *Tellus,* 44B:186–207.

Singer, F., Revelle, R., and Starr, C. (1991) "What to do about greenhouse warming," *Cosmos*, pp. 28–33.

Taylor, K.E. and Penrose, J.E. (1994) "Response of the climate system to atmospheric aerosols and greenhouse gases," *Nature,* 369:734–738.

Wigley, T.M.L. and Barnett, T.P. (1990) "Detection of the greenhouse effect in the observations," *Nature,* 369:239–255.

Wigley, T.M.L. and Roper, S.C.B. (1991) *Climate Change Science: Impacts and Policy,* Jager, J. and Fergusen, H.L. (Eds.), Cambridge University Press, New York.

QUESTIONS

1. What is the place of risk assessment and risk management in the risk analysis of climate change?
2. List key uncertainties in climate change assessment.

CHAPTER II.6

Computer Software Programs, Databases, and the Use of the Internet, World Wide Web, and Other Online Systems

P. J. (Bert) Hakkinen

SUMMARY

The main purpose of this chapter is to provide readers with guidance on how to obtain information about potentially useful risk assessment and risk management computer software programs and databases. The software programs include ones available on floppy disks and/or compact discs (also called CD-ROMs) and also ones available from downloading via online connection to other computers and computer networks. The databases include those accessible either as software or via an online connection to another computer or computer networks. Also included is information about previous publications discussing toxicology and risk assessment software and databases and a discussion of the key scientific journals, magazines, and computer databases to consult for announcements, listings, and reviews of software and databases.

Further, this chapter discusses the current and likely future uses by risk assessors and risk managers of CD-ROMs, global electronic communications networks including the Internet and its World Wide Web (WWW), "personal assistant" electronic books, "information appliance" and "intelligent assistant" devices, and "intelligent agent" software. The chapter is organized as follows: Warnings to Readers; Previous Publications Discussing Software and Databases; Recommended Best Approach to Identify Useful Software and Databases; Scientific Journals to Consult for Announcements, Listings, and Reviews of Software and Databases; Magazines and Catalogs to Consult for Announcements, Listings, and Reviews of Software and Databases;

1-56670-130-9/97/$0.00+$.50

Databases to Consult for Announcements, Listings, and Reviews of Software and Databases; Databases to Consult for Toxicology and Risk Assessment Data and Related Information; Professional Society Activities Relating to Software and Databases; Current and Likely Future Uses of Global Electronic Communications Networks, Including the Internet and Its World Wide Web; Current and Likely Future Uses of CD-ROMs; The Emergence of "Information Appliances," "Personal Assistants" and "Intelligent Assistants"; The Emergence of "Intelligent Agents" and "Java"; More on What to Expect in the Future; Final Comments; References; and Questions.

Key Words: browser, CD-ROM, e-mail, home page, information appliance, intelligent agent, intelligent assistant, Internet, Java, online, personal assistant, search engine, World Wide Web

1. WARNINGS TO READERS

The author is hesitant to recommend specific software programs or databases to anyone who might be reading this chapter years after it is published. Software "ages," and what is state of the art now may not be months or years after this is written. For example, some software programs periodically get revised (and hopefully improved) as new "release versions," e.g., going from Version 1.0 to Version 1.1 or Version 2.0. Also, some software and databases may no longer be available months or years after this chapter is written or may be renamed. Also, some current software will not function as new computer operating systems (e.g., Windows 95) replace operating systems the software was originally developed to function on. Further, even some of the journals, magazines, and databases noted may not be the current best sources of information months or years after this chapter is written.

Another general warning is that some software programs contain outdated exposure assessment and risk assessment values, e.g., "risk potency values" for carcinogens or no-observable-effect levels (NOEL) for other chemicals, that have been superseded by values from more recent studies and/or from additional review and understanding of the original results. Also, although perhaps not noted, some software programs may contain default exposure parameter values and/or distributions (e.g., human body weight distributions) for a specific country such as the United States, which may not be ideal or very appropriate to use in an exposure or risk assessment for another country or area or for a particular population group being assessed. Thus, it is very much "user beware" or "user be wise" when using a software program.

2. PREVIOUS PUBLICATIONS DISCUSSING SOFTWARE AND DATABASES

There have been numerous published reviews and a book dealing with computerized information resources (e.g., software and online databases) for human and environmental toxicology, industrial (occupational) health, and risk assessment.

Readers should note that many of these publications are at least several years old, and all but the most recent do not even mention the development and use of CD–ROMs as sources of software programs and information. The publications are listed in Table 1. Given the rapid changes in software and online databases, readers interested in assembling a useful collection of publications may want to focus their attention on the more recent publications by Wexler (1990a,b), Marnicio et al. (1991), Johnson (1992), Moskowitz et al., (1992), Alston (1994), Anonymous (1994), Kumar and Sahore (1994), Abeytunga (1995), Gardner (1995), and Johnson (1995). Also, as noted in Table 1, expected to be published in 1997 is the third edition of the *Information Resources in Toxicology* (Wexler, chief editor), which will include information on software and databases of interest to toxicologists and others and may also be available on a CD-ROM.

3. RECOMMENDED BEST APPROACH TO IDENTIFY USEFUL SOFTWARE AND DATABASES

The recommended best approach to identify useful software and databases for risk assessment and risk management work is for readers to consult at least some of the publications (e.g., those listed in Table 1), scientific journals, databases, and other potential sources of information described in this chapter. For example, reading the more recent reviews noted in Table 1 and searching for recent risk assessment publications in the journals noted will help identify key current software used for various applications. (Other journals readers have access to may also help identify useful software and databases, and the author welcomes suggestions about other useful journals.) More searching may identify published reviews of a particular software program or database, perhaps comparing its capabilities to other available software programs and databases. Also, contacting the authors of these publications (e.g., by letter or fax message or by e-mail) may bring valuable advice in return.

Further, the author has seen excellent use of the "RISKANAL" Internet mailing list to ask questions and receive answers about software programs and databases and for other exchanges of information. This mailing list was created by the "Columbia-Cascades" chapter of the Society for Risk Analysis and the Pacific Northwest Laboratory and further information about it is provided in Section 9.2.

4. SCIENTIFIC JOURNALS TO CONSULT FOR ANNOUNCEMENTS, LISTINGS, AND REVIEWS OF SOFTWARE AND DATABASES

There are relatively few scientific journals risk assessors and/or risk managers might typically read that contain sections of pages regularly dedicated to announcements, listings, and reviews of software and databases. They include *Risk Analysis, Toxicology, Food and Chemical Toxicology, Journal of the American Medical Association, and Toxicology Modeling.*

Table 1 Previous Publications Listing or Discussing Software and Databases for Human and Environmental Toxicology, Industrial Health, and Risk Assessment

- *Information Resources in Toxicology* (Wexler 1982)
- Toxicology information systems: A historical perspective (Kissman and Wexler 1985)
- Environmental software review (Young 1985)
- Computerized information resources in toxicology and industrial health — A review (Halton 1986)
- Water quality modeling software available from U.S. EPA (Barnwell et al. 1987)
- Environmental software review (Cheremisinoff 1987)
- *Information Resources in Toxicology,* Second Edition (Wexler 1988)
- Strategy for computer-aided searches for information about chemicals (Benz et al. 1989)
- Computer databases for carcinogenicity and risk assessment (Sidhu 1989)
- Environmental software at the U.S. Environmental Protection Agency's Center for Exposure Assessment Modeling (Ambrose and Barnwell 1990)
- Toxicological Information Series, I. Toxicological Information (Cosmides 1990)
- The framework of toxicology information (Wexler 1990a)
- Toxicological information series, II. A survey of toxicology information (Wexler 1990b)
- Risk analysis software and databases: Review of Riskware '90 conference and exhibition (Marnicio et al. 1991)
- 1992 environmental software survey (Johnson 1992)
- Computer models used to support cleanup decision-making at hazardous and radioactive waste sites (Moskowitz et al. 1992)
- Electronic highways: A road map for environmental health (Alston 1994)
- Listings of suppliers of numerous categories of human and environmental software and databases = *ES&T Environmental Buyer's Guide Edition 1995* (Anonymous 1994)
- Public domain software for exposure assessment modeling (Kumar and Sahore 1994) (*Note:* This publication also notes previous reviews from this group, e.g., on use of electronic bulletin boards, and on software for study of environmental effects of hazardous sites, air modeling, and for regulatory compliance.)
- An overview of the electronic health and safety services of the Canadian Centre for Occupational Health and Safety (Abeytunga 1995). [*Note:* This publication discusses numerous diskette, CD-ROM, and online databases, including those developed by major organizations in several countries (Canada, France, U.S., and Switzerland), and also mentions two interactive, multimedia education and training products that are available on chemical and workplace safety.]
- Toxicology abstracts — An electronic journal on the Internet (Gardner 1995)
- U.S. Agency for Toxic Substances and Disease Registry's information databases to support human health risk assessment of hazardous substances (Johnson 1995)
- *Note:* Due to be published in 1997 will be the third edition of the *Information Resources in Toxicology* book (chief editor: Wexler, co-editors: Hakkinen, Kennedy, and Stoss. IOS Press). The contents of this book may also be available as a CD-ROM.

[a] The publications are listed in order of oldest to most recent year of publication.

4.1 *Risk Analysis*

This journal contains risk assessments performed using software and databases, regularly published reviews of software, and occasionally contains listings and summaries of software and databases.

4.2 *Toxicology*

This journal contains a section created in 1994 entitled "Toxicology Information and Resources." This section is devoted to advances in the handling and management of toxicological and environmental data, with an emphasis on computerized methods. For example, hypothetical titles in the publisher's announcement included "CD-ROM software for risk exposure assessment" and "Laser disks as self-instructional tools in undergraduate toxicology courses." Two examples of articles in this journal are "An overview of the electronic health and safety services of the Canadian Centre for Occupational Health and Safety" (Abeytunga 1995) and "Toxicology abstracts — an electronic journal on the Internet" (Gardner 1995). Also to be included are reviews of software and "highlight" stories about a particular software program, database, or organization.

4.3 *Food and Chemical Toxicology*

This journal has contained a "Software Survey Section" to "encourage the open exchange of information on software programs unique to our professional field." However, this section has been at least temporarily discontinued, effective with the 1994 issues (Volume 32).

4.4 *Journal of the American Medical Association*

This journal contains regularly published reviews of software useful to physicians and others and has also contained articles on how databases like the U.S. National Library of Medicine's "Medline" database can best be used, e.g., when seeking quick and helpful information in the diagnosis of various illnesses and diseases. An example of a "how to" article in this journal is one on "Users' guides to the medical literature. VI. How to use an overview" (Oxman et al. 1994).

4.5 *Toxicology Modeling*

The Carfax Publishing Company planned to start publication of this journal in 1995–1996. An announcement for this journal states that it will be "a unique journal bridging toxicology and computer science," covering computer usage in toxicology and ecotoxicology for hazard and risk assessment, retrieval of literature information, evaluation and quality assurance of databases, etc., while also including reviews of books, software, and databases.

5. MAGAZINES AND CATALOGS TO CONSULT FOR ANNOUNCEMENTS, LISTINGS, AND REVIEWS OF SOFTWARE AND DATABASES

Various magazines are devoted to computer topics, including ones focusing on online databases and on the use of the Internet. These include *Database: The Magazine of Electronic Database Reviews*, *Online*, *Online Access*, and *Online Review* (effective 1993 with Volume 17, *Online Review* became *Online & CD ROM Review*). An example of the type of useful information available in magazines is an article on "An Internet navigation tool for the technical and scientific researcher" by Skinder and Gresehover (1995). Also of possible interest to readers is the *Software for Science* catalog published several times a year (see References).

6. DATABASES TO CONSULT FOR ANNOUNCEMENTS, LISTINGS, AND REVIEWS OF SOFTWARE AND DATABASES

Perhaps the best "all-in-one place" source of information about software is DIALOG's several hundred databases, accessible via online connection to DIALOG, or via DIALOG's "ONDISC®"* CD-ROMs. For information about DIALOG's databases, readers can contact DIALOG Customer Services (phone numbers are 1-800-3-DIALOG or 415-858-3810.** WWW home page: *http://www.dialog.com*).

DIALOG contains five online databases of potential interest to people searching for a particular type of risk assessment and risk management software.

6.1 Buyer's Guide to Micro Software

This database is File #237 in DIALOG and is a directory of business and professional software available in the United States. It contains over 6000 records, with each record including technical specifications, a product description, and, when available, reviews of the product. The publisher indicated in 1995 that this database is being converted to a directory of CD-ROM titles.

6.2 Business Software Database (also called Softbase: Reviews, Companies, and Products)

This database is File #256 in DIALOG and is a directory of business, professional, technical, and system software available. It contains over 49,000 records, including basic product information, and either more in-depth product information and an abstract or a bibliography and abstracts of reviews.

6.3 Computer Database

This database is File #275 in DIALOG and provides a wide range of information related to computers, including software information. It contains over 700,000 records, indexed from over 110 English-language journals and magazines on computers. All articles from these publications, except book reviews, notes, editorials, and event announcements, are indexed and abstracted. Each record includes technical specifications, a product description, and, when available, reviews of the product.

* Registered trademark of DIALOG Information Services, Inc., Palo Alto, California.

** All phone numbers provided in this chapter are for U.S. direct access. Readers outside the United States may want to try these numbers using international access codes to the U.S. phone system or may want to try to obtain numbers specific to their country or region of the world from advertisements or from their information services specialist.

6.4 Microcomputer Software Guide

This database is File #278 in DIALOG and claims to contain "information on virtually every microcomputer software program and hardware system available and produced in the U.S." It contains over 13,000 records indexed from information from 4000 software publishers, as well as from books, periodicals, and press releases. Included is a brief description of the software, along with technical specifications and ordering information.

6.5 Datapro Software Directory

This database is File #751 in DIALOG and provides detailed descriptions of business, professional, and system software. It covers over 27,000 software packages from over 7000 vendors. The information is obtained from journals and magazines, press releases and other information from vendors, and elsewhere.

How well do the above DIALOG databases do in identifying risk assessment and risk management software and databases judged to be of possible interest to readers of this chapter? In trial searches by this chapter's author using "software," "databases," and "risk" together as key words, none of the above databases was found to have information about a key recent review of risk-related software (Marnicio et al. 1991; listed as a "Software Review/Listings" in *Risk Analysis*). Also, none of the above databases was found to have information on the U.S. Environmental Protection Agency's (EPA) "Exposure Models Library," described in a "Software Listing" in a 1993 issue of *Risk Analysis*, or about a commercially available risk assessment software program ("SmartRisk") randomly chosen from among those exhibited at the 1993 Annual Meeting of the Society for Risk Analysis. However, all but the Microcomputer Software Guide (File #278) were found to have some information about key Monte Carlo software ("@Risk" and "Crystal Ball") used a great deal in exposure and risk assessments.

Based on the results of these sample searches, none of the DIALOG databases cited here are judged to be ideal for trying to identify software and databases for risk assessment and risk management purposes; however, the author of this chapter has shared these results with the organizations in charge of the databases in an effort to allow these organizations to make these databases more useful in the future for risk assessors and risk managers.

7. DATABASES TO CONSULT FOR TOXICOLOGY AND RISK ASSESSMENT DATA AND RELATED INFORMATION

7.1 The U.S. National Library of Medicine's Toxline Database

DIALOG also provides online and CD-ROM access to the U.S. National Library of Medicine's Toxline®* database. At first glance, an apparent advantage to searching Toxline is that several of the recommended scientific journals to consult for

* Registered trademark of the U.S. National Library of Medicine, Bethesda, Maryland.

announcements, listings, and reviews of software and databases noted previously (*Risk Analysis, Toxicology, Food and Chemical Toxicology,* and the *Journal of the American Medical Association)* are indexed and searchable in Toxline. However, Toxline does not appear to make an attempt to index announcements, listings, and perhaps none of the reviews of software and databases from these journals (see next paragraph). Thus, the author of this chapter strongly recommends that, if at all possible, readers review the hard copies of these journals, or at least review the tables of contents via some means (see Section 7.8).

How well does Toxline do in identifying risk assessment and risk management software and databases? As was done for the DIALOG databases noted earlier, a trial search using "software," "databases," together with "risk" as key words was performed and did not find information about a key recent review of risk-related software (Marnicio et al. 1991) noted previously. Also, Toxline did not have information on the U.S. EPA's "Exposure Models Library" or about the "SmartRisk" software program. However, as was also the case for all but one of the previously mentioned databases, Toxline did have information about Monte Carlo software.

Based on the previously mentioned sample searches, Toxline, like the other databases, is not ideal for trying to identify software and databases for risk assessment and risk management purposes; however, the author of this chapter has shared these results with the National Library of Medicine in an effort to make Toxline more useful.

Although Toxline is not as useful as it could be in identifying software and databases for risk assessment and risk management uses, it is important to note that Toxline does contain references covering the pharmacological, physiological, biochemical, and toxicological effects of chemicals, including risk assessment and risk management publications. Toxline currently covers publications from 1965 to the present. The sources of the references indexed for Toxline are listed in Table 2.

7.2 The U.S. National Library of Medicine's TOXNET Databases

Some other very useful National Library of Medicine databases for risk assessors and risk managers that are available by use of a computer are the TOXNET (TOXicology Data NETwork) databases. TOXNET's databases are described in Table 3.

7.3 Other Ways to Access U.S. National Library of Medicine Databases

Another way to access Toxline and TOXNET is by use of the U.S. National Library of Medicine's "Grateful Med" software program. Grateful Med provides easy online searching of multiple National Library of Medicine databases, along with easy printing or writing of the search results to a file and easy ordering of publications. Information about National Library of Medicine databases and Grateful Med can be obtained by calling 301-496-6193 or 1-800-638-8480 or by sending e-mail to *GMHELP@gmedser.nlm.nih.gov* As indicated in Table 3, not all of TOXNET's 12 databases are searchable via the Grateful Med software.

Further, the CompuServ online service (see Section 9.1), Chemical Information Systems (CIS) (call 1-800-CIS-USER for information), Scientific Technical Network

Table 2 Sources of References for Toxline[a]

- Aneuploidy File, 1970–1986
- Developmental and Reproductive Toxicology File, 1989–present
- Environmental Mutagen Information Center File, pre-1950–1990
- Environmental Teratology Information Center File, pre-1950–1989
- Epidemiology Information System, pre-1940–1988
- (U.S.) Federal Research in Progress, 1984–present
- Hazardous Materials Technical Center, pre-1981–1988
- International Labor Office, CIS Abstracts, 1981–present
- International Pharmaceutical Abstracts, 1970–present
- NIOSHTIC®,[b] U.S. National Institute for Occupational Safety and Health, 1984–present
- Pesticides Abstracts, 1966–present
- Poisonous Plants Bibliography, pre-1976 for most records
- U.S. EPA Toxic Substances Control Act Test Submissions, pre-1988–present
- Toxicity Bibliography, Toxicology Publications in the Medical Literature, as Contained in NLM's Medline Database, 1965–present
- Toxicological Aspects of Environmental Health, 1972–present
- Toxicology Document and Data Repository, U.S. National Technical Information Service, October 1979–present
- Toxicology Research Projects, U.S. National Institutes of Health, three most recent federal fiscal years

[a] These "subfiles" are listed in alphabetical order, with years of publication coverage listed for each subfile.
[b] Registered Trademark of U.S. National Institute of Occupational Safety and Health, Cincinnati, OH.

(STN) online service (call 1-800-753-4227; WWW home page: *http://info.cas.org*), BRS online service (call 1-800-456-7248), the Internet (see Sections 9.2, 9.3, and 9.4), SilverPlatter CD-ROMs (call 1-800-521-0574), and some direct dial numbers (via TYMNET and SPRINTNET) to the National Library of Medicine also provide access to the National Library of Medicine's databases and other databases containing toxicology and other information useful to risk assessors and risk managers. Also, public institutions in numerous countries have agreements with the National Library of Medicine to provide either online or tape access to the National Library of Medicine's databases, and, with an account with the National Library of Medicine, access to the databases can also be obtained via the National Library of Medicine's WWW home page (see Section 9.3). Since the listing of commercial services offering online, CD-ROM, or other access to the National Library of Medicine's databases is incomplete and can change, interested readers might be able to obtain a current listing of country public institutions and commercial services affiliated with the National Library of Medicine from the National Library of Medicine using the phone numbers or e-mail address shown in the previous paragraph.

7.4 Accessing U.S. EPA Information

A useful U.S. EPA online system for risk assessors and risk managers is the "On-Line Library System" (OLS). OLS is free, except for telephone connection charges, and provides the ability to search a wide variety of U.S. EPA publications and document collections (call 919-554-2777 for information). OLS and other U.S.

Table 3 The TOXNET Databases Managed by the U.S. National Library of Medicine

- Hazardous Substances Data Bank (HSDB). Contains peer-reviewed summaries of the toxicology of over 4,000 chemicals.
- Registry of Toxic Effects of Chemical Substances (RTECS). Contains acute and chronic toxic effects data for over 110,000 chemicals. This database is built and maintained by the U.S. National Institute for Occupational Safety and Health (NIOSH).
- Chemical Carcinogenesis Research Information System (CCRIS). Contains carcinogenicity, mutagenicity, tumor promotion, and tumor inhibition data for over 3,500 chemicals. This database is sponsored by the U.S. National Cancer Institute.
- Integrated Risk Information System (IRIS). Contains U.S. Environmental Protection Agency (EPA) health risk and regulatory information for over 650 chemicals, including both carcinogenic and noncarcinogenic risk assessment data for the oral and inhalation routes of exposure. This database also contains U.S. EPA drinking water health advisories and information on environmental standards and regulations.
- GENE-TOX. Contains expert-reviewed genetic toxicology (mutagenicity) data for over 3,000 chemicals.
- Environmental Mutagen Information Center — Front and Back Files (EMIC/EMICBACK). These bibliographic databases contain information on chemical, biological, and physical agents that have been tested for genotoxic activity.
- Developmental and Reproductive Toxicology Database/Environmental Teratology Information Center Backfile (DART/ETICBACK). These bibliographic databases contain information on teratology and developmental toxicology publications.
- Toxic Chemicals Release Inventory (TRI) Series. These U.S. EPA databases contain U.S. data from industry on the estimated releases of toxic chemicals to the environment (air, water, land, or underground injection), along with data on the amounts of these chemicals that are transferred to waste sites, and the methods used for waste treatment.
- Toxic Chemicals Release Inventory Facts (TRIFACTS). This database supplements the TRI environmental release data with information on the health effects and the safety and handling of the chemicals.

Note: As of the writing of this chapter, the only TOXNET databases accessible via Grateful Med are HSDB, RTECS, CCRIS, IRIS, and TRI.

EPA information can also be accessed via the Internet (see Sections 9.2 and 9.3, and Table 5).

Electronic Handbook Publishers, Inc., publishes diskettes containing U.S. EPA risk assessment information. Hyper Text-IRIS™* contains the complete text files of the U.S. EPA's "Integrated Risk Information System" (IRIS), and The Electronic Handbook of Risk Assessment Values (EHRAV™*) contains information from IRIS and the Health Risk Information System (HEAST) (call 206-836-0958 for information).

7.5 Accessing U.S. FDA Information

A useful U.S. Food and Drug Administration (FDA) online system can be accessed 24 hours a day by calling 1-800-222-0185 (call 301-443-7318 for information). The FDA system contains press releases, medical bulletins, proposed regulations, and other information. FDA information can also be accessed via the Internet (see Sections 9.2 and 9.3, and Table 5).

* Registered trademarks of Electronic Handbook Publishers, Inc., Redmond, Washington.

7.6 Accessing the U.S. Agency for Toxic Substances and Disease Registry Information

Lewis Publishers (Boca Raton, Florida), in cooperation with the U.S. Agency for Toxic Substances and Disease Registry (ATSDR), publishes "Toxicological Profiles" on CD-ROM. The CD-ROM version contains the equivalent of over 14,000 pages of text on more than 100 of the most important chemicals found at hazardous waste sites (call 1-800-272-7737). ATSDR information can also be accessed via the Internet (see Sections 9.2 and 9.3 and Table 5).

7.7 Accessing the International Chemical Regulatory Information

The International Chemical Regulatory Monitoring System (ICRMS) from Ariel Research Corporation (call 301-907-7771 for information) is a series of databases (or "modules") containing regulatory information for chemicals. The current or planned modules include ones covering North America, the U.S. Department of Transportation, Western Europe, chemical inventories for various countries, air toxics (hazardous air pollutants), and the Asia/Pacific Rim. This database is available on disk, magnetic tapes, and cartridge, with a CD-ROM version being developed.

The Lexis®-Nexis®* databases are from Lexis-Nexis (call 1-800-227-4908; WWW home page: *http://www.meaddata.com*). These databases are available online and contain many "libraries" or "files." Among the numerous ones are those that can provide worldwide chemical regulatory information and information on various other topics (e.g., complete texts of proposed and existing U.S. government regulations).

7.8 Accessing Abstracts from Scientific Journals

As indicated earlier and in Table 2, Toxline can be searched for abstracts from scientific journals. Also, abstracts from scientific journals can be searched via hard copies of the Institute for Scientific Information's *Current Contents*®,** via searching of *Current Contents on Diskette*®,** or via the online (DIALOG File #440 = Current Contents Search®***) or CD-ROM (SilverPlatter) versions of *Current Contents*. Of the seven editions of *Current Contents*, the Life Sciences, Clinical Medicine, and Agriculture, Biology, and Environmental Sciences editions will be of particular usefulness to most risk assessors and risk managers.

Also, Reed-Elsevier, in cooperation with SilverPlatter, publishes a series of CD-ROMs (EMBASE CD, EMBASE Alert CD, and POLTOX II: EMBASE CD) of possible interest to risk assessors (call 1-800-457-3633). These include abstracts from human and environmental toxicology journals.

Further, Cambridge Scientific Abstracts has developed Internet and Internet WWW access to the contents of its various journals. These journals themselves include abstracts from recent publications that have been abstracted by Cambridge

* Registered trademark of Lexis-Nexis, a division of Reed-Elsevier, Dayton, Ohio.
** Registered trademarks of the Institute for Scientific Information, Inc., Philadelphia, Pennsylvania.
*** Registered trademark of DIALOG Information Services, Inc., Palo Alto, California.

Scientific. The journals include *Toxicology Abstracts, Health and Safety Science Abstracts,* and *Risk Abstracts*. The Internet and World Wide Web access is so up to date that the Cambridge Scientific journal containing the abstracts may not have gone to press before being available on the Internet and WWW. Further information can be obtained by e-mail from Cambridge Scientific (e-mail address: *market@csa.com*). Noteworthy is that Cambridge Scientific Abstracts' Internet Journal Service received the 1994 Information Industry Association's "Hot Shots Award" in the category of best science/technology database.

7.9 Section Summary

The author suggests that anyone interested in learning more about the various databases noted should contact the National Library of Medicine, U.S. EPA, CompuServ, CIS, STN, SilverPlatter, Ariel Research Corporation, Electronic Handbook Publishers, Lewis Publishers, Lexis-Nexis, and Cambridge Scientific Abstracts for more information, including up-to-date listings and descriptions of the databases that are available. Steps toward learning more about how to use the Internet to access information, such as from the National Library of Medicine's databases, are discussed in Section 9.

8. PROFESSIONAL SOCIETY ACTIVITIES RELATING TO SOFTWARE AND DATABASES

8.1 Society of Toxicology Meetings

Annual meetings of the Society of Toxicology have included presentations of software programs and on the use of the Internet. For example, the 1995 Annual Meeting included a session on "Educator's Forum: Use of Computers in Teaching Toxicology."

8.2 International Congress of Toxicology Meetings

Meetings of the International Congress of Toxicology have also featured presentations of software programs and on the use of the Internet. For example, the 1995 "International Congress of Toxicology — VII" meeting included a session on "Education/Information Sources," featuring a number of software programs and ways to use the Internet for information retrieval, communications, and teaching.

8.3 Society for Risk Analysis Meetings

The Society for Risk Analysis has had "Riskware" exhibitions of software and databases as part of its annual meeting since 1991. This exhibition started as an activity of the Ohio Chapter of the Society for Risk Analysis in 1990. Thus far, one review describing Riskware software and databases has been published (Marnicio et al. 1991), with more possible.

Any future Riskware publications are expected to contain cross-referenced tables enabling readers interested in software and/or databases to easily identify a particular software product or database capable of performing a particular type of task or serving as a source of a given type of information. The types of software and databases covered in the Riskware exhibitions so far include those shown in Table 4.

Readers can contact the chapter's author to learn the current status of Riskware publications. Also, the 1991, 1992, and 1993 Riskware exhibitions provided Society for Risk Analysis meeting attendees with a "Softstracts" document, containing approximately one-page descriptions of each piece of software and each database or each related group of software and databases. Additional copies of the 1991–1993 "Softstracts" may be available from the author of this chapter. Finally, anyone interested in exhibiting their software and/or databases at a future Society for Risk Analysis meeting can contact the Society for Risk Analysis for details about becoming a software and/or database exhibitor (Society for Risk Analysis, 1313 Dolley Madison Boulevard, Suite 402, McLean, Virginia 22101. Phone number: 703-790-1745. Fax number: 703-790-2672. e-mail address: *sraburkmgt@aol.com*).

9. CURRENT AND LIKELY FUTURE USES OF GLOBAL ELECTRONIC COMMUNICATIONS NETWORKS, INCLUDING THE INTERNET AND ITS WORLD WIDE WEB

9.1 Commercial Online Services

America Online, CompuServ, Delphi, the Microsoft Network, and Prodigy are among the major commercial online services that allow users to send e-mail, download software programs or files, search for information (including from encyclopedias, magazines, and newspapers), and possibly arrange the retrieved information with the help of software to create personalized electronic newspapers and other documents. These online services also allow users to interact with others via general or special interest group "forums" and to access other computer networks, including the Internet. Often, these commercial online services offer several hours of free online access to allow users to judge the usefulness of the service, including access to the Internet.

Information about Microsoft Corporation's Microsoft Network online service, and the software for it, are part of Microsoft Corporation's Windows 95 operating system software package, or information can be obtained via the WWW as *http://www.msn.com*. Information about other online services can be obtained from advertisements in many magazines or as follows:

America Online:	Call 1-800-827-6364 or via the WWW as *http://www.aol.com*
CompuServ:	Call 614-529-1349 or 1-800-848-8199 or via the WWW as *http://www.compuserv.com*
Delphi:	Call 1-800-695-4005 or send a message to *INFO@delphi.com*
Prodigy:	Call 1-800-Prodigy or via the WWW as *http://www.prodigy.com*

Table 4 An Overview of the Types of Software and Databases Presented at Society for Risk Analysis Meetings

Biological data use
- Access health effects data on mixtures of chemicals
- Access key U.S. EPA risk assessment estimates and the physical and chemical properties of particular chemicals
- Analyze data from studies with a no-observed-effect level (NOEL) using the benchmark dose (BD) method
- Dose-response relationships projection for health effects, generally cancer, from animal data
- Expert system evaluating chemicals for potential to cause cancer
- Identify disease clusters
- Model and assess risks of reproductive/developmental toxicity
- Model the dynamics of wildlife and ecological populations
- Uptake, distribution, metabolism, and excretion simulation of chemicals in animals and humans via physiologically based pharmacokinetic (PB-PK) modeling

Chemical properties
- Access key U.S. EPA risk assessment estimates and the physical and chemical properties of particular chemicals
- Chemical property estimation of compounds and mixtures

Educational
- Tool for teaching toxicology and risk assessment
- Exposure assessment
- Assess daily and/or long-term food consumption or exposure to chemical residues in food
- Assess food consumption at a given time of the day or per eating occasion
- Internal dose assessment for intake of radionuclides
- Perform quantitative exposure and risk assessments for various chemicals and sites

Hazard assessment
- Assess the risks of manufacture, storage, and transport of chemicals
- Hazard assessment modeling alone or with risk assessments for radionuclides released into air
- Hazard assessment modeling for accidental and normal effluent or emission releases from industrial sites, and estimation of their resulting impacts
- Perform fate and transport modeling of chemicals released into air, soil, groundwater, and surface water

Miscellaneous
- Assess process-chemical combinations relative to inhalation risk, analyze usage in various locations, and determine relative risk of a chemical release
- Data center and organization contingency or resumption planning
- Document and report the results of hazard identification and risk analysis studies, including "HAZOPs," "FMCEAs," "PHAs," and "What-If" analyses
- Evaluate various protective action strategies for chemical weapons accidents
- Human reliability analysis (HRA) tool
- Integrated risk model manager for large engineered systems
- Interactive risk assessment product, using expert knowledge base and questionnaire-style interface
- Material and waste tracking
- Material Safety Data Sheet (MSDS) management
- Perform probabilistic risk or safety assessments
- Perform time-dependent risk assessments and reliability analysis of complex engineered systems
- Probabilistic network analyzer, addressing both cost and schedule uncertainties of a project
- Quantitative risk assessment (QRA), including event-tree construction and analysis, fault-tree construction and analysis, human reliability analysis, equipment reliability database, uncertainty and sensitivity analysis, and QRA documentation

Table 4 An Overview of the Types of Software and Databases Presented at Society for Risk Analysis Meetings (continued)

- Risk time trends in system unavailability vs. age
- System test and maintenance strategies risk impacts

Models
- Build, model, and evaluate reliability block diagram (RBD) models
- General modeling environment for constructing, analyzing, and communicating models for risk analysis, decision analysis, and scientific and engineering applications
- Hazard modeling, including consequence analysis of pure components and mixtures
- Prioritizing environmental problems using a multimedia model

Online databases
- Chemical, toxicological, and regulatory data

Process safety
- Chemical process safety management
- Managing chemical process safety tool

Regulations
- Ability to keep track of chemical regulations and advisories
- Assist in the compliance with regulations pertaining to the safe use, fire protection, storage, shipment, and disposal of materials

Uncertainty analysis
- Perform uncertainty analysis (an alternative technique to Monte Carlo simulation) using true mathematical manipulation of probability functions
- Perform uncertainty analysis via Monte Carlo and Latin Hypercube simulation.

9.2 The Internet

After being created for defense purposes in the United States to protect computer networks in the event of war, the Internet has evolved into a large, worldwide collection of computer networks being used for many purposes. Uses include sending e-mail messages worldwide and accessing various databases (including the U.S. National Library of Medicine's Toxline, TOXNET, and U.S. EPA's OLS). Of interest to risk assessors and risk managers is use of the Internet and the commercial online services mentioned previously for information exchange among newsgroups and mailing lists (also called user groups or interest groups), with the worldwide network of these groups called the "Usenet."

For example, the "Columbia-Cascades" chapter of the Society for Risk Analysis and the Pacific Northwest Laboratory have created a "RISKANAL" Internet mailing list for use by the international risk analysis community. This mailing list, as of the middle of 1996, had about 750 members in 25 or more countries. Users ("subscribers") are invited to ask and answer questions and to send news items and announcements of meetings, publications, new software, job openings, etc. Examples of the usefulness of this mailing list include the solicitation and compilation by one mailing list member of a rather comprehensive listing and descriptions of various types of risk analysis software. This compilation was then shared with the mailing list. Also noteworthy is that this mailing list was used to discuss possible topics to be covered in a U.S. EPA Monte Carlo analysis exposure assessment workshop. This led to an often spirited exchange of global communications over several months between some members of the workshop panel and others, while also helping to educate others about technical issues concerning the use of Monte Carlo analysis and related approaches. To join this mailing list, the following message should be sent via the Internet to *listserv@listserv.pnl.gov* (messages about any questions or problems with this mailing list can be sent to *js_dukelow@pnl.gov*): *subscribe riskanal first_name last_name* (e.g., subscribe riskanal Bert Hakkinen)

The Internet is now also used as a way to publish and distribute textbooks, journals, newsletters, weekly bibliographic updates, and other texts. Noteworthy is that some of the texts (e.g., journals) accessible via the Internet do not have hard-copy published counterparts.

9.3 The Internet's World Wide Web

The Internet includes a World Wide Web (WWW). The WWW is the fastest growing part of the Internet (even in 1994, one estimate was that it was then growing at the rate of about 1% a day, while another estimate was that the WWW was six times larger at the end of 1995 than at the start of 1995). Included is a collection of "home pages" that users can navigate through via use of "hypertext." Hypertext enables users to highlight certain pictures or words and then move to a linked picture or page of information. Various companies, other organizations, and individuals have established home pages on the WWW to provide information about themselves, with some home pages linked to related home pages. There were over 30,000 home pages in the WWW as of late 1995. Readers of this chapter may also read or hear elsewhere about a "Gopher" aspect of the Internet to find and retrieve information; however, the direct uses of Gopher are being largely replaced by use of the WWW and will not be discussed in this chapter.

Accessibility to the WWW is becoming very easy for anyone with a computer and modem, with many of the online services noted earlier now providing "browser" software for easy navigation of the WWW. WWW access is also available to anyone with access to the Internet, provided they have the appropriate software for doing so, e.g., the widely available "Mosaic" browser developed by the University of Illinois' National Center for Supercomputing and "Netscape Navigator" marketed by Netscape Communications Corporation. For anyone interested in statistics about the Internet and WWW, including the rate of yearly growth of WWW home pages, a clearinghouse for statistics about the Internet and WWW can be accessed via the WWW at *http://www.zilker.net/swg/*

Some key WWW home pages of possible interest to risk assessors are listed in Table 5, including ones from companies, global and private organizations, professional societies, universities, research institutes, and government agencies.

Home pages are being added to the WWW every day, and other potentially useful WWW home pages related to risk assessment, risk management, or other topics can be identified from use of the WWW home page "search engines" and directories listed in Table 6. These home pages serve as a means of cataloging what WWW home pages exist and what they contain, and they can be used to identify useful home pages and other information by typing in key words or phrases, and, in some cases, questions. Also, some of the search engines listed in Table 6 allow for searching of Internet Usenet newsgroups and mailing lists.

Table 5 Some World Wide Web "Home Pages" and Other WWW Sites of Possible Interest to Risk Assessors and Risk Managers

Companies

3M Company
http://www.3M.com

Chemical Industry Institute of Toxicology (CIIT)
http://www.ciit.org/HOMEP/ciit.html

Dow Chemical Company
http://www.dow.com

du Pont (E. I. du Pont de Nemours and Company)
http://www.dupont.com

Rohm & Haas Company
http://www.rohmhaas.com

Tec.Com Inc.'s RiskWorld
http://www.riskworld.com
(Provides linkages to home pages that deal with human health risks. Also includes "news & reports," a calendar of events, classified ads, and information on workshops, academic courses, professional organizations, government programs, university programs, fellowships, grants, new books, new software, etc.)

Global Organizations

Biological Effects of Low-Level Exposure (BELLE) Advisory Committee
http://www.BELLEonline.com

Central European Environmental Data Request Facility (CEDAR)
http://www.cedar.univie.ac.at
(Also provides access to other international environmental organizations and programs)

Greenpeace International
http://www.greenpeace.org

International Agency for Research on Cancer (IARC)
http://www.iarc.fr

International Occupational Safety & Health Information Centre (ILO-CIS)
http://turva.me.tut.fi/cis/home.html
(Includes listings of online and CD-ROM databases, useful Internet resources, etc.)

International Organization for Standardization (ISO)
http://www.iso.ch/

International Toxicity Estimates for Risk (ITER)
http://www.tera.org

United Nations Environment Programme (UNEP)
http://www.unep.ch
[Includes access to information about the International Register of Potentially Toxic Chemicals (IRPTC)]

Table 5 Some World Wide Web "Home Pages" and Other WWW Sites of Possible Interest to Risk Assessors and Risk Managers (continued)

World Health Organization (WHO)
http://www.who.ch

Journals

Nature
http://www.nature.com

New Scientist
http://www.newscientist.com

Risk: Health, Safety & Environment
http://www.fplc.edu/tfield/rskindx.html

Science
http://www.aaas.org/science/science

Professional Societies

American Association for the Advancement of Science (AAAS)
http://www.aaas.org

American Chemical Society (ACS)
http://www.acs.org

American Industrial Hygiene Association (AIHA)
http://www.aiha.org

American Medical Association (AMA)
http://www.ama-assn.org

International Society of Exposure Analysis (ISEA)
http://www.isea.rutgers.edu/isea/isea.html

Risk Assessment & Policy Association's Health, Safety & Environment on the Internet
http://www.fplc.edu/tfield/links.html

Society for Risk Analysis (SRA)
As of the time this chapter was being finalized, SRA was planning to establish a WWW site, **perhaps** as *http://www.sra.org*

Society of Toxicology (SOT)
http://www.toxicology.org

Canada

Canadian Centre for Occupational Health and Safety
http://www.ccohs.ca

Central Europe

See listing above for Central European Environmental Data Request Facility (CEDAR)

Table 5 Some World Wide Web "Home Pages" and Other WWW Sites of Possible Interest to Risk Assessors and Risk Managers (continued)

Finland

See listing above for International Occupational Safety & Health Information Centre (ILO-CIS) — this WWW home page/server is maintained in collaboration with the Tampere University of Technology in Finland

Japan

National Institute of Health Sciences (Ministry of Health and Welfare)
http://www.nihs.go.jp

Singapore

National University of Singapore BioMed Server
http://biomed.nus.sg

Sweden

National Institute for Working Life, formerly the
Swedish National Institute of Occupational Health (NIOH)
http://www.nioh.se

The Netherlands

National Institute of Public Health and Environmental Protection (RIVM)
http://deimos.rivm.nl/about.html

United States

Agency for Toxic Substances and Disease Registry (ATSDR)
http://atsdr1.atsdr.cdc.gov:8080/atsdrhome.html
(Provides access to a newsletter, listing of related Internet resources, toxicological profiles for various chemicals, etc.)

Centers for Disease Control and Prevention
http://www.cdc.gov/cdc.html

Chemical Industry Institute of Toxicology (CIIT)
See listing above under "Companies"

Environmental Information Center (EIC)
http://www.eic.org

Environmental Protection Agency (EPA)
http://www.epa.gov

EPA's Center for Exposure Assessment Modeling
http://www.epa.gov/software.html

EPA's THERdbASE (Total Human Exposure Risk database and Advanced Simulation Environment) modeling and software tool can be accessed and downloaded at
http://eeyore.lv-hrc.nevada.edu/therdbase.html

Table 5 Some World Wide Web "Home Pages" and Other WWW Sites of Possible Interest to Risk Assessors and Risk Managers (continued)

EXtension TOXicology NETwork (EXTOXNET)
http://www.oes.orst.edu:70/1/ext/extoxnet/
(or stop after "*edu*" and choose EXTOXNET from the menu)
(Cooperative effort among various universities to stimulate dialog on toxicology issues and to make toxicology information available. Includes newsletter, fact sheets, chemical profiles, etc.)

FedWorld
http://www.fedworld.gov
(A gateway into U.S. government information)

U.S. Food and Drug Administration (FDA)
http://www.fda.gov/fdahomepage.html

Internet Disaster Information Network
http://www.disaster.org
(Nongovernment, but is intended to be a clearinghouse for information on large-scale disasters like earthquakes)

Lawrence Berkeley National Laboratory's Risk-Related Research at LBNL
http://www.lbl.gov/LBL-Programs/Risk-Research.html

Massachusetts Department of Environmental Protection
http://www.magnet.ma.us/dep

National Institute for Occupational Safety and Health (NIOSH)
http://www.cdc.gov/niosh/homepage.html

National Institute of Environmental Health Sciences (NIEHS)
http://www.niehs.nih.gov

National Institutes of Health (NIH)
http://www.nih.gov

National Library of Medicine (NLM)
http://www.nlm.nih.gov

National Technical Information Service (NTIS)
http://www.fedworld.gov/ntis/ntishome.html

National Toxicology Program (NTP)
http://ntp-server.niehs.nih.gov/

Natural Hazards Research and Applications Information Center at the University of Colorado
http://adder.colorado.edu/~hazctr/Home.html
(This center's WWW site includes listings of publications and global information sources about hazards and disasters, listings of useful periodicals and Internet resources, and other information)

Oak Ridge National Laboratory
http://www.hsrd.ornl.gov/ecorisk/ecorisk.html
(This site contains ecological risk analysis tools and applications, including databases and completed risk assessments)

Table 5 Some World Wide Web "Home Pages" and Other WWW Sites of Possible Interest to Risk Assessors and Risk Managers (continued)

Occupational Safety and Health Administration (OSHA)
http://www.osha.gov

Resources for the Future (RFF)
http://www.rff.org
(This Washington, D.C. organization includes a Center for Risk Management, and this WWW site provides access to RFF publications and other RFF information, and linkages to the home pages of other sources of risk assessment and other information)

Toxicology News Group via the BIOSCI/bionet Electronic Newsgroup Network for Biology
http://www.bio.net (choose "toxicology" from the newsgroups listed in the "archives")

University of Delaware
http://www.udel.edu/panar/ccr/ccrcat.html
(Sponsored by the Council for Chemical Research, this site is intended to be a source of lecture materials, including information on risk assessment of chemicals)

9.4 Other Information About the Internet

The Internet is also being used by companies to send users updates of software, for providing user-support information, and for establishment of electronic newsgroups/user groups/interest groups to discuss topics related to individual software programs.

There have been many guidance and resource books published about the Internet, and many scientific journals and magazines also contain information about using the Internet. In addition to the scientific journals and magazines mentioned in earlier sections of this chapter, *Science*, *The Scientist*, and *New Scientist* are good sources of information about the Internet's uses, new developments, and issues such as security of access to computer systems and data (e.g., Hoke 1994a,b, Germain 1995). Also, very good short reviews of specific WWW sites are in the "Netropolitan" column of *New Scientist*, e.g., the column in the June 10, 1995 issue (page 21) discusses WWW sites dealing with environmental information.

Of interest to risk assessors is the launching in 1996 of an Internet database summarizing peer-reviewed toxicity values for use in performing human risk assessments. These values will be from government agencies, public organizations, and health groups around the world. The International Toxicity Estimates for Risk (ITER) database is accessible via a WWW home page at *http://www.tera.org*

Finally, as noted previously, some journals such as *Toxicology Abstracts* and *Risk Abstracts* from Cambridge Scientific Abstracts are now available by way of the Internet and its WWW (Gardner 1995).

10. CURRENT AND LIKELY FUTURE USES OF CD-ROMs

CD-ROM technology has become very popular since the early 1990s. Key features of CD-ROMs that have made them so popular in a short period of time

Table 6 Some World Wide Web "Home Pages" To Search To Find Other WWW Sites and Other Internet Information of Possible Interest to Risk Assessors and Risk Managers

---

http://altavista.digital.com (Digital Equipment Corporation's "Alta Vista"). This WWW search engine was said to be indexing over 16 million WWW pages in early 1996. A 1996 search by the author of this chapter found using "risk assessment" and "toxicology" as key words found 200,000 and 20,000 matches or "hits," respectively.

http://gnn.gm/gnn/wic/index.html ("The Whole Internet Catalog"). A 1995 search by the author of this chapter found rather extensive listings of home pages for U.S. and international "health" organizations.

http://lycos.cs.cmu.edu/ (Carnegie Mellon University's "Lycos"). This WWW search engine was said to be indexing over 18 million WWW pages in early 1996.

http://webcrawler.com ("WebCrawler"). A 1996 search by the author of this chapter using "risk assessment" and "toxicology" as key words found 555 and 356 matches or "hits," respectively.

http://www.dejanews.com (DejaNews). This WWW search engine performs full-text searches of various Internet Usenet newsgroups.

http://www.hotbot.com ("HotBot"). The designers of this WWW search engine claim that every WWW page, Usenet newsgroup, and Internet mailing list will be indexed.

http://www.infoseek.com (InfoSeek). This WWW search engine, available for a monthly subscription charge, covers periodicals, Internet Usenet newsgroups, WWW pages, and other publications and indexes the full texts of what it covers. Also offers a search engine for free searches of WWW pages.

http://www.mckinley.com/ (McKinley Internet Directory).

http://www.netmind.com (Netmind Free Services). *Note:* This home page includes access to the "URL-Minder," which can be arranged to notify users by e-mail when the contents of specified WWW home pages change, and which also can be used to perform searches of specified WWW search engines, including some of those noted in this table. The searches are conducted on a regular basis, with notification of the user by e-mail when the results of the searches have changed. This home page also includes access to the "SIFT Home Page," which allows the user to set up a personal search profile for searching Internet Usenet newsgroups for information. The SIFT Home page program will then monitor Usenet information and will send information matching the search profile to the user.

http://www.opentext.com:8080/ This home page allows full-text searches of WWW pages.

http://www.search.com This home page provides access to several hundred search engines.

http://www.wais.com (WAIS, Inc.). *Note:* This home page can handle natural language questions like "What is ___?"

http://www.yahoo.com (The "Yahoo List"). A 1996 search by the author of this chapter using "risk assessment" and "toxicology" as key words found 11 and 36 matches or "hits," respectively.

Note: This is a selected current listing; many other similar WWW search engines and directories are available. Others with differing indexing, search, and output capabilities will be developed and available in future years. It is best to try at least several search engines and directories to see how they meet the individual needs of the user.

Quarterdeck Corporation *(http://www.quarterdeck.com)* markets WebCompass, a software program that, used with a WWW browser, allows users to run simultaneous keyword searches of InfoSeek, Lycos, WebCrawler, Yahoo, and other search engines. The search results are arranged by topic, and WWW linkages are provided to the information obtained in the searches.

include the ability to have one CD-ROM contain the equivalent of several hundred floppy disks and thousands of printed pages and to be able to search and access this information very quickly. It has become very popular to share CD-ROM databases among several sites by using a CD-ROM server located at one site. Also, it is likely in the future that most large (multimegabyte) software programs will be available only as a CD-ROM, rather than as a set of floppy disks.

Popular CD-ROMs now available include numerous games; teaching tools (e.g., combinations of text and sound to learn a foreign language); and resource materials such as encyclopedias, dictionaries, and atlases. CD-ROM versions of printed documents can contain "extras." These extras are possible due to the very large storage capacity and quick access to data capability of CD-ROMs and can include spoken text and other sounds, photographs, interactive maps, "movies," or animation of events. Also, some CD-ROM encyclopedias already offer the ability for users to obtain online updates and other new information that when stored on the user's computer hard drive will create versions of articles and other information that update and add to the information retrieved from the CD-ROM version. It is likely that the online ability to update and revise the content of CD-ROMs will eventually be used for information of particular value to exposure assessors and risk assessors, e.g., to update any future CD-ROM versions of exposure factor compilations and other information used in risk assessments.

Also being marketed are CD-Rs (Compact Disc–Recordable), which allow users to record data on compact discs. Further, although they will not be covered in this chapter, readers likely will be hearing more and more in years to come about other ways to store and access large amounts of data, including via magneto-optical technology, "MiniDisc Data" (MD Data), and various optical storage systems.

As noted earlier, good current examples of the risk assessment and risk management uses of CD-ROMs are the ability to search the U.S. National Library of Medicine's Toxline and TOXNET databases. One drawback to CD-ROM databases is that they themselves cannot be completely up to date, whereas online databases can be updated frequently and accessed as soon as they are updated. However, as noted earlier, it is likely that the rather new online ability to update and revise the content of CD-ROMs via information downloaded to the user's computer hard drive will be used for at least some CD-ROMs of particular value to exposure assessors and risk assessors

If risk assessment and risk management software goes the current way of other software, it is very likely that future risk assessment and risk management programs available on both floppy disks and CD-ROM will have the CD-ROM version contain "extras." These extras will include spoken text and other sounds, "movies" (videos), or animation of events and the ability to easily interact with the program for education and entertainment purposes. This could perhaps include interaction as a "virtual reality" participant in a chemical plant accident or other type of exposure to chemicals much like virtual reality is currently being used by airline companies to train pilots on how to handle emergencies.

An example of how CD-ROMs are currently being used for training purposes is a new CD-ROM series for primary care physicians issued four times a year for use in obtaining continuing education credits (Hogan 1995). This CD-ROM series for physicians contains updates on various subject areas; presents medical procedures as videos together with audio explanations; and also allows the user to go through case studies complete with audio histories, in which the patients can be "examined" by clicking on various quasi-three-dimensional, near-photo-quality images, and even by "listening" to the sounds of the patient's heart.

11. THE EMERGENCE OF "INFORMATION APPLIANCES," "PERSONAL ASSISTANTS," AND "INTELLIGENT ASSISTANTS"

"Information appliances" are computers that are expected to be introduced by several companies in 1996. They will cost much less than a standard computer and have little or no data storage capability, but perhaps can do word processing and use CD-ROMs, while providing access to the Internet.

"Personal assistants" are already available as reference books such as the *Merck Manual* in pocket-size electronic format. Books in this format are easily searchable by key words, with the search results quickly available on the screen of the assistant.

"Intelligent assistants" are also already available as small, portable electronic devices that can send and receive e-mail and faxes, sort incoming messages, and handle other functions such as daily schedules and listings of addresses and phone and fax numbers. One example is Sony's "Magic Link™"* communicator using AT&T's "PersonaLink℠"** software and communication services.

12. THE EMERGENCE OF "INTELLIGENT AGENTS" AND "JAVA"

"Intelligent agents" (also called Smart Agents or Software Agents or Good Viruses or Web Robots) are software programs that are told or essentially learn what information a user likes to see and then search through e-mail, databases, networks, WWW sites, and Internet Usenet newsgroups and mailing lists on an ongoing basis to retrieve that information. In a portable device, they are what is used by intelligent assistants to sort incoming messages based on what the user has looked at first in the past. Intelligent agents also already exist that can interact with other computer networks to check their status and to negotiate the best routes of communication or which agent should be in charge based on the current conditions. For example, intelligent agents now exist that investigate the capabilities and weather conditions of telescopes around the world to select which telescope should be used for a particular astronomy observation. Other existing intelligent agents interact with each other to schedule meetings, based on the length of past meetings with a particular person, and even whether a meeting with that person has already occurred sometime within the past week (Germain 1994).

Finally, "Java" is a computer programming language from Sun Microsystems, Inc. Introduced in 1995, Java is expected to be widely used to share software programs called miniapplications (or "applets") over the Internet (e.g., from WWW sites). The applets can run on any computer and, once downloaded to a computer, can process data (e.g., spoken text or other sounds, animation, etc.) transmitted from a WWW site. Also available is "JavaScript," a simpler version of Java. Applets are also expected to make the WWW more interactive by allowing for "live" updates of data and direct two-way interactions between users.

* Registered Trademark of Sony, Park Ridge, New Jersey.
** Registered Servicemark of AT&T, Parisippany, New Jersey.

13. MORE ON WHAT TO EXPECT IN THE FUTURE

Software and databases accessible via software or online connections are already widely used by toxicologists and others to perform risk assessments, and for risk management purposes. The future is bright for increases in the use of the present types of software and databases, along with further development and use of software like "intelligent agents" and new types of information devices like "information appliances" and "personal assistants."

Aiding in the use of software and databases are the "all-in-one" packages that are being developed. One example is the package recently developed by the U.S. EPA and the University of Nevada at Las Vegas called THERdbASE (Total Human Exposure Risk Database and Advanced Simulation Environment). THERdbASE allows easy access to key software models and databases useful for performing total human exposure assessments (multiple agents present in multiple media, and coming in contact with humans via multiple pathways), and it contains numerous software programs and data files. Information about THERdbASE, and THERdbASE itself, is available via the WWW at *http//eeyore.lv-hrc.nevada.edu/therdbase.html*

Also, the future is certainly likely to see more use by risk assessors and risk managers of the ability to automatically receive personalized electronic magazines, journals, newspapers, and eventually personalized WWW pages. These would be based on the user's general and scientific interests and would be assembled using software program-driven search and assembly of information from different databases. Risk assessors and risk managers will also be able to communicate more and more easily via WWW sites (e.g., via some of those listed in Table 5). There will also be more use by risk assessors and risk managers of personalized Internet e-mail and Usenet newsgroups and mailing lists to collaborate and to ask about or share information (e.g., research plans and results) with one another in far-reaching parts of the world.

The future is likely to see increased use of the Internet for teaching. Already, toxicology and risk assessment are being taught via the simultaneous transmission of voice and computer data between Michigan State University in the United States (Michael Kamrin, Ph.D.; e-mail address: *kamrin@msu.edu*) and the University of Turku in Finland. Students in Finland can hear and question the U.S. instructor and see what the instructor sees on his/her computer monitor. Also, the Internet is already being used for group discussions and scientific meetings held at "virtual facilities." Data can be shared and discussed, and one can even take a "walk" through the virtual facility being used. A discussion of such Internet-based virtual meetings is available (Anderson 1994), as is another view of what the next 10 to 20 years will bring to scientists involved in information exchange and searching via computers (Pool 1993).

The future will be accompanied by further development of so-called wireless computing (e.g., allowing portable computers to communicate with other computers via cellular telephone networks), the further miniaturization of personal computers (likely down to versions that can be easily worn on the body), and perhaps creation of new devices that combine the current features of a computer with other current devices. Descriptions of such devices include (1) a wallet-sized personal computer to be used for information storage and retrieval, to send and receive e-mail and faxes, and for other purposes such as electronic money and even telling the user exactly where they are on earth; (2) personal computers made of soft plastic that could be strapped onto the body with Velcro; and (3) a personal computer for medical emergency technicians that includes a body sensor and a video camera for transmitting data to a doctor at a hospital, a CD-ROM medical encyclopedia with a small display, and a phone. Some of these types of devices were predicted to reach the marketplace by 1996.

14. FINAL COMMENTS

The author of this chapter has not been able to identify commercially available or government online or CD-ROM databases that index listings and reviews of risk assessment and risk management software and databases from key scientific journals. Thus, the current recommended best approach for risk assessors and risk managers to identify useful software and databases is to consult and monitor some or all of the scientific journals described in this chapter (*Risk Analysis, Toxicology, Food and Chemical Toxicology, Journal of the American Medical Association,* and *Toxicology Modeling),* along with reviewing at least some of the references described in this chapter.

Reviewing recent risk assessment publications in these journals and other scientific journals will help identify key current software being used for various applications. Searching these journals and/or some of the DIALOG databases described in this chapter may identify published reviews of any potentially useful software programs, perhaps comparing the capabilities of any software programs under consideration to other available software programs. Also, contacting the authors of any risk assessment studies or software reviews by letter or fax message, or via e-mail, may bring valuable advice in return. Further, the author has seen excellent use of the "RISKANAL" Internet mailing list described in Section 9.2 to ask and answer questions about software programs and databases and for other exchanges of information.

The author of this chapter wishes readers good luck in their identification and use of software and databases for risk assessment and risk management purposes and in their use of the Internet and other online systems and would be very interested in receiving any comments readers might have about this chapter.

REFERENCES

Abeytunga, P. K., An overview of the electronic health and safety services of the Canadian Centre for Occupational Health and Safety, *Toxicology*, 96, 71, 1995.

Alston, G., Electronic highways: A road map for environmental health, *Health & Environment Digest,* 7, 1, 1994.

Ambrose, R. B. and Barnwell, T. O., Environmental software at the U.S. Environmental Protection Agency's Center for Exposure Assessment Modeling, NTIS/PB90-140716, 1990.

Anderson, C., Electronic networks. Cyberspace offers chance to do "virtually real science," *Science*, 264, 900, 1994.

Anonymous, Software and Publications, *Environmental Science & Technology (ES&T Environmental Buyer's Guide Edition 1995),* 28, 167G, 1994.

Barnwell, T. O., Vandergrift, S. B., and Ambrose, R. B., Water quality modeling software available from U.S. EPA, NTIS/PB87-175311, 1987.

Benz, J., Voigt, K., and Mucke, W., Strategy for computer-aided searches for information about chemicals, *Online Review*, 13, 383, 1989.

Cheremisinoff, N. P., Environmental software review, *Pollution Engineering*, 19, 30, 1987.

Cosmides, G. J., Toxicological information series, I. Toxicological information, *Fundamental and Applied Toxicology*, 14, 439, 1990.

Gardner, R. A., Toxicology abstracts — An electronic journal on the Internet, *Toxicology*, 99, 219, 1995.

Germain, E., Software's special agents, *New Scientist*, 142, 19, 1994.

Germain, E., Guarding against Internet intruders, *Science*, 267, 608, 1995.

Halton, D. M., Computerized information resources in toxicology and industrial health — A review, *Toxicology and Industrial Health*, 2, 113, 1986.

Hogan, R., Software review of "PrimePractice: A CD-ROM quarterly for primary care physicians, volume 1, number 1: cardiology," *Journal of the American Medical Association,* 274, 510, 1995.

Hoke, F., Scientists predict Internet will revolutionize research, *The Scientist*, 8, 1, May 2, 1994a.

Hoke, F., Publication by Internet, *The Scientist*, 8, 8, May 2, 1994b.

Johnson, B. L., ATSDR's (U.S. Agency for Toxic Substances and Disease Registry's) information databases to support human health risk assessment of hazardous substances, *Toxicology Letters*, 79, 1, 1995.

Johnson, S. C., 1992 environmental software survey, *Journal of the Air & Waste Management Association*, 42, 1516, 1992.

Kissman, H. M. and Wexler, P., Toxicology information systems: A historical perspective, *Journal of Chemical Information & Computer Sciences*, 25, 212, 1985.

Kumar, A. and Sahore, S., Public domain software for exposure assessment modeling, *Environmental Progress* 13, F14, 1994.

Marnicio, R. J., Hakkinen, P. J., Lutkenhoff, S. D., Hertzberg, R. C., and Moskowitz, P. D., Risk analysis software and databases: Review of Riskware '90 conference and exhibition, *Risk Analysis*, 11, 545, 1991.

Moskowitz, P. D., Pardi, R., DePhillips, M. P., and Meinhold, A. F., Computer models used to support cleanup decision-making at hazardous and radioactive waste sites, *Risk Analysis,* 12, 591, 1992.

Oxman, A. D., Cook, D. J., and Guyatt, G. H., Users' guides to the medical literature. VI. How to use an overview, *Journal of the American Medical Association,* 272, 1367, 1994.

Pool, R., Beyond databases and e-mail, *Science*, 261, 841, 1993.

Sidhu, K. S., Computer data bases for carcinogenicity and risk assessment, *Environmental Toxicology and Chemistry*, 8, 1217, 1989.

Skinder, R. F. and Gresehover, R. S., An Internet navigation tool for the technical and scientific researcher, *Online,* 19, 38, 1995.

Software for Science catalog, free and published quarterly by Scitech Software for Science, 2525 N. Elston Avenue, Chicago, Illinois 60647-2003 (Phone: 1-800-622-3345; Internet World Wide Web home page is *http://www.scitechint.com/scitech/*).

Wexler, P., *Information Resources in Toxicology,* Elsevier-North-Holland, New York, 1982, 333.

Wexler, P., *Information Resources in Toxicology* (Second Edition), Elsevier Science Publishing, Inc., New York, 1988, 510.

Wexler, P., The framework of toxicology information, *Toxicology*, 60, 67, 1990a.

Wexler, P., Toxicological information series, II. A survey of toxicology information, *Fundamental and Applied Toxicology*, 14, 649, 1990b. (A published erratum appears in *Fundamental and Applied Toxicology*, 15, 631, 1990; a published comment appears in *Fundamental and Applied Toxicology*, 16, 210, 1991.)

Young, R., Environmental software review, *Pollution Engineering,* 17(1), 30, 1985.

QUESTIONS

1. Name some scientific journals that could be reviewed to help identify potentially useful risk assessment and risk management software and databases.
2. Provide some reasons for why CD-ROMs have become so useful for storing and accessing information.
3. Provide a key reason why a CD-ROM version of a database may not be the preferred choice over its online version.
4. What is the Internet?
5. What is the World Wide Web?
6. What are some examples of how the Internet and the Internet's World Wide Web can be used by risk assessors and risk managers?
7. Describe one or more of the following: personal assistant, intelligent assistant, or intelligent agent.

Section III

Risk Perception, Law, Politics, and Risk Communication

CHAPTER **III.1**

Risk Perception and Trust

Paul Slovic

INTRODUCTION

Perceived risk can best be characterized as a battleground marked by strong and conflicting views about the nature and seriousness of the risks of modern life. The paradox for those who study risk perception is that, as people have become healthier and safer on average, they have become more — rather than less — concerned about risk, and they feel more and more vulnerable to the risks of modern life. Studies of risk perception attempt to understand this paradox and to understand why it is that our perceptions are so often at variance with what the experts say we should be concerned about. We see, for example, that people have very great concerns about nuclear power and chemical risks (which most experts consider acceptably safe) and rather little concern about dams, alcohol, indoor radon, and motor vehicles (which experts consider to be risky).

Perceptions of risk appear to exert a strong influence on the regulatory agenda of government agencies. In 1987, a U.S. Environmental Protection Agency (EPA) task force of 75 experts ranked the seriousness of risk for 31 environmental problems. The results showed that (1) the EPA's actual priorities differed in many ways from this ranking and (2) their priorities were much closer to the public's concerns than to the experts' risk assessments. In particular, hazardous waste disposal was the highest priority item on EPA's agenda and the area of greatest concern for the public as well, yet this problem was judged only moderate in risk by the experts.

It is important to understand why the public is so greatly concerned today about risks from technology and its waste products. This author does not have the answer, but has several hypotheses about factors that might contribute to the perceptions that such risks are high and increasing. One hypothesis is that we have greater ability

1-56670-130-9/97/$0.00+$.50

than ever before to detect minute levels of toxic substances. We can detect parts per billion or trillion or even smaller amounts of chemicals in water and air and in our own bodies. At the same time, we have considerable difficulty understanding the health implications of this new knowledge. Second, we have an increasing reliance on powerful new technologies that can have serious consequences if something goes wrong. When we lack familiarity with a technology, it is natural to be suspicious of it and cautious in accepting its risks. Third, in recent years, we have experienced a number of spectacular and catastrophic mishaps, such as Three Mile Island, Chernobyl, Bhopal, the Challenger accident, and the chemical contamination at Love Canal. These events receive extensive media coverage which highlights the failure of supposedly "fail-safe" systems. Fourth, we have an immense amount of litigation over risk problems, which brings these problems to public attention and pits expert against expert, leading to loss of credibility on all sides. Fifth, the benefits from technology are often taken for granted. When we fail to perceive significant benefit from an activity, we are intolerant of any degree of risk. Sixth, we are now being told that we have the ability to control many elements of risk, for example, by wearing seatbelts, changing our diets, getting more exercise, and so on. Perhaps the increased awareness that we have control over many risks makes us more frustrated and angered by those risks that we are not to be able to control, such as when exposures are imposed on us involuntarily (e.g., air and water pollution). Seventh, psychological studies indicate that when people are wealthier and have more to lose, they become more cautious in their decision making. Perhaps this holds true with regard to health as well as wealth. Finally, there may be real changes in the nature of today's risks. For example, there may be greater potential for catastrophe than there was in the past, due to the complexity, potency, and interconnectedness of technological systems (Perrow 1984).

Key Words: perceived risk, trust, risk communication, risk assessment, risk management

1. PSYCHOMETRIC STUDIES

Public opinion polls have been supplemented by more quantitative studies of risk perception that examine the judgments people make when they are asked to characterize and evaluate hazardous activities and technologies. One broad strategy for studying perceived risk is to develop a taxonomy for hazards that can be used to understand and predict responses to their risks. The most common approach to this goal has employed the *psychometric paradigm* (Slovic 1986, 1987, Slovic et al. 1985) which produces quantitative representations or "cognitive maps" of risk attitudes and perceptions. Within the psychometric paradigm, people make quantitative judgments about the current and desired riskiness of various hazards. These judgments are then related to judgments of other properties, such as the hazard's status on characteristics that have been hypothesized to account for risk perceptions (e.g., voluntariness, dread, catastrophic potential, controllability). These characteristics of risk tend to be correlated highly with each other across the domain of hazards. For example, hazards judged to be catastrophic also tend to be seen as uncontrollable and involuntary. Investigation of these relationships by means of factor analysis has shown that the broad domain of risk characteristics can be reduced to a small set of higher-order characteristics or "factors."

The factor space shown in Figure 1 has been replicated often. Factor 1, labeled "Dread Risk," is defined at its high (right-hand) end by perceived lack of control, dread, catastrophic potential, and fatal consequences. Factor 2, labeled "Unknown Risk," is defined at its high end by hazards perceived as unknown, unobservable, new, and delayed in their manifestation of harm. Nuclear power stands out in this (and many other) study as uniquely unknown and dreaded, with great potential for catastrophe. Nuclear waste tends to be perceived in a similar way. Chemical hazards such as pesticides and polychlorinated biphenyls (PCBs) are not too distant from nuclear hazards in the upper-right-hand quadrant of the space.

Research has shown that laypeople's perceptions of risk are closely related to these factor spaces. In particular, the further to the right that a hazard appears in the space, the higher its perceived risk, the more people want to see its current risks reduced, and the more people want to see strict regulation employed to achieve the desired reduction in risk (Slovic et al. 1985). In contrast, experts' perceptions of risk are not closely related to any of the various risk characteristics or factors derived from these characteristics. Instead, experts appear to see riskiness as synonymous with expected annual mortality. As a result, conflicts over "risk" may result from experts and laypeople having different definitions of the concept. Expert recitations of risk probabilities and statistics will do little to change people's attitudes and perceptions if these perceptions are based on nonprobabilistic and nonstatistical qualities.

Another important finding from risk perception research is that men and women have systematically different risk perceptions (see Figure 2). Some have attributed this to men's greater knowledge of technology and risk (i.e., science literacy). But a study by Barke et al. (1995) found that risk judgements of women scientists differed from the judgements of male scientists in much the same way as men and women nonscientists differed. Women scientists perceived higher risk than men scientists for nuclear power and nuclear waste.

Recently, Flynn et al. (1994) examined risk perception as a function of both race and gender. Surprisingly, nonwhite men and women differed rather little in their perceptions and differed little from white women. It was white males who stood apart from the rest in seeing risks as less serious than others (see Figure 3). Subsequent analysis showed that this "white male effect" was due to the response of 30% of the white male subgroup of relatively high education and income.

Why do a substantial percentage of white males see the world as much less risky than everyone else sees it? Perhaps white males see less risk in the world because they create, manage, control, and benefit from so much of it. Perhaps women and nonwhite men see the world as more dangerous because in many ways they are

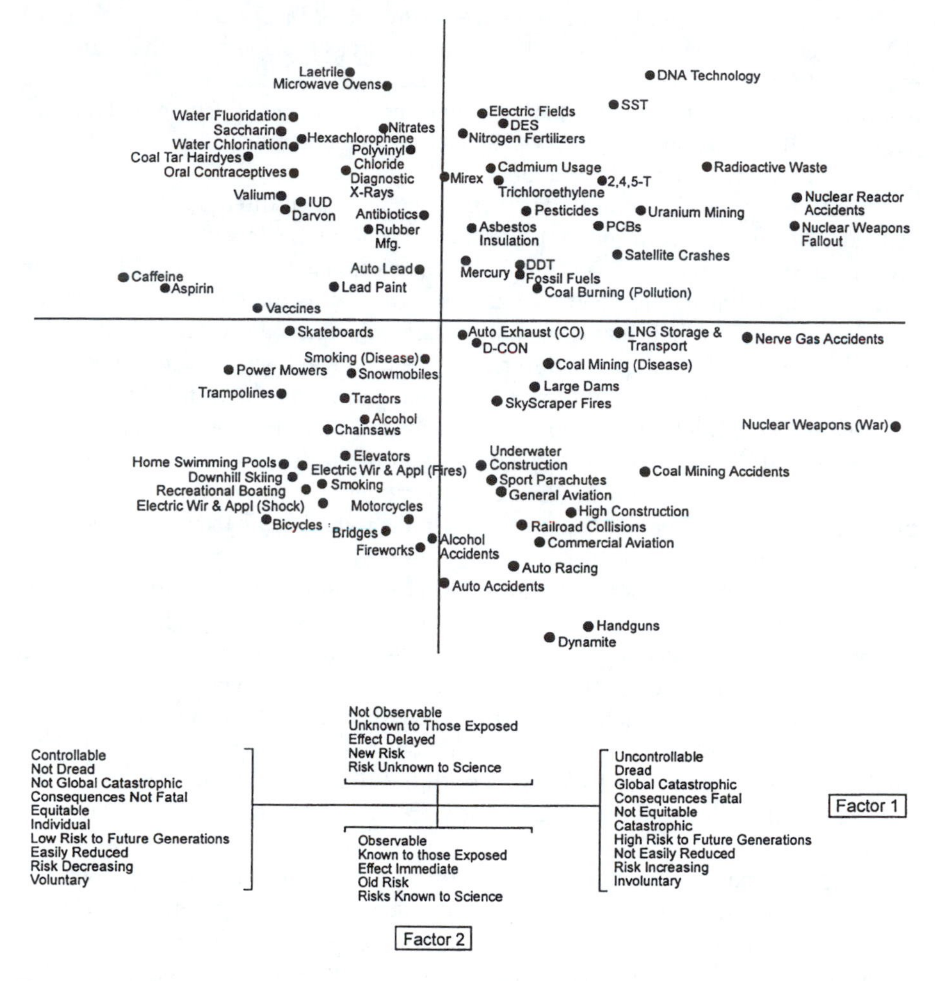

Figure 1 Location of 81 hazards on Factors 1 (Dread Risk) and 2 (Unknown Risk) derived from the interrelationships among 15 risk characterisitics. Each factor is made up of a combination of characteristics, as indicated by the lower diagram. (From Slovic, P. (1987). *Science,* **236**, 280. Copyright American Association for the Advancement of Science. With permission.)

more vulnerable, because they benefit less from many of its technologies and institutions, and because they have less power and control.

Inasmuch as these sociopolitical factors shape public perception of risks, we can see yet another reason why traditional attempts to make people see the world as white males do, by showing them statistics and risk assessments, are unlikely to succeed. The problem of risk conflict and controversy clearly goes beyond science. It is deeply rooted in the social and political fabric of our society. This analysis points to the need for a fairer and more equitable society, as well as for fairer processes for managing risk.

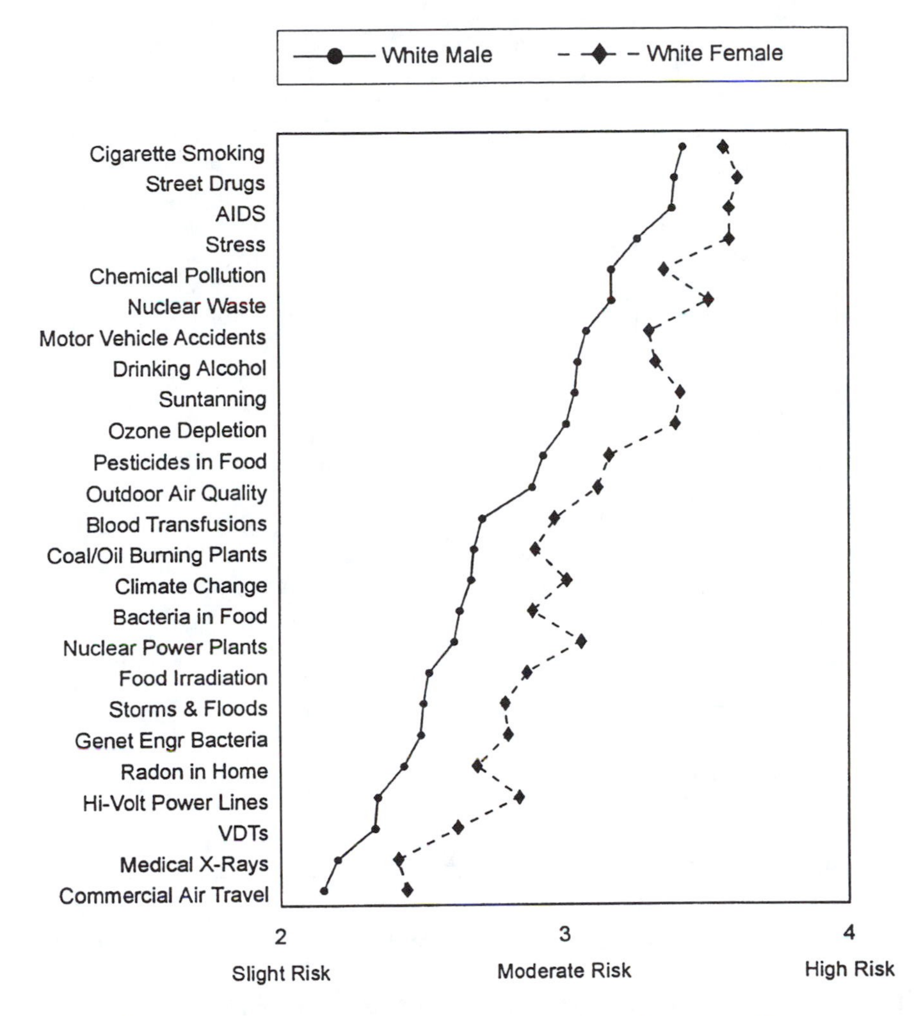

Figure 2 Mean risk perception ratings by white males and white females. (From a survey conducted by P. Slovic and co-workers.)

2. RISK COMMUNICATION AND TRUST

2.1 The Importance of Trust

The research described previously has painted a portrait of risk perception influenced by the interplay of psychological, social, and political factors. Members of the public and experts can disagree about risk because they define risk differently, have different worldviews, or different social status. Another reason why the public often rejects scientists' risk assessments is lack of trust.

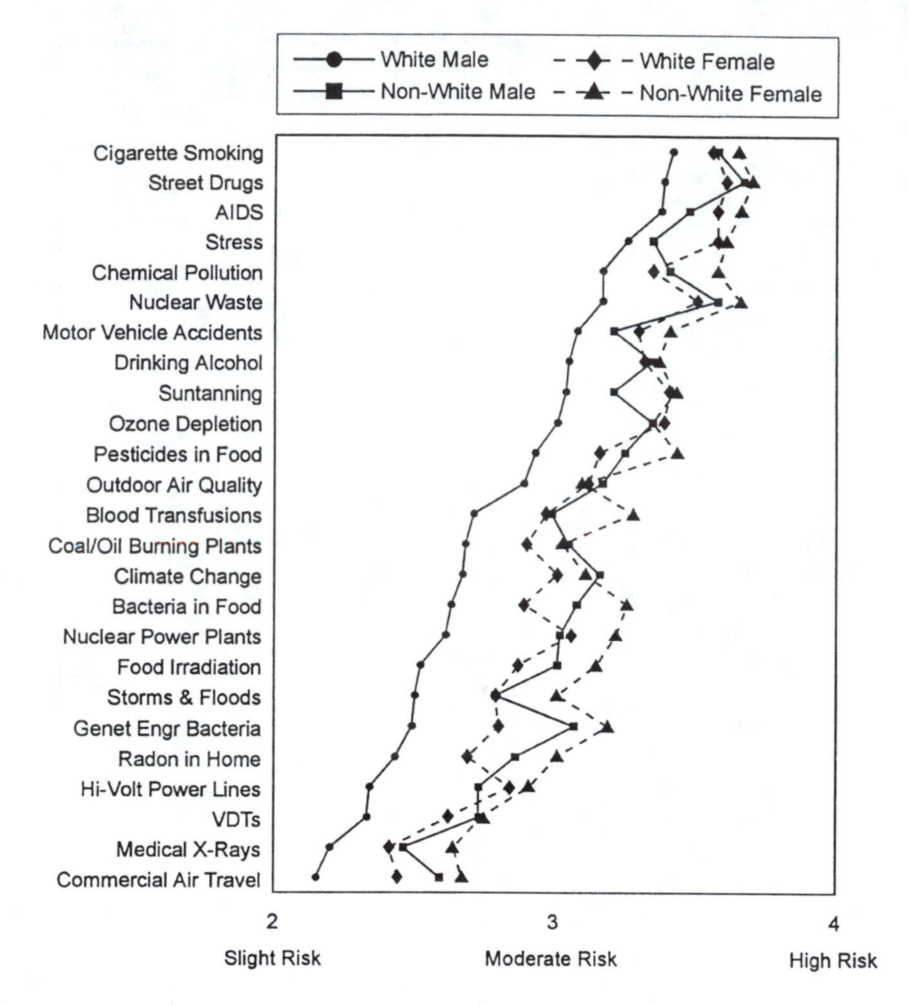

Figure 3 Mean risk perception ratings by race and gender. (From Flynn, J., Slovic, P., and Mertz, C. K. (1994). *Risk Analysis,* **14**(6), 1104. With permission.)

Social relationships of all types, including risk management, rely heavily on trust. Indeed, much of the contentiousness that has been observed in the risk management arena has been attributed to a climate of distrust that exists between the public, industry, and risk management professionals (e.g., Slovic 1993, Slovic et al. 1991). To appreciate the importance of trust, it is instructive to compare those risks that we fear and avoid with those we accept casually. Starr (1985) has pointed to the public's lack of concern about the risks from tigers in urban zoos as evidence that acceptance of risks is strongly dependent on confidence in risk management. Risk perception research (Slovic 1990) documents that people view *medical* technologies based on use of radiation and chemicals (i.e., X-rays and prescription drugs) as high in benefit, low in risk, and clearly acceptable. However, people view *industrial* technologies involving radiation and chemicals (i.e., nuclear power, pesticides, industrial chemicals) as high in risk, low in benefit, and unacceptable. Although X-rays and medicines pose significant risks, our relatively high degree of trust in the physicians who manage these devices makes them acceptable. Numerous polls have shown that the government and industry officials who oversee the management of nuclear power and nonmedical chemicals are not highly trusted (Flynn et al. 1992, McCallum et al. 1990, Pijawka and Mushkatel 1992, Slovic et al. 1991).

Because it is impossible to exclude the public in a highly participatory democracy, the response of industry and government to this crisis of confidence has been to turn to the young and still primitive field of risk communication in search of methods to bring experts and laypeople into alignment and make conflicts over technological decisions easier to resolve. Although attention to communication can prevent blunders that exacerbate conflict, there is rather little evidence that risk communication has made any significant contribution to reducing the gap between technical risk assessments and public perceptions or to facilitating decisions about nuclear waste or other major sources of risk conflict. The limited effectiveness of risk communication efforts can be attributed to the lack of trust. If you trust the risk manager, communication is relatively easy. If trust is lacking, no form or process of communication will be satisfactory (Fessenden-Raden et al. 1987). Thus, trust is more fundamental to conflict resolution than is risk communication.

2.2 How Trust Is Created and Destroyed

One of the most fundamental qualities of trust has been known for ages. Trust is fragile. It is typically created rather slowly, but it can be destroyed in an instant, by a single mishap or mistake. Thus, once trust is lost, it may take a long time to rebuild it to its former state. In some instances, lost trust may never be regained. Abraham Lincoln understood this quality. In a letter to Alexander McClure he observed: "If you *once* forfeit the confidence of your fellow citizens, you can *never* regain their respect and esteem" (italics added).

2.3 The Impact of Events on Trust

The fact that trust is easier to destroy than to create reflects certain fundamental mechanisms of human psychology called here "the asymmetry principle." When it comes to winning trust, the playing field is not level. It is tilted toward distrust for each of the following reasons:

1. Negative (trust-destroying) events are more visible or noticeable than positive (trust-building) events. Negative events often take the form of specific, well-defined incidents such as accidents, lies, discoveries of errors, or other mismanagement. Positive events, while sometimes visible, more often are fuzzy or indistinct. For example, how many positive events are represented by the safe operation of a nuclear power plant for 1 day? Is this one event, dozens of events, hundreds? There is no precise answer. When events are invisible or poorly defined, they carry little or no weight in shaping our attitudes and opinions.

2. When events do come to our attention, negative (trust-destroying) events carry much greater weight than positive events. This important psychological tendency is illustrated by a study in which 103 college students rated the impact on trust of 45 hypothetical news events pertaining to the management of a large nuclear power plant in their community (Slovic et al. 1993). The following events were designed to be trust increasing:
   - There have been no reported safety problems at the plant during the past year.
   - There is careful selection and training of employees at the plant.
   - Plant managers live nearby the plant.
   - The county medical examiner reports that the health of people living near the plant is *better* than the average for the region.

   Other events were designed to be trust decreasing:
   - A potential safety problem was found to have been covered up by plant officials.
   - Plant safety inspections are delayed in order to meet the electricity production quota for the month.
   - A nuclear power plant in another state has a serious accident.
   - The county medical examiner reports that the health of people living near the plant is *worse* than the average for the region.

   The respondents were asked to indicate, for each event, whether their trust in the management of the plant would be increased or decreased upon learning of that event. After doing this, they rated how strongly their trust would be affected by the event on a scale ranging from 1 (very small impact on trust) to 7 (very powerful impact on trust).

   The percentages of Category 7 ratings, shown in Figure 4, dramatically demonstrate that negative events are seen as far more likely to have a powerful effect on trust than are positive events. The data shown in Table 1 are typical. The negative event, reporting plant neighbors' health as *worse* than average, was rated 6 or 7 on the impact scale by 50% of the respondents. A matched event, reporting neighbors' health to be *better* than average, was rated 6 or 7 by only 18.3% of the respondents.

   There was only one event perceived to have any substantial impact on increasing trust. This event stated that: "An advisory board of local citizens and environmentalists is established to monitor the plant and is given legal authority to shut the plant down if they believe it to be unsafe."

   This strong delegation of authority to the local public was rated 6 or 7 on the impact scale by 38.4% of the respondents. Although this was a far stronger showing than for any other positive event, it would have been a rather average performance in the distribution of impacts for negative events.

   The importance of an event is related, at least in part, to its frequency (or rarity). An accident in a nuclear plant is more informative with regard to risk than is a day (or even a large number of days) without an accident. Thus, in systems where we are concerned about low-probability/high-consequence events, problematic events will increase our perceptions of risk to a much greater degree than favorable events will decrease them.
3. Adding fuel to the fire of asymmetry is yet another idiosyncracy of human psychology; sources of bad (trust-destroying) news tend to be seen as more credible than sources of good news. For example, in several studies of what we call "intuitive toxicology" (Kraus et al. 1992), we have examined people's confidence in the ability of animal studies to predict human health effects from chemicals. In general, confidence in the validity of animal studies is not particularly high. However, when told that a study has found that a chemical is carcinogenic in animals, people

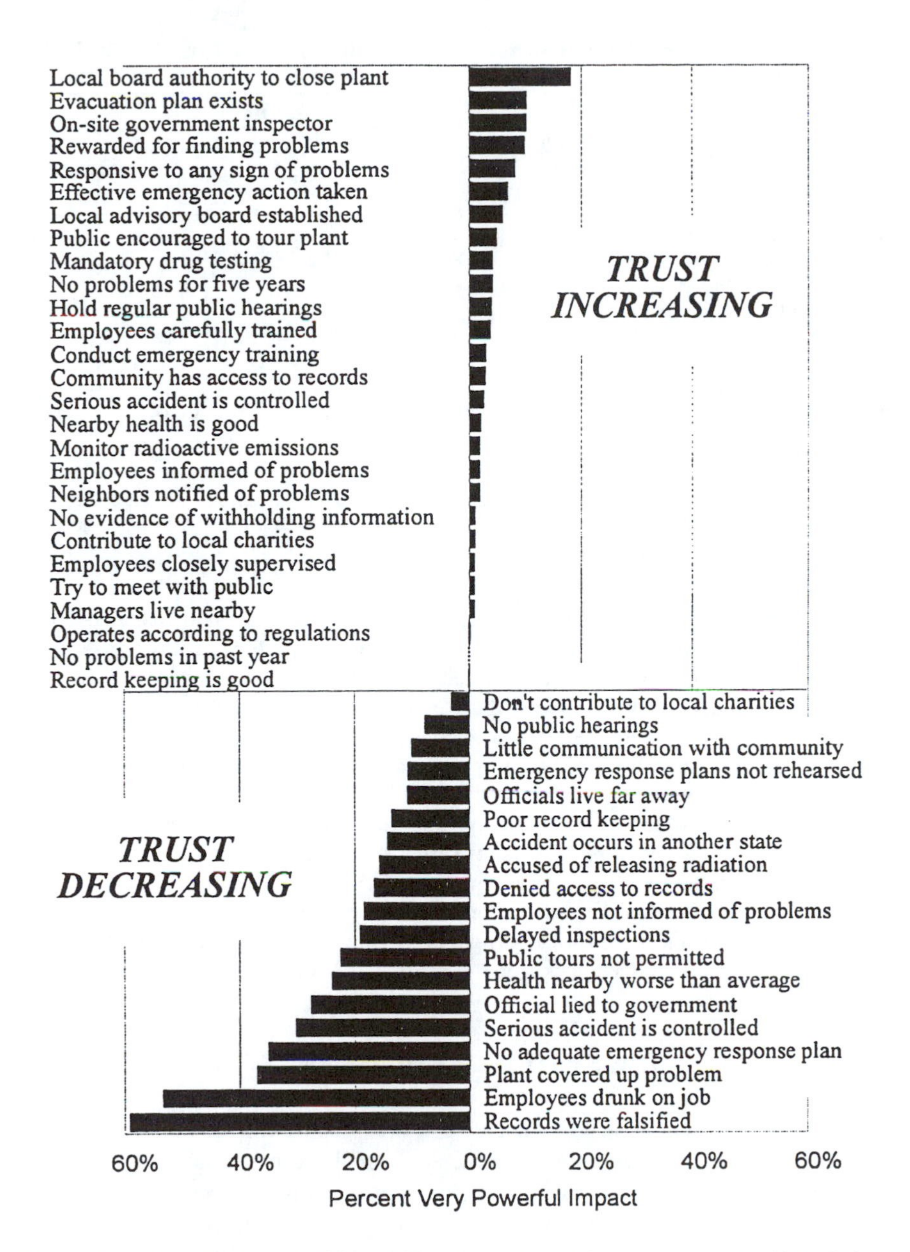

Figure 4 Differential impact of trust-increasing and trust-decreasing events. *Note:* Only percentages of Category 7 ratings (very powerful impact) are shown here. (From Slovic, P. (1993). *Risk Analysis,* **13,** 675. With permission.)

express considerable confidence in the validity of this study for predicting health effects in humans. Regulators respond like the public. Positive (bad news) evidence from animal bioassays is presumptive evidence of risk to humans; negative evidence (e.g., the chemical was not found to be harmful) carries little weight (Efron 1984).

Table 1 Judged Impact of a Trust-Increasing Event and a Similar Trust-Decreasing Event

	Impact on trust						
	Very small 1	2	3	4	5	6	Very powerful 7
Trust-increasing event							
The county medical examiner reports that the health of people living near the plant is *better* than average.	21.5	14.0	10.8	18.3	17.2	16.1	2.2
Trust-decreasing event							
The county medical examiner reports that the health of people living near the plant is *worse* than average.	3.0	8.0	2.0	16.0	21.0	26.0	24.0

Note: Cell entries indicate the percentage of respondents in each impact rating category.
From Slovic, P. (1993). *Risk Analysis,* **13,** 675. With permission.

4. Another important psychological tendency is that distrust, once initiated, tends to reinforce and perpetuate distrust. This occurs in two ways. First, distrust tends to inhibit the kinds of personal contacts and experiences that are necessary to overcome distrust. By avoiding others whose motives or actions we distrust, we never come to see that these people are competent, well meaning, and trustworthy. Second, initial trust or distrust colors our interpretation of events, thus reinforcing our prior beliefs. Persons who trusted the nuclear power industry saw the events at Three Mile Island as demonstrating the soundness of the defense-in-depth principle, noting that the multiple safety systems shut the plant down and contained most of its radiation. Persons who distrusted nuclear power prior to the accident took an entirely different message from the same events, perceiving that those in charge did not understand what was wrong or how to fix it and that catastrophe was averted only by sheer luck.

3. THE SYSTEM DESTROYS TRUST

Thus far, the psychological tendencies that create and reinforce distrust in situations of risk have been discussed. Appreciation of those psychological principles leads us toward a new perspective on risk perception, trust, and conflict. Conflicts and controversies surrounding risk management are not due to public irrationality or ignorance, but, instead, can be seen as expected side effects of these psychological tendencies, interacting with a highly participatory democratic system of government, and amplified by certain powerful technological and social changes in society. Technological change has given the electronic and print media the capability (effectively utilized) of informing us of news from all over the world, often right as it happens. Moreover, just as individuals give greater weight and attention to negative events, so do the news media. Much of what the media reports is bad (trust-destroying) news (Lichtenberg and MacLean 1992). This is convincingly demonstrated by Koren and

Klein (1991), who compared the rates of newspaper reporting of two studies, one providing bad news and one good news, published back to back in the March 20, 1991 issue of the *Journal of the American Medical Association*. Both studies examined the link between radiation exposure and cancer. The bad news study showed an increased risk to leukemia in white men working at the Oak Ridge National Laboratory in Oak Ridge, TN. The good news study failed to show an increased risk of cancer in people residing near nuclear facilities. Koren and Klein found that subsequent newspaper coverage was far greater for the study showing increased risk.

The second important change, a social phenomenon, is the rise of powerful special interest groups, well funded (by a fearful public) and sophisticated in using their own experts and the media to communicate their concerns and their distrust to the public in order to influence risk policy debates and decisions (Wall Street Journal 1989). The social problem is compounded by the fact that we tend to manage our risks within an adversarial legal system that pits expert vs. expert, contradicting each other's risk assessments and further destroying public trust.

The young science of risk assessment is too fragile, too indirect, to prevail in such a hostile atmosphere. Scientific analysis of risks cannot allay our fears of low-probability catastrophes or delayed cancers unless we trust the system. In the absence of trust, science (and risk assessment) can only feed distrust by uncovering more bad news. A single study demonstrating an association between exposure to chemicals or radiation and some adverse health effect cannot easily be offset by numerous studies failing to find such an association. Thus, for example, the more studies that are conducted looking for effects of electric and magnetic fields or other difficult-to-evaluate hazards, the more likely it is that these studies will increase public concerns, even if the majority of these studies fail to find any association with ill health (MacGregor et al. 1994, Morgan et al. 1985). In short, risk assessment studies tend to increase perceived risk.

In sum, the failures of risk management point strongly to the erosion of trust, both in government and in many of our social institutions, as an important causal factor in the conflicts that exist between the community of risk experts and the public. Proposed solutions to the distrust of risk management tend to follow two directions. One path that has been advocated by a number of researchers is to work toward increasing public trust in risk management. This chapter has discussed research that has been conducted in this spirit. While it is much too soon to express either optimism or pessimism about the likely success of this strategy, it is a significantly challenging problem that at the moment appears to have no easy answers.

A second path leads in the direction of developing risk management processes that do not rely on trust or rely on it only minimally. Though it is seldom acknowledged explicitly, many of the steps currently being taken by government and industry to involve the public through community advisory panels and the like are, in effect, establishing layers of oversight such that the checks-and-balances principles inherent in democratic governments are instituted within technological risk management. This may be a fruitful avenue to pursue, and research along these lines is certainly needed.

ACKNOWLEDGMENT

Preparation of this paper was supported by the Electric Power Research Institute and the National Science Foundation under Grant No. SES-91-10592.

REFERENCES

Barke, R., Jenkins-Smith, H., and Slovic, P. (1995). *Risk Perceptions of Men and Women Scientists,* Report No. 95-6. Eugene, OR: Decision Research.

Efron, E. (1984). *The Apocalyptics*. New York: Simon & Schuster.

Fessendon-Raden, J., Fitchen, J. M., and Heath, J. S. (1987). Providing risk information in communities: Factors influencing what is heard and accepted. *Science Technology and Human Values*, **12**, 94–101.

Flynn, J., Burns, W., Mertz, C. K., and Slovic, P. (1992). Trust as a determinant of opposition to a high-level radioactive waste repository: Analysis of a structural model. *Risk Analysis*, **12,** 417–430.

Flynn, J., Slovic, P., and Mertz, C. K. (1995). Gender, race, and perception of environmental health risks. *Risk Analysis*, **14**(6), 1101–1108.

Koren, G., and Klein, N. (1991). Bias against negative studies in newspaper reports of medical research. *Journal of the American Medical Association*, **266**, 1824–1826.

Kraus, N., Malmfors, T., and Slovic, P. (1992). Intuitive toxicology: Expert and lay judgments of chemical risks. *Risk Analysis*, **12**, 215–232.

Lichtenberg, J., and MacLean, D. (1992). Is good news no news? *The Geneva Papers on Risk and Insurance*, **17**, 362–365.

MacGregor, D., Slovic, P., and Morgan, M. G. (1994). Perception of risks from electromagnetic fields: A psychometric evaluation of a risk-communication approach. *Risk Analysis*, **14** (5), 815–828.

McCallum, D. B., Hammond, S. L., Morris, L. A., and Covello, V. T. (1990). Public knowledge and perceptions of chemical risks in six communities (Report No. 230-01-90-074). Washington, D.C.: U. S. Environmental Protection Agency.

Morgan, M. G., Slovic, P., Nair, I., Geisler, D., MacGregor, D., Fischhoff, B., Lincoln, D., and Florig, K. (1985). Powerline frequency electric and magnetic fields: A pilot study of risk perception. *Risk Analysis*, **5**, 139–149.

Perrow, C. (1984). *Normal Accidents: Living with High-Risk Technologies*. New York: Basic Books.

Pijawka, D., and Mushkatel, A. (1992). Public opposition to the siting of the high-level nuclear waste repository: The importance of trust. *Policy Studies Review*, **10**(4), 180–194.

Slovic, P. (1986). Informing and educating the public about risk. *Risk Analysis*, **4,** 403–415.

Slovic, P. (1987). Perception of risk. *Science,* **236,** 280–285.

Slovic, P. (1990). Perception of risk from radiation. In W. K. Sinclair (Ed.), *Proceedings of the Twenty-Fifth Annual Meeting of the National Council on Radiation Protection and Measurements. Vol 11*: *Radiation protection today: The NCRP at sixty years* (pp. 73–97). Bethesda, MD: NCRP.

Slovic, P. (1993). Perceived risk, trust, and democracy: A systems perspective. *Risk Analysis*, **13**, 675–682.

Slovic, P., Fischhoff, B., and Lichtenstein, S. (1985). Characterizing perceived risk. In *Perilous Progress: Technology as Hazard*. R. W. Kates, C. Hohenemser, and J. X. Kasperson (Eds.), (pp. 91–123). Boulder, CO: Westview.

Slovic, P., Flynn, J., Johnson, S., and Mertz, C. K. (1993). The dynamics of trust in situations of risk (Report No. 93-2). Eugene, OR: Decision Research.

Slovic, P., Flynn, J., and Layman, M. (1991). Perceived risk, trust, and the politics of nuclear waste. *Science*, **254**, 1603–1607.

Starr, C. (1985). Risk management, assessment, and acceptability. *Risk Analysis*, **5**, 97–102.

Wall Street Journal. (1989, October 3). How a PR firm executed the Alar scare, pp. A1–A3.

QUESTIONS

1. Name three factors that may be causing perceptions of risk to increase in recent years.
2. According to psychometric studies, how do experts and laypersons tend to differ in their perceptions of risk?
3. What data suggest the influence of sociopolitical factors on perceptions of risk?
4. Why is it unlikely that an agency such as DOE could restore the public's trust in its ability to manage the nation's radioactive waste?

CHAPTER III.2

The Insurability of Risks*

Howard Kunreuther and Paul K. Freeman

SUMMARY

This chapter examines two broad conditions for a risk to be **insurable**. Condition 1 requires the insurer to set a pure premium by quantifying the frequency and magnitude of loss associated with specific events associated with the risk. Condition 2 specifies a set of factors, such as adverse selection, moral hazard, and degree of correlated risk, that need to be taken into account when the insurer determines what premium and type of coverage (maximum limits, nature of deductible) it wants to offer. Finally, a risk is *not* insurable unless there is sufficient demand for the product at some price to cover the upfront costs of developing the product and the expenses associated with marketing policies.

Key Words: insurance, environmental risk, insurability conditions

1. INTRODUCTION

What does it mean to say that a particular risk is insurable? We must address this question from the vantage point of the potential supplier of insurance. We will be focusing on a standard contract between buyer and seller; the insurer offers coverage against a specific risk at some premium R and the insured is protected against a prespecified set of losses defined in the contract.

* The material on which this chapter is based draws heavily on Chapter 4 of a larger study by Paul Freeman and Howard Kunreuther on "Insuring Environmental Risks," to be published. Support from NSF Grant #5-24603 to the Wharton Risk Management and Decision Processes Center, University of Pennsylvania, Philadelphia, PA, is gratefully acknowledged.

1-56670-130-9/97/$0.00+$.50

2. TWO INSURABILITY CONDITIONS

Two conditions must be met before insurance providers are willing to provide coverage against an uncertain event. Condition 1 is the ability to identify and, possibly, quantify the risk. The insurer must know that it is possible to estimate what losses they are likely to incur when providing different levels of coverage. Condition 2 is the ability to set premiums for each potential customer or class of customers. This requires some knowledge of the customer's risk in relation to others in the population of potentially insureds.

If Conditions 1 and 2 are both satisfied, a risk is considered to be insurable. But, it still may not be profitable. In other words, it may not be possible to specify a rate where there is sufficient demand to yield a positive profit from offering coverage. In such cases, there will be no market for insurance.

2.1 Condition 1: Identifying the Risk

To satisfy this condition, estimates must be made of the frequency of specific events occurring and the magnitude of the loss should the event occur. Three examples illustrate the type of data that could be used to identify the risk. In some cases, this may enable the insurer to specify a set of estimates on which to base an insurance premium. In other cases, the data may be much less specific.

2.1.1 Fire

Rating agencies typically collect data on all the losses incurred over a period of time for a particular risk and an exposure unit. Suppose the hazard is fire and the exposure unit is a well-defined entity, such as $300,000 wood-frame homes of similar design, to be insured for 1 year in California. The typical measurement is the pure premium (PP), which is given by

$$\text{PP} = \text{Total Losses/Exposure Unit*} \quad \textbf{(1)}$$

Assume that the rating agency has collected data on 100,000 wood-frame homes in that state and has determined that the total annual losses from fires to these structures over the past year is $20 million. If these data are representative of the expected loss to this class of wood-frame homes in California next year, then, using Equation 1, PP is given by

$$\text{PP} = \$20{,}000{,}000 \;/\; 100{,}000 = \$200$$

This figure is simply an average. It does not differentiate between locations of wood-frame homes in the state, the distance of each home from a fire hydrant, or

* The pure premium (PP) normally considers loss adjustment expenses for settling a claim. We will assume that this component is part of total losses. For more details on calculating PPs see Launie et al. (1986).

the quality of the fire department serving different communities. All of these factors are often taken into consideration by underwriters who set final rates by calculating a premium that reflects the risk to particular structures.

2.1.2 Earthquakes

If there were considerable data available on annual damage to wood-frame homes in California from earthquakes of different magnitudes, then a similar method to the one described for fire could be used to determine the probability and magnitude of loss.

Due to the infrequency of earthquakes and the relatively few number of homes that have been insured against the earthquake peril, this type of analysis is not feasible at this time. Insurance providers have to turn to scientific studies by seismologists, geologists, and structural engineers to estimate the frequency of earthquakes of different magnitudes, as well as the damage that is likely to occur to different structures from such earthquakes.

Table 1 is a template indicating the type of information that would have to be collected to determine the PP for a wood-frame house subject to earthquake damage in California. The first column (Event) reflects one way of calculating the severity of an earthquake occurring, i.e., the modified Mercalli intensity scale. The second column (Probability) specifies the annual probability (p_i) of a wood-frame home in California being damaged in an earthquake. The third column (Loss) is the amount of damage an earthquake might cause to a wood-frame home.

Table 1 Calculating Annual Pure Premium from Scientific Data for Earthquake Damage to Wood-Frame Homes in California

Event[a]	Probability (p_i)	Loss (L_i)
IV		
V		
VI		
VII		
VIII		
IX		
X		

[a] Based on the modified Mercalli Intensity Scale.

If all these data are available from scientific studies, the PP in this case would be equivalent to the expected loss ($E(L)$) which is given by

$$E(L) = p_i \times L_i \tag{2}$$

Over the past 20 years, seismologists have determined certain factors that will influence the probability of an earthquake in a specific area, but they are still uncertain as to how they interact with each other and their relative importance.* At the same time, there has been considerable damage data collected by engineers since the Alaskan earthquake of 1964, which has increased our understanding of the performance of various types of buildings and structures in earthquakes of different magnitudes.*

While seismologists and geologists cannot predict with certainty the probability of earthquakes of different magnitudes occurring in specific regions of California, they can provide conservative estimates of the risk. For example, it is possible to develop worst-case scenarios for determining $E(L)$ using Equation 2 by computing

$$E(L^*) = p^*_i \times L^*_i \tag{3}$$

The factor p^*_i is the maximum credible probability assigned by seismologists to an earthquake of intensity i. The factor L^*_i represents engineers best estimates of the maximum likely damage to a wood-frame house in such an earthquake. Using the estimate from Equation 3 as a basis for calculating a PP, the damage to wood-frame homes from earthquakes becomes a quantifiable risk.

2.1.3 Underground Storage Tanks (USTs)

Suppose that an insurer was attempting to estimate the PP for a new technological advance, such as an improved design for USTs. Since there are no historical data associated with the risk, the insurer would have to rely on scientific studies to estimate the probabilities (p_i) and cleanup costs (L_i) associated with a particular type of defect i in the tank that causes a leak.

To the extent that the insurer has confidence in these scientific estimates of the performance of the tank and the costs of the cleanup from leaks of different magnitudes, it should be able to quantify the risk and calculate a PP. If, on the other hand, the insurer is uncertain about the frequency or loss estimates, it may conclude that the risk cannot be quantified and hence is uninsurable.

2.2 Condition 2: Setting Premiums for Specific Risks

Once a PP is determined using one of the methods specified, the insurer can determine what rate it needs to charge in order to make a profit by providing coverage against specific risks. There are a number of factors that come into play in determining this dollar figure.

2.2.1 Ambiguity of Risk

Not surprisingly, the higher the uncertainty regarding the probability of a specific loss and its magnitude, the higher the premium will be. As shown by a series of empirical studies, actuaries and underwriters are so ambiguity averse and risk averse

* Some of these factors are the time elapsed since the last earthquake, tilting of the land surface, fluctuations in the magnetic field, and changes in the electrical resistance of the ground.
* An Office of Technology Assessment (1995) report provides a detailed discussion on the state of the art of earthquake risk assessment and a comprehensive set of references.

that they tend to charge much higher premiums than if the risk were well specified.* A questionnaire was mailed to 896 underwriters in 190 randomly chosen insurance companies to determine what PPs** they would set for either an earthquake or leaking UST risk. The earthquake scenario involved insuring a factory against property damage from a severe earthquake. The UST scenario involved liability coverage for owners of a tank containing toxic chemicals against damages if the tank leaks. A neutral risk scenario acted as a reference point for the two context-based scenarios. It simply provided probability and loss estimates for an unnamed peril.

For each scenario, four cases were presented, reflecting the degree of ambiguity and uncertainty surrounding the probability and loss as shown in Table 2. A well-specified probability (*p*) refers to a situation in which there are considerable past data on a particular event that enable "all experts to agree that the probability of a loss is *p*." An ambiguous probability (*Ap*) refers to the case where "there is wide disagreement about the estimate of *p* and a high degree of uncertainty among the experts." A known loss (*L*) indicates that all experts agree that if a specific event occurs, the loss will equal *L*. An uncertain loss (*UL*) refers to a situation where the experts' best estimate of a loss is *L*, but estimates range from L_{min} to L_{max}.

Table 2 Classification of Risks by Degree of Ambiguity and Uncertainty

	Loss	
Probability	**Known**	**Unknown**
Well specified	Case 1 *p, L* Life, auto, fire	Case 3 *p, UL* Playground accidents
Ambiguous	Case 2 *Ap, L* Satellite, new products	Case 4 *Ap, UL* Earthquake, USTs

Case 1 reflects well-known risks for which large, actuarial databases exist, e.g., life, automobile, and fire insurance. Satellite accidents are an example of a Case 2 risk, since there is normally considerable uncertainty regarding the chances of their occurrence. If they do happen, the satellite is destroyed and the loss is well specified. Playground accidents illustrate Case 3 since there are good data on the chances of an accident occurring, but considerable uncertainty as to the magnitude of the liability award should a person be injured or killed. Finally, there is considerable ambiguity and uncertainty related to earthquakes and UST risks, so they are illustrative of Case 4.

In the questionnaire to the underwriters, Case 1 was represented by providing a well-specified probability (e.g., $p = .01$) and a well-specified loss (e.g., $L = \$1$ million). The other three cases introduced ambiguity and uncertainty into the

* For more details on the survey and the analysis of findings, see Kunreuther et al. 1995.

** The questionnaire instructions stated that PPs should exclude "loss adjustment expenses, claims expenses, commissions, premium taxes, defense costs, profits, investment return and the time valuation of money."

picture. For the case where L = \$1 million, the uncertain estimates ranged from L = \$0 to L = \$2 million.

One hundred and seventy-one completed questionnaires (19.1% of the total mailed) were received from 43 insurance companies (22.6% of those solicited). Table 3 shows the ratio of the average PP that underwriters would want to charge for each of the three cases where there is uncertainty and ambiguity in either p and/or L in relation to the average PP they specified for a risk that is well specified (Case 1). The data reveal that underwriters will want to charge a much higher premium when there is ambiguity and uncertainty regarding probabilities and/or losses. For example, as shown in Table 3, the premium for the Case 4 earthquake scenario was 1.5 times higher than for the well-specified Case 1 scenario.

Table 3 Ratio of Average Pure Premiums Specified by Underwriters Relative to a Well-Specified Case (Case 1) (p = .01, L = \$1 million)

Scenario	p, L Case 1	Ap, L Case 2	p, UL Case 3	Ap, UL Case 4
Neutral (N = 24)[a]	1	1.5	1.1	1.7
Earthquake (N = 23)	1	1.2	1.3	1.5
UST (N = 32)	1	1.5	1.4	1.8

[a] N = Number of respondents.

Data from Kunreuther et al. (1995).

Why do actuaries and underwriters price uncertain and ambiguous risks higher than well-specified risks? In two very insightful papers, Stone (1973a,b) indicates that, in setting premiums for any particular risk, insurers are motivated by the impact that their actions will have on the stability and solvency of their firm. Stability is measured by the loss ratio (LR), i.e., paid losses/written premiums, for a particular risk. Stability requires a probability less than some specified level p' (e.g., p' = .05) that the loss ratio exceeds a certain target level LR^* (e.g., LR^* = 1).

Solvency is measured by the survival constraint that relates aggregate losses for the risk in question to the current surplus plus premiums written. It requires that the probability of insolvency be less than p'' (e.g., p'' = 1 in 100,000). Berger and Kunreuther (1995) have shown that, if underwriters and actuaries are mindful of the two constraints of stability and solvency, they will set higher premiums as specific risks become more ambiguous and uncertain.

2.2.2 Adverse Selection

If the insurer cannot distinguish between the probability of a loss for good and bad risk categories, it faces the problem of adverse selection. What this means is that, if the insurer sets a premium based on the average probability of a loss using the entire population as a basis for this estimate, only the bad risks will want to purchase coverage. As a result, the insurer will expect to lose money on each policy that is sold.

The assumption underlying adverse selection is that purchasers of insurance have an informational advantage by knowing their risk type. Insurers, on the other hand, must invest considerable expense to collect information to distinguish between risks. A simple example illustrates the problem of adverse selection for a risk where the probabilities of a loss are $p_G = .1$ (good risks) and $p_B = .3$ (bad risks). For simplicity, assume that the loss is $L = \$100$ for both groups and that there are an equal number of potentially insurable individuals ($N = 50$) in each risk class. Table 4 summarizes these data.

Table 4 Data for Adverse Selection Example

Good risks	$p_G = .1$	$L = 100$	$N = 50$
Bad risks	$p_B = .3$	$L = 100$	$N = 50$

In the example in Table 4, the expected loss for a random individual in the population is 20.* If the insurer charged an actuarially fair premium across the entire population, only the bad risk class would normally purchase coverage, since their expected loss is 30 [.3(100)] and they would be pleased to pay only 20 for insurance. The good risks have an expected loss of 10 [.1(100)], so they would have to be extremely risk averse to be interested in paying 20 for coverage. When only the poor risks purchase coverage, the insurer would suffer an expected loss of –10 (20 – 30) on every policy it sold.

There are two principal ways that insurers can deal with this problem. If the company knows the probabilities associated with good and bad risks, but does not know the characteristics of the individuals, it can raise the premium to at least 30 so that it will not lose money on any individual purchasing coverage. In reality, where there is a spectrum of risks, the insurer may only be able to offer coverage to the worst risk class in order to make a profit. Hence, raising premiums is likely to produce a market failure in that very few of the individuals who are interested in purchasing coverage to cover their risk will actually do so at the going rate.

A second way for the insurer to deal with adverse selection is to offer two different price-coverage contracts. Poor risks will want to purchase contract 1 and good risks will purchase contract 2.** For example, contract 1 could be offered at price = 30 and coverage = 100, while contract 2 could be price = 10 and coverage = 40. If the good risks preferred contract 1 over contract 2 and the poor risks preferred contract 2 over contract 1, this would be one way for the insurers to market coverage to both groups while still breaking even.

Finally, the insurer could require some type of audit or examination to determine the nature of the risk more precisely. In the case of property, the audit could take the form of an inspection of the structure and its contents. For individuals, it could be some type of an examination, e.g., a medical exam if health insurance were being

* The expected loss for a random individual in the population is calculated as follows: [50(.1)(100) + 50(.3)(100)] / 100 = 20.

** This solution has been proposed by Rothschild and Stiglitz (1976).

offered. Certain types of coverage may not lend themselves to an exam, however, due to the nature of the risk. It is difficult to test a person for driving ability, for example, although past records and experience may be useful indicators as to whether a person is a good or bad risk.

Finally, it is important to remember that the problem of adverse selection only emerges if the persons considering the purchase of insurance have more accurate information on the probability of a loss than the firms selling coverage. If the customers have no better data than the underwriters, both groups are on an equal footing. Coverage will be offered at a single premium based on the average risk, and both good and poor risks will want to purchase policies.

2.2.3 Moral Hazard

Providing insurance protection to an individual may serve as an incentive for that person to behave more carelessly than before he/she had coverage. If the insurer cannot predict this behavior and relies on past loss data from uninsured individuals to estimate rates, the resulting premium is likely to be too low to cover losses.

The moral hazard problem is directly related to the difficulty in monitoring and controlling behavior once a person is insured. How do you monitor carelessness? Can you determine when a person decides to collect more on a policy than he/she deserves, e.g., making false claims or moving old furniture to the basement just before a flood hits the house?

The numerical example used previously to illustrate adverse selection can also demonstrate moral hazard. With adverse selection, the insurer cannot distinguish between good and bad risks. Moral hazard is created because the insurer must estimate the premium based on the probability of a loss before insurance is purchased, but the actual probability of a loss is much higher after a policy is sold. Table 5 depicts these data for the case in which there are 100 individuals, each of whom face the same loss of 100. The probability of a loss, however, increases from $p = .1$ before insurance to $p = .3$ after coverage has been purchased.

Table 5 Data for Moral Hazard Example

Before insurance	$p = .1$	$L = 100$	$N = 100$
After insurance	$p = .3$	$L = 100$	$N = 100$

If the insurance company does not know that moral hazard exists, it will sell policies at a price of 10 to reflect the estimated actuarial loss ($.1 \times 100$). The expected loss will be 30, since p increases to .3. Therefore, the firm will lose 20 (10 – 30) on each policy it sells.

One way to avoid the problem of moral hazard is to raise the premium to 30 to reflect the increase in the probability (p) that occurs once a policy has been purchased. In this case, there will *not* be a decrease in coverage as there was in the adverse selection example. Those individuals willing to buy coverage at a price of 10 will still want to buy a policy at 30 since they know that their probability of a loss with insurance is .3.

Another way to avoid moral hazard is to introduce deductibles and coinsurance as part of the insurance contract. A deductible of D dollars means that the insured party must pay the first D dollars of any loss. If D is sufficiently large, there will be little incentive for the insureds to behave more carelessly than prior to purchasing coverage because they will be forced to cover a significant portion of the loss themselves.

A related approach is to use coinsurance — the insurer and the firm share the loss together. An 80% coinsurance clause in an insurance policy means that the insurer pays 80% of the loss (above a deductible) and the insured pays the other 20%. As with a deductible, this type of risk sharing encourages safer behavior because the insureds want to avoid having to pay for some of the losses.*

A fourth way of encouraging safer behavior is to place upper limits on the amount of coverage an individual or enterprise can purchase. If the insurer will only provide $500,000 worth of coverage on a structure and contents worth $1 million, then the insured knows he/she will have to incur any residual costs of losses above $500,000.**

Even with these clauses in an insurance contract, the insureds may still behave more carelessly than if they did not have coverage, simply because they are protected against a large portion of the loss. For example, they may decide not to take precautionary measures that would otherwise have been adopted had they not purchased insurance. The cost of these measures may now be viewed as too high relative to the dollar benefits that the insured would receive from this investment.

If the insurer knows in advance that an individual will be less interested in loss reduction activity after purchasing a policy, then it can charge a higher insurance premium to reflect this increased risk or require specific protective measure(s) as a condition of insurance. In either case, this aspect of the moral hazard problem will have been overcome.

2.2.4 Correlated Risk

By correlated risks we mean the simultaneous occurrence of many losses from a single event. Natural disasters, such as earthquakes, floods, and hurricanes, illustrate cases where the losses in a community are highly correlated: many homes in the affected area are damaged and destroyed by a single event.

If a risk-averse insurer faces high correlated risks from one event, it may want to charge a higher premium to protect itself against the possibility of experiencing catastrophic losses. An insurer will face this problem if it has too many eggs in one basket, such as mainly providing earthquake coverage to homes in Los Angeles county rather than diversifying across the entire state of California.

To illustrate the impact of correlated risks on the distribution of losses, assume that there are two policies sold against a risk where $p = .1$ and $L = 100$. The actuarial

* More details on the role of deductibles and coinsurance to reduce the chances of moral hazard can be found in Pauly (1968).

** We are assuming that the firm will not be able to purchase a second insurance policy for $500,000 to supplement the first one and, hence, be fully protected against a loss of $1 million (except for deductibles and coinsurance clauses).

loss for each policy is 10. Table 6 depicts the probability distribution of losses for the two policies when the losses are independent of each other and when they are perfectly correlated.

Table 6 Data for Correlated Risk Example

Risks	$L = 0$	$L = 100$	$L = 200$
Independent	$p = .81$	$p = .18$	$p = .01$
Perfectly correlated	$p = .9$		$p = .1$

The expected loss for both the correlated and uncorrelated risks is 20. However, the variance will always be higher for correlated than uncorrelated risks which have the same expected loss. Risk-averse insurers will always want to charge a higher premium for the correlated risk.

Empirical data on the impact of correlated risks on premium-setting behavior comes from a mail survey of professional actuaries who were members of the Casualty Actuarial Society. Of the 1165 individuals who were sent questionnaires, 463 (or 40%) returned valid responses. Each of the actuaries evaluated several scenarios involving hypothetical risks, where the probability of a loss was either known or ambiguous.

One of these scenarios involved a manufacturing company that wants to determine the price of a warranty to cover the $100 cost of repairing a component of a personal computer. Each actuary was asked to specify premiums for both nonambiguous and ambiguous probabilities when losses were either independent or perfectly correlated and $p = .001$, .01, and .10. Table 7 presents the ratios of premiums for correlated risks to independent risks for well-specified and ambiguous probabilities using median estimates of the actuaries' recommended premiums. If the actuaries perceived no differences between the independent and correlated risks, the ratios would all be 1.

Table 7 Ratio of Premiums for Correlated Risks to Independent Risks for Scenarios with Nonambiguous (*p*) and Ambiguous Probabilities (*Ap*)[a]

	Probability level		
Nature of probability	.001	.010	.100
Well specified (*p*)	.910	1.160	1.250
Ambiguous (*Ap*)	2.000	5.560	2.000

[a] 100,000 units insured; $L = \$100$.

Data from Hogarth and Kunreuther (1992).

The data reveals a very different story. The median premiums were always higher for the correlated risks except for the case where $p = .001$ and the probability is well specified. The ratios were noticeably higher when the probabilities were ambiguous. In fact, when $p = .01$, the ratio of median premiums was more than 5.5 times larger for a correlated risk than for an independent risk.

2.2.5 *Administrative Costs*

The insurer must also be able to recover the costs of analyzing, underwriting, selling and distribution, paying claims, and meeting the regulatory requirements of issuing insurance policies. Generally speaking, these costs, collectively referred to as administrative expenses, are calculated as a percentage of premium dollars paid by an insured.

Administrative costs are also incurred in the process of quantifying risk which involves the following steps:

1. Obtaining a statistical database for estimating the risk
2. Underwriting cost associated with setting the premium using the statistical database
3. Obtaining the necessary regulatory approvals to market a policy
4. Marketing and distribution costs — determining the nature of the demand for the product and then using a sales force to promote the product

2.2.6 *Marketability*

Even if an insurer determines that a particular risk meets the first two insurability conditions, it will not invest the time and money to develop a product unless it is convinced that there is sufficient demand to cover these costs. An insurer must be able to cover development and marketing costs through its premiums. These costs include upfront costs for product development, as well as the expenses associated with marketing and distribution. The higher these costs, the higher the premium will have to be for a fixed number of customers. The final premium will be a function of the administrative costs and the elasticity of demand with respect to price.

REFERENCES

Bainbridge, John (1952) *Biography of an Idea: The Story of Mutual Fire and Casualty Insurance* (Garden City, N.Y.: Doubleday & Company).

Berger, Larry and Kunreuther, Howard (1995) "Safety First and Ambiguity" *Journal of Actuarial Practice* 2:273–291.

Hogarth, Robin and Kunreuther, Howard (1992) "Pricing Insurance and Warranties: Ambiguity and Correlated Risks" *The Geneva Papers on Risk and Insurance Theory* 17:35–60.

Kunreuther, Howard, Meszaros, Jacqueline, Hogarth, Robin and Spranca, Mark (1995) "Ambiguity and Underwriter Decision Processes" *Journal of Economic Behavior and Organization* 26:337–352.

Launie, J., J. Lee, and N. Baglini (1986) *Principles of Property and Liability Underwriting* (Third Edition) (Malvern, PA: Insurance Institute of America).

Office of Technology Assessment (1995) *Reducing Earthquake Losses* (Washington, D.C.: USGPO).

Pauly, Mark (1968) "The Economics of Moral Hazard: Comment" *American Economic Review* 58:531–536.

Rothschild, Michael and Stiglitz, Joseph (1976) "Equilibrium in Competitive Insurance Markets: An Essay on the Economics of Imperfect Information" *Quarterly Journal of Economics* 90:629–650.

Stone, John (1973a) "A Theory of Capacity and the Insurance of Catastrophic Risks" *Journal of Risk and Insurance (Part I)* 40:231–243.

Stone, John (1973b) "A Theory of Capacity and the Insurance of Catastrophic Risks" *Journal of Risk and Insurance (Part II)* 40:339–355.

QUESTIONS

1. Why are insurers more comfortable providing coverage against fire risks than they are against earthquake risks?
2. Suppose you are interested in buying medical insurance because you know that your health is below average of those individuals in your age group. What steps is your insurer likely to take so that they do not sell you coverage unless you pay a premium above the average for your age group?
3. What are the reasons that private insurers have not been interested in providing coverage against risks such as floods? Why can the federal government offer protection against this risk? *Note:* Federal flood insurance has existed since 1968 because private insurers refused to offer coverage.
4. Suppose a private insurer was interested in providing coverage to protect private contractors who clean up asbestos against those exposed to asbestos fibers who might contract cancer. How would the insurer determine whether such a risk is insurable?

CHAPTER **III.3**

Setting Environmental Priorities Based on Risk

Paul F. Deisler, Jr.

SUMMARY

This chapter briefly describes the background, development, and current status of the relatively new art of *comparative risk analysis*, its use, and where this new art fits into the general field of environmental risk analysis. It also offers insights on how to organize and prosecute comparative risk analysis projects, and it offers cautions as to the uncertainties involved in the process and as to the meaning of the results. It is not intended to be a complete guide to comparative risk analysis; other sources of information on the subject are given.

Key Words: risk, comparative, analysis, assessment, risk criteria, ranking, consensus building, management, prioritization, policy making, uncertainty

1. COMPARATIVE RISK ANALYSIS: INTRODUCTION AND BACKGROUND

The comparison of risks to make decisions is at least as old as the human race and is not limited to it. We have all witnessed our dogs or other pets, faced with choices between what are to them dangerous or frightening alternatives, hesitate with apparent uncertainty before finally making a choice. This is risk management, using risk comparison, at its simplest level.

1-56670-130-9/97/$0.00+$.50

Many risks are easily compared, for example, human deaths in a population per year caused by various types of accidents or diseases. Here, the comparison is based on well-defined, accessible, recorded numbers of incidents.

In the last few years, a new, very broad and far from simple art has sprung up within the wide field of risk analysis, *comparative risk analysis*. Applied to environmental risks, it is intended to be an instrument of governmental environmental prioritization, policy making, and policy implementation. It had its beginnings in the attempt by the U.S. Environmental Protection Agency (EPA) to answer a practical, seemingly simple, question: "Are the funds and effort allocated to the abatement of environmental risks in keeping with the actual levels of the risks to be abated?" The result was the U.S. EPA's well-known report, "Unfinished Business" (1987), and the answer, in brief, was "not necessarily."

The U.S. EPA undertook other, internal comparative risk projects, but it was not until the publication in 1990 of the U.S. EPA Science Advisory Board's (SAB) report "Reducing Risk" (1990), which examined, validated, and extended "Unfinished Business," that the new art began to enjoy wider public use. A major reason for this was the strong support and publicity given to it by then-EPA Administrator William K. Reilly, at whose personal request the SAB had prepared the report.

The use of this art has spread rapidly. Today, seven states have completed comparative risk ranking studies, and most are in the implementation stage; fourteen states have studies in progress; nine states are in the planning stages before commencing their studies; and several other entities (various cities, groups of counties, Native American tribes, and others) have completed or are carrying out their studies (Northeast Center for Comparative Risk 1995).

Once environmental risks are assessed and ranked, the consideration of additional factors such as the feasibility of risk reduction; the benefits of risk reduction; public risk perception; special risks to subgroups or ecosystems; and political, economic, and social factors, is needed to develop a prioritization of the same risks for attention and, finally, to develop policy options leading to legislation, regulation, or other types of risk abatement possibilities. It is important to note, here, that a ranking according to risk is not synonymous with a ranking according to priority.

The entire span of the environmental *comparative risk analysis* (CRA) process consists of the following two major stages, as is the case with other forms of risk analysis:

1. Comparative risk assessment (CRASS), in which the risks associated with specific environmental issues or problems are assessed and compared, usually by being ranked against each other
2. Comparative risk management (CRM), in which there are three stages:
   a. Risk reduction prioritization
   b. Risk reduction policy option development
   c. Implementation of risk reduction policy options, including monitoring of the results

Carrying out this full process requires several different groups working together and/or in sequence. Thus, *comparative risk studies are very labor intensive*. Much of the detailed, hard work is often supplied by large numbers of volunteers (50 or 100 or more) working within several types of committees or workgroups over a time span of 1 to 2 years or more, taking valuable time away from their career pursuits. Moreover, the work, if it is to get done at all in the face of gaps in data and theory and of large uncertainties, often requires scientists and other specialists to use a degree of "guesstimation" and extended — even speculative — judgment they would not ordinarily use when working in their respective fields. Also, it requires close, reciprocal, and effective communication between the many kinds of specialists and the many diverse nonspecialists that a typical study involves.

Nonetheless, the interest this new art holds is sufficient to attract otherwise busy people to do the work. And from the fact that so many states and other entities have entered into and sponsored such projects, that funding has been forthcoming from the EPA and others, and that implementation is going forward in most places where studies have been completed, it is clear that policy makers find this new art, with all its uncertainties and difficulties, to be exciting and potentially useful.

This chapter deals primarily with stage 1, CRASS, alluding as appropriate to the remaining stages. It is based, in part, on the experience of the author as a participant in three comparative risk projects and another having many of the characteristics of a comparative risk project: the U.S. EPA's reducing risk study (1990), two comparative risk studies in progress in Texas at this writing*,** and the U.S. EPA SAB's studies (1995a,b) on the environmental future.

2. GENERAL FEATURES OF THE CURRENT PRACTICE OF COMPARATIVE RISK ANALYSIS

In the familiar processes for assessing environmental risks to human health (National Research Council 1983) and to ecological health (U.S. EPA 1992) caused by exposures to specific types of stressors, the processes are scientifically based, even though they are filled with assumptions and fraught with serious data gaps and uncertainties, and they are typically carried out by technically trained people who know risk assessment. Also, risk assessment, as an activity, is supposed to be kept unaffected by risk management considerations, although communication between risk managers and risk assessors is necessary (Deisler 1988). Some forms of risk assessment processes, such as for cancer risks for regulatory purposes (U.S. EPA 1986), have been specified to the point where, in some cases, a single, knowledgeable, scientific risk assessor can carry out the assessment.

The first CRASSs carried by and within the U.S. EPA were by groups of U.S. EPA technical employees without public input, paralleling as closely as possible the other forms of risk assessment just mentioned insofar as process was concerned.

* State of Texas Environmental Priorities Project (STEPP): a statewide comparative risk study conducted by the Texas Natural Resources Conservation Commission, Austin, Texas, as part of a statewide, environmental priority-setting project.

** Houston Environmental Foresight Project: a regional, full comparative risk analysis study embracing the city of Houston, Texas, and eight contiguous counties, within one of which the city is located, conducted by the Center for Global Studies of the Houston Advanced Research Center, The Woodlands, Texas.

The many judgments such assessments must necessarily incorporate were those of the scientific and technical experts involved, alone. Although there was considerable support for this approach, it was also subjected to considerable criticism, as was CRASS in general. Questions were raised, such as whether national or regional assessments might not reduce the attention needed by more local problems, whether CRASS could be valid if done on a "scientific" basis without public input, and whether assessing such broad areas of environmental risk was a valid activity in the first place (Finkel and Golding 1994).

CRASS, as currently practiced, is carried out by processes involving groups of both trained specialists of many types and of nonspecialists. Members of the public, not necessarily trained specialists (although some may be), are drawn from both private and public segments of the affected community and are deeply involved in the CRASS. While the specialists, in carrying out much of their work, usually do so in separate groups, much of the scope of their work and the definition of how their output is to be presented to the public group for inclusion in their deliberations is set by the public group itself in consultation with the specialist groups. Communication between the two types of groups must be close and frequent throughout the risk assessment process, with the final characterization — the ultimate, overall ranking — of the risks being the responsibility of the public group with inputs and assistance from the specialist groups. This kind of process is intended to provide as good a scientific and scholarly basis to the characterization of comparative risks as possible, to be responsive to the concerns and perceptions of the public, and to make the results as understandable and acceptable as possible to members of the affected community.

3. ORGANIZING A COMPARATIVE RISK ANALYSIS STUDY AND GENERAL DESCRIPTION OF THE PROCESS

There are important resources available to those planning, initiating, or participating in CRAs. The U.S. EPA has published a very useful guidebook (U.S. EPA 1993) containing information on the organization of the committees and work groups, the selection of the problems or issues to be assessed and ranked, the kinds of criteria that might be used to assist in ranking (qualitatively or by scoring), and other useful information. The EPA was also instrumental in establishing two centers, the Northeast Center for Comparative Risk (NCCR) (P.O. Box 96, Chelsea Street, South Royalton, Vermont 05068) and the Western Center for Comparative Risk (624 Concord Ave., Boulder, Colorado 80304), which provide valuable services: workshops; intensive short courses; information on the literature and the types of resources and tools available; information on comparative risk studies completed, in progress, or being planned; and consultation, advice, and on-location assistance. They also publish, together with the Northeast Center, a bimonthly bulletin (NCCR 1995). There are numerous other publications dealing with CRA and related topics (for example, Cothern [1992] and Wernick [1995]). With these resources readily accessible, only a few practical thoughts on organizing and implementing the CRA process will be described here.

There are two kinds of committees of basic importance in a comparative risk study: a *public committee* and a series of *specialist committees*. These committees are given different names in different studies, but these will be used here. Typically, three types of specialist committees are established (although more can be added if needed) and filled with a variety of specialists needed to address a list of environmental issues from the perspectives of each of the three committees. These three types of specialist committees typically deal, respectively, with human health, ecological, and socioeconomic risks (this last risk has been given different names in different studies, including "welfare," for example [U.S. EPA 1987]). According to the type of risk considered, each of these committees assesses and ranks, comparatively, a list of environmental issues which has been agreed upon by the public committee, with input from the specialist committees (what are called "issues" here have been called "problems," "concerns," etc. . . . elsewhere).

In addition to the risk-based issue rankings (one for each specialist committee), each specialist committee also produces documents which collect, examine, and give scientific information on the risks associated with each of the issues considered. Because these will form part of the background of the public committee when it deliberates on a final, overall ranking, they should be readable by nonspecialists and should be *aimed specifically at providing information, judgments, and expert opinion describing the risks associated with each issue and the uncertainties involved in such a way as to assist in the comparative assessment of risks*. The specialist committee rankings, the issue documents, and oral discussions by and with representatives of the specialist committees are all input to the public committee. It is advisable for representatives of the specialist committees to serve as resource persons during the public committee's assessment and ranking sessions; these representatives may or may not be members of the public committee.

Not all members can attend all meetings, and numerous experts are needed for each one; therefore, larger committees of both kinds are to be preferred to smaller ones. Redundancy of expertise, viewpoint, and participation is a benefit, not a detriment, in CRASS, and, especially in the public committee, inclusiveness of community viewpoints, not exclusiveness, is needed for success. Public committees of 100 members are more desirable than those of 50 if the latter would leave some community interest out. Moreover, in each specialist committee, experts should be present, representing not only different specializations, but different career experiences in their fields. And, although members of the requisite specializations probably have assessed risks in their respective fields, their perspectives on risk and what it is will be very different. Therefore, having members on, or consultants available to, both kinds of committees, public and specialist, who are well versed in risk assessment is advisable. While national experts have their place as members on the specialist committees, the bulk of the membership should come from the region under study so as to bring their understanding of local problems to bear.

The public committee has the task, always with the assistance of and in close communication with the specialist committees, of guiding the study, of considering all inputs to it (including inputs which outside experts or its own members bring to the table), and of producing the final statement on the assessment of the risks associated with the different issues. This latter task often includes developing a final, overall, risk-based ranking of the list of issues according to the several kinds of risks, taken together, that the specialist committees have considered, plus their own considerations. This final statement, assessment, and ranking is the first, major product of a CRA study.

CRA stages 2a, 2b, and 2c defined earlier may be undertaken by the public committee for the sponsoring organization, by the study's sponsoring organization with input and assistance from the public committee, or by still other committees set up by the sponsoring organization to track the study and make use of its results with input and assistance from the public committee, depending on how the study is designed to fit the needs and organizational pattern of the sponsoring organization. In Louisiana, for example, the assessment (stage 1) has long been complete, but members of the Public Advisory Committee of the study "LEAP to 2000" continue to work with Louisiana's Department of Environmental Quality "to develop strategies for action on six of the top ten highest risk priorities" at this time (NCCR 1995, p. 3).

One other type of committee is highly desirable as a means of facilitating the entire process. This committee is a small, coordinating committee composed of at least the chairs of the specialist committees and some advisors who are not members of the specialist or the public committees. This committee can serve as a communication device between the specialist committee chairs; as a way for them to obtain expert and diverse, even independent, views on the work of their committees and what it might include, exclude, or otherwise consider; as a way for them to plan their work jointly and to be, so far as is possible, consistent with each other and with the goals of the study; as a focal point for generating joint proposals to be put before the public committee; and as a means for arranging for outside review of the issue documents, selecting members for the specialist committees, and other functions. Such a committee may be included in the original design of the study (an advisable course of action) or, if not, it may arise as an ad hoc group. In any case, this kind of committee may also include specially interested members of the public committee. One thing this coordinating committee should not do is act for any of the other committees or usurp their functions in any way.

The committees must be supported by a highly competent, knowledgeable, and hard-working staff. The staff are not members of any of the committees, but they perform essential functions ranging from clerical tasks to highly sophisticated editorial and writing tasks in turning out all kinds of reports. This includes assisting with the issue reports and writing the final reports, from the initial planning of the study to the intermediate planning and management of the study process. What the staff does not do is enter into the debates within the committees or in any way "manage" the conclusions reached by the committees. Staff can only be truly effective if they are seen by the committees as assisting the process to proceed, but not influencing the direction of the deliberations. There is technical assistance the staff can and should render, such as making suggestions that help the committees, reminding them of principles they have agreed to when they seem to be straying, securing outside expertise when needed, and so on. A very important function of the staff is to assist in the full documentation of the processes and rationales followed by each committee in their deliberations to produce a clear record of how the results were obtained; this is very important in ensuring the credibility of the results of a study, in answering questions about them, and in utilizing them with understanding. Without good staff, the process is not possible.

4. PREPARATIONS NEEDED BEFORE CONDUCTING A COMPARATIVE RISK ASSESSMENT

One of the first things the U.S. EPA did in starting the work which resulted in the "Unfinished Business" report was to agree on a list of environmental problem areas of interest to the agency to be ranked. They chose a list of 31 problems (for example, hazardous waste, pesticides, global warming) in which the agency had a direct or indirect interest.

It is easy to list large numbers of issues to be ranked. One may list "drinking water quality," for example, or more specific issues such as "nitrates in drinking water" and so on. What is listed depends on two things: (1) that the issues be of importance for the region which is the subject of study, and (2) that the issues be clearly definable and there be as little overlap as possible (some overlap may be unavoidable). As a part of selecting issues and getting started, a number of other matters need some attention and agreement (which may be altered as learning takes place during the process). (1) *The establishment of what is included as an "environmental issue."* "Environmental" can cover every kind of stressor to which human beings and ecosystems can be exposed, from pollutants to earthquakes. For example, it can include social environment, it can be limited more narrowly to pollutants of human origin, or it can include microbial contamination of food and water and yet exclude the workplace. In brief, the public committee, assisted and advised by the specialist committees, must decide, at the start, what the environmental issue will and will not cover. (2) *The establishment of a framework within which issue ranking will take place.* In addition to defining the characteristics of the region within which the study is taking place and its subregions, time span is a very important part of the framework for risk ranking. Issues can be ranked according to immediate risks, longer-term risks, or a consideration of both, and the rankings can be very different as a result. An explicit discussion of the framework is important in getting the effort started on a reasonable footing, recognizing that not all problems can be resolved in advance, but will often have to be dealt with on a case by case basis. The framework may also include the importance that the committees might give to such concepts as sustainability of the environment over the long term, from which the related risks to human society (and its members) and the ecologies upon which it depends can be considered. This concept, carefully developed throughout the study, can help to provide a unifying principle when the public committee undertakes its final ranking. To give the best consideration possible to future trends or scenarios, including a professional futurologist as an advisor to all committees, or as a member of the public committee, or both is highly advisable and will help to produce a first-class product. The EPA SAB's environmental futures study's annex (1995b) and the book by Schwartz (1991) are recommended reading. (3) *The establishment of some common understanding of what risk consists of.* In its simplest terms, risk has two components: the *probability* that an adverse consequence (or effect) will occur and the *severity* of the consequence if it occurs. In formal cancer risk assessment, the effect of any cancer is considered so severe that probability is the only factor considered; in ecological risk assessment, one often finds that probability is not considered very much — exposure usually leads to a stated consequence — and risk assessment becomes largely a matter of consequence assessment, although the probability of exposure is sometimes considered. Socioeconomic risk assessment is the most difficult of the risk areas in CRA, as well as the least developed and understood; it is the most derivative and yet the most integrative form of risk assessment, since the risks it deals with are contingent on the realization of human health and ecological risks, plus other risks that might be considered such as risks to aesthetic values. Careful application of basic risk concepts in all CRASS areas will lead to the development, as the study proceeds, of risk criteria for each type of risk useful in the ranking of the issues according to risk. Members of, or advisors to, the committees who are versed in risk assessment will be useful in committee discussions early in the process and later on. (4) *What is meant by "residual" risk and its use in CRA?* Residual risks are the risks estimated to exist given whatever current and continuing risk management, regulation, and control may (or may not) exist within the framework of the study. These are the risks to be compared, since what is called "residual" risk is the actual risk faced by the populace or by ecosystems within the scope of the study. The concept of residual risk is an easy one from which to stray in the course of a long and complex project; participants must remind themselves and each other what it is that is sought. Also, very importantly, prioritizers, policy makers, and other users of the results of a CRA must understand the use of residual risk: because an issue is in a lower comparative risk ranking category, a cut in budget or effort relating to that issue is not necessarily called for. The issue may rank lower only because there are substantial programs in place to keep risk low and these must be maintained.

Later in the process, the committees will find it useful to develop criteria for describing comparative risks appropriately in each of their areas. Examples are probability of occurrence of an adverse effect, intensity of the effect, extent of the effect, timing of the effect, and so on. These may be used qualitatively or given quantitative scores to assist in ranking; this is described in the EPA's guidebook (U.S. EPA 1993), and an example of the development and use of criteria similar to those needed for comparative risk analysis is given in the U.S. EPA SAB's environmental futures study (U.S. EPA 1995b).

5. ACHIEVING COMPARATIVE RISK-BASED RANKINGS OF THE ENVIRONMENTAL ISSUES

The risk-based ranking of a list of environmental issues is not a defined quantitative, analytical process. The large gaps in available data and information of all kinds, the major uncertainties in what information there is and in its application to risk assessment, and the need to use assumptions and judgments (and the uncertainty as to the degree to which values and perceptions enter into the judgments needed to bridge the gaps and cope with the uncertainties), all of which plague the other, more specific forms of health and ecological risk assessment, are magnified in the much broader field of comparative risk assessment, where even the relative severities of effects and other factors which might help to make different kinds of risks comparable to each other are often matters of educated guesswork at best. If, as Weinberg (1985) has written, the usual health risk assessments are not science, but "trans-science," then comparative risk assessment is at the extreme of trans-science, almost to the point of being science fiction. Satisfactory analytical procedures do not exist which will cope with this situation, and reaching a consensus on risk ranking among the well-prepared members of the committees involved in a comparative risk study is necessary for a credible result to be achieved.

There are many methods for reaching a consensus on ranking: jointly examining the information about, and the importance of, the selected risk-describing criteria to achieve separate rankings by members of the committees and then comparing and debating these; the joint examination of matrices of issues and risk criteria; the assignment of quantitative scores to the criteria and, by an agreed-upon formula, the combination of these into total scores for each issue, thus ranking them according to their scores; the use of various forms of voting to achieve a ranking; and the use of various forms of issue-to-issue comparison. Any ranking achieved by any of these systems should be treated as an initial ranking, subject to examination, debate, and change before reaching a final ranking.

A process which has worked well, in the author's experience, is for the committees to follow the following five-step sequence: (1) decide what constitutes a consensus (a final majority vote, a lack of objection to what may be a final ranking, the freedom to enter minority opinions or not, etc.); (2) review the agreed risk principles to be followed one more time (residual risk, selected criteria, and so on) and modify them, if necessary, one last time; (3) fully absorb written and then orally presented basic data on each issue, with discussion; (4) make a trial or "straw" ranking by any technique agreeable to the committee members and discuss it in depth, especially where there are differences; and, (5) meet to make a final consensus ranking. Enough time should elapse between steps 4 and 5 for staff to send out summaries of the first steps and what was achieved in them and for the members of committees to have time to rethink their positions, possibly doing their own ranking exercises, but using some common, consistent technique. But, not too much time should elapse as to allow the members to go stale and forget too much of what was done in the first four steps. Two weeks is barely enough time for the interim work to be done, and more than 1 month may allow committee members to grow cold. Any rankings achieved by step 4 should be considered tentative, at best; the main benefit is as a "warm up" for the final session (step 5) and to bring out different viewpoints.

At the last meeting, step 5, an agreed upon method for achieving the final ranking should be used, and it is especially important for a trained, neutral, firm and perceptive facilitator to conduct the meeting, however, it has been chaired up to that point. Also, having the facilitator at least observe the preliminary ranking effort and possibly discuss the final ranking method he/she recommends is highly desirable.

One particularly effective method for achieving a final ranking, which the author has observed in action, is one of facilitated, computer-assisted voting to compare and rank sequential pairs of issues. In this method (Dominus 1995), developed by Saunders Consulting of Toronto, Canada, each committee member is given a pad on which he/she may vote "yes" or "no" to questions of the form, "Does issue A rank higher than issue B?". The individual votes are tallied by a computer and displayed both numerically and as a bar chart. Discussion can then ensue as to why individuals voted as they did, and another vote can then be taken, which often is accepted as the consensus on that pair of issues. Not infrequently, major shifts in voting can occur on the second vote, indicating that the exchange of ideas has been effective. Rarely is a third vote needed.

The computer can also display, when desired, the ranking achieved after a series of paired rankings (issue A with B, B with C, and so forth), sometimes showing that a pair ranked one way now ranked the opposite when all rankings achieved at that point in the process were displayed. Further discussion often yields an important point previously missed.

This method of paired voting permits the ranking of lists of approximately 20 issues in about half a day. If any member has a truly serious disagreement with the ranking achieved, he/she can voice it and defend a different ranking, then the committee can decide whether, and how, to change the ranking. While another equally prepared committee addressing the same list of issues might arrive at some different rankings, it would be a surprise if there were gross differences. However, this point has not yet been tested.

At times, one or another of the committees might not be able to rank an issue because it is not relevant to that committee (e.g., "indoor exposures to radon" might not be a relevant issue for an ecological committee) or because data are too sparse or nonexistent. It is necessary for the specialist committees to state why they are not ranking an issue and to offer whatever information or opinions they have on the subject to assist the public committee members as much as possible in their final ranking effort.

6. SOME REFLECTIONS ON COMPARATIVE RISK ANALYSIS

There is scope, here, only for a few final comments and reflections on the CRA process. A critique of the art, as it now stands, would be highly desirable, but would occupy at least another chapter. However, Andrews (1995), in his review of California's study (California EPA 1994), offers useful comments on this new field, and the reader is referred to his article. Four final reflections are offered here.

1. ***It is necessary to keep in front of everyone, participants in the study and users of its results alike, the fact that a comparative risk ranking is but one input to ranking by priority; a comparative risk ranking is not, by itself, a priority ranking or even an indicator of priorities.***
2. A comparative risk ranking for a given region represents an overall assessment for that region and, unless pains are taken to address them, risk "hot spots" affecting particular segments of the region's populations or specific ecosystems can get lost in the "bigger picture." With a broad enough representation on the committees,

these hot spots can be identified and, although it may not be possible to rank them within the overall ranking, they can be pointed out in the final report, and their importance made clear. Given the state of the art, little more than this can be done, but policy makers should be made fully aware of such special points.

3. ***For a comparative risk ranking to succeed, participants, whatever their background or special interests may be, must enter into it from the very beginning with a commitment to achieve the rankings; to listen as well as to be heard; to be open in their communication with the others (a thought unsaid cannot be heeded); to maintain an attitude of mutual, personal goodwill, respect, and attentiveness; and to work hard.***
4. Sponsors must understand from the very beginning that participants commit themselves and their time and energies to an effort of the magnitude of a comparative risk study because they have a deep interest in the environment and its future; thus, they incur an obligation to the participants. Thanking them is not payment enough: following up vigorously on the results of the study, being seen to follow up vigorously, and feeding back to the participants on the utilization of their work and on what effects it has had is the only real payment possible. Without this, the impulse to volunteer when a new study is needed will not exist. A CRA study is, in the final analysis, the product of a thinking, feeling, perceptive, multihuman computer.

REFERENCES

Andrews, R. N. L., Report on reports: toward the 21st century: planning for the protection of California's environment (a review), *Environment*, **37,** 25, 1995.

California Environmental Protection Agency, Office of Environmental Health Hazard Assessment, California Comparative Risk Project, Toward the 21st Century: Planning for the Protection of California's Environment, May, 1994.

Cothern, C. R. (Editor), *Comparative Environmental Risk Assessment*, Lewis Publishers, Boca Raton, Florida, 1992.

Deisler, P. F., Jr., The risk management-risk assessment interface, *Environmental Science and Technology*, **22,** 15, 1988.

Dominus, M., Director, Organizational Development and Training Office, Texas Natural Resources Conservation Commission, Austin, Texas, personal communication on the group decision making system of the Saunders Consulting Group (1210 Sheppard Avenue East, Suite 300, (Postal Box 36) Toronto, Ontario, Canada M2K 1E3), July, 1995.

Finkel, A. M. and Golding, D. (Editors), *Worst Things First? The Debate Over Risk-Based National Priorities*, Resources for the Future, Washington, D.C., 1994.

National Research Council, *Risk Assessment in the Federal Government: Managing the Process*, National Academy Press, Washington, D.C., 1983.

Northeast Center for Comparative Risk (NCCR), Project news, *The Comparative Risk Bulletin*, 7 (May/June), 1995.

Schwartz, P., *The Art of the Long View: Planning for the Future in an Uncertain World*, Doubleday, New York, 1991.

U.S. Environmental Protection Agency, A Guidebook to Comparing Risks and Setting Environmental Priorities, EPA-230-B-93-003, September, 1993.

U.S. Environmental Protection Agency, Framework for Ecological Risk Assessment, EPA-630-R-92/001, February, 1992.

U.S. Environmental Protection Agency, Guidelines for carcinogen risk assessment, in *Federal Register*, **51,** U. S. Printing Office, Washington, D. C., p. 33992, 1986.

U.S. Environmental Protection Agency, Office of Policy Analysis, Office of Policy, Planning and Evaluation, Unfinished Business: A Comparative Assessment of Environmental Problems, February, 1987.

U.S. Environmental Protection Agency, Science Advisory Board (1400), Reducing Risk: Setting Priorities and Strategies for Environmental Protection, EPA-SAB-EC-90-021, September, 1990.

U.S. Environmental Protection Agency, Science Advisory Board, *Beyond the Horizon: Using Foresight to Protect the Environmental Future*, EPA-SAB-EC-95-00, January, 1995a.

U.S. Environmental Protection Agency, Science Advisory Board, SAB Report: Futures Methods and Issues: A Technical Annex to "Beyond the Horizon: Protecting the Future with Foresight," EPA-SAB-EC-007A, January, 1995b.

Weinberg, A. N., Science and its limits: the regulator's dilemma, *Issues in Science and Technology*, 59, 1985.

Wernick, I. K. (Editor), *Community Risk Profiles: A Tool to Improve Environment and Community Health*, The Rockefeller University, New York, 1995.

QUESTIONS

1. What is the difference between risk analysis and risk assessment?
2. What is the difference between risk analysis and risk management?
3. What is the difference between comparative and other forms of risk analysis?
4. Why are comparative risk analysis and comparative risk assessment described as arts and not as sciences, despite the fact that scientific information and judgment are used in carrying them out?
5. What is residual risk, and why is it used in comparative risk assessment? Why is it so easy to misinterpret? What should policy makers understand about the use of residual risk in comparative risk assessment?
6. What uses do defined criteria of risk serve in comparative risk assessment? Should uncertainty be used as a criterion of risk or as a characterization of an assessment of risk?
7. Why is a ranking of environmental risks not a ranking in order of priority for action to reduce the risks, starting with the worst?
8. Comparative risk assessments are generally carried out by relatively large groups of people of various backgrounds. Can a single, skilled, well-informed individual not carry out such an analysis? If not, why not?
9. Since risk, in principal, is a mathematically defined, scientifically measurable parameter, why does a comparative risk analysis project involve nonscientifically trained members of the public?
10. Is a final, single ranking of environmental risks, which includes risks to human health, to ecosystems, and to economic and social well-being, truly a comparative *risk* ranking? In any event, how would you describe it?
11. Local groups at times object to the use of broad statewide, regional, or national comparative risk analyses in setting risk-reduction priorities. Why is this?
12. Why is "guesstimation" or speculation required of the scientists participating in comparative risk analysis exercises? Why not just use good science?

13. What makes the practice of this new art of comparative risk analysis sufficiently attractive to cause otherwise very busy individuals to be willing to volunteer large amounts of their time and much hard work to carrying it out?
14. Would two equally skilled comparative risk assessment projects, ranking the same risks within the same framework and using the same definitions, achieve the same rankings? If different, how and why might they be different?
15. Why is comparative risk assessment uncertain, and what are the sources of uncertainty?
16. How should uncertainty affect the use of the results of comparative risk assessment?
17. In setting up priorities for dealing with environmental risks, what alternatives are there to using comparative risk analysis? Do you see any that are better than the use of comparative risk analysis? If so, explain what they are and why they are better.

CHAPTER III.4

Comparative Risk Analysis: A Panacea or Risky Business?

Vlasta Molak

SUMMARY

Comparative risk analysis (CRA) is a term used to describe a rapidly growing number of projects performed around the United States by state and local groups as a promised new cure for "irrational" environmental management. CRA is supposed to combine the "science of risk analysis" with "community values" (Stone 1994). However, close examination of several CRAs performed recently by the states and local groups, indicates that neither scientific risk analysis or community values are properly incorporated. The name "comparative risk analysis" is a misnomer and a distortion of the term risk analysis. Moreover, attempting to use such CRAs for setting environmental priorities may do more harm than good, because the public is led to believe that decisions are based on "objective risk science," while, in fact, it is only the opinions of the groups involved in conducting CRA (often industry and government dominated). CRAs seem often to be used as a means of speeding up the process to meet unreasonable deadlines and satisfy officials eager to embrace CRA as a cost-cutting measure in environmental management (mostly for polluters).

Although performing CRA adequately may be difficult, requiring integration of many different disciplines, it can be done so long as proper precautions are taken, the limitations of the analyses are clearly spelled out, and the results are used with caution. The U.S. Environmental Protection Agency's (U.S. EPA) goal to manage environmental issues, based in part on the magnitude of risks to human health and ecosystems, can be achieved only if good science is practiced and proper procedures are followed in performing risk analyses (U.S. EPA 1987, 1990, 1993). While procedures for simpler types of risk analyses (such as estimating health effects of a

1-56670-130-9/97/$0.00+$.50

particular chemical or process) have been developed, methods that would enable comparing various types of problems and multiple stakeholders values attached to those problems and/or their solutions still need to be developed. A good starting point may be a multiattribute utility theory, which provides a mathematical framework for analyzing choices involving multiple competing outcomes (Kadvany 1995). ***However, any model one may conceive for such a purpose must be validated if it is to be useful. Also, one should not forget that some complex questions are unanswerable and that only time will show the validity of our predictions.*** True CRA could become an ongoing process in which an interdisciplinary and long-term approach could be applied in evaluating and solving environmental problems using the advances in the fields of risk analysis and decision analysis. Ironically, this type of environmental problems analysis and management is very close to the medicinal circle approach of some native American tribes where the knowledgeable elders of the tribe (interdisciplinary group of experts) would discuss the impacts of important decisions on seventh generation (long-term approach to environmental management).

Key Words: comparative risk, environmental priorities, state studies, risk perception, multiattribute utility theory, community values, future generations, environmental management

1. INTRODUCTION

Risk analysis is a scientific method used to evaluate adverse effects (a value judgment) of a particular agent (physical, chemical, or other) on human health, ecological health, or economic well-being (in case of business risk analysis). Although there are many types of risk analysis, some common elements are necessary to qualify the process as "risk analysis." Those elements are (NAS 1983)

1. Hazard (agent) identification
2. Dose–response relationship (how is quantity, intensity, or concentration of a hazard related to adverse effect)
3. Exposure analysis (who is exposed? to what and how much? for how long? other exposures?)
4. Risk characterization (reviews all of the previous items and makes calculations based on data, with all the assumptions clearly stated, or with the conclusion that more data and/or improvement in methodology is needed)

Deciding WHAT is an adverse effect (and to some extent hazard identification) is a value judgment that can be made by well-informed citizens. The consideration of other components of risk analysis is a complex process, which in order to be properly conducted requires extensive training. Just as one would not want to have a surgery performed by an untrained layman, risk analysis may be a risky business if performed by untrained people. Because of its interdisciplinary nature and complexity, risk analysis requires an appropriate amount of time to evaluate all pertinent data, even when one deals with problems of lesser complexity.

2. SHORTCOMINGS OF CURRENTLY PERFORMED CRAs

2.1 Composition and Expertise of the Committees Performing CRA

In a process titled Comparative Risk Analysis (CRA) for Ohio, neither training nor time is given to laypeople (with a few exceptions) to perform valid risk analysis. In addition, if one observes closely how CRAs are conducted in other states and localities, it appears that neither community values nor scientific risk analysis were properly used, and the resulting conclusions are based on someone's (often unqualified and/or biased) opinion, frequently clouded by obvious conflicts of interest. This is the case for various group representatives (industry, agriculture, developers, transportation) who do not want to believe that their activity may present any environmental risks or government employees (most often state EPA) who are put into a bind of trying to please industry (their jobs are threatened by budget cuts), while the mission of their agency is to protect public. While we may not doubt the honest intentions of those individuals whose employers are paying for their participation in CRA, their bias is inevitable. Studies show that even scientists in the risk analysis field have different views about toxicity of chemicals and validity of toxicological studies depending on their work affiliation (Slovic et al. 1993). Additionally, since very few volunteers or staff in the environmental organizations or concerned citizens can afford time to participate in such an extensive effort, the views of environmental organizations, which have better than average knowledge of environmental problems in their communities and concerned citizens are not properly represented. Thus, the composition of the various committees performing CRAs are skewed toward government and industry representatives who have more resources to donate their time and effort to such an endeavor. ***An additional difficulty in getting environmental organizations to participate in CRA is a distrust that many environmentalists have in an endeavor that has been perceived as yet another tool for dismantling already endangered environmental regulations.*** Their concern is that the "priority list" of environmental problems derived by CRAs will be exploited by special interest groups to justify further expansion of their activities and profits without regard for environmental well-being and health of the citizens and ecosystems. Since very limited resources of the environmental groups are currently focused on preventing the dismantling of environmental regulations, which are happening rapidly both nationally and statewide, they are reluctant to participate in a project that promises very little return in protecting the environment.

2.2 Community Values and Environmental Problem Ranking

The data used to define "environmental problems in Ohio" were obtained from facilitated public meetings by the EPA's Ohio Comparative Risk Project, in which over 700 environmental problems were identified and eventually grouped into environmental problem categories for scientific analysis. The EPA's Ohio Comparative Risk Project Technical Work Group members developed the final 11 environmental problem categories by consensus (rather than following any type of scientifically derived rules). Thus, many of the original 700 problems were omitted as "unimportant" for further consideration by CRA, and thus, a possibility exists to miss an important environmental problem since risk analysis was not properly applied in eliminating less risky perceived environmental problems. ***Consensus generally reflects cumulative knowledge and perception and biases of a group, which in our case was rather small (less than 100 people altogether), and does not necessarily reflect either the opinions of general population of Ohio or the knowledge and awareness of environmental problems in the state of Ohio.*** "Science by majority opinion of lay people not trained in science" can hardly be considered as "scientific basis" for proper environmental management. The final 11 perceived "environmental problem" categories derived by the Quality of Life Committee or Ohio CRA are

1. Surface and Groundwater Quality
2. Habitat Loss and Degradation
3. Outdoor Air Quality
4. Waste Management
5. Land Use and Development
6. Environmental Awareness and Access to Information
7. Indoor Air Quality
8. Environmental Management
9. Natural Resource Use
10. Food Safety
11. Drinking Water at the Tap

These 11 categories are a mixture of "environmental problems" (items 1, 2, 3, 7, 10, and 11) (as generally defined by the EPA and environmental groups), potential causes of environmental problems (items 4, 5, and 9), and potential solutions of environmental problems (items 6 and 8). Such mixing of "apples" and "oranges" does not easily render itself to scientifically valid risk analysis.

Additional data used by the Ohio CRA were by the EPA's Comparative Risk Project Survey at the August 1994 Ohio State Fair. Data were collected from 2626 individual respondents. Eight quality-of-life indicators (criteria) were pictorially presented on various displays, and each survey was grouped into one of six geographic locations in Ohio. The survey asked various demographic information and the question "if your quality of life was threatened, which of the following concerns would be most important to you":

1. Peace of mind — safety, happiness, and health
2. Mobility — ease of getting from one place to another
3. Aesthetics — visibility, noise, odors, and any visual impacts
4. Future generations — impact on our children, availability of alternatives, reversibility of effects
5. Sense of community — neighborhoods and personal growth
6. Economic impacts — maintaining a comfortable standard of living, achieving personal goals, costs
7. Fairness — sense of equity, respect of individuals or property owners rights, number of effected persons, severity of effects on different groups
8. Recreation — access to and quality of recreational lands, opportunities for solitude

It was hoped that by asking these questions, one could establish the value that the citizens of Ohio assign to various quality-of-life indicators, thereby fulfilling the promise of the Ohio CRA to "combine scientific risk analysis with community values." This survey focused on obtaining public input for the Quality-of-Life Subcommittee to use in evaluating the eight quality-of-life indicators. In order to ascertain how each indicator (criterion) is affected by the 11 environmental problem categories, one should apply principles of risk analysis to each perceived environmental problem category.

In order to have community values incorporated into the CRA, one needs to have community involvement. Analysis of participants in the CRA, however, reveals that most participants were either government or industry employees. Those participants were getting paid by their institutions to participate in the CRA, while representatives of communities potentially effected both by industrial operations and/or governmental agency decisions could not afford to take unpaid time from work to participate. Thus, those who are currently performing the CRA are NOT representing properly various community values. A clear conflict of interest is obvious. Those who pollute or are involved in governmental regulation promotion are not in a position to make objective judgments, since their livelihood may be effected by the conclusions derived in the CRA. It is in their interest to dismiss or underestimate potential risks stemming from their companies or organizations' activities and to overestimate risks of other potential hazards.

An example from the Columbus CRA indicates that "effects of street lights on ecological health" was considered an important enough topic for discussion, while "health effects from industrial emissions" were deemed unimportant, based on the opinion of an industrial engineer, who just said so, without any support from data and without performing proper risk analysis. Thus, a CRA performed in this way is NOT going to properly reflect "community values," and the resulting conclusions may do more harm than good by lulling the community into complacence about potentially serious environmental problems. In addition, a large number of community surveys, where various people were polled in focus groups, it is clear that the agricultural concerns were predominant.

The resulting number one "environmental problem" perceived by those polled indicates that those people are not representatives of the broader community or that they misunderstood the meaning of risk analysis. While "environmental education" may be a problem in the larger context of society (or may be reflecting that those polled felt unqualified in the environmental field), it certainly is NOT a meaningful "environmental problem" as defined in any kind of risk analysis. Mixing environmental education with a host of true environmental problems obstructs proper conduct of the CRA and misleads the public. According to such "analysis" and priority setting, if the state of Ohio invests money on environmental education, the true environmental problems would be considerably reduced without solving the problem of industrial pollution, poor planning and development, fossil fuels derived energy use, transport, and other problems that are degrading environments of Ohio. While environmental education may raise the awareness of pressing environmental problems (mostly coming from industrial pollution, energy production, transportation, and poor planning and development) and the citizens may put political pressure for a solution, without directly addressing those problems, the environment of Ohio is still in danger.

2.3 Science of Risk Analysis

It appears that risk and risk analysis have been terms redefined by many groups. Risk is defined as "hazard + outrage" and not the usually accepted definition of "probability of adverse effect." This definition would place an environmental problem into a high-risk category, even if scientifically conducted risk analysis showed no or negligible risk if sufficient outrage was created. If outrage is considered as a major contributor to risk, one could spend lots of money on fixing nonexisting problems, as is the case in several outbreaks of mass psychogenic illness in the workers and communities where the cause was erroneously attributed to "chemical exposures" (Boxer 1985, 1990a,b). Moreover, such a definition may miss some real risks which the general public may not be aware of, similar to the outbreak of cholera caused by the erroneous decision to stop chlorinating water "to avoid cancer risk posed by chlorinated byproducts in water" (Anderson 1991). Also, equating "hazard" with risk is a fallacy that any risk assessor is aware of, since risk does not exist if there is no potential exposure to the hazard (chemical or physical agent, natural event, process, technology, etc.). ***Thus, following such a definition of risk (hazard + outrage) would defeat the very purpose of CRA: to establish some rational means of approaching environmental problems, rather than the response to perceived risks which may not be real.***

After committees were formed and several brainstorming sessions resulted in long lists of potential hazards, volunteers were requested to search for data and establish the risks from a particular hazard. Whether or not the person had any qualifications in the area he/she was examining or had any familiarity with the issues was irrelevant. In order to perform a risk analysis properly, one has to be able to evaluate obtained information critically, search for other relevant information, and follow generally verifiable methods (hazard identification, dose–response relationship, exposure assessment, and risk characterization). Most of the problems (hazards) being reviewed by the various committees of the Ohio CRA are of such a complex nature that an in-depth evaluation and search for data, which would be sufficient for several Ph.D. theses, is required. One cannot throw partial data at a group of well-meaning volunteers and expect them to have the expertise and time necessary for such an ambitions project.

It has taken many expert risk analysts years of study to be able to make any valid conclusions even in much simpler cases! It is impossible for laypeople to critically evaluate masses of data and derive defendable conclusions without being thoroughly trained in risk analysis and without having sufficient time to obtain and analyze the data. While citizen meetings may be useful to find out what "community values" are in order to define adverse effects, without proper scientific examination of data by proper scientific risk analysis, the whole process is flawed and even fraudulent if the CRA is promoted as a true-science-based ranking of environmental concerns. The CRA projects currently promoted certainly do not fulfill this promise of integrating community values and science.

3. CONCLUSION

CRA could be a worthwhile goal (U.S. EPA 1987, 1990) if the factors addressed in this chapter were taken into consideration (Molak 1990). Instead, it appears that CRA is being used rather to speed up processes, to meet unrealistic deadlines and satisfy needs of politicians who are eager to embrace CRA as a cost-cutting measure in environmental management (mostly to save costs to polluters). ***Recently passed and currently pending laws, both on a federal level and in Ohio state, indicate that the term "risk analysis" is used indiscriminately, thus endangering the scientific credibility of the whole discipline of risk analysis. The scientists involved in risk analysis should educate the American public about the meaning of the term "risk analysis" and prevent current misuse of this term. In particular, the assumptions and uncertainty associated with risk analysis and the resulting risk numbers should be clearly spelled out in order to prevent simplistic solutions to current environmental problems (Breyer 1993).***

There are many qualified risk analysts who could perform valid analyses and come to reasonable conclusions in which the community values are truly taken into account. The U.S. EPA's attempt to manage environmental problems based in part on the magnitude of risks to human health and ecosystems is a reasonable goal. However, this can be achieved only if good science is practiced and proper procedures are followed to perform risk analyses and reflect community values. Although performing CRA may be a difficult task, requiring integration of many different disciplines, it can be done if proper precautions are taken, the limitations of the risk analyses are clearly spelled out, and the results are used with caution. The results of several CRAs conducted so far by the states and local groups could be used to design better approaches to real CRA, which would truly serve the purpose of solving environmental problems in a most efficient and equitable way.

To put it in the language of risk analysis, CRAs currently performed by states and local groups in the United States may present a hazard with a clearly indicated dose–response relationship (the more CRAs are done, the more potential damage if they are used for decision making), and all citizens are potentially exposed to this risk. Such massive exposure of citizens to CRA may result in severe adverse effects from lessening already weakened environmental regulations. This would place citizens and the environment in a high environmental risk category! If improperly conducted, various CRAs, which only pretend to apply scientific methods and community values when if fact they may only reflect the current political power's goals, are very risky for Americans! The value of "environmental priorities" lists derived in such CRAs is extremely limited and should not be used at this stage for management of environmental problems. Since lots of time and effort has been expanded in performing CRAs, and those CRA reports contain lots of valuable data, they could be used as a starting point for an ongoing process of stakeholders' dialogue, where all the cards are brought on the table and are discussed within a newly developed framework of multiattribute utility theory and decision analysis. It is important to get ALL the stakeholders included in this process (rather than having some groups boycott the process because of distrust in its purpose and hidden agendas).

True CRA could then become an ongoing process in which an interdisciplinary and long-term approach could be applied in evaluating and solving environmental problems using the advances in the fields of risk analysis and decision analysis. Ironically, this type of environmental problems analysis and management is very close to the medicinal circle approach of some native American tribes where the knowledgeable elders of the tribe (interdisciplinary group of experts) would discuss the impacts of important decisions on seventh generation (long-term approach to environmental management). Chapter IV.5 deals in depth with possible risk management options among native American tribes, which may illustrate the concepts of medicinal circle and seventh generation.

REFERENCES

Anderson, C. 1991. Cholera epidemic traced to risk miscalculation. *Nature* 354: 255.

Boxer, PA. 1985. Occupational mass psychogenic illness: history, prevention and management. *J. Occupational Med.* 27: 867–872.

Boxer, PA. 1990a. Pollution panic reduction. *Indoor Pollution Law Report* 4(1): 6–7.

Boxer, PA. 1990b. Indoor air quality: a psychosocial perspective. *J. Occupational Med.* 32: 425–428.

Breyer, S. 1993. *Breaking the Vicious Circle — Toward Effective Risk Regulation.* Harvard University Press, Cambridge, MA.

Kadvany, J. 1995. From comparative risk to decision analysis (in print).

Molak, V. 1990. Pollution as a terminal disease? *Risk Analysis* 10(4): 605–607.

Moynihan, P. 1993. Risk reduction. *Congressional Record* 139.

NAS 1983. Risk Assessment in Federal Government: Managing the Process. U.S. Department of Health and Human Services, Washington, DC.

Slovic, P. 1993. Perceived risk, trust, and democracy: a systems perspective. *Risk Analysis* 13: 675–682.

Stone, R. 1994. Comparative risk analysis. *Science* October.

U.S. EPA 1987. Unfinished Business: A Comparative Assessment of Environmental Problems. Office of Policy Analysis, U.S. EPA, Washington DC.

U.S. EPA 1990. Reducing Risk: Setting Priorities and Strategies for Environmental Protection. Science Advisory Board, U.S. EPA, Washington DC (SA-E-90-021).

U.S. EPA 1993. A Guidebook to Comparing Risks and Setting Environmental Priorities. Washington, DC (EPA 230-B-93-003).

QUESTIONS

1. What is comparative risk analysis?
2. What is the stated goal of the comparative risk analysis?
3. What are the major dangers of current comparative risk analyses?
4. How can one avoid a conflict-of-interest problem?
5. What are some of the community values ascertained by surveys in Ohio?
6. What is wrong in defining "risk = hazard + outrage" within a comparative risk analysis study?

CHAPTER III.5

Environmental Justice

Rae Zimmerman

SUMMARY

The environmental justice movement arose in the early 1980s from social and environmental movements of the previous decade or two. Prior to Federal Executive Order 12898 of February 11, 1994, environmental laws did not directly address environmental justice issues. Environmental justice studies most commonly have been conducted with respect to waste sites or waste-generation facilities. Although there is considerable debate as to what population characteristics to measure in an environmental justice study, most studies use population characteristics defined in data from the U.S. Bureau of the Census. Environmental justice studies have been conducted using population characteristics that are defined at many different scales and distances from the facilities. Because many subjective judgments are involved in evaluating whether or not an environmental justice issue exists, alternative approaches should be used in order to examine a range of results.

Key Words: environmental equity, environmental justice, environmental racism, hazardous waste, health risks

1. ORIGINS OF THE ENVIRONMENTAL JUSTICE MOVEMENT

Environmental equity and environmental justice were recognized as a public issue around the early 1980s. The environmental justice movement appears to have originated from three separate traditions: the overall diffusion of environmental consciousness in the population during the late 1960s and early 1970s, changes in the organization of environmental activism from one dominated by national organi-

1-56670-130-9/97/$0.00+$.50

zations to one with more grassroots involvement, and the civil rights movement in general which moved from emphasizing social causes to encompassing environmental concerns as well (Zimmerman 1993).

Although it is usually difficult to place the beginning of a movement, the environmental justice movement is considered to be an outgrowth of the public sensitivity to and outrage about hazardous waste disposal in the late 1970s and early 1980s. The hazardous waste site at Love Canal, New York is credited as the hazardous waste issue that led to the passage of one of the leading hazardous waste management statutes — the Comprehensive Emergency Response Compensation and Liability Act of 1980 (CERCLA) and its 1986 amendments — the Superfund Amendments and Reauthorization Act (SARA). The linkage between hazardous waste concern and minority exposures was not immediate, however. In spite of the considerable attention paid to Love Canal, little is mentioned of racial or ethnic disparities in documentation of the case, with the exception that Levine's (1982) notable study mentions that a public housing project allegedly affected by Love Canal wastes had a majority of black residents living in it.

Soon after Love Canal, attention to racial and ethnic disparities escalated. In 1983, the characteristics of the population near four hazardous waste facilities in the southeastern United States were investigated in a congressional study (U.S. GAO 1983). The study was initiated by protests in Warren County, North Carolina against a landfill siting in an area with a majority of African-Americans and low-income people (U.S. EPA 1992). The environmental justice movement is considered by many to have originated with the Warren County protests (U.S. EPA 1992).

By the early 1990s, several dozen cases involving racial or ethnic controversies were already identified, primarily involving hazardous waste disposal facilities, municipal waste disposal facilities, or other industrial-type operations (Zimmerman 1993). An extensive literature on various aspects of the environmental justice movement and its concerns began appearing in the late 1980s and early 1990s (Bryant and Mohai 1992, Bullard 1990, 1993).

Inactive hazardous waste sites regulated under CERCLA's Superfund program were particularly the target of environmental justice concerns addressed, for example, in an article by Lavelle and Coyle (1992). This led to numerous investigations of Superfund sites from the perspective of environmental equity and justice. In 1995, the U.S. General Accounting Office (U.S. GAO) expanded investigations of justice issues to municipal solid waste landfills in a nationwide study (U.S. GAO 1995a,b). At the same time, however, environmental concerns of minority populations expanded from concerns over waste processing and disposal facilities to the location of "Not-In-My-Backyard" facilities (NIMBYs) in general and to the distribution of public services to rectify a wide range of adverse environmental conditions.

The concern for environmental justice became formalized for federal government actions by Executive Order 12898, signed on February 11, 1994. The order required that each federal agency consider justice issues in its decisions and in carrying out its responsibilities. Prior to that order, the requirements to consider environmental justice in federal and state laws covering the siting and permitting of facilities were largely ad hoc and discretionary. The criteria for siting and operational decisions rested primarily on environmental locational or performance standards and guidelines irrespective of the population profile near the sites. Consideration of equity issues directly in hazardous waste legislation was largely absent, with the exception of some Department of Energy regulations (Greenberg 1995). At most, the magnitude of the population was considered, but not its characteristics. Environmental impact statements (EISs) conducted under the National Environmental Policy Act (NEPA) and similar state and local initiatives could have incorporated environmental justice issues in the context of socioeconomic analyses; however, environmental justice considerations were largely absent in both the decisions to conduct an EIS on a project and in the choice of subjects within an EIS for those projects requiring one (Zimmerman 1995). The environmental justice issue, however, could and often did arise in connection with public hearings and commentary required for these facility decisions. Few direct actions were likely, other than agency responses to the concerns, however, since specific regulatory mandates to consider the nature of the population affected were largely absent. Thus, prior to the executive order, few environmental laws directly addressed environmental justice.

The environmental justice movement was a grassroots movement. Local and regional organizations such as the Mothers of East Los Angeles, the Southwest Network for Environmental and Economic Justice, and the People for Community Recovery of Southside Chicago (U.S. EPA 1992) became common in the 1980s, representing the concerns of racial and ethnic minorities. The groups converged initially at two early conferences: "The First National People-of-Color Environmental Leadership Summit" in 1990 at the University of Michigan and the "Environmental Leadership Summit" in 1991 in Washington, D.C. (U.S. EPA 1994). After that time, a number of meetings were organized under the auspices of the U.S. Environmental Protection Agency (EPA), and a formal environmental workgroup was set up in connection with the executive order. The Office of Environmental Equity established in 1992 was changed to the Office of Environmental Justice in 1994.

2. ISSUES, CONCEPTS, AND DEFINITIONS

2.1 Basic Terminology: Equity, Justice, Fairness, and Racism

How inequity and injustice are addressed largely depends upon how the problems or the issues are defined and identified. Some of the concepts that have emerged in the environmental justice area that reflect different perspectives and, in fact, reflect the evolution of the movement include equity, justice, fairness, and racism.

According to the U.S. EPA (1992), *environmental equity* refers to "the distribution and effects of environmental problems and the policies and processes to reduce differences in who bears environmental risks." It refers to equality in the distribution of environmental risks across different sectors of the population, as well as the distribution of activities or procedures to mitigate these risks. This important distinction between distributional (spatial) and procedural (institutional) equity, which underlies the concept of environmental equity, was introduced by Kasperson and Dow (1991).

The concept of *environmental justice* introduces a normative dimension and goes beyond distributional issues to affording all groups in the population equal protection and access to healthy and environmentally sound conditions. The term was actively used in 1992, when two bills introduced into the U.S. House of Representatives (HR 5236) and the U.S. Senate (S 2806) used the term environmental justice.

The related concept of *fairness* is considered a key component of environmental justice. It has been used to connote equal treatment of groups with respect to the siting and location of existing facilities, such as New York City's "Fair Share" doctrine has advocated. This doctrine requires that the distribution and concentration of existing facilities in a community be considered before siting new ones. Been (1993) has extensively discussed the concept of fairness with respect to hazardous waste facilities.

The various dimensions of the concept of *environmental racism*, which actually appeared earlier than the other concepts, have been discussed by Hamilton (1995), and its initial use dates to the mid-1980s (United Church of Christ 1987). The U.S. EPA (1992) notes that in some circles the term has come to mean "disproportionate environmental risks in racial minority communities."

2.2 Defining Health and Environmental Risk

The linkage of population characteristics to environmental and health risks is of paramount importance to the environmental equity and justice movements. Justice can theoretically only be addressed in the context of some measure of environmental or health risk. The need to characterize these risks in equity research has grown, and a number of symposiums have begun to address the issue (Johnson et al. 1992, Sexton and Anderson 1993).

In spite of the apparent newness of environmental equity and justice, precedents for associations between population characteristics and actual and perceived environmental conditions or risks have had a very long history in the fields of sociology (including public opinion polling) and epidemiology. For example, most textbooks in epidemiology today contain data that disaggregate disease rates by race and ethnicity and other population characteristics, such as age, and the *American Journal of Public Health* periodically publishes articles and symposiums on the subject of patterns of disease among minority populations. Nevertheless, a great deal of work still has to be done to identify and quantify health and environmental risks to the point where their association with population characteristics can be explored.

Although it is often difficult to characterize health risks, especially for subpopulations without extensive and lengthy health studies, some issues have surfaced as having a high priority. First, health risk assessment guidance uses values for exposure assessment, such as rates and types of food consumption, that underestimate what minority populations consume and how these substances may be assimilated in the body. The example commonly used is estimates for fish consumption. Since certain minority groups tend to consume more fish than nonminorities, and fish often absorb pollutants, it is argued that using nonminority averages underestimates the risks of contaminants in fish to minorities (U.S. EPA 1992, 1994). A second area of concern is that the focus of current environmental justice analyses on the distribution of pollutants may be drawing attention from pollution prevention or pollution reduction initiatives.

2.3 Defining Minority Groups

There is no absolute consensus on what constitutes a minority or disadvantaged group nor on the terms used to characterize the disparities or disproportional impacts that such groups may experience. In the continental United States, the focus has been on the following racial and ethnic minority groups: African-Americans, Haitians, Asians, Native Americans, and Hispanics, as well as low-income groups. Each of the broad ethnic categories are further subdivided into often numerous subcategories; the category of Native Americans, for example, is subdivided into several hundred different tribes (Zimmerman 1994). Different regions will be concerned with different groups, since minority groups often differ in how they are distributed geographically.

Regardless of which groups are of interest in an environmental justice analysis, many common issues arise in connection with the classification of individuals into such groups, including (1) in the case of mixed heritages, whether or not one classifies an individual on the basis of what the individual regards as his/her identification or on the basis of some biological criteria (U.S. DOC 1993); (2) given a general definition of minority categories, what threshold constitutes the emergence or existence of an environmental justice issue; and (3) the extensiveness of racial and ethnic misclassification and its impacts on the accuracy of mortality and morbidity rates (Hahn et al. 1992, Sugarman et al. 1993, U.S. Congress 1986, U.S. DOC 1993). The classification issue has largely been addressed through consistent application of criteria, rather than trying to achieve consensus about which categories or classification criteria to use.

3. ANALYZING THE ENVIRONMENTAL JUSTICE ISSUE

The number of studies being conducted and the analytical frameworks for environmental justice have exploded over the past decade, concentrated particularly after the appearance of the very rich database afforded by the U.S. Bureau of the Census' *1990 Census of Population and Housing* and some initial court cases in which population data played a key role in the decisions. Although environmental justice studies have focused upon waste sites, other contaminant conditions have been studied as well, such as air pollution (Gelobter 1992), pesticides, and the presence of chemicals in the human body (which was the subject of a recent literature review [Goldman 1993]). Within the subject area of waste sites, the focus has been primarily on the association between population characteristics and both site location and the pattern of remediation or regulation of the site. The intensity of this activity has escalated along side the continued growth of environmental justice as a grassroots-based social movement. The United Church of Christ report (1987), for example, brought census data to bear upon the issue of justice on a national scale, although earlier case studies had also drawn on census data (U.S. GAO 1983). Analysis and activism often become intertwined.

The availability and accessibility of indicators of population characteristics and the ability to aggregate this data at very fine geographic scales has enhanced the ability to analyze environmental issues quantitatively. Yet basic analytical problems remain, which are discussed in the next sections.

Environmental equity/justice analyses typically serve two purposes. One is to map entire land surfaces according to some equity characteristic or set of characteristics such as race, ethnicity, and poverty. Given this spatial mapping, the potential impact on equity of locating a facility at any given point on such a surface can be ascertained. A second more common purpose is to define a given site or point first, describe population characteristics around it, and compare them to characteristics of some reference group in order to determine if disparities exist. The issues in the next sections pertain primarily to this second purpose, starting with a particular location and defining the population around it, although they are generally applicable to spatial mapping as well.

3.1 Locating the Site

First are *locational issues* relating to how finely a data point can, in fact, be located, which is the first step in defining the population around it. Satellite imaging and the availability of geographic positioning systems (GPS) that can be taken into the field are increasing the accuracy of site or area locations. Even if locations could be known with certainty, however, very large sites pose problems since they are not easily characterized by a single point. If the waste source cannot be located well within the site or if it is very spread out, a choice is usually made between using the site's geographic center or taking into account the entire area of the property. Although existing locational databases are becoming more refined, there is still a lack of consistency in how location is defined and the criteria used to define location. Ultimately, the criteria used to define a site's location depend more upon what the purpose is. The typical objective is to define population near some exposure point on the site. Since knowledge of such points usually requires extensive site investigation work, an approximation such as the center of the site is usually used. Results will not be affected very much by this assumption for sites with small areas relative to the population area that is being described around them.

3.2 Selecting Populations

Second is the set of choices pertaining to (1) what *population* should be the focus of the equity analysis (defined as who is frequently near the alleged source of the risk), (2) the *distance* from a site population where characteristics should be defined, and (3) the *size of the population data units* that should be aggregated.

The first related issue — what population is of interest — ideally should reflect who is potentially exposed to the environmental conditions at the site in question. Examples of alternatives are residential populations, workers, recreationists, and persons in institutions such as hospitals or schools. Since the exposed population is often never known with any certainty and databases are usually most commonly available only for resident populations, residents are the most common target population analyzed. This is a reasonable assumption, since it captures exposures that occur around the clock, such as constant air emissions from an abandoned hazardous waste disposal landfill or waste pile and any water supply contaminants that may enter the immediate water supply used by those residents. Of course, it will not capture exposures that are more transient or intermittent, i.e., of persons who may enter the area during the day for work, shopping, or recreation trips. It will also not capture exposures of more distant persons who may be using water supplies that may be contaminated by the source, but not reside near the site. Only more information on the nature of the exposures and a direct survey of the population exposed will yield such information.

The second related issue — distance from the site or area of interest — also depends upon the geographic extent of the exposure area which is often unavailable or so highly variable (temporally and by route of exposure) that it is not useful for spatial analysis. In order to compensate for this uncertainty, analyses are usually conducted at various distances from sites, typically about 1 to 4 miles from the defined site location.

The third related issue — unit of analysis — pertains to the geographic unit at which data should be collected. Data that originate from the census are aggregated in fixed geographic units, and a selection among them has to be made for an environmental justice analysis. The alternative census units that have been commonly used are (listed approximately in order of size) counties (Hird 1993, Perlin et al. 1995), zip codes (Goldman and Fitton 1994, Hamilton 1995, United Church of Christ 1987), census tracts (Anderton et al. 1994a,b), and block groups or blocks (Zimmerman 1994). Court cases have introduced another set of choices in determining whether environmental equity issues have arisen in connection with existing and proposed waste sites (Collin 1992, Colquette and Robertson 1991), although they have tended to use census tracts.

Related to the distance and the data unit selected (tract, block) is how data should be extracted. One approach is to employ some distance criterion (e.g., a 1-mile radius) as a guide for drawing whole data units based on the location of the centroids of the data units falling within the distance criterion (Zimmerman 1994). Another approach is to aggregate only those data units whose entire boundaries fall within the distance criterion (U.S. GAO 1983). Both of these approaches have the advantage of being very rapid. The outer boundaries of these areas are usually uneven, since they consist of the boundaries of the data units. Still another, far more common approach is to use a geographic information system (GIS) to intersect the outer data units in order to adhere to a strict distance criterion, that is, data are aggregated within a circle of a given radius. The choice of the unit of analysis and method of aggregation may alter the results in some cases and not in others, as shown in a recent test of sensitivity of race, ethnicity, and poverty results to geographic unit used by Glickman (Glickman and Hersh 1995, Perlin et al. 1995).

3.3 Selecting a Comparison Area

Third, and perhaps the most critical point, is the *comparison area or reference group* — the area to which population characteristics defined for the site or area should be compared. Here, the options vary widely, including the nation, a geographic region, the state, county, municipality, or an adjacent area such as a census tract or neighboring county or municipality. Where rings or circles of a certain distance are used, a ring or circle closer to a site might be compared with a ring or circle further away. All of these comparisons have been used and provide different perspectives. Another approach that borrows from case-control studies in epidemiology is to compare characteristics within a certain distance from a set of sites (aggregated for all sites) to characteristics of all other areas that lie outside of these proximate areas. Using the example of a 1-mile radius circle to define populations near each site in a set of sites and the county in which the sites are located as a comparison area, the total population in the circles around all of the sites would be compared to the population in the rest of the area within the county outside those circles. This is usually only an effective approach if the size of the area surrounding the site or sites (in aggregate) is large compared with the size of the comparison area.

3.4 Temporal Patterns

Fourth is the *temporal dimension* — whether or not a snapshot of population characteristics at one point in time is used vs. time trends in those characteristics. Most equity research emphasizes the former. A couple of studies have recently underscored the need to do time trend analyses in order to determine whether waste sites moved to where minorities were or vice versa (Anderson et al. 1994, Been 1994). It can be argued, however, that the time dimension is not important to whether or not a justice issue exists, since both circumstances — people moving in before the waste site or afterward — constitute inequities, though the solutions to the problem may differ.

3.5 Selecting Statistical Protocols

Finally, where the characteristics of a set of sites or areas is analyzed (as distinct from a single site), the implicit and explicit criteria used to interpret the patterns in the data are as significant as the choice of geographic units, boundaries, and statistical methods. Various statistical protocols are common, such as means, both weighted and unweighted (Zimmerman 1993, 1994); medians; quartiles; quintiles (Greenberg 1995); computations that take into account the standard deviation around the means; attention to statistical significance at alternative levels of significance; etc. The issue of indexing the raw data (classifying the data into just a few categories) for simplification is still another aspect of this issue. All of these issues are complex, but some simple rules are to examine the distribution of the underlying data prior to choosing criteria for interpreting it and evaluate patterns in alternative ways.

4. FUTURE DIRECTIONS

Few, if any, predictions of the rise of the environmental justice movement existed before it arose in full force during the 1980s, and few predictions of its current directions exist, especially in light of the radical changes pending in the environmental area in general.

As is typical of many social movements, analysis is playing a large role along with activism to determine how environmental justice concerns should be conceptualized and institutionalized. A number of guidelines and precautions emerge from the foregoing discussion that influence how an environmental justice issue is identified in a particular location. Many of these have already been summarized or argued in the literature (Greenberg 1993, Zimmerman 1994).

More work is needed to identify and quantify the risk element (whether environmental/ecological or health based) that different subpopulations confront. Spatial associations between risk and population characteristics and living conditions in general are an important guide in identifying areas of concern; however, for health issues, conducting an epidemiologic investigation is a necessary follow-up for more definitive results.

There is an underlying subjectiveness in many environmental justice criteria pertaining to what threshold constitutes a disparity, how comparisons are drawn, etc. Given those problems, multiple criteria should be used, e.g., different types of data and different comparison areas (municipalities, counties, states, regions, and the nation). Sensitivity analyses should be conducted on alternative criteria to see what impact the differences have on determinations that environmental justice issues exist. The measures used for environmental justice and environmental equity should be expanded to accommodate some of the more subtle distinctions, such as alternative concepts of poverty and persons from mixed racial and ethnic backgrounds.

REFERENCES

Anderson, A. B., Anderton, D. L., and Oakes, J. M., Environmental Equity: Evaluating TSDF Siting Over the Past Two Decades, *Waste Age*, 25, 83, 1994.

Anderton, D. L., Anderson, A. B., Rossi, P. H., Oakes, J. M., Fraser, M. R., Weber, E. W., and Calabrese, E. J., Hazardous Waste Facilities. "Environmental Equity" Issues in Metropolitan Areas, *Evaluation Review*, 18, 123, 1994a.

Anderton, D. L., Anderson, A. B., Oakes, J. M., and Fraser, M. R., Environmental Equity: The Demographics of Dumping, *Demography*, 31, 229, 1994b.

Been, V., Locally Undesirable Land Uses in Minority Neighborhoods: Disproportionate Siting or Market Dynamics?, *The Yale Law Journal*, 103, 1383, 1994.

Been, V., What's Fairness Got to do With It? Environmental Equity in the Siting of Hazardous Waste Facilities, *Cornell Law Review*, 78, 1001, 1993.

Bryant, B. and Mohai, P., Eds., *Race and the Incidence of Environmental Hazards*, Westview Press, Boulder, CO, 1992.

Bullard, R. D., *Dumping in Dixie. Race, Class and Environmental Quality,* Westview Press, Boulder, CO, 1990.

Bullard, R. D., ed., *Confronting Environmental Racism. Voices from the Grassroots*, South End Press, Boston, MA, 1993.

Collin, R. W., Environmental Equity: A Law and Planning Approach to Environmental Racism, *Virginia Environmental Law Journal*, 11, 495, 1992.

Colquette, K. M. and Robertson, E. A. H., Environmental Racism: The Cases, Consequences, and Commendations, *Tulane Environmental Law Journal*, 5, 153, 1991.

Gelobter, M., Toward a Model of Environmental Discrimination, in *Race and the Incidence of Environmental Hazards*, Bryant, B. and Mohai, P., Eds., Westview Press, Boulder, CO, 1992.

Glickman, T. S. and Hersh, R., *Evaluating Environmental Equity*, Resources for the Future, Washington, D.C., 1995.

Goldman, B. A., Not Just Prosperity, Achieving Sustainability with Environmental Justice, National Wildlife Federation, Washington, D.C., 1993.

Goldman, B. A. and Fitton, L., *Toxic Wastes and Race Revisited*, Center for Policy Alternatives, Washington, D.C., 1994.

Greenberg, M. R., Proving Environmental Inequity in Siting Locally Unwanted Land Uses, *Risk — Issues in Health & Safety*, 4, 235, 1993.

Greenberg, M. R., Separate and Not Equal: Health-Environmental Risk and Economic-Social Impacts in Remediating Hazardous Waste Sites, in *Environmental Contaminants, Ecosystems and Human Health*, Majumdar, S. K., Brenner, F. J., Willard Miller, E., and Rosenfeld, L. M., Eds., Pennsylvania Academy of Science, Philadelphia, (in press), Chap. 32.

Hahn, R. A., Mulinare, J., and Teutsch, S. M., Inconsistencies in Coding of Race and Ethnicity Between Birth and Death in U.S. Infants, *JAMA*, 267, 259, 1992.

Hamilton, J. T., Testing for Environmental Racism: Prejudice, Profits, Political Power?, *Journal of Policy Analysis and Management*, 14, 107, 1995.

Hird, J. A., Environmental Policy and Equity: The Case of Superfund, *Journal of Policy Analysis and Management*, 12, 323, 1993.

Johnson, B. L., Williams, R. C. and Harris, C. M., *National Minority Health Conference. Focus on Environmental Contamination*, Princeton Scientific Publishing Co., Inc., Princeton, NJ, 1992.

Kasperson, R. E. and Dow, K. M., Developmental and Geographical Equity in Global Environmental Change. A Framework for Analysis, *Evaluation Review*, 15, 149, 1991.

Lavelle, M. and Coyle, M., Unequal Protection. The Racial Divide in Environmental Law, *The National Law Journal*, Special Investigation 12, September 21, 1992.

Levine, A. G., *Love Canal: Science, Politics and People*, Lexington Books, Lexington, MA, 1982.

Perlin, S. A., Setzer, R. W., Creason, J., and Sexton, K., Distribution of Industrial Air Emissions by Income and Race in The United States: An Approach Using the Toxic Release Inventory, *Environmental Science & Technology*, 29, 69, 1995.

Sexton, K. and Anderson, Y. B., Equity in Environmental Health: Research Issues and Needs, *Toxicology and Industrial Health*, 9, 679, 1993.

Sugarman, J. R., Soderberg, R., Gordon, J. E., and Rivara, F. P., Racial Misclassification of American Indians: Its Effect on Injury Rates in Oregon, 1989 through 1990, *American Journal of Public Health*, 83, 681, 1993.

United Church of Christ, Commission on Racial Justice, *Toxic Wastes and Race in the United States*, United Church of Christ, New York, 1987.

U.S. Congress, Office of Technology Assessment, *Indian Health Care*, Office of Technology Assessment, Washington, D.C., 1986.

U.S. Department of Commerce, Bureau of the Census, *Challenges of Measuring an Ethnic World. Science, Politics, and Reality*, U.S. Government Printing Office, Washington, D.C., 1993.

U.S. Environmental Protection Agency, *Environmental Equity,* Reducing Risk for All Communities. Volume 1: Workgroup Report to the Administrator; Volume 2: Supporting Document, U.S. EPA, Washington, D.C., June 1992.

U.S. Environmental Protection Agency, Environmental Justice Initiatives, 1993, U.S. EPA, Washington, D.C., 1994.

U.S. General Accounting Office, *Siting of Hazardous Waste Landfills and Their Correlation with Racial and Economic Status of Surrounding Communities,* GAO/RCED-83-168, U.S. GAO, Washington, D.C., June 1, 1983.

U.S. General Accounting Office, *Hazardous and Nonhazardous Waste, Demographics of People Living Near Waste Facilities,* GAO/RCED-95-84, U.S. GAO, Washington, D.C., June 1995a.

U.S. General Accounting Office, *10 Studies on Demographics Near Waste Facilities,* GAO/RCED-95-158R, U.S. GAO, Washington, D.C., June 1995b.

Zimmerman, R., Social Equity and Environmental Risk, *Risk Analysis: An International Journal*, 13, 649, 1993.

Zimmerman, R., Issues of Classification in Environmental Equity: How We Manage Is How We Measure, *Fordham Urban Law Journal*, XXI, 633, 1994.

Zimmerman, R., An Environmental Equity Study for Inactive Hazardous Waste Sites, Superfund Program for Inactive Hazardous Wastes Sites on the NPL, U.S. Environmental Protection Agency, Region 2, New York, February 9, 1994.

Zimmerman, R., Integrating Environmental Justice (EJ) Methodologies into Environmental Impact Assessment, in *Environmental Challenges: The Next 20 Years,* National Association of Environmental Professionals 20th Annual Conference Proceedings, Washington, D.C., 1995.

QUESTIONS

1. What is the difference between environmental equity, environmental justice, and environmental racism?
2. What are the origins of the environmental justice movement?
3. To what extent is the requirement to evaluate environmental justice issues currently present in environmental laws and regulations?
4. Why is the definition or quantification of health risks particularly difficult in environmental justice analyses of hazardous waste sites?
5. What are some of the problems involved in defining minority groups?
6. At what spatial levels have environmental justice studies been conducted?
7. What are the alternative methods for aggregating census data units in order to define total population characteristics around a location?
8. After data is obtained for a particular site, how is the data used to determine whether or not a justice issue exists?
9. What aspects of environmental justice evaluations are subjective, and how can these be addressed?

CHAPTER III.6

Law and Risk Assessment in the United States

Peter Barton Hutt

SUMMARY

Since recorded history, all human activity has included both risk assessment and risk management. Even animals exhibit these characteristics. Only very recently have humans begun to quantify risks and thus to improve their management of the daily risks that we all face.

Government control of risks is not a recent phenomenon. Every civilized country has adopted some form of government regulation to reduce societal risk. The oldest recorded governmental laws and regulations that reflect a collective approach to risk assessment and risk management were adopted to protect the food and drug supply.

This chapter focuses on development in the United States of various forms of government regulation, based on risk assessment and risk management, to protect the public health and safety. Because much of our approach to risk assessment and risk management is derived from our rich heritage of food and drug regulation, the chapter deals more extensively with this area than with more recent regulation of other consumer products, occupational safety, and the environment, all of which became subject to organized federal regulation only very recently, in the last 25 years.

As this chapter demonstrates, during most of our country's history, we had available only very primitive tools to assess risk and to manage the risk that was identified. Nonetheless, our early regulatory systems did a remarkable job in establishing basic public health protection. Only in the very recent past have more sophisticated quantitative methods of risk assessment become available. Our biggest problem today is that we can find risk and quantify it far more readily than we can understand it or know whether or how to manage it.

1-56670-130-9/97/$0.00+$.50

1. BEFORE 1900

The people who left England and the European continent to colonize America brought with them an extensive heritage of statutory and common law intended to protect the integrity of the food and drug supply. This heritage was reflected in the laws and regulations adopted in colonial America. Statutes were enacted to prevent food and drug adulteration, and trade guilds enforced them as a condition of membership. Citizens could resort to the courts to obtain damages for the sale of adulterated food and drugs, and the courts also determined that the sale of unwholesome food and drugs was a crime under both statutory law and common law.

By modern standards, detection methods for determining adulteration were very limited. Nonetheless, these early laws were enforced and clearly prevented what would otherwise have been a completely chaotic deterioration of the food and drug supply. Just as detection methods were relatively simple, adulteration methods were also relatively simple. It did not require sophisticated methodology to find stones in bread or burnt charcoal in pepper.

During this era, the statutes and the literature almost invariably referred to impure food and drugs as "adulterated," not as "unsafe." Our concept of "safety" did not emerge until this century. Use of the term adulterated, moreover, covered two quite distinct problems. First, it referred to economic adulteration, i.e., substitution of a cheap substance (e.g., water) for a more expensive and valuable substance (e.g., milk). Second, it referred to safety problems, such as the use of harmful ingredients. Because economic adulteration often involved the use of contaminated or harmful ingredients, these two meanings of the word adulterated often merged. Frederick Accum, in his famous *Treatise on Adulterations of Food and Culinary Poisons* published in England in 1820, was one of the first to explicitly point out that economic adulteration often resulted in danger to health as well as the pocketbook.

Following the American Revolution, the colonial food and drug laws were reenacted as city, county, and, ultimately, state statutes. In the mid-1800s, two events occurred that had major impacts on our laws to protect the food and drug supply. In 1846, Dr. Lewis C. Beck, Professor of Chemistry at Rutgers and Albany Medical College, published a treatise demonstrating widespead adulteration of food and drugs in the United States. In 1850, Lemuel Shattuck of Boston published the *Report of the Sanitary Commission of Massachusetts*, documenting a startling reduction in average life expectancy at birth for United States citizens living in large urban areas and attributing this problem to public insanitation and to adulterated food and drugs. Both of these reports were early examples of risk assessment. The Shattuck report recommended, as a risk management approach, the enactment of state and local food and drug laws to deal with adulteration. By the end of the 19th century, cities, counties, and states throughout the country had enacted some form of statute to prohibit adulterated food and drugs.

At that time, Congress and the Supreme Court were of the opinion that regulation of food and drugs was a local matter, not within the legislative authority of Congress under our Constitution. Throughout the 19th century, therefore, Congress studied the problem of food and drug adulteration and enacted statutes dealing with the import and export of adulterated food and drugs, but never enacted a statute establishing federal requirements for pure food and drugs on a nationwide basis.

2. THE FEDERAL FOOD AND DRUGS ACT OF 1906

In 1906, Congress enacted two statutes designed to prevent food and drug adulteration: the Federal Food and Drugs Act, which applied to all food, and the Federal Meat Inspection Act, which was designed specifically to cover the slaughter of animals used for human food. These statutes incorporated risk assessment and risk management into federal regulatory law. They began the tradition in our country, now 90 years old, of direct involvement by the U.S. Food and Drug Administration (FDA) in assuring the safety of the entire food and drug supply. Because FDA regulates meat as well as all other food, this chapter does not specifically discuss the narrower function of the U.S. Department of Agriculture (USDA) in regulating meat, and subsequently poultry and eggs as well.

The 1906 act prohibited, as adulterated, any food that contained an added poisonous or deleterious substance that may render the food injurious to health. As interpreted by the Supreme Court, this provision prohibited any substance that resulted in a reasonable possibility of human harm, based on actual conditions of use in the food supply. The FDA was given substantial court enforcement powers to implement this provisions.

It is clear that the safety standard in the 1906 act was not written narrowly and was not well defined. A "poisonous or deleterious substance," even under the Supreme Court interpretation of that phrase, could be applied broadly or narrowly, depending upon the FDA interpretation.

In 1900, Congress had appropriated funds for the FDA to investigate the use of preservatives and colors in the food supply. At that time, the concept of animal testing was still not developed. There were no colonies of inbred laboratory animals for toxicity testing. Therefore, the FDA tested five of the most widely used food categories of preservatives of that day — boric acid and borax, salicylic acic and salicylates, sulfurous acid and sulfites, benzoic acid and benzoates, and formaldehyde — by feeding them to 12 young USDA employees from 1902 to 1904. The results of this human feeding study were published in 1904 to 1908. These reports listed table after table of the observed clinical chemistry results in the 12 subjects involved, but contained little useful analysis of the type that would be expected from any such experiment today. The sciences of pharmacology and toxicology had not yet begun to develop, and, thus, the reports could only state in general terms that the substances tested appeared to be injurious and harmful to health.

Following enactment of the 1906 act, the FDA banned all five of these categories of preservatives from food. After an extraordinary appeal by industry directly to President Theodore Roosevelt and a subsequent review by a scientific advisory committee, one of the preservatives — sodium benzoate — was found not to be a poisonous or deleterious substance and thus was allowed to remain in soft drinks. It is still used in food products today.

This was a remarkable early experiment in risk assessment and risk management. Reading these early reports, however, one is struck by the lack of precision or detail of the risk assessment, even in qualitative terms, and by the quick resort to the most draconian form of risk management, a complete ban of the substances involved.

By the time that the status of sodium benzoate was resolved, animal feeding studies had begun to replace human feeding studies as the basis for risk assessment. Inbred strains of laboratory animals began to be raised in various centers throughout the country. Previously, ad hoc testing of substances and products was done on single animals, but could not be reproduced in a colony of animals because of their genetic diversity. Once the technique of raising inbred animals was perfected, following the rediscovery of Mendel's laws of heredity, animal experimentation began in earnest.

The use of animal testing had two effects. First, it greatly accelerated toxicity testing. Animals were obviously easier to control than humans, and they could be sacrificed and examined in infinitely greater detail. Second, the availability of data on large numbers of interchangeable animals for the first time allowed scientists to begin to attempt to establish the relative potency of compounds, the first step toward quantification of risk. For the most part, however, the results of animal toxicity testing were simply regarded as one factor to be considered in determining the safety of a food or drug. The relationship between animal toxicity and human toxicity was still unknown, and, thus, animal testing could not be used as a direct surrogate for human safety.

In spite of these obstacles, however, the FDA was extremely successful in taking regulatory action under the 1906 act to prevent and punish the adulteration of food and drugs. Thousands of cases were successfully brought by the agency, and as a result the industry responded by substantially improving its products.

By 1917, the FDA had concluded that the 1906 act needed substantial expansion and modernization. Nothing concrete was done about this until 1933, when President Franklin Roosevelt took office. At that time, the FDA proposed the new legislation that ultimately became the 1938 act.

3. THE FEDERAL FOOD, DRUG, AND COSMETIC ACT OF 1938

While the 1938 act was pending enactment, a national drug tragedy occurred that was to have a major impact on the field of risk assessment. In the fall of 1937, the Massengil Company developed, and immediately marketed without animal testing, a product labeled Elixir Sulfanilamide, the first successful attempt to put the new miracle antibacterial drug, sulfanilamide, into solution. Unfortunately, the solvent used was diethylene glycol, a frank poison now used as antifreeze, and the product killed more than 100 people within a few days.

Two FDA scientists were determined not to let the matter rest there. They undertook what became one of the most important series of experiments in the history of risk assessment. First, they obtained all available records on the individuals who consumed Elixir Sulfanilamide and either lived or died, thus determining a rough LD_{50} for diethylene glycol in humans. Then they conducted animal feeding experiments in a wide variety of laboratory animals, also determining the LD_{50} in those species. They concluded that there was roughly a tenfold variation in toxicity among humans and among test animals. Multiplying 10 by 10, they arrived at the 100 to 1 safety factor that has been widely accepted and used for acute toxicity ever since.

The importance of this development cannot be overstated. For centuries, no one understood the relationship between animal toxicity and human toxicity, because it was clearly impossible to conduct an LD_{50} study, or anything even close to it, in humans. It was only the Elixir Sulfanilamide tragedy, and the extraordinary scientific initiative of the FDA scientists that lead to this landmark work. Neither of these scientists has ever received the recognition that should have resulted from this brilliant work.

The Elixir Sulfanilamide tragedy also provided the impetus necessary for enactment of the 1938 act. This statute built upon the foundation of the 1906 act and carried it further. For new drugs, it required premarket notification relating to safety, based upon both animal and human testing. For food, however, it continued to rely on FDA policing the marketplace to assure an unadulterated and pure supply of food. For the first time, the word "safe" was used in the statute.

Having invented the 100 to 1 safety factor in the early 1940s, the FDA proceeded to implement it. As new chemicals entered the food and drug supply, they were subjected to animal toxicity testing, and the lowest no-observed-effect level (NOEL) was determined. That NOEL was then divided by 100 to determine the safe level for humans.

It has often been stated that the FDA initially applied the 100 to 1 safety factor to carcinogens as well as noncarcinogens. Recent research has shown, however, that this never happened. From the 1940s on, FDA scientists concluded that carcinogens acted differently than noncarcinogens and that no safe threshold could be determined for carcinogens. Accordingly, the FDA never applied the 100 to 1 safety factor to a chemical that had been determined to be carcinogenic in test animals.

There are a number of examples that document this policy. In the mid-1940s, the FDA banned a color additive, butter yellow, because of animal carcinogenicity. In the early 1950s, the FDA banned two nonnutritive sweeteners, dulcin and P-4000, because of findings of animal carcinogenicity. In effect, the FDA adopted a zero-tolerance standard for carcinogens under the general safety provisions of the 1938 act. The agency concluded that, since no carcinogen could be shown not to be poisonous or deleterious at any level, they all should be banned.

4. THE FOOD ADDITIVES AMENDMENT OF 1958

In 1950, Congress established a Select Committee to Investigate the Use of Chemicals in Food. The idea of a select committee came from Representative Frank B. Keefe (R, WI), but he persuaded Representative James J. Delaney (D, NY) to chair the select committee. The select committee recommended legislation to assure the safety of new food additives. Late in the consideration of legislation to implement this recommendation, Representative Delaney added a provision that has since been called the Delaney Clause, prohibiting FDA approval on any food additive found to induce cancer in humans or animals after ingestion or other appropriate testing. The FDA initially opposed the Delaney Clause, but later accepted it when it became clear that the legislation otherwise could not be enacted. The agency stated that the clause simply codified the policy that it had adopted administratively more than a decade earlier. Subsequently, the Delaney Clause was also added to the Color Additive Amendments of 1960 and the Animal Drug Amendments of 1968.

5. IMPLEMENTATION OF THE DELANEY CLAUSE

Implementation of the Delaney Clause was difficult from the very beginning. First, it applied only to substances explicitly defined as food additives. Thus, it does not apply to food itself or to food substances determined to be generally recognized as safe (GRAS substances) or subject to an FDA or USDA approval during 1938 to 1958 (prior-sanctioned substances). Therefore, the Delaney Clause does not apply to most substances that comprise the food supply. Second, when confronted with a difficult problem under the Delaney Clause in 1962, Congress enacted an exemption to the Delaney Clause for carcinogenic drugs used in food-producing animals if no residue could be found in the human food using methods of analysis approved by the FDA. This exemption was enacted to allow the FDA to continue to approve the carcinogenic animal drug, diethylstilbestrol (DES), for use to promote growth in cattle and sheep.

For the decade of the 1960s, the FDA simply implemented the Delaney Clause as it was written. Two unimportant indirect food additives were banned under the Delaney Clause during that time, but otherwise it had no significant impact.

Nonetheless, three related scientific developments were occurring that, by the early 1970s, created a major regulatory crisis. First, more and more chemicals used in food and drugs were being tested in animal bioassays, and thus more were found to be carcinogens in laboratory animals. Second, the protocols used for this testing were gradually made more stringent, thus resulting in a larger percentage of the tested substances being determined to be carcinogenic. For example, roughly half the chemicals tested in the National Toxicology Program protocols have been found to be carcinogenic. Third, the sensitivity of analytical detection methodology plummeted from parts per million to parts per quadrillion, and now even parts per quintillion, thus allowing the detection of carcinogenic substances throughout the food and drug supply. It gradually became apparent that the FDA could not ban all carcinogenic food substances or all food containing carcinogenic constituents, because this would require a ban of virtually all food in the country.

6. THE DEVELOPMENT OF QUANTITATIVE RISK ASSESSMENT FOR CARCINOGENS

In mid-1972, it was determined that, even after 1 week of withdrawal, DES was still detected in meat obtained from cattle for human consumption. The FDA immediately banned DES. It initially acted under the general safety provisions of the law, not under the Delaney Clause, but, after a court-ordered administrative hearing, the agency later relied upon the Delaney Clause and the ban was upheld.

More important than DES itself, this incident dramatized to the FDA that its implementation of the Delaney Clause for the past decade could no longer be sustained. If DES could be found in the meat of animals to whom small amounts were administered, even after 1 week of withdrawal, it was apparent that the same problem would arise with all other carcinogenic animal drugs. The FDA chief counsel therefore declined to approve any other carcinogenic animal drug until a new approach to regulating these products under the Delaney Clause could be found and adopted.

Beginning in the mid-1930s, academic scientists had already begun to consider how the risk of carcinogenic compounds could be quantified, based upon animal testing results. Numerous approaches were explored in the scientific literature. One of the more influential papers was published by Mantel and Bryan in 1961, advancing a mathematical model for determining carcinogenic risks through low-dose extrapolation and recommending an acceptable risk level of 10^{-8} (1 in 100 million). In 1970, Gross, an FDA scientist, published a paper with Mantel, applying that model to the regulation of a commonly used flavoring substance that was thought to be a potential reproductive toxicant rather than a carcinogen.

Accordingly, at the time of the DES crisis in mid-1972, there was a substantial body of theoretical literature to support the use of quantitative risk assessment in determining an acceptable risk level for a carcinogen like DES. Gross advocated this approach for DES within the FDA prior to mid-1972, but his views were not accepted. After the DES crisis occurred, however, the matter was reconsidered. Lehman, another FDA employee, took the initiative in explaining the entire matter to the FDA chief counsel, resulting in a decision to adopt quantitative risk assessment in the regulation of animal carcinogens under the Delaney Clause.

The new FDA policy became known as the "sensitivity of the method" policy. Simply put, the FDA took the position that the sponsor of carcinogenic new animal drug must develop an analytical method sufficiently sensitive to detect in the food obtained from the animal the level of the carcinogenic drug that represents a risk of 10^{-6} (1 in 1 million). This represented the first use of quantitative risk assessment by any regulatory agency in the world for any purpose. It was refined in a series of proposed and final regulations published in the *Federal Register* beginning in 1973.

This was, however, only the beginning of the matter. Carcinogenic animal drugs represented only a small part of the problem. Carcinogenic food additives and color additives, and carcinogenic constituents throughout the food and drug supply, represented a far more formidable challenge.

Undoubtedly, the largest part of the problem was represented by carcinogenic constituents throughout the food supply. The FDA immediately took the position that only the complete food additive, and not its constituents, were subject to the Delaney Clause. The constituents were subject only to the more general "poisonous or deleterious" standard, which permitted quantitative risk assessment. This was announced by the FDA in a report to Congress in 1974, was first relied upon to approve an additive in a decision published in the *Federal Register* in 1982, and was upheld by a reviewing court in 1984.

The agency then turned its attention to carcinogenic food additives and color additives themselves. Faced with three color additives found to be carcinogenic in test animals, but for which quantitative risk assessments demonstrated a human risk of far below 1 in 1 million, the FDA announced in 1986 that these color additives would be approved on the basis of the applicable risk assessments. In 1987, however, the reviewing court overturned the FDA decision. The court concluded that the Delaney Clause must be interpreted literally to ban any carcinogenic risk, however small, represented by a color additive. The same decision was later reached by another court, applying the Delaney Clause to pesticide residues that concentrated in processed food and thus were subject to regulation as food additives. As a result, the regulated industry has requested Congress to revise the statute to impose an insignificant or negligible risk standard to replace the current zero-tolerance approach in the Delaney Clause. The outcome of the congressional debate on this matter remains uncertain.

7. REFINEMENT OF RISK ASSESSMENT

As a result of the FDA adoption of quantitative risk assessment, scientists throughout the world have reconsidered and refined existing approaches to quatifying the risk of both noncarcinogenic and carcinogenic substances and have developed new approaches. It is not the purpose of this chapter to relate or evaluate any of the risk assessment methodology. It is important to note, however, that the initiative taken by the FDA to use quantitative risk assessment directly in product regulation has fostered substantial scientific inquiry and insight.

It is apparent to all who work in this field that both the quantitative risk assessment methodology pioneered by the FDA, and that used today, remain primitive and crude. Current techniques represent the best available tools to assess risk at this time, but enormous progress must be made before confidence can be obtained in the validity and accuracy of these techniques. It is certain that in years to come our successors will look back on our efforts as highly inaccurate and immature, just as we look back on the work of our predecessors in this field.

8. THE PROLIFERATION OF QUANTITATIVE RISK ASSESSMENT IN OTHER AREAS OF THE LAW

Following the development of quantitative risk assessment as a regulatory tool by the FDA, it has been used widely by other government agencies and in private litigation in the courts.

Beginning in 1976, the U.S. Environmental Protection Agency (EPA) published general cancer assessment principles to govern regulation of pesticides and contaminants in air and water. Since then, the U.S. EPA has applied quantitative risk assessment throughout all of its regulatory responsibilities.

The Consumer Product Safety Commission (CPSC) and the Occupational Safety and Health Administration (OSHA) were slow to adopt quantitative risk assessment. Following a major Supreme Court decision in 1980 holding that OSHA may ban only significant risks to public health, OSHA has adopted the use of the quantitative risk assessment. Similarly, an adverse court of appeals decision in 1983 has stimulated the use of quantitative risk assessment by CPSC.

Issues relating to the toxicity of chemical substances that comprise consumer products, that once were resolved by regulatory agencies, have now spilled over into the courts in the form of so-called "toxic torts" litigation. These lawsuits typically involve hundreds or thousands of individual plaintiffs suing one or more corporations for damages, alleging real or feared injury from the chemical or product involved. Examples include such chemicals as bendectin, asbestos, dioxin, and formaldehyde and such products as the Dalcon Shield and breast implants. Quantitative risk assessment is the major focal point on which many of these cases rest.

9. CONCLUSION

Without question, quantitative risk assessment represents a major advance over the qualitative judgments that formerly had to be relied upon in making important decisions about the health consequences of chemicals in our food and drug products and in the environment. Equally without question, these techniques remain unsatisfactory because of the large lack of precision and resulting enormous uncertainty in the ultimate calculated level of potential risk. While these techniques are the best we have today, one can only look forward to the inevitable progress that will be made in this field as the science progresses, thus leading toward far more accurate and reliable risk assessments in the future. As this occurs, the importance of quantitative risk assessment in all aspects of the law will undoubtedly increase.

REFERENCES

Hutt, P. B. and Merrill, R. A., *Food and Drug Law: Cases and Materials*, 2d ed., Foundation Press, Westbury, New York, 1991.

Hutt, P. B. and Hutt, P. B., II, A History of Government Regulation of Adulteration and Misbranding of Food, *Food Drug Cosmetic Law Journal,* 39, 2, 1984.

CHAPTER III.7

Science, Regulation, and Toxic Risk Assessment*

Howard Latin

SUMMARY

Regulation of toxic substances is an extremely complex, uncertain, and controversial enterprise. The regulatory process is customarily divided into two discrete functions: risk assessment and risk management. Risk assessment ostensibly is a scientific activity that develops estimates of human health hazards or environmental hazards at varying pollutant exposure levels, whereas risk management is a political activity that balances competing interests and values to determine whether identified toxic risks should be considered unacceptable or tolerable (U.S. NRC 1983). Risk management is the process of weighing policy alternatives and selecting the most appropriate regulatory action, integrating the results of risk assessment with social, economic, and political concerns. This sharp distinction between the scientific and social policy dimensions of toxics regulation is embodied in the U.S. Environmental Protection Agency's (EPA) guidelines (1986a), which state that risk assessments must "use the most scientifically appropriate interpretation" and should "be carried out independently from considerations of the consequences of regulatory action." The U.S. EPA and other federal agencies stress the need for scientifically credible risk assessments and presume that their analyses should be grounded exclusively on the best available scientific theories and data, even if the resulting predictions do not achieve the degree of reliability ordinarily required for valid scientific conclusions.

We challenge this view that only scientific perspectives should dominate the risk assessment process. Risk assessment is too important and too uncertain to be

* The comments in this chapter have been condensed from a much longer treatment written two years after the last major revision of the EPA Carcinogen Risk Assessment Guidelines.

left exclusively to the risk assessors. Instead, social policy considerations must play as prominent a role in the choice of risk estimates as in the ultimate determination of which predicted risks should be deemed unacceptable.

Key Words: social policy, regulations, toxic substances, risk assessment, risk management, laws, politics, legislation

1. INTRODUCTION

When harm will be substantially irreversible, as in the cases of carcinogenic exposures, extinction of species, or acid-rain contamination of lakes and forests, the problem of how long regulators should wait for "enough" information to enable reliable scientific judgements is likely to be controversial.

Risk assessment suffers from fundamental uncertainties about causal mechanisms for cancer and other hazards, extrapolative relationships between high-dose and low-dose responses and between animal test data and human risks, latent effects and latency periods, special sensitivities in exposed subpopulations, synergistic or co-carcinogenic effects of various substances, past and present exposure levels, dispersion patterns for contaminants, and virtually every other area of required knowledge (Ruckelshouse 1983). These uncertainties generally preclude reliable assessments of relevant effects, and there is no scientific consensus on how they should be resolved. For example, conflicting risk estimates submitted in U.S. Food and Drug Administration (FDA) proceedings on saccharin varied by more than a millionfold (Leape 1980), and predictions of the hazards posed by TCE, a drinking-water contaminant, varied by many millions (Cothern et al. 1986). The same discussion of TCE regulation noted that the "estimates provide a range of uncertainty equivalent to not knowing whether one has enough money to buy a cup of coffee or pay off the national debt."

Part 2 of this chapter evaluates the risk assessment principles generally used by federal agencies. We demonstrate that the risk assessment efforts in regulatory proceedings seldom achieve professionally accepted standards of scientific validity and inevitably entail implicit or explicit policy judgements.

Part 3 describes the social ramifications of the EPA's current emphasis on "good science." In practice, this risk assessment focus is likely to result in reduced public protection against potential toxic hazards, increased regulatory decision-making costs, and expanded opportunities for obstructive behavior by agency bureaucrats or private parties hostile to toxics regulation. ***These consequences might be acceptable if they were the product of an explicit, well-formulated, and open political decision, but they should not arise unintentionally from the EPA's single-minded pursuit of "good science" in an area where reliable scientific conclusions are difficult if not impossible to attain.***

Part 4 describes social policy criteria that agencies could use to supplement scientific evidence on toxic hazards. These factors include the hierarchy of legislative priorities in particular regulatory statutes, the expense and time requirements associated with individualized assessments of recurring scientific issues, the potential for catastrophic results from erroneous risk estimates, and the likelihood that specific uncertainties can or cannot be resolved in the near future. Contrary to the EPA's carcinogen guidelines, analysis of regulatory purposes and possible social consequences — not attempts at "good science" alone — should shape risk assessment efforts. ***Statutory preferences for safety from toxic substances should not be undermined by low-visibility adoption of speculative risk assessment practices that cannot be grounded in reliable science. Instead, we should examine a range of social policy criteria that could be incorporated in the risk assessment process after appropriate public discussion.***

2. EVALUATION OF FEDERAL RISK ASSESSMENT

2.1 Tensions between Risk Assessment Science and Risk Regulation

Under current regulatory practices, EPA scientists produce risk assessments that seldom approach the level of reliability normally expected of scientific findings; indeed, many estimates are little more than educated guesses (U.S. EPA 1984a, Goldstein 1985). Yet, the choice among competing estimates — a prediction of only a minuscule hazard or one a million times greater — can determine whether toxic exposures are characterized as "acceptable" or "unacceptable" irrespective of any values in the risk management process. Absent a scientific consensus on which risk assessment principles should be applied, an agency's choice among competing risk estimates should not be exclusively a result of provisional scientific judgements. If substantial uncertainty exists about the extent of toxic hazards and the possible benefits from risk reduction, social consequences and political values must play an integral role in determining which speculative risk estimates are adopted.

Unlike in pure scientific research, where the proper response to uncertainty is reservation of judgement pending the development of adequate data and testable hypotheses, the risk assessment process cannot be suspended without significant social consequences. A finding that a vital issue is currently indeterminate would be entirely consistent with the practice of "good science," but "no decision" on a possible toxic hazard inescapably is a decision that promotes interests which benefit from the regulatory status quo (Bazelon 1981, Latin 1982). ***Risk assessment is not driven by the pursuit of knowledge for its own sake, but by the need to decide whether potentially severe health hazards should be allowed to continue or whether control costs should be imposed with potentially severe economic consequences. Thus, scientists in regulatory proceedings are expected to produce "answers" in a timely manner even if their predictions are highly speculative. Any reluctance to relax the standards of proof and certainty generally required of valid science may introduce a bias in favor of regulatory inaction.***

Science aims at the dispassionate pursuit of truth. In contrast, scientists in risk assessment proceedings frequently represent industries, labor unions, consumers, environmentalists, or agency bureaucracies with great interests at stake. These affiliations may often explicitly or unintentionally color interpretations of available evidence (Latin 1985, Stewart 1981). Scientists seldom base conclusions on data and experiments that cannot be reproduced, but information in regulatory hearings is routinely submitted by affected parties and frequently cannot be replicated or effectively challenged by other participants (U.S. EPA 1984a). Scientists are no more immune to cognitive dissonance and wishful thinking than are nonscientists. Scientists tend to design research studies in light of which data are available and which experiments may be feasible, whereas the critical questions in risk assessment proceedings are usually determined by statutory or judicial requirements that need not be responsive to the state of scientific knowledge (*Industrial Union Dept.* 1980, *Gulf* 1983, *Texas Indep. Ginners* 1980). Budgetary and time limitations often influence the scientific research agenda, but no good scientist would feel that definitive answers must be produced irrespective of resource constraints. The opposite predisposition may be appropriate for good regulators (Latin 1982, 1985). These comments are not intended to call into question the competence or ethics of all scientists who participate in risk assessments. Rather, the point is that the risk assessment process is fundamentally shaped by the requirements, constraints, and adversarial climate of regulation, not by the disciplinary norms of science.

The illusion that risk assessment is a purely scientific activity reduces the visibility and political accountability of policy judgements, which often guide regulatory decisions on toxic hazards. Federal agencies have employed controversial risk assessment assumptions to justify inaction on some hazardous substances. Regulators have also attempted to make determinations based on "good science" without considering the implications of this approach for decision-making costs, regulatory delays, and opportunities for obstructive or strategic behavior by affected parties. Risk assessors often respond to scientific uncertainties by adopting conservative safety-oriented positions on some important issues while using best-current-scientific-guess, middle-of-the-range, methodological-convenience, or least-cost treatments on other material issues. The EPA and other agencies have never explained the scientific or policy rationales underlying these inconsistent treatments of uncertainty, and risk managers may not recognize that substantial inconsistency exists. In light of these diverse risk assessment practices, regulatory policy judgements and scientific judgements must be applied coherently, explained forthrightly, and tested actively through public debate.

After unsuccessful attempts to achieve environmental deregulation, the Reagan administration adopted a strategy purportedly designed to improve the efficiency of pollution control programs (Latin 1985). One EPA assistant administrator contended that efficient standards must be based on "scientific evidence and not on rumor and soothsaying" (Eidsness 1982), and another official noted that the cancer guidelines "hopefully will add to the scientific credibility" of agency decisions (Shabecoff 1982). In a *Science* symposium on risk assessment, two EPA regulators claimed that the guidelines were intended "to reduce possible confusion by dealing consistently and openly with the assumptions and extrapolations that are required to bridge the gap between scientific findings and the risk assessments derived from them" (Russell and Gruber 1987).

Critics of this approach regard administration prescriptions for "good science" as a subterfuge designed to accomplish de facto deregulation. Then congressman Al Gore, stated, "The upper echelon science policy-makers have made a crass, calcu-

lated, cynical change in the traditional policy of seeking to prevent cancer" (Marshall 1982). He claimed that the administration had "reached way down into the processes of government to control the science. They think that if you control the science you can control the conclusions about whether to control this or that substance."

This "good science" orientation, whatever its initial purposes, has become entrenched in a myriad of regulatory programs as agencies increasingly rely on quantitative risk assessment, risk–benefit analysis, or cost–benefit analysis (NRDC 1986) to justify pollution control decisions and to establish staff priorities (Russell and Gruber 1987, Lave 1987). The EPA carcinogen guidelines (1986a), for example, have been the most influential statement of federal risk assessment practices for years, and yet they have not been thoroughly scrutinized from public policy and legal perspectives. It must be stressed that thousands of lives and billions of dollars in regulatory costs may depend on an agency's choice of controversial risk assessment principles.

The primary purpose of this chapter is to encourage agency officials, legislators, and other legal decision makers to examine critically the scientific limitations and broader public policy implications of alternative risk assessment treatments.

2.2 Risk Assessment Principles in Federal Agencies

Social policy judgements have always been perceived as central to the risk management process, and regulatory agencies have assigned different weights to competing factors in response to changing political or economic conditions. Under the Carter administration, risks above one fatality per million exposed people were usually treated as "unacceptable" if feasible control measures were available (Cross 1986). Reagan administration agencies concluded that risks as high as one in ten thousand, or even one in a hundred in some settings, were tolerable (Cross 1986, Russell and Gruber 1987, U.S. EPA 1986a). These risk management decisions reflect different ideological preferences and different assumptions about the economic and political effects of toxic substances regulation. Similar considerations implicitly influence risk assessment practices and resulting estimates of toxic hazards. Yet, social policies and values adopted in risk assessment proceedings typically have not been made explicit nor have they been applied in a consistent manner (Latin 1985).

During the Carter administration, the EPA, OSHA, the FDA, and the Consumer Product Safety Commission (CPSC) formed the Interagency Regulatory Liaison Group (IRLG) to develop a common set of risk assessment principles. The IRLG guidelines, which were intended to achieve consistent resolutions of recurring scientific issues, emphasized the need for safety-oriented protective treatments under conditions of uncertainty (IRLG 1979). OSHA and the EPA also created generic cancer policies partly motivated by their desire to prevent repetitive submissions of scientific theories and supporting data that the IRLG had rejected as unreliable (OSHA 1980a, U.S. EPA 1979, CPSC 1982a). For example, OSHA noted that industry representatives in every proceeding on toxic substances had argued for the existence of a threshold exposure level below which cancer risks are negligible (OSHA 1978). The IRLG guidelines and OSHA generic cancer policy found this contention was unproven and implausible in biological terms. The agencies instead adopted a protective nonthreshold causation theory in recognition of continuing scientific uncertainty (IRLG 1979, OSHA 1980b). In the absence of any fundamental advance in the state of scientific understanding, OSHA and other IRLG agencies concluded that there was little reason to debate the threshold-level issue for every potential carcinogen. This was an explicit social policy judgement incorporated into the risk assessment phase of toxics regulation.

The later EPA carcinogen guidelines (1986a) may be examined at face value as an attempt to improve the quality and consistency of risk assessments. The specified practices usually conform to recommendations made by politically independent scientific organizations; the guidelines were widely reviewed by outside scientists; and in some instances the guidelines adopt conservative treatments similar to those in the IRLG guidelines (IARC 1982, U.S. NRC 1983, OSTP 1985). There was, however, a subtle but important shift in emphasis. Although the current guidelines are intended to encourage some degree of analytical consistency, EPA experts must assess risks independently on the "weight of evidence" for each substance under review. The guidelines make clear that risk assessments will be conducted on a case-by-case basis, giving full consideration to all relevant scientific information (U.S. EPA 1986a). The "weight of evidence" approach requires "an overall, balanced judgement of the totality of the available evidence" that "should be dealt with on an individual basis." This regulatory philosophy implies that risk assessors must examine any potentially relevant scientific theories and data that any party may choose to submit (Preuss and White 1985, OSHA 1980a). The guidelines never consider additional decision-making and administrative costs, regulatory delays, and opportunities for obstructive private behavior that may arise from implementation of this individualized "weight of evidence" treatment. It is fair to say that, in comparison with the IRLG approach, the EPA now places considerably more emphasis on attempts to ground regulatory decisions on "good science" than on the need to provide effective pollution control under conditions of scientific uncertainty.

Consider the following examples. The EPA's carcinogen guidelines follow the widely held view that "risks at low exposure levels cannot be measured directly either by animal experiments or by epidemiologic studies" (U.S. EPA 1986a). Analysts must therefore extrapolate from observed effects at high dosages to predict risks at low exposure levels. They also must frequently extrapolate from results in high-dosage animal tests to animal risks and long-term human hazards at significantly lower doses. Scientists have developed a number of competing extrapolative models during the past two decades, but none has achieved general acceptance (OSHA 1980a, DHS 1984). Although all of the models fit the observed high-dosage data reasonably well, their estimates of low-dosage hazards can vary by several orders of magnitude (Luken and Miller 1981, U.S. EPA 1986a). The EPA guidelines (1986a) candidly acknowledge, "Goodness-of-fit to the [high-dose] experimental observations is not an effective means of discriminating among models." In other words, there is usually no reliable experimental basis for selecting one extrapolative model over another (Environ Corp. 1986).

The IRLG agencies adopted a "one-hit" linear extrapolative theory that assumed the absence of safe threshold levels; they did not, however, choose this approach simply because it was an "uncomplicated methodology." The linear one-hit model is the most conservative credible theory in the sense that it generates the highest risk estimates at low exposure levels. In an explicit policy judgement made in response to persistent uncertainties, the Carter administration agencies chose to maximize safety at the possible cost of overly stringent regulation by adopting the most protective extrapolative model with significant support in the scientific community (OSHA 1978, 1980a).

The current EPA guidelines recommend adoption of a linearized multistage model in most carcinogenic risk assessments. This extrapolative theory is quite conservative and produces risk estimates at low exposure levels similar, though not usually equal, to the results of the one-hit linear model (Environ Corp. 1986). The EPA selected the multistage model on the grounds that it provides a better fit with the available experimental evidence than the one-hit model and also appears more compatible with current knowledge about some biological processes related to cancer causation (DHS 1984). Thus, the agency adopted a protective, but not worst-case, extrapolative theory because it considered the multistage model most plausible based on the present state of scientific understanding. It is not, however, apparent why the agency should prefer marginally greater scientific plausibility to marginally greater public protection given the EPA's recognition that no extrapolative model is demonstrably correct and that goodness of fit for high-dose results does not prove a model's value in predicting low-dose effects (Environ Corp. 1986, Preuss and White 1985). ***The multistage theory may be tenable science in light of our imperfect knowledge about carcinogenesis mechanisms, but the EPA's selection of this provisional extrapolative model in pursuit of "good science" represents an implicit social policy judgement. Moreover, the guidelines make clear that the linearized multistage model is a default methodology to be used "in the absence of adequate information to the contrary." Agency experts or regulated parties may now argue for adoption of competing models on the basis of individualized circumstances. The guidelines provide no selection criteria for competing extrapolative theories in specific circumstances, and simply state: "When a different model is chosen, the risk assessment should clearly discuss the nature and weight of evidence that led to the choice" (U.S. EPA 1986a). This treatment gives broad, if not unlimited, discretion to agency analysts and encourages regulated parties to present any extrapolative theories and data that support the outcome they desire.***

3. RAMIFICATIONS OF THE EMPHASIS ON "GOOD SCIENCE"

These risk assessment practices indicate that an attempt to base risk estimates on "the most scientifically appropriate interpretation" entails several controversial social ramifications.

3.1 Trade-Offs between the Pursuit of "Good Science" and Effective Protection under Uncertainty

Although most guidelines embody some conservative risk assessment principles, the individualized "weight of evidence" approach coupled with agency attempts to tailor all analyses in light of changing scientific knowledge will often reduce the degree of protection previously afforded. Few if any of the revised treatments in the carcinogen guidelines have achieved general scientific acceptance, and the EPA does not contend that most uncertainties can be resolved with reasonable scientific assurance. Given the imperfect state of the risk assessment art, regulators must decide how much potential but uncertain public protection should be traded for some potential but uncertain improvement in the accuracy of scientific judgements that the EPA clearly recognizes are far from reliable. The present guidelines assume that every tentative step, however provisional, in the direction of "good science" is warranted regardless of its possible effect on the scope of protection. The wisdom of this presumption is surely a public policy issue rather than a purely scientific question. The guidelines reflect a relative shift in the EPA's emphasis on two recurring questions in toxic substances regulation: Is there sufficient reliable evidence that a chemical produces "toxic" effects at high or unknown past exposure levels, and is there enough evidence to derive reliable quantitative risk assessments at specific exposure levels (Ruckelshouse 1983, U.S. NRC 1977, Latin 1982)? ***If the EPA delays regulation until the "weight of evidence" enables predictions about specific dose–response relationships, as the guidelines presume, then the EPA may allow years of continued exposures to a known toxic substance because the precise level of toxicity cannot be reliably estimated. The agency decision to wait until regulators can meet the particularized evidentiary requirements of the guidelines is equally a decision to stress scientific validity rather than safety after an indeterminate toxic hazard has been qualitatively identified. This preference is neither inevitable nor consistent with past practices.***

As one illustration of how a requirement for "good science" in regulatory determinations can affect the scope of public protection, the Clean Water Act initially provided that the EPA must control toxic pollutants based on their degree of toxicity. This harm-based regulatory strategy presupposed that the EPA could produce particularized assessments of the hazards created by specific substances. After the agency's failure to issue any toxic water pollutant standards was challenged in litigation, the EPA adopted a "technology-based" approach in which it imposed strict standards based on qualitative proof that a substance is "toxic" and that controls are technologically and economically feasible (Rodgers 1977, Latin 1985). The EPA Assistant Administrator for Water and Hazardous Materials testified before Congress that the original approach was "technically impractical" because the agency could not "demonstrate the cause and effect relationship between pollutants and public health." Administrator Costle similarly testified in 1977 that "experience with the alternative approaches . . . leave[s] us firmly convinced that for the bulk of known or suspected toxics of concern, technology-based standards established on an industry-by-industry basis are by far the most feasible to implement and administer" (House Committee 1972). In short, the EPA adopted the technology-based regulatory approach because the agency lacked the information necessary to perform quantitative risk assessments and because it decided that protective regulatory action was necessary despite scientific uncertainty (Latin 1985, Stewart 1981). The EPA has promulgated more toxic substances standards under this one technology-based program than it has under all of its programs that require quantitative risk assessments based on "good science."

The carcinogen guidelines, in contrast, require precisely the kind of individualized evidence that the EPA had previously found difficult to obtain, and in recent years the agency has promulgated few harm-based standards for toxic water pollutants. Of special concern, the carcinogen guidelines do not indicate what regulatory actions are appropriate during the often lengthy period between the time a substance has been identified qualitatively as "toxic" and the time quantitative risk estimates become practicable. To the extent administrators are now required to support regulation of carcinogens with the kind of "weight of evidence" assessments envisioned in the guidelines, this position clearly places the burden of scientific uncertainty on exposed populations.

3.2 Effects on Agency Behavior

The pursuit of "good science" based on individualized circumstances is likely to increase the decision-making costs and time requirements associated with the risk assessment process. With respect to animal tests, for example, the guidelines state that the "weight of evidence" for potential human hazards rises "with the increase in number of animal species, strains, sexes, and number of experiments and doses showing a carcinogenic response" (U.S. EPA 1986a). With respect to data from epidemiological studies, the guidelines similarly observe that the "weight of evidence increases rapidly with the number of adequate studies that show comparable results on populations exposed to the same agent under different conditions." Both types of studies are expensive, may take years to complete, and are frequently inconclusive. The carcinogen guidelines, however, never address EPA budgetary restrictions or the time lag — with accompanying irreversible health effects — that may occur while regulators wait for sufficient data to make reliable scientific judgements.

A more subtle ramification is that the guidelines invite EPA officials to evaluate their own performance, and that of their subordinates, in terms of scientific competency rather than regulatory competency. If the primary decisional criterion is whether regulators select the "most scientifically appropriate interpretation to assess risk," officials may be reluctant to choose speculative treatments that increase public safety under conditions of uncertainty but cannot be identified as the most plausible scientific theories among a constellation of competing hypotheses. Moreover, the majority of interveners in regulatory proceedings are sponsored by affected industries or trade associations (Bayley 1972–73, U.S. EPA 1984b,c), which means the scientific performance of agency officials will regularly be monitored and challenged by industry scientists who advocate less conservative risk assessment practices. Agency bureaucrats, like other people, are sensitive to criticism and may deliberately or subconsciously seek to placate persistent critics.

3.3 Increased Opportunities for Obstructive Behavior by Affected Parties

Even if EPA risk assessors are assumed to be motivated solely by a desire to conduct the best possible scientific analyses based on the available evidence, a comparable assumption cannot be applied to the goals of interveners who espouse conflicting private interests (Latin 1985). The primary incentive of industry representatives is to minimize regulatory costs, not to promote "good science." The primary interest of environmentalist interveners is to minimize health and ecological risks irrespective of regulatory costs, not to promote "good science." The "weight of evidence" approach embodied in the carcinogen guidelines allows parties in each proceeding to make any conceivable scientific argument — and some inconceivable ones if past practices are any guide — which may affect agency decision making directly through the force of debatable scientific arguments or indirectly through increased delays and costs.

3.4 Increased Opportunities for Abuse of Discretion by Agency Decision Makers

Emphasis on individualized "weight of evidence" judgements may enable regulators to make ideologically motivated decisions under the guise that they represent "good science." In 1982, for example, the EPA Assistant Administrator for Pesticides and Toxic Substances, John Todhunter, concluded that formaldehyde poses only a low carcinogenic risk, which need not be regulated under the Toxic Substances Control Act (Ashford 1983a,b). This decision ostensibly was predicated on the agency's risk assessment, not on regulatory cost considerations or political values incorporated in the risk management process. Todhunter's formaldehyde risk assessment, however, incorporated many questionable analytical assumptions. He presumed that a safe threshold level exists for low exposures, that only body-site-specific tumors should be counted in the test results, that positive animal tests do not fairly indicate human hazards, and that vulnerable people will protect themselves because high exposures may cause unpleasant irritant effects (Latin 1985). The EPA adopted these assumptions on an ad hoc basis despite their inconsistency with previous agency practices and with risk assessment principles widely held in the scientific community. Indeed, the contemporaneous scientific literature sharply criticized Todhunter's analytical positions and conclusions (Ashford 1983a,b, Hileman 1982, Marshall 1982, Perera and Petito 1982). After Todhunter left office, the EPA reopened the formaldehyde issue and decided in 1984 that two categories of emissions sources should be regulated (U.S. EPA 1984a–d). The public policy problem with this degree of quasi-scientific discretion is that regulatory judgements expressed as individualized findings of "good science" are likely to be less visible and more immune from effective judicial or legislative review than decisions clearly based on economic concerns or controversial political values.

3.5 Susceptibility to Intrusive Judicial Review

Agency contentions that toxic controls are grounded on "good science" may increase the vulnerability of regulations to hostile judicial review. The CPSC, for example, tried to regulate urea–formaldehyde foam insulation on the basis of one experiment in which more than 40% of the animals contracted cancer within 24 months (*Gulf* 1983, Ashford 1982a,b). This finding showed an unusually high degree of carcinogenic potency in comparison with the animal data on other toxics in widespread use (DHS 1984, Luken and Miller 1981). In *Gulf South Insulation v. Consumer Product Safety Commission* (1983), the Court of Appeals for the Fifth Circuit overturned the formaldehyde regulation because the judicial panel decided: "It is not good science to rely on a single experiment, particularly one involving only 240 subjects, to make precise estimates of cancer risk. To make precise estimates, precise data are required." The opinion provided no intimation of how much precision is required for risk estimates nor how much "precise data" are necessary to constitute substantial evidence in support of regulatory judgements. The Fifth Circuit judges, however, apparently were prepared to make this decision themselves rather than defer to agency determinations.

The court's opinion reflects insensitivity to the protective goals of the organic regulatory legislation and a fundamental misunderstanding of the limited evidence on which most risk assessments of carcinogens are based. The court did not consider the social consequences of allowing the toxic hazard to continue while the EPA tried to accumulate precise data on formaldehyde cancer risks. Urea–formaldehyde foam insulation was a relatively new product, which precluded the acquisition of human epidemiologic data because of the long latency periods for many forms of cancer. Moreover, virtually all regulatory discussions of toxic hazards agree that test results from the most sensitive species and exposure conditions should receive special weight in risk assessments of carcinogens. Thus, findings of lower potency in subsequent studies would not necessarily negate the significance of the initial finding of high toxic potency. Many, if not most, quantitative risk assessments based on animal test data have relied on findings from one experiment or one series of related tests conducted by a single group of experimenters. ***The Court of Appeals opinion seemed to assume that valid science requires agencies to average the results of several positive tests before developing a quantitative risk assessment or finding that a substance causes cancer in humans. This judicial conclusion is not generally accepted in the scientific community nor warranted from the viewpoint of good regulation.***

Unrealistic judicial requirements for comprehensive agency assessments of all potentially relevant factors and for a high degree of scientific precision have substantially decreased efficiency of environmental control programs in the past decade (Mashaw and Harfst 1987, Latin 1985, Stewart 1981). Yet, the EPA's current "good science" orientation exacerbates this problem. Regulated industries and other interveners invariably can challenge the scientific bases of carcinogen risk assessments because uncertainty is pervasive and agency officials must adopt many debatable procedures in response to resource constraints and limited data. If regulators explicitly rely on quasi-legislative policy choices under conditions of scientific uncertainty, rather than pretending that their risk assessment decisions are predicated on reliable scientific judgements, appellate courts might be less prone to accept arguments that agency analyses are irrational or flawed from a scientific perspective. There is no perfect way for administrators to protect their decisions against unsympathetic appellate review, but the current agency emphasis on "good science" invites judicial criticism of toxic risk assessments on grounds where the assessments are sure to be especially vulnerable.

4. INTEGRATION OF SCIENCE AND SOCIAL POLICY JUDGEMENTS

The EPA's carcinogen guidelines (1986a) and an influential NRC study (U.S. NRC 1983) maintain that risk assessors should strive to make the best possible scientific judgements based on current knowledge and ordinarily should divorce these judgements from the economic, political, and ethical dimensions of regulation. ***No one favors bad science, but this "good science" perspective is simplistic and potentially harmful in situations where the best available science is unreliable. When no consensus exists on how to resolve fundamental scientific uncertainties, policy considerations should and must influence agency choices on which provisional risk estimates to adopt.*** Explicit incorporation of social policy judgements into the risk assessment process raises two related problems: Which types of policy criteria should be considered in risk assessment analyses as well as risk management decisions, and should the distinction between risk assessment and risk management be maintained once risk assessors employ social policy criteria to resolve scientific uncertainties?

4.1 Applicable Social Policy Criteria

Several types of policy criteria can guide risk assessments when an agency decides that the best available science is insufficient to yield reliable risk estimates. Some of these criteria may be evaluated once for each regulatory program and can provide the basis for generic treatments of recurring issues, while other material factors are linked to the particular characteristics of each toxic substance and must receive individualized treatments.

4.1.1 Interpretation of Legislative Policies

An agency choice among competing treatments of uncertainty on any scientific issue should be shaped by the policies in the applicable regulatory legislation. Legislatures have often recognized scientific uncertainties associated with toxic hazards and nonetheless required agencies to impose effective regulatory controls (Latin 1983). For example, a California Department of Health Services benzene assessment was conducted pursuant to a statutory mandate that provides "while absolute and undisputed scientific evidence may not be available to determine the exact nature and extent of risk from toxic air contaminants, it is necessary to take action to protect public health" (DHS 1984). These legislative prescriptions do not offer a complete program for defining how agencies should resolve scientific uncertainties, but they do provide guidance that may help regulators develop their own systematic social policy responses. Explicit legislative mandates for protection against toxic substances despite the recognized presence of uncertainty should caution against agency adoption of "good science" requirements that in effect preclude control of most known or probable toxic hazards.

The Clean Air Act's treatment of hazardous air pollutants may provide another illustration of how regulators can shape risk assessment practices in light of specific legislative policies. The Act requires national ambient air quality standards (NAAQS) to maintain an "adequate margin of safety," while standards for hazardous air pollutants must provide an "ample margin of safety to protect the public health" (Environ Corp. 1986). The unequivocal language on hazardous air pollutants indicates that Congress intended to place a high, and possibly absolute, priority on assurance of public protection in this regulatory context (Currie 1981). The legislative history of the 1977 amendments expressed congressional dissatisfaction with the failure of the NAAQS to include safety margins equal to those in other environmental control programs, such as radiation standards (Congressional Research Service 1978). The committee comments justified "adequate" safety margins of one to two orders of magnitude in response to scientific uncertainty about the health effects of widespread air pollutants. It seems reasonable to conclude that hazardous air pollutant standards, which are at least equally subject to scientific uncertainty and are supposed to include "ample" safety margins, should be even more biased in the direction of protection than the NAAQS limits. This interpretation is consistent with Congress's special concern for protection against toxic hazards, as expressed in an array of regulatory statutes enacted during the past three decades. Yet, the EPA has not attempted to resolve the many uncertainties presented by regulation of toxic air or water pollutants systematically in light of this congressional preference. Even if the EPA may consider costs in setting hazardous air pollutant standards, the risk management analysis should be performed after risk estimates are adjusted to reflect ample safety margins under conditions of scientific uncertainty.

4.1.2 Cost-Effectiveness of Individualized Analyses

If the effectiveness of toxic substances control programs is considered on a synoptic level, regulatory agencies must acknowledge that their achievements have fallen far short of legislative intentions. Indeed, these programs often suffer from bureaucratic paralysis and are invariably more expensive and time consuming than Congress or the agencies themselves expected. Notwithstanding its criticisms of the administrative process, the NRC rightly concluded that ***"the basic problem in risk assessment is the sparseness and uncertainty of the scientific knowledge of the health hazards addressed, and this problem has no ready solution" (U.S. NRC 1983). Yet, agencies have seldom examined the effectiveness of their risk assessment procedures in light of this fundamental problem. Given the inherent complexity of toxic hazards and severe constraints on agency resources, regulators must consider which analytical procedures are cost-effective and which scientific issues are worth assessing repeatedly in individualized proceedings.***

The EPA carcinogen guidelines (1986a) are almost entirely lacking in this form of self-analysis. The guidelines provide no indication of which risk assessment issues are especially difficult or expensive to address. They do not specify which analytical issues cannot now be resolved in a reasonably reliable manner due to the absence of any scientific consensus and which issues have been raised repetitively but inconclusively in prior regulatory proceedings. They do not identify which risk assessment issues and procedures are likely to enable obstructive behavior by regulated parties. ***Given the great difficulty in regulating any toxic substance, it is doubtful that agencies should assess in each instance whether safe***

threshold levels exist, whether one speculative extrapolation theory should be preferred over competing models, and whether benign tumors should be excluded from findings of animal studies. No doubt these and many other recurring issues are relevant to attainment of the best possible science, but individualized assessments of all material scientific issues in all toxic control proceedings may preclude achievement of adequate and timely protection for exposed populations.

Regulators might choose an intermediate position in which generic presumptions against certain kinds of theories or evidence could be rebutted by a credible showing that a scientific consensus has emerged on a previously contested issue. Risk assessors should recognize that their treatments of recurring issues and uncertainties have important implications for the scope, cost, and timing of toxic substances regulation. It is unclear whether the EPA's disregard of such factors in the carcinogen guidelines represents an instance of scientific tunnel vision or a deliberate attempt to impede effective regulation, but the guidelines and accompanying explanatory statements never question the utility of the agency's "good science" focus.

4.1.3 Potential for Catastrophic Miscalculations

Regulators could increase the conservative bias in their risk estimates when a particular toxic substance may have catastrophic effects if it proves more potent than the agency assessment anticipates. The presence of several individualized circumstances might support this form of social policy judgement.

4.1.3.1 Widespread Population Exposures

Some chemical usage and dispersion patterns entail significant exposures for only a relatively small number of workers or residents near pollution sources, whereas other hazardous substances may endanger millions of people. ***If the "best-current-guess" prediction underestimates actual risks by more than two orders of magnitude for pollutants where exposure is limited or localized, the result may be "only" a few dozen unexpected fatalities. In contrast, similar mistakes in estimation of the risks from widespread toxic exposures may have catastrophic effects.*** It is widely recognized that susceptibilities of individuals and population subgroups to toxic hazards vary widely, but there is currently no accepted methodology for tailoring risk estimates in response to those differences (U.S. EPA 1984a–d, DHS 1984). Present risk estimates are usually based on dose–response data derived from epidemiological studies of the entire population or of white male workers. Agency decision makers may, however, choose to adjust risk estimates in order to provide additional protection for unusually vulnerable subgroups, as in the cases of exposure of children to high lead concentrations or of pregnant working women to certain hazardous industrial chemicals. Again, this precautionary judgement would reflect social policy considerations in light of the possible consequences of agency mistakes under conditions of scientific uncertainty, rather than the current practice of treating risk assessment purely as a function of "good science."

4.1.3.2 Absence of a Long Historical Record of Exposures

Some toxic substances, such as benzene, have been in common use for decades at higher exposure levels than are now prevalent. This historical record reduces the chance of catastrophic risk assessment errors because hazards of epidemic proportions presumably would already have manifested themselves. In contrast, many substances are introduced each year that may eventually have toxic effects, and long latency periods may conceal those hazards for decades. Regulators might increase the conservative bias in their risk estimates for substances that lack a long historical record of exposures and related health effects. Yet, agencies seldom consider this factor in their scientific risk assessment deliberations.

4.1.3.3 Evidence of Unusual Potency

Regulators may occasionally receive evidence that a substance is unusually hazardous prior to their acquisition of sufficient data for a reliable dose–response assessment. In some instances, the substance under investigation may have a close chemical resemblance to another substance whose hazardous effects have been clearly documented, or it might yield positive results in short-term *in vitro* tests. In other cases, data from animal studies may reveal an especially high degree of toxic potency. Risk assessors could decide on the basis of these preliminary but suggestive indications of severe toxic hazards to increase the conservative bias in risk estimates derived from incomplete data. For example, CPSC tried to regulate formaldehyde after one animal study found that the substance may be an unusually potent carcinogen; the Fifth Circuit Court of Appeals' contrary decision, however, left millions of people exposed to a hazard of unknown but potentially serious dimensions.

4.1.4 Ability to Resolve Uncertainty

In practice, agencies seldom commence regulatory proceedings until considerable evidence has accumulated that a substance may be hazardous. When sufficient information or public controversy exists to justify an expensive risk assessment hearing, agency experts usually consider whatever data happen to be available at the time. In some instances, however, agencies may be able to identify ongoing scientific studies or to sponsor collection of data on acute toxic effects, prevailing exposure patterns, or other material issues. In such cases, regulators might adopt interim strategies on the assumption that specific uncertainties can be resolved in the near future. They might, for example, allow a substance under investigation, such as a newly developed drug, to be used when no substitute is available but not if its primary advantage is marginally lower costs. Because risk assessors typically cannot predict the outcome of scientific research with assurance, this type of hedging strategy clearly entails a problematical policy choice to accept some risks on a tentative basis in return for the social benefits associated with use of the toxic substance (U.S. NRC 1983, DHS 1984, U.S. EPA 1986a). This criterion could, however, facilitate abuse of discretion by agency officials because it may allow amorphous trade-offs influenced by political or economic pressures. Moreover, the history of environmental control programs indicates that pollution control standards often remain in effect indefinitely as a result of agency inertia and higher regulatory priorities.

The NRC study of risk assessment problems advised regulatory agencies to adopt generic approaches for risk assessment problems on the grounds that uniform science policy guidelines "could help separate risk assessment from risk management considerations, improve public understanding of the process, foster consistency, and prevent oversights and judgements that are inconsistent with current scientific thought" (U.S. NRC 1983). It is often advisable for regulatory agencies to rely on generic treatments of recurring scientific issues, but generic policies cannot resolve all scientific and social policy questions in each toxic risk assessment proceeding. Particularized circumstances, such as those pertaining to the potential for catastrophic errors or the likelihood that specific uncertainties can be resolved, would preclude resolution of all uncertainties in a consistent fashion. Requiring agencies to provide cogent reasons for their treatments of various types of scientific uncertainty may be more realistic and more important than a high degree of uniformity in risk assessment outcomes.

4.2 Interaction of Risk Assessment and Risk Management

Most discussions of risk assessment stress the need for scientists or regulators to identify significant uncertainties and to explain the assumptions used to resolve them. The EPA carcinogen guidelines (1986a) acknowledge "in every quantitative risk estimation that the results are uncertain" and then provide that "whichever method of presentation is chosen, it is critical that the numerical estimates not be allowed to stand alone, separated from the various assumptions and uncertainties upon which they are based."

For example, analysis of whether adequate substitutes exist for a potentially toxic product or process is usually regarded as a risk management function. The social value of a toxic chemical is determined in part by the availability of safer alternatives, and the existence of reasonable product substitutes would clearly be an important element in the risk–utility balancing comparison that often forms the heart of risk management deliberations. If the risk manager is provided with a prediction that the substance under investigation poses only a minuscule hazard, the administrator would be unlikely to regulate that substance although safer substitutes are in common usage. Yet, the agency's risk estimate may be unreliable from a scientific perspective, even if it is the best current guess, and adoption of different risk assessment assumptions could suggest a much greater danger. Because risk managers are seldom equipped or disposed to modify risk estimates, it may be appropriate for risk assessors to increase the conservative bias in their estimates when available substitutes could achieve reasonably equivalent functions. For example, after the Fifth Circuit decision on urea–formaldehyde foam insulation, millions of people were subjected to low-level formaldehyde exposures despite the availability of other forms of nontoxic insulation. When a toxic substance's primary benefit is a marginal cost advantage over substitute products, agencies should not necessarily adopt the same risk assessment treatments that they employ in the context of new drugs, pesticides, or other hazardous materials that offer distinctive benefits. Yet, most risk-assessment treatments do not consider this type of distinction because it is based on social policy, not scientific, considerations.

We contend that agencies cannot wait until the risk management stage of toxic regulatory proceedings to address the social ramifications of scientific uncertainty, but this position does not suggest that the distinction between risk assessment and risk management should be abandoned. In some toxic contexts, risk assessors can provide reliable estimates based on generally accepted scientific principles. In the case of some noncarcinogenic toxic substances, such as cotton dust and lead that have been in use for many decades and produce chronic effects after long-term exposures, scientists may be able to obtain reliable epidemiologic data and identify reasonably accurate dose–response relationships (*American Textile* 1981). There is no reason why risk assessments in such contexts should be modified in response to policy criteria when "good science" judgements can be grounded on valid science. Moreover, risk managers must address economic, political, and ethical factors relevant to each toxic substance even if the social policies applied in the risk assessment stage are incorporated in generic treatments of recurring scientific issues. To the extent the conventional distinction represents a real rather than symbolic division of decision-making responsibilities, risk managers should retain the ultimate authority to determine the scope of toxic regulations. Nevertheless, risk assessment inescapably plays a central role in the toxic substances regulatory process and this function is too uncertain to be treated exclusively as an exercise in "good science."

5. CONCLUSION

Environmentalists attack the risk assessment process because they believe it frequently produces unreliable estimates of toxic hazards and because it is subject to manipulation by industrial dischargers and government bureaucrats. However sympathetic one may be to these objections, which surely have ample basis in past regulatory experience, society cannot feasibly eliminate all carcinogenic risks nor enjoin use of all toxic substances. Society must therefore develop some rational method for deciding which risks are unacceptable and for allocating scarce regulatory resources. Notwithstanding the risk assessment uncertainties and analytical shortcomings emphasized in this chapter, it is unlikely that regulators should or could eliminate attempts to estimate the dimensions of diverse toxic hazards. Moreover, after more than a decade of intrusive appellate decisions and political emphasis on cost-effectiveness justifications, risk assessment procedures are firmly embedded in the federal regulatory agencies responsible for toxic substances control.

Because predictions of toxic effects generally cannot be grounded on reliable scientific judgements, social policy criteria must play an influential role in the choice among competing risk estimates. Once we recognize that toxic substances regulation requires a panoply of policy determinations to supplement provisional

scientific judgements, it is essential that risk assessment agencies explicitly consider the social ramifications of scientific uncertainty, strive for analytical coherence in their treatments of currently indeterminate issues, and clearly explain the principles, practices, and values underlying particular estimates of toxic hazards.

NOTES ADDED IN PROOF

The preceding material was written two years after the last major revision of the EPA Guidelines for Carcinogen Risk Assessment (U.S. EPA 1986a). In mid-1996, the EPA issued new proposed guidelines for "reevaluating" carcinogen risk assessments (U.S. EPA 1996a, U.S. EPA 1996b). The proposed treatments would retain and even increase the prominence of individualized "weight of evidence" determinations on the rationale that, "The intent of this proposal is to take account of knowledge available now and to provide flexibility for the future in assessing data and employing default inferences, recognizing that the guidelines cannot always anticipate future research findings." (U.S. EPA 1996a). The proposed new approach would focus just as much as the prior guidelines on utilization of the best available science even if this "good science" falls far short of conventional standards of scientific reliability. The new guidelines generally would not incorporate social-policy judgements in determining how scientific uncertainties should be resolved. Thus, all of the observations in this chapter are equally applicable to the EPA proposed guidelines. In a recent statement, EPA Administrator Carol Browner noted that children face "special risks from toxic chemicals" and that the EPA would begin to regulate pollutants in light of the risks presented for children and other vulnerable subpopulations (Cushman 1996). Such an agency treatment, if it is ever implemented, would represent one application of the kind of explicit social policy-based risk assessments advocated in this chapter.

REFERENCES

Ackerman, B. and Hassler, W. Clean Coal/Dirty Air, 79–103 (1981).

Ackerman, B. and Stewart, R. Reforming Environmental Law, 37 *Stanford Law Rev.* 1333, 1357 (1985).

Albert, R. Carcinogen Assessment Group's Final Report on Population Risk to Ambient Benzene Exposures (Jan. 10, 1979).

Albert, R. U.S. Environmental Protection Agency Revised Interim Guideline for the Health Assessment of Suspect Carcinogens, in D. Hoel, R. Merrill and F. Perera, Eds., 19 *Banbury Report, Risk Quantitation and Regulatory Policy,* 307, 308 (1985).

American Textile Mfrs. Inst. v. Donovan, 452 U.S. 490, 505 n.25 (1981).

Ames, B.N. Identifying Environmental Chemicals Causing Mutations and Cancer, 204 *Science* 587, 589 (1979).

Ames, B.N., Magaw, R., and Gold, L.S. Ranking Possible Carcinogenic Hazards, 236 *Science* 271, 275 (1987).

Asbestos Information Association v. OSHA, 727 F.2d 415, 424–426 (5th Cir. 1984).

Ashford, N., Ryan and Caldart. A Hard Look at Federal Regulation of Formaldehyde: A Departure from Reasoned Decisionmaking, 7 *Harvard Environ. Law Rev.* 297, 298–299, 330–331 (1983a).

Ashford, Ryan and Caldart. Law and Science Policy in Federal Regulation of Formaldehyde, 222 *Science* 894 (1983b).

Bayley, N. Memoirs of a Fox, 2 *Environ. Affairs* 332 (1972–73).

Bazelon, D. Science and Uncertainty: A Jurist's View, 5 *Harvard Environ. Law Rev.* 209, 213 (1981).

California Air Resources Board and Dept. of Health Services, Report to the Scientific Review Panel on Benzene: Overview and Recommendation 8 (Nov. 1984).

Congressional Research Service. 95th Cong., 2d Sess., 4, *A Legislative History of the Clean Air Act Amendments of 1977,* at 2573–2595, 2674–2678 (Comm. Print 95-16, 1978).

Cothern, Coniglio and Marcus. Estimating Risk to Human Health, 20 *Environ. Sci. Technol.* 111, 113–115 (1986).

CPSC. Benzene-Containing Consumer Products; Proposed Withdrawal of Proposed Rule, 46 *Fed. Reg.* 3034 (1982a).

CPSC. Ban of Urea-Formaldehyde Foam Insulation, 47 *Fed. Reg.* 14,366 (1982b).

Cross. Beyond Benzene: Establishing Principles for a Significance Threshold on Regulatable Risks of Cancer, 35 *Emory Law J.* 1, 17 (1986).

Currie, D. *Air Pollution: Federal Law and Analysis,* § 7.13, 10.01 (1981).

Cushman, J. Children's Health Is to Guide E.P.A., *N.Y. Times,* Sept. 12, 1996, at A14 col. 4.

DHS Benzene Report, 1984. California Dept. of Health Services, Report to the Scientific Review Panel on Benzene: Part B — Health Effects of Benzene, 68–80 (Nov. 1984).

Douglas, M. and Wildavsky, A. *Risk and Culture,* University of California Press, Berkeley (1982).

Eidsness. An Administration Sold on Clean Water, *N.Y. Times*, Nov. 9, at A30, col. 4 (1982).

Environ Corp. *Elements of Toxicology and Chemical Risk Assessment: A Handbook for Nonscientists, Attorneys and Decision Makers* 37–41 (1986).

Goldstein, B. Risk Assessment and Risk Management of Benzene by the Environmental Protection Agency, in D. Hoel, R. Merrill and F. Perera, Eds., 19 *Banbury Report, Risk Quantitation and Regulatory Policy,* 293, 295 (1985).

Gulf South Insulation v. Consumer Prod. Safety Commission, 701 F.2d 1137, 1141, 1146 (5th Cir. 1983).

Hileman, R. Formaldehyde; How Did EPA Develop Its Formaldehyde Policy?, 16 *Environ. Sci. Technol.* 543 (1982).

House Committee on Public Works and Transportation. 95th Cong., 1st Sess., *Implementation of the Federal Water Pollution Control Act: Summary of Hearings on the Regulation and Monitoring of Toxic and Hazardous Chemicals Under the Federal Water Pollution Control Act* [P.L. No. 92-500, 86 Stat. 816 (1972)] 26 (Comm. Print 1977). [House (1972)].

Huber, P. Safety and the Second Best: The Hazards of Public Risk Management in the Courts, 85 *Columbia Law Rev.* 277 (1985).

Industrial Union Dept., AFL-CIO v. American Petroleum Inst., 448 U.S. 607, 617–628, 656–657, and n.64. (1980).

Interagency Regulatory Liaison Group (IRLG). Scientific Bases of Identification of Potential Carcinogens and Estimation of Risks, 44 *Fed. Reg.* 39,858, 39,872–39,875 (1979).

International Agency for Research on Cancer (IARC). 29 *IARC Monographs on the Evaluation of the Carcinogenic Risk of Chemicals to Humans,* Suppl. 4, 121 (1982).

Krier, J. The Irrational National Air Quality Standards: Macro- and Micro-Mistakes, 22 *UCLA Law Rev.* 323–330 (1974).

Latin, H. The "Significance" of Toxic Health Risks: An Essay on Legal Decisionmaking under Uncertainty, 10 *Ecology Law Q.* 339 (1982).

Latin, H. The Feasibility of Occupational Health Standards: An Essay on Legal Decision-making under Uncertainty, 78 *Northwest. Univ. Law Rev.* 583, 605–611 (1983).

Latin, H. Ideal versus Real Regulatory Efficiency: Implementation of Uniform Standards and "Fine-Tuning" Regulatory Reforms, 37 *Stanford Law Rev.* 1267, 1282–1297 (1985).

Latin, H. Good Science, Bad Regulation, and Toxic Risk Assessment, 5 *Yale J. on Regulation* 89 (1988).

Lave, L. Health and Safety Risk Analyses: Information for Better Decisions, 236 *Science* 291 (1987).

Leape, J. Quantitative Risk Assessment in Regulation of Environmental Carcinogens, 4 *Harvard Environ. Law Rev.* 86, 103 (1980).

Luken and Miller. The Benefits and Costs of Regulating Benzene, 31 *J. Air Pollution Control Ass'n.* 1254, 1256–1257 (1981).

Marshall. EPA's High-Risk Carcinogen Policy, 218 *Science* 975 (1982).

Mashaw, J. and Harfst, D. Regulation and Legal Culture: The Case of Motor Vehicle Safety, 4 *Yale J. Regulat.* 257, 312–313 (1987).

McGarity, T. Substantive and Procedural Discretion in Administrative Resolution of Science Policy Questions: Regulating Carcinogens in EPA and OSHA, 67 *Georgetown Law J.* 729 (1979).

National Toxicology Program. Report of the Ad Hoc Panel on Chemical Carcinogenesis Testing and Evaluation of the National Toxicology Program (1984).

Natural Resources Defense Council, Inc. v. EPA. 25 E.R.C. 1105 (D.C. Cir. 1986) (No. 85-1150). [The Court of Appeals sitting en banc vacated the panel's opinion and heard oral arguments on April 29, 1987. On July 28, 1987, the en banc court remanded the case. 824 F.2d 1146 (D.C. Cir. 1987).] [NRDC (1986)].

Office of Science and Technology Policy (OSTP). Chemical Carcinogens: Review of the Science and Its Associated Principles, 50 *Fed. Reg.* 10,372 (1985).

Olson, E. The Quiet Shift of Power: Office of Management and Budget Supervision of Environmental Protection Agency Rulemaking under Executive Order 12,291, 4 *Va. J. Nat. Resources Law* 1 (1984).

OSHA. Identification, Classification and Regulation of Toxic Substances Posing a Potential Occupational Carcinogenic Risk, 42 *Fed. Reg.* 54,146, 54,148, 54,155–54,156 (1977).

OSHA. Occupational Safety and Health Standards, Occupational Exposure to Benzene, 43 *Fed. Reg.* 5918, 5928–5931, 5946–5947 (1978).

OSHA. Generic Cancer Policy, 45 *Fed. Reg.* 5002, 5023–5024, 5131 (1980).

OSHA. Identification, Classification and Regulation of Potential Occupational Carcinogens; Proposed Amendments, 46 *Fed. Reg.* 7402 (1981).

OSHA. Identification, Classification and Regulation of Potential Occupational Carcinogens, 45 *Fed. Reg.* 5002, 5200 (1980b) [codified at 29 C.F.R. § 1900.101–1990.152 (1987)].

Pedersen, J. Why the Clean Air Act Works Badly, 129 *Univ. Pa. Law Rev.* 1059 (1981).

Perera, F. and Petito. Formaldehyde: A Question of Cancer Policy?, 216 *Science* 1287 (1982).

Preuss, P. and White. The Changing Role of Risk Assessment in Federal Regulation, in D. Hoel, R. Merrill and F. Perera, Eds., 19 *Banbury Report, Risk Quantitation and Regulatory Policy,* 331, 335 (1985).

Rodgers, W. *Environmental Law,* 486–487 (1977).

Ruckelshouse, W. Science, Risk, and Public Policy, 221 *Science* 1026, 1027 (1983).

Russell, M. and Gruber, M. Risk Assessment in Environmental Policy Making, 236 *Science* 286–287 (1987).

Shabecoff, P. Administration Drafting New Policy on Regulating Cancer-Causing Agents, *N.Y. Times*, Dec. 4, at 32, col. 2 (1982) (Quoting Dr. Denis Prager, Assistant Director of the Office of Science and Technology).

Stewart, R. Regulation, Innovation, and Administrative Law: A Conceptual Framework, 69 *Calif. Law Rev.* 1256, 1274–1275, 1338–1353 (1981).

Sullivan, N. The Benzene Decision: A Contribution to Regulatory Confusion, 33 *Admin. Law Rev.* 351 (1981).

Texas Indep. Ginners Association v. Marshall. 630 F.2d 398 (5th Cir. 1980).

The Odds on Cancer: EPA's Recent Bets, 218 *Science* 976 (1982).

U.S. EPA. National Emission Standards for Hazardous Air Pollutants; Policy and Procedures for Identifying, Assessing, and Regulating Airborne Substances Posing a Risk of Cancer, 44 *Fed. Reg.* 58,642 (1979).

U.S. EPA. National Emission Standards for Hazardous Air Pollutants; Benzene Fugitive Emissions, 46 *Fed. Reg.* 1165 (1981).

U.S. EPA. National Emission Standards for Hazardous Air Pollutants; Regulation of Benzene; Response to Public Comments, 49 *Fed. Reg.* 23,478–23,480, 23,484, 23,493 (1984a).

U.S. EPA. National Emission Standards for Hazardous Air Pollutants; Proposed Standards for Benzene Emissions from Coke By-Product Recovery Plants, 49 *Fed. Reg.* 23,522, 23,527–23,528 (1984b).

U.S. EPA. National Emission Standards for Hazardous Air Pollutants; Benzene Emissions from Maleic Anhydride Plants, Ethylbenzene/Styrene Plants, and Benzene Storage Vessels; Withdrawal of Proposed Standards, 49 *Fed. Reg.* 8386, 8389, 23,558, 23,562 (1984c).

U.S. EPA. Formaldehyde; Determination of Significant Risk, 49 *Fed. Reg.* 21,870 (1984d).

U.S. EPA. Guidelines for Carcinogen Risk Assessment, 51 *Fed. Reg.* 33,992–33,993, 33,996 (1986a).

U.S. EPA. Standards for Radon-222 Emissions from Licensed Uranium Mill Tailings, 51 *Fed. Reg.* 34,056, 34,057 (1986b).

U.S. EPA. National Emission Standards for Hazardous Air Pollutants; Benzene Equipment Leaks (Fugitive Emission Sources), 40 C.F.R. § 61.110–61.112 (1987).

U.S. EPA. Proposed Guidelines for Carcinogen Risk Assessment, 61 *Fed. Reg.* 17960 (1996a).

U.S. EPA. Proposed Process for Reevaluating Cancer Assessments, 61 *Fed. Reg.* 32799 (1996b).

U.S. National Research Council (NRC). *Decision Making in the Environmental Protection Agency* (1977).

U.S. National Research Council (NRC). *Risk Assessment in the Federal Government: Managing the Process* 3 (1983).

United Steelworkers v. Marshall. 647 F.2d 1189 (D.C. Cir. 1980), Cert. denied, 453 U.S. 913 (1981).

Section IV

Risk Management

CHAPTER IV.1

Risk Management of the Nuclear Power Industry*

B. John Garrick

SUMMARY

It is clear from the other chapters of this book that risk assessment and risk management means different things to different groups. While there are many different groups involved in the risk field, including engineers, health scientists, social scientists, and environmental scientists, I would like to divide them into just two groups and refer to the two as engineers and environmentalists. The engineer group sees risk assessment as principally a quantification of the "source term" (i.e., a release condition), while the environmental group's concept of risk assessment is principally pathway analysis and exposure assessment. This arbitrary division is not to suggest that engineers are not environmentalists and environmentalists do not include engineers, but is done only to provide a more convenient framework for discussing two different approaches to risk assessment and risk management.

Engineers and environmental groups had very different beginnings in the risk assessment and risk management field. The environmental group, for the most part, had its start with the U.S. Environmental Protection Agency (EPA) cancer risk assessment guidelines in the mid-1970s and the National Academy of Science paradigm on risk assessment in 1983 (Barnes 1994). The engineering community, on the other hand, made its biggest jump into the risk assessment field in 1975 with the release of the reactor safety study (U.S. Nuclear Reg. Com. 1975). Even before the Reactor Safety Study, there was research going on to change our way of thinking

* Some of the material of this chapter uses the same source material as a similarly titled chapter written by the author in the reference: Garrick, B. J., Risk management in the nuclear power industry, in *Engineering Safety*, David I. Blockley, Ed., McGraw-Hill International (UK) Limited, 1992, Chap. 14.

1-56670-130-9/97/$0.00+$.50

about safety in general and nuclear safety in particular (Garrick 1968). Since this chapter is devoted to the nuclear power industry, the principles of risk assessment and risk management practiced follow those advocated by such investigators in the field as Rasmussen, Garrick, and Kaplan and as generally practiced in the engineering field.

Key Words: probabilistic risk assessment (PRA), nuclear power, radiation, nuclear waste, risk-based regulation, nuclear accidents, source term, defense in depth

1. INTRODUCTION

It is important to point out that the early applications of probabilistic risk assessment (mid-1970s to mid-1980s) in the nuclear power industry were the best examples of full-scope risk assessments that integrated both the engineering and environmental considerations into the basic analysis models. Full scope implies both front- and back-end detailed analyses. The front end refers to the engineering modeling necessary to quantify the source term of a health and safety threat, and the back end includes exposure pathways and the analysis of health and property effects. Had the practice of full-scope risk assessments for nuclear power plants been continued, then it is most likely that the differences between the engineering group and the environmental group would not be great, if even significant, because it forced the two groups to work together. However, the nuclear industry, driven by changing regulatory practices, chose not to continue supporting the full-scope approach to risk assessment, but rather to focus on the new requirements of the U.S. Nuclear Regulatory Commission, starting with the individual plant examination program (U.S. Nuclear Reg. Com. 1988), which emphasized the assessment of core damage frequency. While there was logic to the argument that a damaged core was necessary to have a release, it terminated the important work of quantifying pathways and health effects, not to mention property damage, and allowed the two groups in many respects to go their separate ways. The end result is that the knowledge base for risk management in the nuclear power industry is not as complete as it might have been, had the emphasis not changed with respect to risk assessment.

2. THE NUCLEAR POWER INDUSTRY

While there continues to be uncertainty about the future of nuclear power, its present status is that of a very significant industry. Currently, nuclear energy is about 5.3% of the world primary energy production and about 17% of its electrical generation (Häfele 1994). This represents a very major industry as energy is the most capital-intensive industry in the world. There is somewhat of a standstill in nuclear power in the United States and Europe, although there are locations of high usage. For example, in France and Belgium, approximately 70% of the electricity comes from nuclear generation; the number is 50% in Sweden and Switzerland and greater than 40% in Korea and Taiwan. In the United States, approximately 20% of the electricity is from nuclear power plants. While there may be a standstill in nuclear power in Europe and the United States, there continues to be a buildup in Japan, South Korea, Taiwan, China, and elsewhere. In terms of the number of nuclear plants, the United States leads all nations, with 109 plants, followed by France and the former Soviet Union, with between 50 and 60 plants each. There are between 425 and 450 nuclear plants operating worldwide. These plants are generating approximately 350,000 MW of electricity, of which over 100,000 MW come from the U.S. plants.

3. THE RISK OF NUCLEAR POWER PLANTS

The evidence is strong that nuclear power is among the safest of the developed energy technologies in spite of the high profile accidents at Three Mile Island and Chernobyl. The problem is that a large segment of the world population is not convinced of the safety of nuclear power, and there is always the chance of a major accident, however unlikely it may be. Unlike most major industries affecting our quality of life, safety has been a first priority of nuclear power since its very beginning. Nevertheless, the "fear anything nuclear" syndrome prevails. This is probably because of the manner in which nuclear fission was introduced to the world, namely, as a devastating weapon of massive destruction. Of course, a nuclear power plant is nothing like a nuclear weapon.

The United States, as discussed later, utilizes light water reactor technology for its power plants. There are two types of light water reactors, pressurized water reactors and boiling water reactors. Simplified flow diagrams of these two reactor types are illustrated in Figures 1 and 2.

The difference in the two concepts is primarily in the thermal hydraulics of the coolant during normal operation. In the pressurized water reactor, the water used to cool the reactor is kept under pressure to prevent boiling and is circulated through secondary heat exchangers, called steam generators, to boil water in a separate circulation loop to produce steam for a standard steam turbine cycle. In a boiling water reactor, the water used to cool the reactor is allowed to boil in the reactor at a lower pressure than in a pressurized water reactor and the resulting steam is routed to the steam turbine to produce electricity.

The distinguishing threats of nuclear power are radiation and something called decay heat. While it is possible to immediately stop the nuclear fission process of a nuclear reactor, it is not possible to immediately shut off all of the radiation in a reactor core. This is because of the existence of large quantities of radioactive fission products — a byproduct of the energy-producing nuclear fission process. The fission products have varying lifetimes that radioactively decay with time and involve different types of radiation. For example, if the reactor has been operating for a long time, say 1 year, the power generated immediately after shutdown (i.e., after stopping the fission process) will be approximately 7% of the level before shutdown. For a 1000-MW(e) nuclear plant, this means about 200 MW of heat will be generated, which is enough heat to cause fuel melt in the absence of decay heat removal. Of course, loss of decay heat removal is guarded against with elaborate and highly

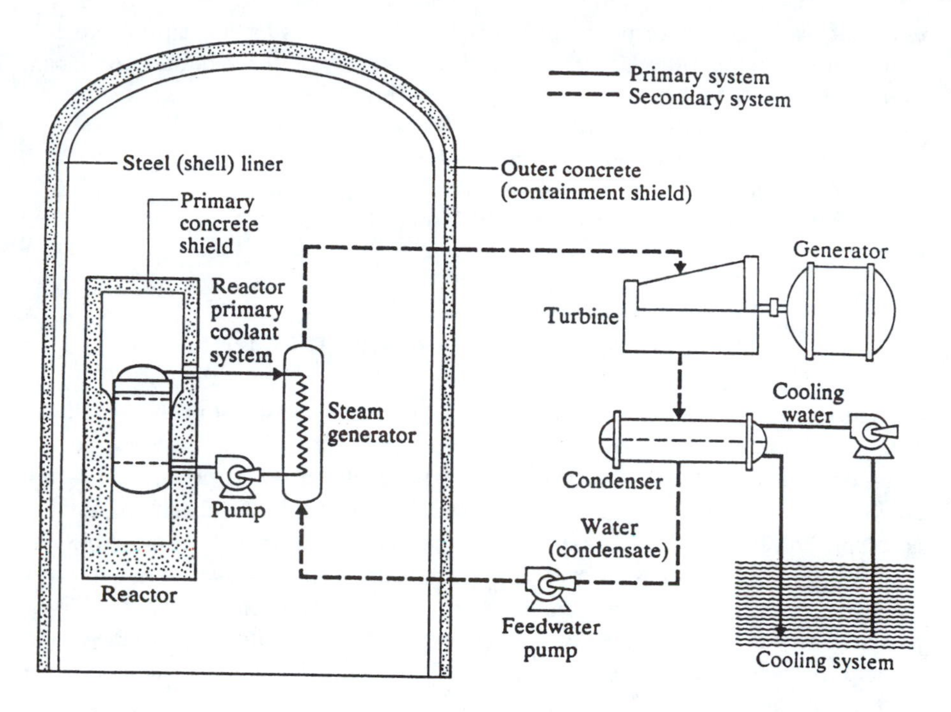

Figure 1 Schematic of a pressurized water reactor power plant (From Nero, A. V., Jr., *A Guidebook to Nuclear Reactors,* University of California Press, Berkeley, 1979. With permission.)

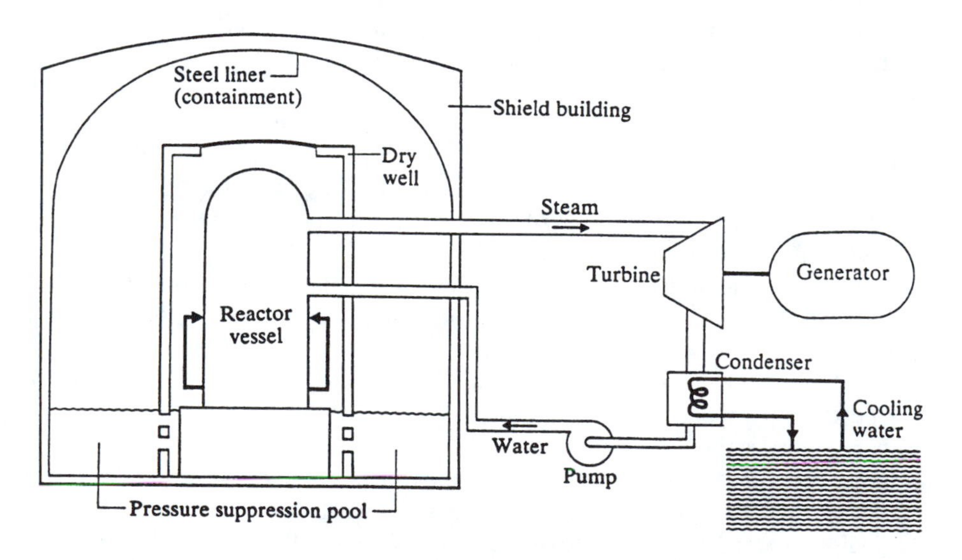

Figure 2 Schematic of a boiling water reactor power plant (From Nero, A. V., Jr., *A Guidebook to Nuclear Reactors,* University of California Press, Berkeley, 1979. With permission.)

reliable decay heat removal systems. Even as reliable as such systems may be, additional protective measures are included in the form of accident mitigating systems to terminate the progression of accidents.

Besides loss of decay heat, there are other risk issues associated with the operation of nuclear power plants. Two accident mechanisms that require intervention should they occur are nuclear transients and loss of coolant. Both mechanisms could lead to serious fuel damage and, should the accident mitigation systems fail (such as containment), could eventually lead to radiation releases from the plant. These are extremely low-probability events and are the reasons for the excellent safety record of commercial nuclear power plants.

While the emphasis on the risk of nuclear power has focused on the nuclear power plant itself, there are other segments of the nuclear fuel cycle that are also in the risk picture of nuclear power. They too have been carefully analyzed and must be a part of the nuclear power risk management agenda. These segments of the fuel cycle include fuel fabrication; fuel reprocessing; and nuclear waste processing, handling, and storage. Most of these steps of the fuel cycle have had quantitative risk assessments performed similar to those performed on nuclear power plants. One of the most difficult challenges is to be able to demonstrate the safety of proposed geologic waste repositories over periods of time corresponding to tens of thousands of years. Much of the assessment effort to demonstrate long-term repository performance is ongoing at the present time. Should these efforts fail, then it may be necessary to consider other alternatives to waste disposal, such as monitored and maintained engineered facilities.

4. NUCLEAR POWER PLANT ACCIDENT HISTORY

As indicated at the beginning of this chapter, the safety record of nuclear power is outstanding and without parallel in the development of a major technology that has advanced to the stage of widespread public use throughout the world. Still, incidents and accidents have occurred. For nuclear power, the accident history is dominated by two accidents: one that did not result in acute injuries or deaths (the Three Mile Island, Unit 2 accident in the United States) and the other much more serious Chernobyl accident in the former Soviet Union, where there were several early deaths and injuries. The full level of damage of the Chernobyl accident has not yet been fully assessed.

Before the Chernobyl and Three Mile Island accidents are described, it is important to put the risk and safety record of nuclear power in perspective. There are some 440 nuclear power plants located throughout the world, 109 of which are in the United States. These plants represent a total cumulative operating experience as of January 1995 of more than 7000 in-service reactor years. Add to this experience base the reactors used in weapon systems (most notably submarines), weapons production, and research, and the actual experience is estimated to exceed 10,000 reactor years. Almost 70% of this experience involves water reactors, the type used in the United States, for which there was only one accident involving a nonmilitary operation. No member of the public or the operating staff was killed or injured in that accident. Considering the complexity of the industry and the extensiveness of application of nuclear power, this is a rather remarkable safety record, as mentioned earlier, not matched by any other of the major energy industries. However, the Three Mile Island and Chernobyl accidents do remind us that accidents can happen, and it is extremely important that we learn as much as possible from these accidents. A brief description of both accidents is given based on descriptions contained in Chapter 14 of *Engineering Safety* (Blockley 1992).

The Three Mile Island, Unit 2 (TMI-2) nuclear power plant, located near Harrisburg, Pennsylvania, went into commercial operation in December 1978. The plant consists of a Babcock & Wilcox pressurized water reactor and generates approximately 800 MW of electricity. The accident occurred on March 28, 1979, at 4:00 a.m.

The early stages of the accident involved events that were quite routine, in terms of the ability of the reactor operators to respond. There was a trip (i.e., an automatic shutdown) of the main feedwater pumps, followed by a trip of the steam turbine and the dumping of steam to the condenser. As a result of the reduction of heat removal from the primary system, the reactor system pressure began to rise until the power-operated relief valve opened. This action did not provide sufficient immediate pressure relief, and the control rods were automatically driven into the core to stop the fission process.

At this point, complications began to develop. First, there was the problem of significant decay heat, which could have been handled straightforwardly had it not been for some later problems with such systems as emergency feedwater. The second, and turning point of the accident, was that a pressure relief valve failed to close, and the operators failed to recognize it. The result was the initiation of the now-famous small loss of coolant accident; i.e., the small LOCA. The stuck-open valve, together with some valve closures that had not been corrected from previous maintenance activities, created a severe shortage of "heat sinks" to control the heat loads of the plant. The events were further complicated by the failure of the operators to recognize that coolant was, in fact, being lost through the stuck-open relief valve.

These events resulted in initiation of high-pressure emergency cooling. Meanwhile, the operator concerned about losing pressure control over the primary system shut down the emergency cooling and transferred slightly radioactive water outside the containment building to the auxiliary building. Fortunately, the transfer was terminated before much radioactivity was involved.

Pump vibration and continued concern about overpressurizing the primary system led to the operators eventually shutting down all of the main reactor coolant pumps. It was at this point that the severe damage to the core took place. The critical events were the overheating of the reactor and the release of fission products into the reactor coolant. The time interval for this most serious phase of the accident was 1 to 3 hours following the initial feedwater trip. At about 2 hours and 20 minutes into the accident, the block valve over the pressurizer was closed, thus terminating the small LOCA effect of the stuck-open relief valve. However, it was almost 1 month before complete control was established over the reactor fuel temperature when adequate cooling was provided by natural circulation.

In terms of the threat to public health and safety, the consequences of the accident were quite minimal. There were measurable releases of radioactivity outside the containment, but not of sufficient magnitude to cause any immediate injuries. The latent effects are very speculative. Of course, the damage to the reactor was essentially total.

The Chernobyl Nuclear Power Station accident was by far the most serious nuclear power plant accident ever to occur. The specific reactor involved in the accident was Unit 4 of the four-unit station. The reactor is a 1000-MW(e), boiling water, graphite-moderated, direct cycle, USSR RBMK type.

The Chernobyl accident occurred on April 26, 1986, and was initiated during a test of reactor coolant pump operability from the reactor's own turbine generators. The purpose of the test was to determine how long the reactor coolant pumps could be operated, using electric power from the reactor's own turbine generator under the condition of turbine coast down and no steam supply from the reactor. One of the reasons for the test was to better understand reactor coolant pump performance in the event of loss of load and the need to bypass the turbine to avoid turbine overspeed. The reactor should have been shut down during the test, but the experimenters wanted a continuous steam supply to enable them to repeat the experiment several times.

At the beginning of the test, half of the main coolant pumps slowed down, resulting in a coolant flow reduction in the core. Because of prior operations leaving the coolant in the core just below the boiling point, the reduced flow quickly led to extensive boiling. The boiling added reactivity to the core because of the positive void coefficient, a property of this particular type of reactor, and caused a power transient. The negative reactivity coefficient of the fuel (i.e., an offsetting effect) was insufficient to counteract the dominance of the positive void coefficient because of the conditions in the core at the time of the test. By the time the operators realized that the reactor was rapidly increasing in power, there was insufficient time to take the appropriate corrective action because of the slow response time of the control system. The power excursion caused the fuel to overheat, melt, and disintegrate. Fuel fragments were ejected into the coolant, causing steam explosions and rupturing fuel channels with such force that the cover of the reactor was blown off. The near-term damage included 30 fatalities from acute doses of radiation and the treatment of some 300 people for radiation and burn injuries.

The off-site consequences are still being investigated, even though the accident occurred almost 9 years ago. To be sure, there will be latent effects from the accident. It is known that 45,000 residents of Pripyat were evacuated the day after the accident, and the remaining population within approximately 20 miles of the reactor were evacuated during the days that followed the accident. The ground contamination continues to be a problem, and it is not known when the nearby areas will be inhabited again.

Nuclear power suffered a severe setback from this accident. Even though this type of reactor is not used outside the former Soviet Union for the production of electricity and even though the consequences from the accident do not rank with major public disasters in our history, at least in terms of the short-term damage, the accident has left a scar from which the nuclear power industry may never recover.

5. THE PRINCIPAL ELEMENTS OF RISK AND SAFETY MANAGEMENT

5.1 Regulatory Practices

Most nuclear-capable nations are similar in their approach to nuclear power plant regulation. The key elements are (1) an independent government regulatory agency that is not responsible for the development or promotion of nuclear energy; (2) a formal licensing process for the siting, construction, and operation of nuclear power plants; and (3) inspection and enforcement powers within the regulatory agency over the nuclear power industry, including the authority to terminate operations in the interest of public safety or environmental impact.

While the regulatory agencies have large staffs of engineers and scientists, advisory groups, and extensive analytical tools for independent licensee compliance verification, one of the most basic principles guiding the regulatory process is "defense in depth." The defense-in-depth principle has been a major driver in the development of such protection concepts as (1) containment systems capable of containing major accidents, (2) very conservative design basis accidents, and (3) the single failure criteria: i.e., the requirement that a plant be able to withstand the failure of any single component without fuel damage. The defense-in-depth concept has been a major player in the promulgation of very specific deterministic regulations.

The defense-in-depth concept has resulted in a very safe industry, but it has also made nuclear power very expensive by requiring extensive equipment redundancy and greatly increasing plant complexity. The concern among many experts is that the safety management process is overemphasizing safety and creating a serious imbalance between safety and societal benefits. The search for better methods for measuring safety performance has resulted in the increased use of probabilistic risk assessment (PRA), a concept based on the reactor safety study sponsored by the NRC (1975). PRA is discussed in the following sections.

5.2 Risk and Safety Assessment Practices

In no other industry has the practice of safety analysis reached the level of sophistication of that in the nuclear power industry. The most advanced form of safety analysis is that embodied in a full-scope probabilistic risk assessment or probabilistic safety assessment (PSA), the preferred label in international circles. PSA is a rigorous and systematic identification of possible accident sequences, which we call scenarios, that could lead to fuel damage, biological damage, or environmental damage, and a quantitative assessment of the likelihood of such occurrences. All nuclear plants in the United States now have some form of a PSA to serve as critical source material for the management of the risks associated with specific plants. In addition to the United States, PSA is practiced at most nuclear plants throughout the world. In fact, in some locations such as Germany, the PSAs are having an even greater influence on the design of their plants than they do in the United States. Other countries such as France, Sweden, and Japan are also now making extensive use of the PSA as the method of choice for in-depth understanding of the safety of their plants. Of course, an in-depth understanding of contributors to risk is the very best basis of all to formulate a meaningful risk management program.

It should be pointed out that the risk and safety analysis methods are far more advanced than the extent of their adoption in the regulatory process. In particular, the regulatory process is not yet risk based. In fact, it may never be totally risk based, but it is clear that there is movement in that direction.

5.3 Future Directions in Risk Management and the Move toward Risk-Based Regulation

In the United States, some form of risk assessment is now a requirement for all nuclear plant licensees. With the expanded use of quantitative risk assessment (QRA), another name often used to describe the same process as PRA and PSA, the NRC has been active in updating the work of the original reactor safety study. One major activity in this regard was the severe accident risk study performed for five U.S. nuclear power plants (NUREG-1150) (U.S. Nuclear Reg. Com. 1990). NUREG-1150 is expected to have a major influence on the NRC's severe accident policy.

The reactor safety study, NUREG-1150, and the Zion\Indian Point risk assessments (Pickard, Lowe and Garrick, Inc. 1981, 1982) were probably the three most influential risk studies affecting the current confidence in the use of risk-based technologies in the nuclear regulatory process. Of course, the other knowledge base important to the future direction of risk-based regulation is the plant-specific risk assessments supplied by the applicants. The lessons learned are many and far-reaching and should be a part of the basis for making future decisions about risk-based regulation. There is no clear cut process in place for maximizing the knowledge base created by the risk assessments submitted by the licensees.

On the surface, with analytical methods available to support risk-based regulation, it appears that it is the only logical direction to take. Why, then, are we making so little progress, and why are there so many obstacles to its implementation? Well, the problems appear to be many, and here are what appear to be but a few:

- *The institutional structure in which regulations are made and enforced is culturally resistant to changes that have the appearance of uncertainty being a part of the process.* The regulatory process has developed a "speed limit" mentality. The answers have to be yes or no, 0 or 1, go or no-go, or above or below some sort of a "limit line." That is, regulators are much more comfortable in a "binary" world. Since, in reality, all issues about the future have uncertainty associated with them, the risk assessment process recognizes this and merely attempts to quantify what the level of uncertainty might be. Therefore, when it comes to performance measures or damage parameters, if we are honest with ourselves, we will admit that there is uncertainty and present our results accordingly. In the nuclear regulatory world, where decisions have been made based on very conservative, deterministically based criteria, the adoption of a point of view that embraces the notion of uncertainty in critical parameter calculations is, to say the least, an extremely difficult concept to accept. Yet it is the only way to tell the truth about the analysts' state of knowledge of any performance measure.

- *There is concern that the price of maintaining a plant-specific risk model is too costly.* The point here is that regulating on the basis of risk would require the plant operators to keep their risk models current, which, it is argued, may be a very expensive undertaking. The idea of risk-based regulation is to have a more or less continuous knowledge of the most important contributors to risk in order to be in the best possible position for their direct positive control. Since risk is a dynamic process, so needs to be the process of risk assessment or risk monitoring.
- *Regulators and operators have concerns that the lack of consistency in different risk models precludes meaningful comparisons between plants and could lead to inconsistencies in regulatory enforcement.* In order for regulators to make decisions for the industry based on risk-based arguments, there must be some consistency among nuclear plant risk models regarding the boundary conditions, completeness, and level of detail at which accident sequences are modeled. Experience has indicated some difficulty in prescribing risk assessment methods and scopes. The problem is that risk-based technologies do not lend themselves to a "best method," and there is great value in remaining flexible to stimulate creative modeling and analysis. The result has often been new and important insights. The other problem is that the industry and the regulators have difficulty in agreeing on what constitutes a suitable scope for a risk analysis on which to base regulatory judgments.
- *The question of quality control and communication of the risk assessment results are a concern to both regulators and licensees.* The question is, "How does one prescribe a quality control system for what is basically an analysis activity that crosses dozens of technical disciplines and thousands of pieces of hardware?" The expansiveness of a risk analysis creates a question and answer (QA) nightmare of detailed knowledge of hardware, software, procedures, personnel qualifications, analysis methods, analysts' qualifications, etc. The communication issue relates to the choice of performance measures and the form of the results. It is becoming increasingly clear that no single performance measure, such as core damage frequency, is adequate to communicate the risk, nor can a single number, curve, table, or graph adequately represent the total risk involved.

So the question is, "Where are we?" Is risk-based regulation even feasible? Should we continue to pursue it as the foundation for the risk management of nuclear power? To the last question, this author believes that, indeed, we should — that some form of risk-based regulation is not only essential for nuclear power, but should be the foundation for all decisions affecting the health, safety, and welfare of all societies.

As to where we now stand on nuclear power and its move toward risk-based regulation, the following situation seems to exist. There now exists an opportunity on the basis of NRC encouragement to perform some pilot applications of risk-based regulation, and industry needs to take the initiative. Early applications on using risk-based arguments to get relief on technical specifications (U.S. Nuclear Reg. Com. 1994) have indicated an interest on the part of the NRC with some, not totally, encouraging results. Furthermore, the applications on tech spec relief have demonstrated the ability to cut maintenance and operating costs without compromising safety.

Early indications from the pilot applications being proposed by industry are that the approach for risk-based regulation most likely to succeed is a mix of probabilistic, deterministic, and mechanistic analysis. It is clear that the transition is going to be very evolutionary and may never be completely probabilistic. It is also clear that it is going to be very difficult to move the regulators off the pass/fail threshold way of thinking. The idea of making a decision on the basis of a series of probability curves, while the better way is to expose the truth, may never happen. In spite of all of the obstacles and problems, there is strong evidence that risk assessment as an aid to decision making in the regulation of nuclear power is becoming increasingly accepted.

Some of the challenges to a more rapid acceptance of risk-based regulation and a resolution of the problems noted earlier are the following:

- There needs to be implemented an effective quality control system for the nuclear plant risk assessments. This is important to reduce the potential for miscommunication, misapplication, and abuse of risk assessment results.
- There needs to be a better definition of risk assessment scopes, terminology, success criteria, boundary conditions, and the form of the results.
- Risk assessment results, including the quantification of uncertainty, are not compatible with legal decisions, the basis of the regulatory process — litigation and legal transactions thrive and prosper when there is uncertainty. This is a fundamental problem that needs to be solved between the technical and legal communities.
- There needs to be developed a consensus for risk-based regulation within industry and the regulatory community while building public confidence.
- The regulators need to be more of a single voice in providing guidance and encouragement on risk-based regulation. While NRC management carries the voice of reason and encouragement, the staff often comes across with business as usual with very little evidence of wanting to change anything. Meanwhile, industry needs to work harder at winning public confidence. The public needs to be convinced that industry really cares about the environment and their health and safety.
- For risk-based regulation to really work, there needs to be a greater commitment from industry to keep their risk models and databases current to reflect as-operated conditions.
- As a form of leadership toward risk-based regulation, the NRC needs to develop a strategy for transistioning into risk-based regulation.
- Finally, it is clear that for risk-based regulation to have broad-based appeal, it needs to be demonstrated that it can accommodate what some people call the "soft science" issues such as human factors and human values.

Considering that these are some of the problems and needs for an effective risk management program, it is interesting to speculate on some of the actions that would push the process along. There are many possibilities. They include initiatives for licensees to submit specific license amendment requests based on risk assessment findings. It would also help for the different industry groups to collaborate, so as to present more of a common front to the regulators. For example, such industry groups as the Electric Power Research Institute (Palo Alto, CA), the Nuclear Energy Institute (Washington, D.C.), and the Institute for Nuclear Power Operations (Atlanta, GA) should work together with industry consultants and suppliers to formulate a unified approach to risk-based regulation. The result of such collaboration would be a much stronger industry partner to collaborate with the NRC in making constructive progress. The further result would be an NRC action plan that reflects reality and, in particular, a plan that takes full advantage of the total knowledge base of industry and government. Such an approach would greatly facilitate the development of a strategy that would result in increased public confidence, something both the NRC and industry greatly needs.

6. SUMMARY AND CONCLUSIONS

The risk management of nuclear power is in a state of transition from deterministically based rules and regulations to greater dependence on probabilistic risk assessments. While the transition is far from complete, nuclear power, perhaps more than any other industry, has used quantitative risk assessment methods and applications to gain insights into the safety of their plants. The safety record of nuclear power is outstanding, with two accidents having the greatest impact on the course of the industry and the safety practices employed. Considering that the experience base for nuclear-generated electricity has reached approximately 7000 reactor years, this is a most impressive record. However, these accidents, the Three Mile Island, Unit 2 plant and Unit 4 of the Chernobyl station, are an important reminder of the need for a comprehensive risk management process to gain the full benefits of nuclear power.

The nuclear power industry is further advanced than any other major industry in having a comprehensive knowledge base of detailed and quantitative risk assessments to support meaningful risk management. This is about the only industry to perform extremely detailed risk assessments that quantify not only the frequencies of releases of radiation (i.e., the source term), but also the likelihood of injuries and property damage off-site. In recent years, there has been less emphasis on off-site consequences and greater emphasis on assessing precursor events such as the likelihood of core damage. Both the owner/operators and the regulators have made extensive use of the risk assessments in making decisions about the safe operation of the plants.

The issue now is whether to change the regulatory process to take greater advantage of the robust amount of information contained in the risk assessments by more formally making regulatory decisions using risk-based arguments of probabilistic risk assessment. There are many obstacles before such a transition is complete, with perhaps the biggest one being the cultural change required in the regulatory agencies. The NRC is encouraging pilot applications of risk-based licensing changes to develop confidence in the process. While risk-based regulation is not yet a reality, what is a reality is that risk assessment arguments are now routine in the risk management process for both the regulators and the owner/operators of the plants. What is also a reality is that the application of risk assessment technologies has added greatly to the understanding of nuclear safety and our confidence in the safety of nuclear power.

REFERENCES

Barnes, D. G., Times are tough — brother, can you paradigm?, *Risk Analysis*, 14(3), 219, 1994.

Blockley, D.I., *Engineering Safety,* McGraw-Hill International (UK) Limited, Chap. 14, 1992.

Garrick, B. J., United systems safety analysis for nuclear power plants, Ph.D. thesis, University of California, Los Angeles, 1968.

Häfele, W., The role of nuclear energy in the global context of the 21st century, presented at the Dave Ross memorial lecture, MIT, Cambridge, Massachusetts, April 20, 1994.

Pickard, Lowe and Garrick, Inc., Westinghouse Electric Corporation, and Fauske & Associates, Inc., Indian Point Probabilistic Safety Study, prepared for Consolidates Edison Company of New York, Inc. and the New York Power Authority, March 1982.

Pickard, Lowe and Garrick, Inc., Westinghouse Electric Corporation, and Fauske & Associates, Inc., Zion Probabilistic Safety Study, prepared for Commonwealth Edison Company, Newport Beach, CA, September 1981.

U.S. Nuclear Regulatory Commission, Reactor Safety Study: An Assessment of Accident Risks in U.S. Commercial Nuclear Power Plants, WASH-1400 (NUREG-75/014), October 1975.

U.S. Nuclear Regulatory Commission, Individual Plant Examination for Severe Accident Vulnerabilities, Generic Letter No. 88-20, November 23, 1988.

U.S. Nuclear Regulatory Commission, Severe Accident Risks: An Assessment for Five U.S. Nuclear Power Plants, NUREG-1150, Volumes 1 and 2, December 1990.

U. S. Nuclear Regulatory Commission, Safety evaluation by the Office of Nuclear Reactor Regulation related to Amendment nos. 59 and 47 to facility operating license nos. NPF-76 and NPF-80, Houston Lighting & Power Company, City Public Service Board of San Antonio, Central Power and Light Company, City of Austin, Texas, docket nos. 50-498 and 50-499, South Texas Project, units 1 and 2, Washington, D.C., February 1994.

QUESTIONS

1. What distinguishes nuclear power plant safety from other engineered facilities?
2. What has been the record for nuclear plant safety?
3. What major accidents have occurred and how have they influenced nuclear power?
4. What are the principal elements of managing the safety of nuclear power?
5. What progress is being made in the transition to risk-based regulation?
6. What distinguishes probabilistic risk assessment from other risk assessment techniques?

CHAPTER **IV.2**

Seismic Risk and Management in California

William E. Dean

SUMMARY

California has high incidences of damaging earthquakes. Eighty percent of the state's population lives in the seismic zone with the greatest probabilities of strong ground motion. Most earthquake-related death and property loss result from damage to structures. Many old buildings remain from the early years before the building codes had significant provisions for seismic resistance. In particular, many unreinforced masonry buildings pose real hazards to human life. Seismic retrofit greatly reduces the life risk at a fraction of the building's replacement cost. Risk analysis provides a basis for deciding if retrofit makes sense as a risk-reduction strategy.

The risk analysis provides estimates of the cost of preventing a quake-related death. Estimates for the typical cost of preventing a death are as follows: for unreinforced masonry bearing wall buildings, $0.6 million; for unreinforced masonry infill wall buildings, $3.7 million; and for nonductile concrete frame buildings, $9.6 million. The uncertainty in these results is about a factor of 10. The building-to-building variability introduces another factor of 50 to the distributions. Surveys of Americans indicate that they value incremental risk reduction at $3 million to $7 million per life saved. On this basis, retrofit of unreinforced masonry bearing wall buildings is a good way to save lives. Retrofit of unreinforced masonry infill wall buildings makes sense where local conditions indicate a high hazard. In light of the uncertainty, requiring the retrofit of all nonductile concrete frame buildings is not a good way to save lives.

Some local governments in California have taken action against the dangers of unreinforced masonry buildings. Long Beach is a pioneer, passing an ordinance in

1-56670-130-9/97/$0.00+$.50

1971 that required retrofit or demolition of buildings. The city of Los Angeles passed a mandatory retrofit ordinance for its 8000 bearing wall buildings in 1981. The city is now gearing up to take on infill wall buildings. The state passed the Unreinforced Masonry Building Law in 1986, which required cities and counties to establish mitigation programs by 1990. Many of these programs require retrofit, but other programs are ineffective.

Key Words: earthquake, California, seismic retrofit, natural hazard, building safety

1. INTRODUCTION

California is unique among the 50 states, in that it is both the most populous state and has high incidences of damaging earthquakes (Gore 1995). Eighty percent of the state's population, including the Los Angeles and Orange County metropolis and the San Francisco Bay Area, lives in Seismic Zone 4, with the greatest probabilities of strong ground motion (see Figure 1).

Figure 1 Seismic Zone 4.

Most earthquake-related death and property loss result from damage to structures. California building codes aim for life safety. In this regard they have been successful. For example, the 1994 Northridge quake caused $40 billion of property damage (Adkisson 1995). Yet the quake killed only 57 people. (Perhaps the state may not fare so well when a M 7 strikes under an urban area, as happened in Kobe. Recent calculations show that such a quake could cause collapse of 20-story, steel-frame buildings over an area of 50 km^2. [Heaton et al. 1995].) So far, California engineers have a good track record in terms of protecting lives in modern buildings.

Many old buildings remain from the early years before the building codes had significant provisions for seismic resistance. In particular, many unreinforced masonry (URM) buildings remain in use, and these buildings pose real hazards to human life. Nonductile concrete-frame buildings are also dangerous.

It is possible to retrofit these buildings, greatly reducing the life risk at a fraction of the building's replacement cost. For example, typical replacement cost is about $80/$ft^2$, whereas typical cost of seismic retrofit is about $17/$ft^2$ (Hart Consultant Group Inc. 1994). While several people were killed by URM in Loma Prieta, nobody died from this cause in the Northridge quake, where most URM buildings had already been retrofitted.

Seismic retrofit is sufficiently expensive that it is not taken lightly. Some party has to pay for it (or else decide not to do it), but no party is eager to do so. The responsibility for making a building safe falls squarely on the owner. Governments issue building codes, license contractors, and inspect their work, but the costs fall on the owner.

Mandatory rehabilitation policies that put the burden on the owners are unpopular with them; voluntary programs fare better politically (Beatley and Berke 1990). The city of Los Angeles has a mandatory program for load-bearing URM buildings. All observers agree that financial considerations have been a major headache in the implementation of this program.

Sometimes the owner can recover part of the cost of seismic retrofit by passing it on to the tenants. If rent is at market rate, the owner must upgrade the building in other ways as well to make it more attractive to tenants. If the building is a rent-controlled housing facility, the city allows the owner to pass through some (but usually not all) of the cost. Although seismic retrofit makes the old building safer, occupants do not seem willing to pay for it. There appears to be no link between seismic work and rent levels (Tyler and Gregory 1990). The market levels for rent after retrofit are not much more than they were before retrofit.

Seismic retrofit does not increase the market value of the building to a level above that which it had before the mitigation program began. In Los Angeles during the 1980s, unstrengthened URM buildings sold at a discount roughly equal to the expected cost of seismic work. The market value of strengthened buildings is higher than unstrengthened buildings only by the approximate cost of strengthening (Tyler and Gregory 1990). As a result, banks will not make loans for seismic rehabilitation, even though the amount is less than the value of building and most owners have no other loans outstanding. Bankers are not willing to make loans for projects that do not increase the value of a building (Jouleh 1992).

Most people seem unwilling to pay now to prepare for the next earthquake, but, after it happens, the same people criticize the government for not doing enough to prepare. Risk analysis provides a basis for deciding if retrofit makes sense as a risk-reduction strategy. The method used in this study involves comparison of two quantities. One is the cost of preventing a death due to building collapse. The other is the monetary value of incremental risk reduction.

The remaining sections are as follows: a description of the classes of dangerous old buildings in California, a discussion of the valuation of risk reduction, a risk analysis of retrofit of the dangerous buildings, a report on what the state is doing to encourage seismic retrofit, and, finally, a report on the Development of a Standardized Earthquake Loss Estimation Methodology.

2. DANGEROUS OLD BUILDINGS

California has four types of buildings that pose threats to life in an earthquake:

- *URM bearing wall buildings* — These are almost always brick buildings, in which the load is borne by the walls themselves. The walls consist solely of bricks and mortar, with no reinforcement rods nor anchors to tie the walls to the roof or to upper-story floors. These buildings were all built before 1934 and are the most hazardous of the four classes. The International Council of Building Officials adopted Appendix 1 of the Uniform Code for Building Conservation in 1991 as a standard for seismic retrofit of bearing wall buildings.
- *URM load-bearing frame buildings (also called infill wall buildings)* — These have concrete or steel frames with URM infill walls. They were built before 1940. The Hazardous Buildings Committee of the Structural Engineers Association of Southern California is developing provisions for seismic retrofit of infill buildings.
- *Nonductile concrete-frame buildings* — These brittle buildings were built before the 1973 building code change. The strength of the building comes from the reinforced concrete. During strong shaking, the concrete fails catastrophically. In contrast, ductile concrete has extra steel reinforcement and can bend without breaking. Nonductile frame buildings are the most expensive to retrofit of the four classes. Engineers understand these buildings qualitatively, but more research is needed to obtain the numbers for guidelines for seismic retrofit.
- *Pre-1976 concrete tilt-up buildings* — For these buildings, the walls are precast and tilted into place, and a roof is added. Earlier building codes permitted the walls and ceiling to be attached by mere nails. More recent codes require anchors. This simple, inexpensive precaution prevents a tilt-up building from becoming tilted down during an earthquake. Los Angeles has adopted a mandatory retrofit ordinance for these buildings, in response to the good performance that voluntarily retrofitted tilt-up buildings displayed during the Northridge earthquake (Dames & Moore 1994).

After the Northridge earthquake, engineers discovered cracks in frames and connections of many steel-frame buildings. Repairs cost $7000 to $22,000 per joint. The city of Los Angeles passed an ordinance requiring inspection and repair for the 100 nonresidential buildings in the vicinity of greatest shaking. Of these buildings, 75% had some broken joints (EERI 1995). The cost boils down to about $14 to $40/ft^2. The threat to life is hard to quantify, because so far no steel-frame building has collapsed in an earthquake in California (Heaton et al. 1995). Most engineers consider it sufficient to repair a *few* hundred steel-frame buildings after a strong earthquake, rather than retrofit all of them now prior to future earthquakes.

3. VALUATION OF RISK REDUCTION

This section discusses the valuation of risk reduction in terms of the cost of preventing a death due to collapse of a dangerous building in an earthquake. Two issues need attention. First, what is a reasonable quantitative estimate of the value of risk reduction? Second, what is a reasonable way to think about discounting future deaths?

3.1 Value of Risk Reduction

The discussion centers on the value of a "statistical life" in contrast to an "identified life." For example, a boy lost on a mountain is an identified life; government agencies will provide lots of resources, and many volunteers will give much time to search for the boy, though there is scant chance of finding him alive. Mitigation of earthquake hazards, on the other hand, saves "statistical lives" because it reduces risk a little bit for many people, and nobody can identify in advance which individuals will be spared from death. The analyst sums up the incremental risks to calculate the number of statistical lives saved.

The willingness-to-pay approach has come into favor in the past 15 years. This approach considers how much people are willing to pay to reduce risks of mortality. This approach does not calculate how much society ought to value risk reduction, but it tries to measure how much people do value risk reduction.

In this context, the phrase "value of life" is misleading. It conjures up the image of a scale balance: a person sits on one pan, and the other pan holds a pile of money; the object is to decide how much money it takes for us to be indifferent between the two. That is not so at all. The analyst really means "the incremental value of incrementally reducing the probability of death from some small level to another yet smaller level." It is less long winded to use the phrase "value of life."

Many people are squeamish about putting a dollar value on a life as a whole. Yet these same people have no qualms about quantifying the value of portions of life. Every employee concedes that at least some of his/her time is less valuable as leisure than the wages earned on the job. Likewise, one can put dollar values on marginal risk reduction.

Normal people do not spend all their resources on safety; they also purchase other goods. They try to make the tradeoff between safety and other goods so that the marginal utility of safety equals the marginal utility of other goods. Then they are indifferent whether their last dollar goes toward safety (instead of going toward other goods) or toward other goods (instead of toward safety).

A recently introduced concept, "willingness to spend," is the income loss expected to induce one premature fatality. This quantity equals willingness to pay divided by the marginal propensity to spend on risk reduction (Lutter and Morall 1994).

If a family or a society spent all its resources on safety in an attempt to achieve the complete elimination of one or two kinds of risk, then it would find itself impoverished and would find itself exposed to the risks associated with poverty (Keeney 1990). People have to tolerate some risk to life.

A recent review of a multitude of studies suggests a range for the value of life of $3 million to $7 million (Viscusi 1993). A recent estimate of willingness to spend suggests a range of $9 million to $12 million in 1991 dollars ($10 million to $13 million in 1994 dollars) (Lutter and Morall 1994).

3.2 Discounting and Pseudo-Discounting of Lives

By seismic retrofit, deaths are prevented in future years. The owner spends the dollars in 1 year, and the risk reduction occurs throughout the remaining life of the building, typically 30 years.

If a policy prevents deaths from some particular cause at some time in the future, how are those lives to be valued in comparison with lives saved in the present? There is not a market by which human lives can be bought, sold, or invested. It is not obvious that lives saved in the future should be discounted the same way, or at the same rate, as monetarized benefits and costs are discounted.

The point is not about deferring risk for individuals. The population (occupants of hazardous buildings) can be thought of as a diverse lot. Individuals may move in and out of the buildings, but presumably the characteristics of the population (age distribution, etc.) change slowly. So, whenever the quake may strike, the same kinds of fatalities are prevented.

Value-of-life calculations usually do not have to take the future into account. Tradeoffs between fatality risks and wages consider industrial accidents, fire-related deaths, homicides, and suicides. These all are near-term causes of death. From a moral perspective, it makes no difference when a life is saved. The prevention of a death 20 years from now is just as valuable as the prevention of a death this year. (The comparison is between someone now and someone else 20 years later. The comparison is *not* between an individual now and the *same individual* 20 years from now.) One should not discount lives, because there are no opportunity costs to saving lives later rather than sooner (MacLean 1990).

The classic argument asserts that the discount rate for life-saving benefits ought to be the same as the discount rate for money, or else analysis produces strange results. If lives are not discounted, the decision maker is paralyzed. Money that could save lives this year sits in the bank until next year, so that it can save even more lives next year. The perversities disappear if one uses a discount rate for lives that equals the discount rate for money (Keeler and Cretin 1983). The Office of Budget and Management recommends a real discount rate of 7% per year, because it approximates the marginal pretax return on an average investment in the private sector in recent years (OMB 1992).

Consider an alternate viewpoint based on these considerations:

- A life is a life, and lives are not discounted.

- For a given policy, the valuation of life is the dollar amount that society would have to pay for each life saved, that amount being the *same* for each life saved.
- The dollar amount is adjusted according to the time at which the life is saved.

Consider the following example. The U.S. Environmental Protection Agency's (EPA) analysis of a uranium mill tailings standard estimated that the short-term costs would be $338 million and that the standard would save 4.9 lives per year. If lives are not discounted, does that mean that the cost-effectiveness is $800,000 per life over a horizon of 100 years or $80,000 per life over a horizon of 1000 years (MacLean 1990)? One is rightly suspicious of a policy analysis with a horizon of 100 years, and the consequences of radioactive waste 1000 years from now are utterly unknowable.

The correct question is this: "How much to charge society for each life saved *at the time that it is saved*, so that these charges all add up to $338 million." Clearly, the arithmetic is the same as the problem of calculating payments on a perpetuity with a non-zero interest rate. In the example of the uranium mill tailings standard, the cost per life saved would be $7 million, given a 10% discount rate. For a 2% discount rate, the cost per life saved would be only $1.4 million.

Consider the same problem from yet another perspective. Suppose that the cost is financed by a loan. Then every year a payment is due. The cost of saving lives each year gets paid that same year.

It is not at all necessary to discount lives. Spending $3 million *now* to save a life *now* is equivalent to spending $3 million *in 20 years* to save a life *20 years from now*. Likewise, spending $3 million to save a life now is equivalent to spending the present discounted value of $3 million to save a life 20 years from now. Notice that it is the *dollars* and *not* the *lives* that are being discounted! Yet the mathematical formalism is identical with that used if lives are discounted, if the discount rate reflects opportunity costs for money. This practice can be called "pseudo-discounting."

4. RISK ANALYSIS

Risk reduction is usually seen as an end in itself. The comparison of benefits to the costs of abatement is almost an afterthought. However, this comparison deserves to be a central concern of any policy analysis of risk reduction. Is seismic retrofit cost-effective? It depends on the objective. This chapter focuses on seismic retrofit rather than on new construction. So the appropriate objective is risk to life, rather than structural damage or content loss.

The cost of preventing a (pseudo-discounted statistical) death depends on various quantities:

- Retrofit cost, in dollars per square foot (Hart Consulting Group Inc. 1994)
- Replacement cost of building, $80/ft^2
- Building occupancy, 0.9 to 3.3 occupants per 1000 ft^2
- Street occupancy, 0 to 62 bystanders per 1000 linear feet

- Length of building footage, 100 ft
- Annual probability of quakes of various intensities
- Lifetime of retrofitted building, 30 years
- Social discount rate, 7% per year.

Further details, such as tables of inputs and formulas for the derived quantities, are in a previously published study (Dean 1993). The cost is partly offset by reduced structural damage. For building classes other than tilt ups, this damage reduction is about 10% of the cost of retrofit. According to an assessment of URM buildings shaken by the Northridge earthquake, 11% of unstrengthened buildings suffered severe damage, in contrast to only 0.3% of retrofitted buildings (Penera 1995).

Here are some "typical" values of the cost of preventing an earthquake-related death for three of the four types of buildings discussed previously (Dean 1993):

- URM bearing wall, $0.6 million
- URM infill wall, $3.7 million
- Moment-resisting, nonductile, concrete frame, $9.6 million

These are median values from a distribution produced by multiple runs of the model, using different combinations of inputs each time, to account for uncertainty and variability. However, the previous study does not draw a distinction between uncertainty and variability. For that reason, the medians are reported here and labeled "typical" values (Dean 1993). (Tilt-up buildings do not require such analysis because of the clear benefits — beyond life safety — of seismic retrofit for tilt ups.)

A first-cut comparison suggests that seismic retrofit of URM bearing wall buildings is cost-effective because the typical cost falls below the range for valuation of risk reduction, which is $3 to $7 million. The infill wall building falls inside the range, so it is not clear whether infill wall buildings are good candidates for risk reduction. The nonductile concrete-frame building falls above the range. So the first-cut comparison suggests that seismic retrofit of these buildings is not cost-effective.

4.1 Uncertainty

A second-cut comparison looks at uncertainty as well as the typical value. What if the cost is really several times the typical value? Or several times less?

Uncertainty in the probability estimates for earthquakes is on the order of a factor of two up or down from the best estimate (Lamarre et al. 1992). The estimates of death rates are also uncertain by roughly the same factor (Holmes et al. 1990).

The equations for the risk analysis consist mainly of multiplication and division of factors. It is appropriate, then, to treat each uncertain factor as if it has a lognormal distribution, so the result from calculation also has a lognormal distribution. This procedure is more complicated than simply multiplying ranges together, but it avoids exaggerating the size of the uncertainty (Bogen 1994). The combination of sources of uncertainty leads to a factor of 10 range. The true value could be three times as high or three times as low as the "typical" values cited earlier.

4.2 Variability

Real buildings differ from a typical building in terms of exposure to earthquakes, cost of retrofit, susceptibility to ground shaking, etc. The distribution of real-world variability among buildings needs to be taken into account in the risk analysis (Hattis and Burmaster 1994). It is important to distinguish between uncertainty caused by ignorance, on the one hand, and variability, on the other hand (Hoffman and Hammonds 1994).

The cost of retrofit varies because buildings come in different sizes, shapes, etc. A recent study concluded that the dispersion factor is 4.07 for a 90% confidence interval for retrofit cost (Hart Consultant Group 1994).

The probability of various levels of ground motion differs greatly throughout Seismic Zone 4 (Algermissen 1991). A given acceleration is roughly five times more likely inside Seismic Zone 4 than at its edge.

The quality of soil has a major role in the extent of building damage. Model runs with poor soil show a death rate 14 times higher than model runs with good soil (Dean 1993).

The combination of the three factors leads to a range of 50. So, for the lowest 5-percentile building, the cost of preventing a death is about 1/7 the cost for the median building. Likewise, for the 95-percentile building, the cost is seven times that of the median building.

4.3 Conclusions of Risk Analysis

One can safely conclude that seismic retrofit of URM bearing wall buildings seems a cost-effective way for society to save lives. Retrofit of the median building saves lives at a cost under $3 million. Even the high-percentile buildings fall within the range of valuation of risk reduction.

For some URM infill wall buildings, seismic retrofit is cost-effective. For others, the cost of preventing a death is too high. Perhaps retrofit of infill wall buildings should be required on a selective basis, such as in Los Angeles, where risk is high, to mandate retrofit of these buildings.

For all but a few nonductile concrete-frame buildings, the question of cost-effectiveness of seismic retrofit has an unclear or negative answer. Life safety justifies seismic retrofit for a few buildings, especially if they have a pattern of higher than average occupancy. Retrofit programs ought to be voluntary, with incentives to encourage retrofit, but with the decision in the hands of the party who will have to pay for it. These buildings are not as dangerous as URM buildings, so retrofit has to be less expensive to get the same risk reduction per dollar. Perhaps engineers will invent new techniques that will drop the cost.

5. ACTIONS TO MITIGATE RISK

This section describes two local mitigation programs and a state law that promotes local programs. The Long Beach and Los Angeles programs are important because they provided examples of mandatory programs that other cities could follow. The state law, SB-547, requires that cities and counties in Seismic Zone 4 start their own programs.

5.1 Long Beach

In 1959, Long Beach (Alesch and Petak 1986) amended its municipal code to define earthquake hazards associated with buildings as nuisances. This allowed the city to take legal action against owners for elimination of hazardous buildings. In 1969, opponents requested a moratorium on condemnations while the city performed a study of the problem. The ordinance committee was still considering the issue when the San Fernando earthquake struck in February 1971. Because of the background work and the concern aroused by the quake, Long Beach passed its Earthquake Hazard Ordinance in June 1971. The original ordinance ranked buildings into four priority groupings. In 1976, the ordinance was amended to simplify the ranking process. The amendment also stipulated an explicit time table for enforcement, with deadlines for the more hazardous buildings by January 1984 and deadlines for the least hazardous by January 1991.

5.2 Los Angeles

After 6 years of debate, the Los Angeles City Council passed a retrofit ordinance in January 1981 (Alesch and Petak 1986). The lateral force standards reflected those in effect from 1940 to 1960. These standards have been incorporated in the state model ordinance. The ordinance applied to bearing wall URM buildings in Los Angeles, except detached residential buildings with fewer than five units. Buildings were assigned to four classifications. Owners had 3 years to comply after official notification. However, owners could choose to install wall anchors within 1 year after notification in exchange for additional time for full compliance. After the 1985 Mexico City earthquake, the Los Angeles ordinance was amended to speed up the mitigation program. The new ordinance is called Division 88. The program was nearly completed in time for the Northridge earthquake. Some of the retrofitted buildings suffered damage, but none collapsed.

5.3 SB-547: The Unreinforced Masonry Building Law

The California legislature passed the Unreinforced Masonry Building Law, SB-547, in 1986. The law requires cities and counties in Seismic Zone 4 to make an inventory of their URM buildings and to develop a program for hazard abatement. The Seismic Safety Commission oversees implementation of SB-547. The state law tells the cities and counties to develop a program, but does not require any particular type of program.

Mandatory programs have been adopted by half of the cities and counties in Seismic Zone 4, affecting about three fourths of the URM buildings (California Seismic Safety Commission 1991). These programs legally remove the do-nothing option for owners. The owners have several years to retrofit or demolish.

A few municipalities have voluntary programs. The owners have to prepare hazard evaluation reports, which are made public. Owners do not *have to* do anything. They face special incentives that make it easier to retrofit or replace the building. Also there are incentives to make it less attractive to do nothing (such as fear of litigation if someone gets hurt in a quake).

About a quarter of the cities and counties in Seismic Zone 4 have "notification-only" programs. Owners receive a letter indicating that their URM building is hazardous, but typically there is no indication of standards that the building should meet nor recommended procedures for making the building safer. No municipality had a notification-only program until the deadline loomed for starting a mitigation program required by SB-547. The Seismic Safety Commission considers such notification-only programs as falling short of complying with the spirit of the law, although they comply with the letter of the law.

Some programs do not fit into the three categories just described. "Other" programs include posting signs in the URM building themselves warning occupants that the building is hazardous. Another example is requiring seismic rehabilitation upon increases in occupancy, alterations, or additions.

The more URM buildings, the more likely the local government opts for a notification-only or "other" program. Cities with more than 200 URM buildings are unlikely to impose a mandatory program. (Notable exceptions are San Francisco, with more than 2000 bearing wall buildings, and Los Angeles.) Even where mandatory programs are in place, building officials report that owners drag their feet in compliance, so mitigation programs fall behind schedule (Turner 1995); see Figure 2.

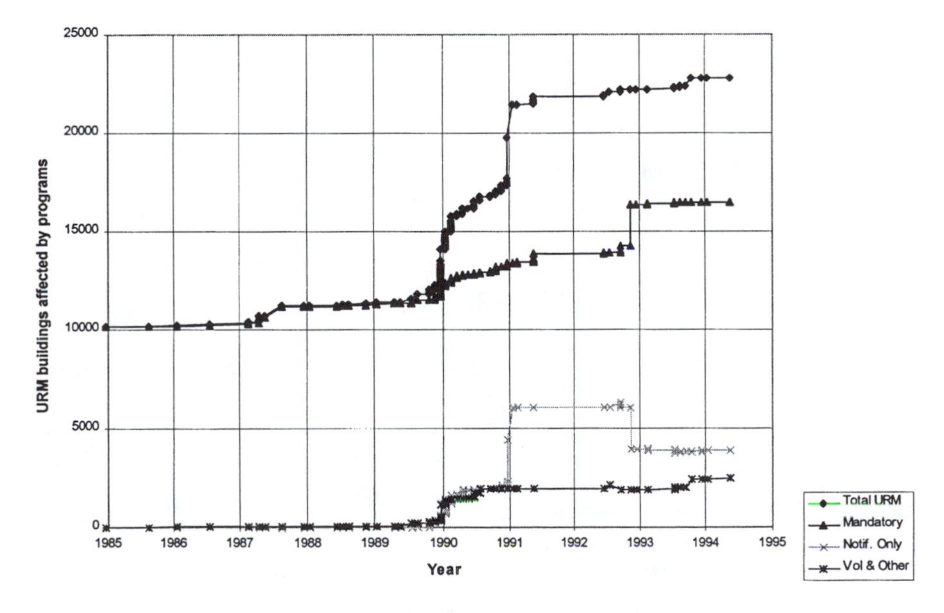

Figure 2 History of compliance with the Unreinforced Masonry Building Law.

6. DEVELOPMENT OF A STANDARDIZED LOSS ESTIMATION METHODOLOGY

The National Institute of Building Sciences is coordinating a major effort on the "Development of a Standardized Earthquake Loss Estimation Methodology." When completed at the end of 1996, the working group will produce a FEMA report. Risk Management Solutions, Inc. will implement the methodology as a PC-based geographic information system coupled to a thorough database. The software is scheduled for release in early 1997. The methodology encompasses

- Potential earth science hazards
- Direct physical damage
- Induced physical damage
- Direct economic and social losses
- Indirect economic losses

This accomplishment will enable researchers to perform detailed, high-quality, credible risk analysis without having to build crude ad hoc models from scratch.

7. CONCLUSION

Risk analysis can show which types of dangerous buildings in California are worth retrofitting. One can safely conclude that seismic retrofit of URM bearing wall buildings seems a cost-effective way for society to save lives. Retrofit of infill wall buildings should be required on a selective basis, such as in Los Angeles, where risk is high, to mandate retrofit of these buildings. Nonductile concrete-frame buildings are poor candidates, at least with current technology.

The National Institute of Building Sciences is coordinating a major effort on the "Development of a Standardized Earthquake Loss Estimation Methodology." This accomplishment will enable researchers to perform detailed, high-quality, credible risk analysis without having to build crude ad hoc models from scratch.

REFERENCES

Alesch, D. J. and Petak, W. J., *The Politics and Economics of Earthquake Hazard Mitigation,* Institute of Behavioral Science, Boulder, CO, 1986.

Algermissen, S. T., Perkins, D. M., Thenhaus, P. C., Hanson, S. L., and Bender, B. L., Probabilistic Earthquake Acceleration and Velocity Maps for the United States and Puerto Rico, MF-2120, U.S. Geological Survey, Reston, VA, 1991.

Adkisson, J., personal communication, 1995.

Beatley, T. and Berke, P., Seismic safety through public incentives: The Palo Alto seismic hazard identification program, *Earthquake Spectra,* 6, 57, 1990.

Bogen, K. T., A note on compounded conservatism, *Risk Analysis,* 14, 379, 1994.

California Seismic Safety Commission, Status of California's Unreinforced Masonry Building Law, SSC 91-04, State of California, Sacramento, 1991, Figures 3 and 4.

Dames & Moore, *The Northridge Earthquake January 17, 1994,* Dames & Moore, Los Angeles, CA, 1994.

Dean, W. E., Earthquakes and Old Buildings: Seismic Risk Permits for California Structures, N-3604-RGSD, RAND, Santa Monica, CA, 1993, Chapter 5 and Appendix E.

EERI White Paper Workshop, Social, Economic and Political Issues Involved in Decisions about Building Safety, Earthquake Engineering Research Institute, Oakland, CA, 1995.

Gore, R., Living with California's faults, *National Geographic,* 187, 2, 1995.

Hart Consultant Group, Inc., Typical Costs for Seismic Rehabilitation of Existing Buildings, Second Edition, FEMA-156, Federal Emergency Management Agency, Washington, D.C., 1994.

Hattis, D. and Burmaster, D. E., Assessment of variability and uncertainty distributions for practical risk analyses, *Risk Analysis,* 14, 713, 1994.

Heaton, T. H., Hall, J. F., Wald, D. J., and Halling, M. W., Response of high-rise and base-isolated buildings to a hypothetical M_w 7.0 blind thrust earthquake, *Science,* 267, 206, 1995.

Hoffman, F. O. and Hammonds, J. S., Propagation of uncertainty in risk assessments: The need to distinguish between uncertainty due to lack of knowledge and uncertainty due to variability, *Risk Analysis,* 14, 707, 1994.

Holmes, W. T., Lizundia, B., Dong, W., and Brinkman, S., *Seismic Retrofitting Alternatives for San Francisco's Unreinforced Masonry Buildings: Estimates of Construction Cost and Seismic Damage,* Rutherford & Chekene, San Francisco, CA, 1990, 6–10.

Jouleh, A., personal communication, 1992.

Keeler, E. B. and Cretin, S., Discounting of life-saving and other nonmonetary effects, *Management Science,* 29, 300, 1983.

Keeney, R. L., Mortality risks induced by economic expenditures, *Risk Analysis,* 10, 147, 1990.

Lamarre, M., Townshend, B., and Shah, H. C., Application of the bootstrap method to quantify uncertainty in seismic hazard estimates, *Bulletin of the Seismological Society of America,* 82, 104, 1992.

Lutter, R. and Morall, J. F., Health-health analysis: A new way to evaluate health and safety regulation, *Journal of Risk and Uncertainty,* 8, 43, 1994.

MacLean, D. E., Comparing values in environmental policies: Moral issues and moral arguments, in *Valuing Health Risks, Costs, and Benefits for Environmental Decision Making,* Hammond, P. B. and Coppock, R., Eds., National Academy Press, Washington, D.C., 1990, 83.

OMB, Guidelines and Discount Rates for Benefit-Cost Analysis of Federal Programs, Circular No. A-94, Revised, Office of Management and Budget, Washington, D.C., 1992.

Penera, V. A., personal communication, 1995.

Turner, F., personal communication, 1995.

Tyler, M. B. and Gregory, P., *Strengthening Unreinforced Masonry Buildings in Los Angeles: Land Use and Occupancy Impacts of the L.A. Seismic Ordinance,* William Spangle and Associates, Inc., Portola Valley, CA, 1990, 60, 99.

Viscusi, W. K., The value of risks to life and health, *Journal of Economic Literature,* 31, 1912, 1993.

QUESTIONS

1. This chapter focuses on earthquake-related threats to life safety. What are some other detrimental effects of earthquakes that are beyond the scope of this chapter?
2. What can you say to an owner of a dangerous building who insists that, because the building has survived 60 years of earthquakes, it must be a safe building?
3. Why is mandatory seismic retrofit of URM buildings unpopular in cities with many of them, even though those cities are the places with greatest potential for death and injury?
4. Suppose that the state has two options: (1) retrofit its buildings now or (2) wait 10 years for development of an improved technique that cuts the cost of retrofit in half. Which would you recommend?
5. The URM buildings in California were constructed prior to 1934. If a building will be demolished in a few years, does it make sense to retrofit it?
6. If the "typical" cost of retrofit for nonductile concrete-frame buildings is $25/ft^2, how low would the cost have to fall before you consider retrofit a cost-effective measure for this class?
7. How does a geographic information system (GIS) improve seismic risk analysis?

CHAPTER IV.3

Sustainable Management of Natural Disasters in Developing Countries

Terence Lustig

SUMMARY

Disaster-management systems have not been very successful. A large part of the problem stems from the tendency for a community's preparedness for the next disaster to decline over time after the previous event; the tendency for newcomers to deny the problem; and the likelihood that the effectiveness of the disaster-management system will deteriorate quite rapidly because of the rapid turnover of key staff in the various agencies making up the disaster-management system.

There are powerful psychological barriers which make it difficult to enforce proper maintenance of disaster-management systems. These stem from the fact that we need to feel in control of our lives, whereas warnings will often be taken as threats to our sense of control.

For a disaster-management system to be sustainable, therefore, it should be designed not only to convey the message to the members of the disaster-prone community that they are in control, but also that the system is actually under their control.

Key Words: developing countries, disaster, disaster management, hazards, natural disasters, preparedness, risk, risk communication, sustainability, sustainable development, sustainable disaster mitigation

1-56670-130-9/97/$0.00+$.50

1. INTRODUCTION

A disaster can be defined as "an unexpected disruption of economic and/or environmental systems, entailing widespread losses which exceed the ability of the affected society to cope using its own resources." The key point is that the disruption is unexpected, and, thus, people are unprepared. The difficulty for any disaster mitigation system is then for the community to prepare for that which the people are not prepared.

The economic losses through worldwide disasters from 1980 to 1989 have been estimated at $35 trillion (U.S. dollars), and the rate of losses apppears to be increasing (Kreimer and Munasinghe 1991). Since population densities in developing countries are increasing, many of their disadvantaged citizens will come under increasing pressure to settle in hazardous areas. This will increase the susceptibility of these countries to disasters as a whole, threatening to impede their economic development.

Disaster-management programs have not frequently been successful. Even after decades of disaster-mitigation works, the annual losses from the disasters can be greater than at the beginning of disaster-mitigation programs (U.S. Water Resources Council 1976). Therefore, if we are to achieve sustainable disaster management, we should first understand the underlying causes for the increase in losses from disasters.

2. THE PROBLEM OF DECLINING COMMUNAL PREPAREDNESS

It is in the nature of disaster-prone communities that their overall capability of coping with a disaster decreases with the time since the previous event. Certainly, once people have experienced a disaster, they will usually be better prepared for the next one (e.g., Lustig and Haeusler 1989). However, as people who have experienced a disastrous event die or move out, those who replace them will not have the experience and thus will tend to discount people's accounts of the severity of former events. Consequently, they normally will be unprepared for it.

Even though these inexperienced people may be told about the hazard, they will not fully appreciate how bad it can be (Schiff 1977). Thus, as new people replace those who went through the last event, the preparedness of a community will tend to decrease over time, as typified in Figure 1. A derivation of the relationship of the theoretical curve in Figure 1 is given by Lustig (1994), Sinclair Knight Merz (1995b), and Lustig and Maher (1996).

On the other hand, if there are frequent disasters, the preparedness of the community will remain high.

2.1 The Problem of Successful Disaster Mitigation

This is why we have a continual problem in disaster mitigation. The more successful we are in mitigating disasters, the less experience people will gain, and the less prepared will the community be.

In addition, as the community becomes less prepared for a disaster, a greater and greater proportion of the households will be unaware of the dangers of settling

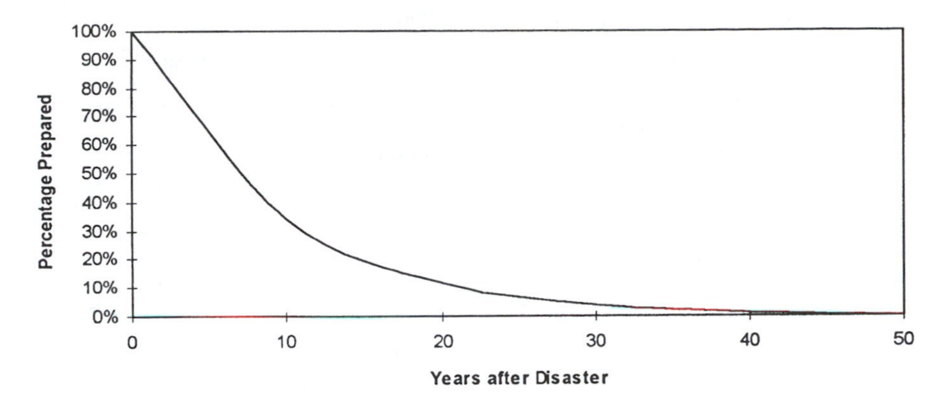

Figure 1 Typical decline in preparedness of a community since the last event.

in hazardous locations and will be more inclined to do so, rendering the community as a whole more prone to disasters.

The less prepared the community, the less political pressure there will be to direct resources toward disaster management. This includes not only resources directed to emergency services, but also to those authorities concerned with forecasting, flood mitigation, catchment management, land-use planning, and environmental conservation.

For example, in Vietnam, in the years 1900 to 1945, there were 18 years in which dykes failed in the Red River Delta. After the war and with the achieving of independence, the system of maintenance improved, and the frequency of dykes breaching steadily declined. Following the massive floods of 1971, when the dykes failed in a number of places, the attention to maintaining the dykes intensified still further. Since then, apart from one near failure in 1986, there have been no more dyke failures in the Red River Delta. Now, however, evidence is accumulating that the dykes and other disaster-mitigation structures of Vietnam are no longer being maintained as diligently as before.

It will be argued in this chapter that there are powerful psychological barriers which make it difficult to enforce proper maintenance of disaster-management systems. This, in turn, has implications for determining the most appropriate institutional arrangements for ensuring a sustainable program of operation and maintenance.

2.2 The Problem of Denial

Why do people deny that they are prone to hazards? Let us consider the distinction between voluntary and involuntary activities as shown in Table 1. A voluntary activity is one which the participant freely chooses to undertake, and an involuntary activity is one which, to a great extent, is imposed.

Most people would rather be involved in voluntary risky activities than involuntary ones, even when the voluntary activities are easily shown to be more hazardous (Slovic et al. 1984). This seems to be because involuntary activities are those over

Table 1 Voluntary and Involuntary Risky Activities

Voluntary activities	Involuntary activities
• Fishing off rocky headlands • Fishing at sea in the typhoon season • Riding a bicycle in heavy traffic • Riding on the side of an overloaded bus or train • Riding on top of an overloaded bus or train • Lighting firecrackers • Smoking	• Riding in a bus of a busline with a poor safety record • Living in a house with a chemical factory with modern safety facilities being built nearby • Living in a house with a dam being built just upstream

which we have no control, while we feel happier undertaking more hazardous activities if we feel we are in control. Just how important it is for people to feel that they are in control can be readily seen if we consider that people will give up their lives for the idea of "freedom."

It is a prerequisite for mental health that we feel in control (Langer 1978). A feeling of helplessness can be debilitating and, in chronic cases, can lead to death (Langer 1975, 1983). Studies on animals (Lefcourt 1973) and humans (Langer 1983, Glass and Singer 1969) show that mental and physical stress can be more readily coped with if the subjects have a sense of control. This does not mean that they *are* in control, merely that they *perceive* they are in control.

Let us imagine that people who have just bought a house are told it is in a hazardous location. This threatens their sense of control, since they cannot eliminate the hazard. The only way they might feel they can retain a sense of control is to deny the problem. (To think how we might behave in this kind of situation, we could imagine ourselves in analogous circumstances. Let us envisage that we have almost completed a project. Then, someone comes along and points out a fatal flaw which would compel us to revise all the work. What is our reaction?)

This denial is an extremely powerful influence. Even after a disaster, people who had not expected it will be telling themselves that it could not happen again. It is also a well-known source of frustration for disaster managers. They go out of their way to provide the community with information about a hazard and see it is largely ignored.

There is a further difficulty. Even if an area is subjected to a disaster, not all the people will be affected. For example, during a moderate flood, some houses prone to flooding only in a large flood would be spared, and many of these householders would be convinced that they would always be above flood level. Langer (1978) explains that in order to rationalize that we are in control, we tend to attribute favorable outcomes of risky circumstances to our skill and unfavorable outcomes to bad luck. Thus, many of those who are flood prone, yet have been above a previous flood, may convince themselves that they are clever enough to have acquired a house above the flood level. As well, some of those who were flooded would have rationalized that another flood could not recur in their lifetime. This idea would also have been reinforced by the "availability bias," whereby we tend to recall small, frequent events more easily than rare, large ones (Saarinen 1990). Thus, the preparedness of a community could decline even more rapidly than shown in Figure 1.

We should recognize that people act not so much to minimize losses, but to minimize distress (Green 1990). Thus, they will only start to reduce losses if they perceive that this is the most effective strategy for minimizing distress. Handmer and Penning-Rowsell (1990a) conclude that this is why people cope with unavoidable threats by ignoring them and devote themselves to matters they perceive they can control.

Miransky and Langer (1978) have documented how different occupants of apartment blocks in New York approached their concerns with burglary in unexpected ways. Those who thought their area was safe used all their locks **more** than those who thought their area unsafe. Further, more than two thirds of respondents thought that it was the responsibility of **others** to prevent burglary in their own dwellings.

The authors suggest that people may be wanting to distance themselves from negative events and taking steps to reduce burglary may make the event seem more likely. They conclude that simply telling people it is their responsibility to reduce the chances of burglary would probably not work, neither would dire warnings.

Macgregor (1991) has come to a similar conclusion. He found that people tend to worry more about matters which they feel they have some control over than those which they perceive as uncontrollable. Thus, simply giving people more information about an event which seems uncontrollable may have little effect on how well they can cope with it.

Saarinen points out that there is very little support from the hazard literature for there being any relationship between awareness and behavior. People may be aware of a hazard, but they tend to underestimate the probability of an unfavorable outcome (Quinnell 1981). This tendency can be found just as easily among disaster-management experts as among lay people (Saarinen 1990).

If we are all prone to deny uncomfortable facts, persons with authority to ensure that a disaster-management system is maintained should not be wholly relied upon to make the correct decision on how best to do so if they are inexperienced or untrained. If these persons are faced with other pressures to allow inappropriate development, they may resolve the dilemma by denying the hazard.

Thus, they may tend to respond more to the demands of the community on which they must rely for reappointment than the warnings of the disaster-management expert. If the community believes that the structure is sufficiently strong to allow harmful developments, the end result often can be that the hazard is denied and the advice of the expert is ignored.

2.3 The Problem of Declining Organizational Readiness

Disaster-management systems are invariably made up of a number of government and nongovernment organizations. All too often, they find it difficult to coordinate their activities so that they function smoothly when there is an emergency.

Part of the problem is that these organizations may be busy with other priorities during times when there is no emergency and may not pay enough attention to preparing for the next event.

Also, the people within an organization change positions or leave, so that gradually those with experience of the last disastrous event are no longer available to pass on their knowledge. The longer the period since the last event, the less the appreciation by the emergency workers of the pitfalls in carrying out their duties and liaising with other organizations.

Unless there is very thorough training, the inexperienced replacements are unlikely to appreciate fully how they should work with others of the disaster-management system. As a result, two inexperienced members of two cooperating organizations may have different understandings of who should do what, so some tasks may be left undone during the next disaster.

Figure 2 indicates that with an average 5-year turnover of staff and perhaps five organizations in a flood-warning system, the chances of it working without too many mistakes could become very small within a few years (Sinclair Knight Merz 1995c). The assumptions made in deriving this figure were that an experienced member of the staff would have a 90% chance of not making a serious error, and an inexperienced member of the staff would have only a 10% chance, while a trained but inexperienced person would have a 50% chance.

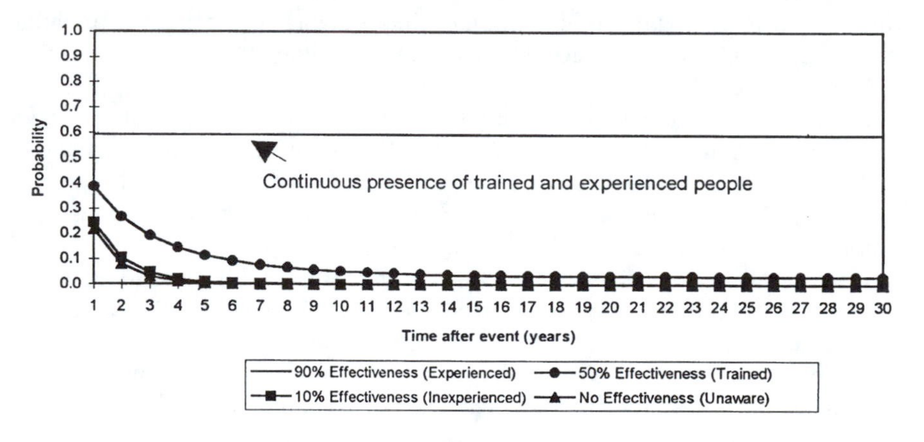

Figure 2 Decline of organizational readiness disaster-management system with five organizations comprising a disaster-management system.

There is an inherent difficulty of coordination between government agencies, even at the best of times. By their very nature, bureaucracies have to take care not to offend their counterparts. Yet quite clearly, coordination during an emergency is highly likely to encounter situations where there is little time for delicacy and subtlety.

Also, we should bear in mind that the tendency to assign responsibility for an accident increases as the consequences become more serious (Walster 1966). This tendency could make it harder to discern the true nature of the problem in a briefing session of coordinating agencies after a disaster.

We would suggest that while strong efforts should be made to improve communication and coordination, we would do well to recognize in designing a disaster-management system that (1) it is in the nature of disaster management for coordination to break down and (2) that interagency rivalry is generally endemic throughout the world.

2.4 Economic Changes

A new approach to managing disasters in developing countries is needed also because of the rapid social, economic, and technical changes they are undergoing.

In deltaic areas in the past, for example, thousands of trained water professionals and millions of workers would donate their labor annually to maintain the flood infrastructure. Today, however, economic pressures are making this level of commitment increasingly difficult. It follows that manual labor and traditional construction methods will need to be substituted by mechanized equipment, new construction techniques, and new ways of mitigating disasters.

Also, it is now recognized that some of the disaster-management infrastructure may have been built to a level of protection which was too low, while others may have been built to a level too high to be economically justified. Given the scarce resources available to governments of developing countries, it will be important to ensure that future disaster-management systems are designed, not only to be economically justified, but also to be economically efficient.

2.5 Environmental Changes

Natural disasters in developing countries have been aggravated by environmental degradation. Destruction of mangroves and coral reefs has increased the vulnerability of coastal settlements to typhonic winds and waves.

In the hills and mountains, the removal of trees, both through war and economic activities, has substantially increased erosion and runoff, so that flood levels are higher than they used to be. At the same time, with less and less water infiltrating into the ground, dry-season flows are reduced, and these can result in severe water shortages and seawater intrusion at the coast.

Also, it is recognized that the frequency of heavy storms and typhoons will increase with global warming.

Therefore, it will become increasingly urgent for disaster-management professionals to develop social practices which foster an ethic of ecologically sustainable development.

2.6 Insufficient Maintenance

Most developing countries are anxious to raise their standard of living as quickly as possible. A frequently preferred strategy is to put resources into an improved infrastructure. Unfortunately, this can often be at the expense of proper operations and maintenance.

A lack of maintenance can be particularly pronounced for an infrastructure that is used only infrequently: for example, disaster-mitigation works. The community which should benefit from this infrastructure would be largely unaware of the poor care being taken and thus be unlikely to press for any improvement. As a result, when the infrastructure is ultimately called upon to play its part, it may no longer be up to it. If it is being relied upon to mitigate disasters, the consequences can be even worse than if it were not there.

2.7 Social Impacts

It is generally accepted that the social effects of disasters are greater after the actual event and are experienced longer than the physical and monetary effects. (Most people will agree that deaths, illnesses, and trauma are worse than the loss of possessions.) If we are to attain sustainable disaster management, we will need to understand the processes which people go through in trying to cope so that we might see where social intervention might be most effective.

Often people in a community will help each other during and just after the disaster, and this will help the community to become closer. It will raise their spirits, as they will gain a tangible sense of achievement. This can be sufficient for people to regain some sense of control. As Solomon and others (1989) hypothesize, it is the activity of solving problems that offers the prospect of eventual control.

People begin to enter what Filderman (1990) refers to as the "teachable moment." Since people can reach a maximum rate of learning under conditions of optimal arousal, a short time after the event is when people can be most receptive to new ideas on preparedness (Wilson 1990).

Prince-Embury (1992), in studying the aftermath of Three Mile Island (TMI), also discusses how people seek control by acquiring information and suggests that there are psychological benefits from information and education after a disaster. She points to an increased sense of loss of control and psychological symptoms after TMI being associated with a lack of adequate information.

Unfortunately, all too often, disaster victims become frustrated as they realize how much they have lost and how difficult it will be to recover. They can suffer shock when they realize that the control they thought they had was not there after all. Therefore, it is vital that the people have access to disaster counseling, support, and recognition within days to help them develop or even maintain a sense of achievement in overcoming their troubles. The earlier such a service is available, the better. Solomon and others have found that those who cope best tend to be people who do not blame themselves, but do accept responsibility to deal with the consequences of a disaster. Victims who blame themselves are unlikely to seek help from relief organizations and are likely to have mental health problems (Solomon et al. 1989). As Lefcourt (1973) reminds us, once subjects have learned helplessness it is difficult to reestablish a sense of control.

According to Ladrido Ignacio and Perlas (1994), a prerequisite for recovery from a disaster is gaining a sense of control. For this reason, the assistance should be based on a strategy of mutual help rather than simply one of charity. For example, if it is feasible, it can be helpful to sell materials and equipment for recovery at a low price rather than for free. Selling the goods should help give the recipients a sense of ownership and encourage their sense of control over their destinies.

If disaster recovery centers are set up, they should run for at least 1 year. This is because some people need time to come to terms with what happened before they seek emotional assistance.

Thus, to mitigate social effects in a sustainable manner, there should be prior planning to

- Ensure that people can participate fully in their own recovery.
- Provide early information on the disaster and ways of reducing the impact the next time.
- Provide early counseling and recognition of the disaster.
- Ensure that assistance is provided in a manner that fosters the victims' sense of control.

3. IMPROVING THE MANAGEMENT SYSTEM

Underlying the attitude of many agencies involved in disaster management appears to be a hierarchical view of what constitutes a disaster-management system. It would seem that many see the management system as comprising only the public agencies, with the public itself consisting of passive recipients of the management services. This view is found to coexist with the belief that it is the effectiveness of the agencies which is the primary determinant of the effectiveness of the system as a whole.

For example, disaster-management systems are typically depicted as in Figure 3, with advice or service going **down** to the people. This contrasts with what many would argue should be the role of such a system, namely, to provide a service **up** to its client, the community. Also, such depictions rarely acknowledge the fact that it is often the people who provide valuable information and services to the disaster-management agencies and that in developing countries it is often the people themselves who play a major role as providers of disaster-mitigation infrastructure.

Maskrey (1989) argues that a feeling of control is necessary for effective communal action in mitigating disasters. If this principle is not adopted, the system is unlikely to be effectively sustained until the next disaster. It is only with the people feeling they have a stake in the mitigation system that they will remain actively involved. Further, unless the people are participating, the political and hence financial support for maintaining the disaster mitigation system will tend to drop away.

This implies that the disaster-management system should be designed to convey a strong message to the occupants of the vulnerable area that it belongs to them and that they are its most effective component.

3.1 Putting the Community in Control

In order to promote the idea that the disaster-management system belongs to the community, there should be a disaster-management board or committee, consisting of representatives of the disaster-prone community. Preferably, the members of this board should have a personal stake in sustaining the disaster-mitigation system.

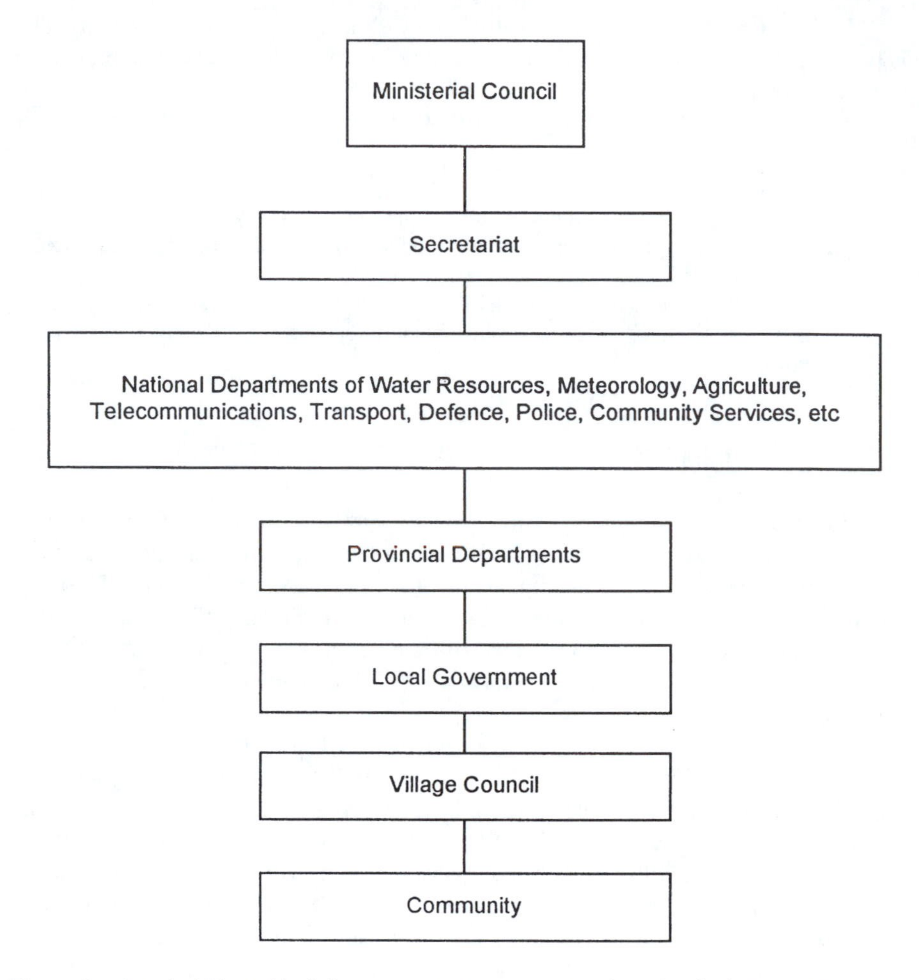

Figure 3 A typical hierarchical disaster-management system in a developing country.

All the agencies with responsibilities for setting up, operating, and maintaining the disaster-management system would report to this board. Perhaps, the organizational arrangements could be as shown in Figure 4.

The members of this board would not be responsible for the day-to-day operation of the disaster mitigation system, but would have oversight of the performance of the various agencies, which would be reporting to it. We might expect that the board would normally delegate the executive task to a local disaster-management coordinator, probably seconded from one of the agencies involved in the system.

The board could have the following tasks:

- Appoint an executive officer and secretariat for terms up to, say, 3 years
- Monitor the capability of each agency with responsibility for some component of the disaster-management system

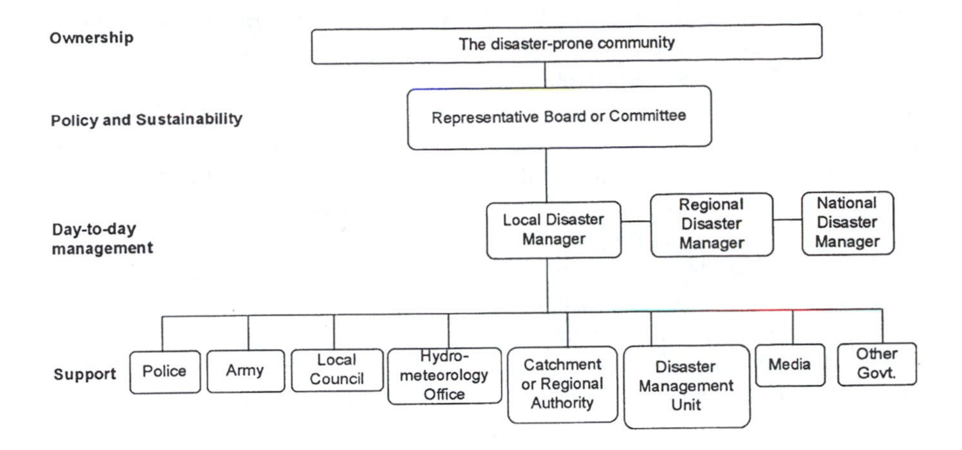

Figure 4 Suggested model for control of a disaster-management system.

- Institute checks on the operational effectiveness of the disaster-management system as a whole
- Initiate regular disaster-preparedness campaigns
- Run regular checks on the state of community preparedness in the whole disaster-prone area, up to the extreme event
- Maintain a high political profile for the disaster-management system
- Act as role models for preparing for the next disaster and for responding to warnings when the disaster comes (Quinnell 1981, Handmer and Penning-Rowsell 1990c)
- Act as personal motivators to prepare for the next disaster (Handmer and Penning-Rowsell 1990b)

There are several reasons for having no representatives of the disaster-management agencies on the board.

- The members of the board must be seen by the community as the stewards of their disaster-management system.
- The members should be directly accessible by the occupants of the disaster-prone area.
- If one agency did not perform to expectations, or was encountering difficulties, the board would not be inhibited in making representations at a sufficiently high level to resolve the matter.
- With the membership of the board being people with a personal stake in the disaster-management system, they could help provide the continuity needed for sustaining it in a state of readiness.

The work of the board would be demanding and would require people with staying power. It would be difficult to keep the issue of disaster management in front of an increasingly unaware and, hence, an increasingly indifferent community. For this reason, consideration should be given to recognizing the contributions by the members of the committee toward sustaining the disaster-management system with appropriate citations and/or decorations. This may help slow the decline in enthusiasm which is often experienced in the years after a disastrous event.

Some funding of the secretariat and local components of the disaster-management system would be needed. It would be politically and psychologically desirable for a significant part of this to be raised by a special levy on people in the hazardous area. This would help make clear that the disaster-management system was going to belong to the community, since they were in fact making an important financial contribution to its continued operation.

The rating structure could be roughly in proportion to the potential benefits of the disaster-management system, and this could also provide an incentive for people to make their properties less vulnerable.

This direct levy would also help ensure that the disaster-management system continued to have control by locals. For as long as the occupants of the floodplain were paying the levy, the board would continue to function. Were there to be no such direct charge, the funding of the board would be subject to the normal political processes which could lead to funds being gradually diverted elsewhere.

3.2 The Disaster-Warning System

For centuries before modern methods of disaster management were developed, people would use other techniques, such as observing the behavior of animals or changes in plants to forecast hazardous events (MWR 1995). Even today, information from official sources would generally appear to play a minor role, and it is frequently the case that, when people receive a warning which causes them to take action, it is more often than not from a friend, relative, or neighbor (Sinclair Knight Merz 1995a, Handmer and Penning-Rowsell 1990d).

If we are to take account of the fact that the members of the community, besides being the owners of the disaster-warning system, are probably also its most effective component, the flow of information in the warning system could be as depicted in Figure 5.

It is essential that the warning be transmitted in several ways at once. There are two reasons for this. The first, which is widely accepted, is that redundancy is important in case some communication links become inoperative during the disaster management cycle. The second is that people need the opportunity to obtain confirmation of the impending hazard or disaster.

The warning message should not consist simply of facts. There is no evidence for the provision of information leading directly to a change in attitude, much less a change in behavior (Handmer and Penning-Rowsell 1990d). In fact, as argued earlier, merely making an announcement about an impending disaster can be counterproductive.

Marks (1990) advises that the message should be designed to be

- Multimodal, using many different forms (e.g., acoustic, visual, graphic) and many modes of communication; this enables people to get confirmation from several sources.

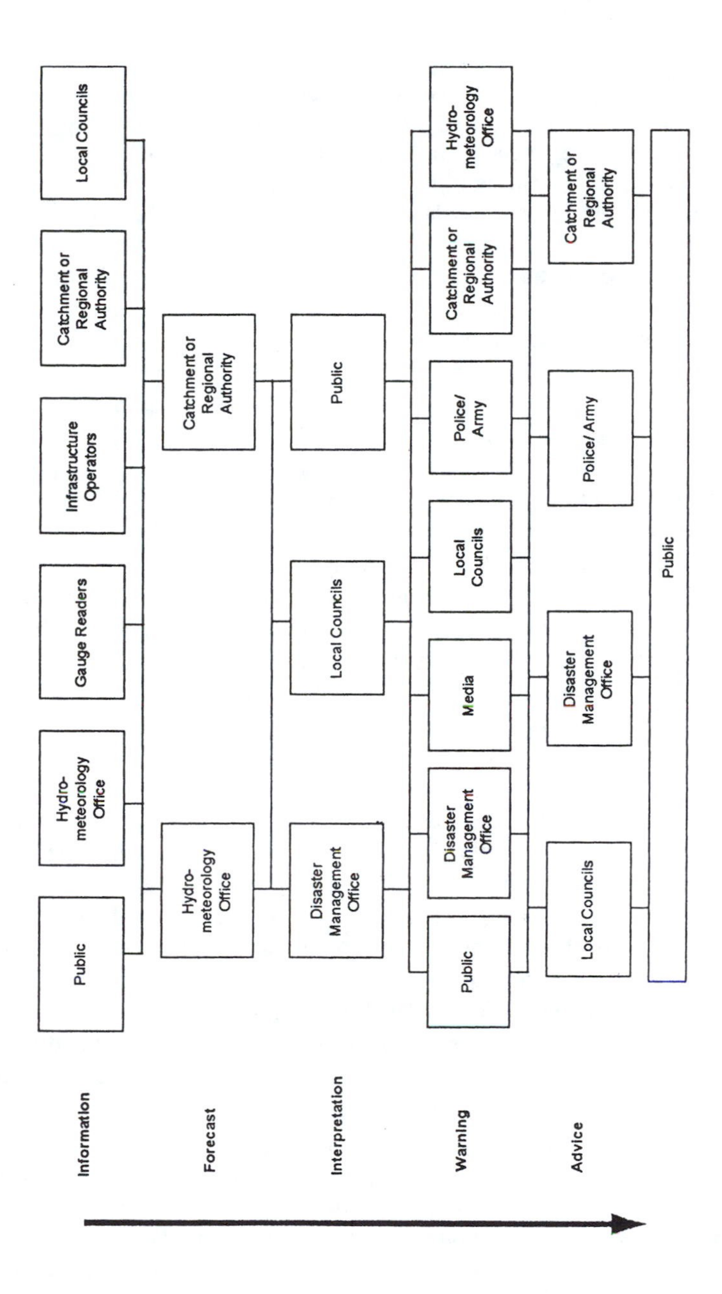

Figure 5 Suggested flow of information in a disaster-warning system.

- Positive, stressing the benefits rather than negatively stressing the adverse possibilities; Langer (1978) states that even the anticipation of success in a risky venture enhances the sense of control.
- Specific.
- Inviting participants' activity rather than passivity; Langer (1978) has shown that people prefer to participate in games with a low probability of winning, but where they feel they are in control, rather than in games with a higher chance of winning, but where they have no control (by seeking their participation, people are being told that they can be in control).
- Inviting sociability rather than isolation (e.g., check that your neighbors are aware).

Often, it could be valuable if the National Hydrometeorological Office (NHO) — assuming the hazard is related to weather — could set up a system of accrediting local organizations for forecasting. These local organizations would often have access to a wide formal and informal network for supplying information on local observations. They would also have a much better idea of which of the formal and informal observers were reliable.

The local forecasters could be responsible for producing forecasts in areas not dealt with by the NHO and for interpolating between the results of the models of the NHO. This local capability can be very helpful. In Central Vietnam, for example, when a typhoon is approaching and there is no reliable forecast available from the National Hydrometeorological Service in Hanoi, the local provincial meteorologists at different locations along the coast compare barometric readings. Where the reading is lowest, it is assumed to be where the typhoon is likely to strike (Lustig et al. 1993).

With such an arrangement, the local disaster-mitigation and emergency agencies should be in a better position to issue warning messages which are meaningful to the community, and this would help foster a greater sense of control within the community.

The NHO should reinforce this perception when issuing forecasts by conveying the message that the warning system is relying on the people's skills. First, it should make sure that the recipient of any warning message is informed in the beginning that detailed interpretations can be obtained from local disaster-mitigation and emergency agencies, the local government authority, the local police, and perhaps some other local agency. This is so that more than one source of the same information can be accessible to the public. Second, it should ask that the people inform others. Third, it should state that any observations of earlier floods (e.g., flooding in the upper catchment) or later records of the event (e.g., markings of peak water levels) would be welcome.

The warnings should be issued in a graduated fashion (known colloquially in the United States as "Ready-Set-Go"). The "Ready" advice would be only a generalized announcement indicating that a disaster-producing event could be expected somewhere within a certain large area. Later, when the forecasts could be much more accurate, the "Set" warning would indicate there was a high likelihood of a disastrous event within a reasonably well-defined area. When the prediction could be very accurate, the "Go" warning could state where and when the disaster was likely to be.

This stepwise approach to providing a warning should be more likely to result in more people responding than at the present time. Langer (1978) and Langer and Roth (1975) found that, even in games which were purely random, people who had a longer time to become involved tended to have a greater sense of control and seemed to believe they had a greater chance of winning. This perception has been observed even with educated people (e.g., university students) who could be certain to appreciate how random the outcomes were (e.g., tossing a coin) when they were looking at the situation objectively. Of course, during an emergency when a warning is being given, the proportion of the community which is likely to remain objective would be low.

There is an opposing argument to this strategy, based on the "cry wolf" syndrome. This argument holds that the "Ready" warnings should not be made public, since more often than not the situation does not worsen and people will tend to ignore them. While we acknowledge that there is this risk, we would suggest that it can be reduced by conveying to people the message that they are in control and that it is their skills which are needed to help reduce losses.

Studies have shown that people prefer to have information about uncertain, unpleasant, yet unavoidable outcomes (Saarinen 1990, Lanzetta and Driscoll 1966). It might be argued also that the problem is not so much the false warnings, but people's prior beliefs. People tend to seek or accept information which is consistent with their beliefs and will interpret new evidence in a way which fits in with those beliefs (Handmer and Penning-Rowsell 1990a).

3.3 Preparing the Community

The issue then is how to prepare or condition the community for these uncertain, but inevitable events. If we accept that people act more to reduce stress than to reduce losses, we need to design our preparedness campaigns so that they will not threaten to cause undue stress and be rejected. If the information can be presented positively, over a long period, people will have an opportunity to become habituated to it in their own way and adapt it to their own needs (Prince-Embury 1992). Prior conditioning can also reduce the stress during and after the event itself (Lazarus and Alfert 1964, Langer et al. 1975).

It will not be easy for this effort and enthusiasm to be sustained at a high level. Indeed, despite all efforts, the preparedness of the community will decline.

This is not to say, however, that the work to improve awareness and to motivate preparedness should be dispensed with as futile. Rather, it is recommended that there be a comprehensive program for heightening and sustaining community awareness and regular professionally designed campaigns to motivate people to gradually prepare for the next disaster. Some strategies for maintaining awareness and sustaining preparedness could be as set out in Appendix A in this chapter.

The campaign to prepare for the next disaster should start during the last one. This is when people will be most receptive to advice and most likely to become motivated.

The information should be provided regularly to facilitate habituation and in a form which conveys the message that the people themselves have important skills to contribute. For example, by having friendly competitions between towns, the emphasis can be on people's skills, and this can enhance their sense of control (Langer 1975).

The message should be positive and advise on how to "beat the odds." As Langer (1978) explains, the lower the probability of success, the more difficult it is to have control. Hence, if people can be helped to "win" in such circumstances, the greater will be their sense of mastery.

The concept of blame should be strenuously downplayed. It will be a continually recurring theme among those who were spared. Especially with those who were not affected by a small event, but who are still disaster prone, there will be an inclination to blame those who were victims, in order to forestall the uncomfortable recognition that it could happen to themselves (Walster 1966).

3.3.1 Coercion vs. Persuasion

Communal preparedness also relates to a matter which is a long-standing source of discussion among emergency services, namely, whether or not they should have the power to enforce evacuations.

It is recommended that this power not be available for two reasons:

- In practical terms, it may not be logistically significant. It is not clear how it will help if emergency personnel are devoting their resources to removing someone who will not move, when they could be more effectively helping others who are more cooperative.
- A more important point is that the emergency agency would tend to rely on such a power, rather than on seeking the cooperation of the community. The existence of this power would convey the strong implication that the disaster-management system is owned and controlled by the government, and, thus, the members of the community are not in control. As we have stressed several times, it is vital that the community be part of the disaster-management system, and the availability of this power would tend to work against their becoming involved.

It would follow from this that the extent that an emergency agency would be relying on coercion to achieve its objectives during a disaster would be an index of its lack of success in preparing the community beforehand.

3.3.2 Disaster Recovery Services

The disaster recovery plan should be a vital component of the preparedness campaign. Key points of this plan should be as follows.

- It should commence operation at the time of the disaster in the evacuation centers. The evacuated victims are at that time suffering from boredom, terror, and frustration. The advice they receive could be a powerful aid to their regaining a sense of control over their destiny (Ladrido Ignacio and Perlas 1995).
- The disaster recovery officers should be trained and be able to supply kits with practical advice on subjects such as

    - Assistance available from government and nongovernment agencies
    - Techniques for making property resistant to disasters
    - Improving the informal disaster-management system
- The appeal for donations should be planned well in advance by marketing experts with experience in promoting public issues (e.g., AIDS, family planning).
- There should be regular liaison between government and nongovernment disaster-relief and recovery agencies to ensure coordination.
- The potential roles of volunteers and nongovernment agencies should be set out.
- There should be an explicit plan for making manifest the continuing recognition of the plight of the victims by those outside the affected areas.
- A contingency plan should be prepared for supplementary programs of assistance for the long-term unemployed, the newly unemployed, and other recipients of welfare services.
- The mental health of those working in disaster relief and recovery should be monitored to ensure there are regular opportunities to relieve stress and ready counseling in case of an impending crises.
- The plan should be reviewed, updated, and drilled regularly, at least every 5 years.
- There should be a disaster-recovery desk at each evacuation center.

3.3.3 The Media

The media's effectiveness can range from extremely good to counterproductive. For this reason, the warning agencies should not take it for granted that the media will perform as desired. We would suggest that, because it is such a powerful medium, local radio and television stations should be regularly contacted to ensure that the current staffs are aware of the very beneficial role they could play, that they are agreeable to playing this role, and that they know what to do. Radio and television stations often have a high turnover of staff, and this regular contact should help sustain the effectiveness of this means of communication.

The help of both the local radio stations and the local newspapers should also be sought in promoting awareness of the disasters at other times through talk-back radio, newspaper articles, information drama, etc.

4. CONCLUSION

To sustain systems for mitigating disasters, it is required that the design of the systems take into account the tendency for people to deny the problem. This tendency stems from the strong need for all of us to feel we are in control of our lives, and this implies that a sustainable disaster-management system must convey the message that it is the people who own and control it.

The nature of disaster management is that there are many agencies with some responsibility for the operation of the disaster-management system. As the time since the last disaster increases, the various agencies will tend to become less and less ready to operate as effectively together. To compensate for this, each agency should be responsible to a board comprised of representatives of the local disaster-prone community, with day-to-day management being delegated to an executive officer, possibly seconded from one of the agencies making a contribution to the system.

The funding of the secretariat and other expenses of the system should rely, to a substantial degree, on a levy upon those benefiting. This would make funding less subject to changing political priorities.

APPENDIX A. SUGGESTED STRATEGIES FOR SUSTAINING COMMUNAL PREPAREDNESS AND ENGENDERING A SENSE OF CONTROL

- Permanent marks in public places indicating the severity of the worst historical disaster, for example, the height reached by the highest previous flood
- Photos in public buildings showing previous disasters
- Disaster advice kits containing simple strategies for protecting property upon receiving the warning and for preparing for the next hazardous event
- Advise on ways of safeguarding very important personal property such as memorabilia
- Kits for schools to teach aspects of the hazard in topics such as science, geography, social studies, etc.
- School and village plays
- Articles in the local newspapers
- Videos on previous disasters and on disaster preparedness, which are available for a nominal fee
- Programs on local radio stations
- Use as a topic for talk-back radio
- Information leaflets in relevant public offices
- Data personalized for each household, relating the effects of the hazard on the property
- Use as a topic for drama on radio and television
- Incorporate information into radio quizzes

STRATEGIES FOR DEVELOPING A SENSE OF CONTROL AND MOTIVATING PEOPLE TO PREPARE FOR THE NEXT HAZARDOUS EVENT

- Disaster-preparedness advice desks at evacuation centers during and after a disaster
- Disaster-preparedness advice stands at fairs, festivals, and other public occasions
- Competitions between villages as part of hazard drills
- Well-publicized subsidies for disaster-proofing strategies

DEDICATION

This chapter is dedicated to Tu Mao, Director of Dyke Management and Flood Control, and to Tran Nhon, Vice-Minister of Construction, who have devoted themselves to developing a sustainable water-disaster management system in the Socialist Republic of Viet Nam.

REFERENCES

Filderman, L. Designing public education programmes: a current perspective, in *Hazards and the Communication of Risk.* Handmer, J. and Penning-Rowsell, E., Eds. Gower, England, 1990, Chap. 12.

Glass, D. C. and Singer, J. E. Psychic cost of adaptation to environmental stressor. *J. Personality Soc. Psychol.,* 12(3), 200, 1969.

Green, C. Perceived risk: past, present and future conditional, in *Hazards and the Communication of Risk.* Handmer, J. and Penning-Rowsell, E., Eds., Gower, England, 1990, Chap. 3.

Handmer, J. and Penning-Rowsell, E., Eds., *Hazards and the Communication of Risk.* Gower, England, 1990a, 70.

Handmer, J. and Penning-Rowsell, E., Eds., *Hazards and the Communication of Risk.* Gower, England, 1990b, 72.

Handmer, J. and Penning-Rowsell, E., Eds., *Hazards and the Communication of Risk.* Gower, England, 1990c, 73.

Handmer, J. and Penning-Rowsell, E., Eds., *Hazards and the Communication of Risk.* Gower, England, 1990d, 313.

Kreimer, A. and Munasinghe, M. Managing environmental degradation and natural disasters: an overview, in *Managing Natural Disasters and the Environment.* Kreimer, A. and Munasinghe, M., Eds. World Bank, Washington, D.C., 1991, 3.

Ladrido Ignacio, L. L. and Perlas, A. P. *From Victims to Survivors: Psychosocial Intervention in Disaster Management.* University of Philippines, IPPAO, Manila, 1995, 57.

Ladrido Ignacio, L. L. and Perlas, A. P. *From Victims to Survivors: Psychosocial Intervention in Disaster Management.* UP Manila Information, Publication and Public Affairs Office, Manila, 1994, 57.

Langer, E. J. and Roth, J. Heads I win, tails it's chance: the illusion of control as a function of the sequence of outcomes in a purely chance task. *J. Personality Soc. Psychol.,* 32(6), 951, 1975.

Langer, E. J. The illusion of control. *J. Personality Soc. Psychol.,* 32(2), 311, 1975.

Langer, E. J. The psychology of chance. *J. Theory Soc. Behav.,* 7(2), 185, 1978.

Langer, E. J. *The Psychology of Control.* Sage, Beverly Hills, 1983, 241.

Langer, E. J., Janis, I. L. and Wolfer, J. A. Reduction of psychological stress in surgical patients. *J. Exp. Soc. Psychol.,* 11, 155, 1975.

Lanzetta, J. T. and Driscoll, J. M. Preference for information about an uncertain but avoidable outcome. *J. Personality Soc. Psychol.,* 3(1), 96, 1966.

Lazarus, R. S. and Alfert, E. Short-circuiting of threat by experimentally altering cognitive appraisal. *J. Abnormal Soc. Psychol.,* 69(2), 195, 1964.

Lefcourt, H. M. The function of the illusions of control and freedom. *Am. Psychologist,* May, 417, 1973.

Lustig, T. L. and Haeusler, T. M. Social and economic effects of floods, *29th Annual Conference of Flood Mitigation Authorities,* Batemans Bay, Australia, 1989, Chap. 4.

Lustig, T. L. Viet Nam's strategy and action plan for mitigating water disasters, in *Proceedings of the International Consultation on Strategy and Action Plan for Mitigating Water Disasters in Viet Nam.* Ministry of Water Resources, United Nations Development Programme and United Nations Department of Humanitarian Affairs, Hanoi, 1994, 19.

Lustig, T. L., Glemarec, Y., Solomatine, N., Silver, M., and Aamodt, T. *Strategy and Action Plan for Mitigating Water Disasters in Viet Nam.* UNDP, UNDHA and Socialist Republic of Vietnam, Hanoi, 1993, 68.

Lustig, T. L. and Maher, M. Sustainable Flood-Plain Management Plans, 36th Annual Flood-plain Management Conference, Grafton, Australia.

MacGregor, D. Worry over technological activities and life concerns. *Risk Anal.,* 11(2), 315, 1991.

Marks, D. Imagery, information and risk, in *Hazards and the Communication of Risk.* Handmer, J. and Penning-Rowsell, E., Eds. Gower, England, 1990, Chap. 2.

Maskrey, A. *Disaster Mitigation: A Community Based Approach.* Oxfam, Oxford, 1989, 40.

Miransky, J. and Langer, E. J. Burglary (non) prevention: an instance of relinquishing control. *Personality Soc. Psychol. Bull.,* 4(3), 399, 1978.

MWR. *Proceedings of International Consultation on the Strategy and Action Plan for Mitigating Water Disasters in Viet Nam.* Ministry of Water Resources, United Nations Development Programme and United Nations Department of Humanitarian Affairs, Hanoi, 1995, 51.

Prince-Embury, S. Information attributes as related to psychological symptoms and perceived control among information seekers in the aftermath of a technological disaster. *J. Applied Soc. Psychol.,* 22(14), 1148, 1992.

Quinnell, A. L. Human information processing of flood risk. *Proceedings of the Floodplain Management Conference.* AWRC, AGPS, 303, 1981.

Saarinen, T. Improving public response to hazards through enhanced perception of risks and remedies, in *Hazards and the Communication of Risk.* Handmer, J. and Penning-Rowsell, E., Eds. Gower, England, 1990, Chap. 16.

Schiff, M. Hazard adjustment, locus of control, and sensation seeking: some null findings. *Environ. Behav.,* 9(2), 233, 1977.

Sinclair Knight Merz. *Review of the Effectiveness of the Flood Warning Service Provided During the Floods of North-East Victoria in October 1993.* Victorian Flood Warning Consultative Committee, Melbourne, Australia, 1995a, 29.

Sinclair Knight Merz. *Review of the Effectiveness of the Flood Warning Service Provided During the Floods of North-East Victoria in October 1993.* Victorian Flood Warning Consultative Committee, Melbourne, Australia, 1995b, Appendix 1.

Sinclair Knight Merz. *Review of the Effectiveness of the Flood Warning Service Provided During the Floods of North-East Victoria in October 1993.* Victorian Flood Warning Consultative Committee, Melbourne, Australia, 1995c, Appendix B.

Slovic, P., Fischoff, B. and Lichtenstein, S. Behavioural decision theory perspectives on risk and safety, *Acta Psychol.,* 56, 183, 1984.

Solomon, S. D., Regier, D. A. and Burke, J. D. Role of perceived control in coping with disaster. *J. Personality Soc. Psychol.,* 8(4), 376, 1989.

U.S. Water Resources Council. *A Unified National Program for Floodplain Management.* Washington, D.C., 1976.

Walster, E. Assignment of responsibility for an accident. *J. Personality Soc. Psychol.,* 3(1), 73, 1966.

Wilson, C. Education and risk, in *Hazards and the Communication of Risk.* Handmer, J. and Penning-Rowsell, E., Eds. Gower, England, 1990, Chap. 4.

QUESTIONS

1. Why will the preparedness of a community for a disaster tend to decline?
2. Why will the effectiveness of a disaster-management system for a community tend to decline?
3. What are the main impediments to sustainable disaster management in developing countries?
4. On whom should the community rely to mitigate the next disaster?
5. Where should the major resources be directed to sustain a disaster-management system?
6. Think about a natural hazard in your area. What is the most important cost-effective way of managing it in a sustainable manner?
7. For class discussion: Is it easier to manage disasters sustainably in developed countries than in developing countries?
8. For class discussion: Imagine you are the head of a village and one day you see a cousin of yours digging into the toe of a levee which protects the village from floods. He says he is only digging in a little way in order to make some more space for growing vegetables. If you were to remonstrate with him, you could alienate your whole extended family on whom you rely for electoral support. What should you do?

CHAPTER IV.4

Risk Analysis, International Trade, and Animal Health

Stuart C. MacDiarmid

SUMMARY

The international trade in animals or commodities derived from them cannot be conducted without some element of risk to importing countries. Significant diseases of livestock have been spread around the world through trade. The application of risk analysis to the regulation of importation of animals and animal products is a relatively recent development intended to provide regulatory veterinarians with a transparent, objective, repeatable, and defensible estimate of the risks posed by a particular import proposal. While those working in the field of quarantine and animal imports have always analyzed risks before making their decisions, it is only in recent years that this process of identifying the hazards, assessing the probability of their actual occurrence, and then formulating risk-reducing measures with which to manage them has been approached in a structured way. This chapter describes this structured approach and presents some examples of actual cases.

Key Words: animal diseases, animal imports, anthrax, embryo transfer, fish diseases, quarantine, rabies, risk analysis, risk assessment, salmonid diseases

1. INTRODUCTION

The establishment of the World Trade Organization (WTO) and the progressive implementation of the commitments negotiated in the Uruguay Round of the General Agreement on Tariffs and Trade (GATT) have led to important changes in international

1-56670-130-9/97/$0.00+$.50

trade. Members of the WTO have agreed to remove barriers to trade in agricultural products, except in situations where a particular import can be demonstrated to constitute a risk to the animal, human, or plant health of the importing country. The application of risk analysis to the trade in animals and their products is being developed to give decision makers the means to assess whether or not particular trade proposals do, indeed, pose a disease risk to the importing county and to demonstrate to interested parties the basis on which decisions are made.

When analyzing the risks associated with a proposed import of animals or animal products, it must be remembered that such imports cannot be made without some element of risk. A "zero risk" import policy has an intuitive appeal to conservative sectors of a nation's livestock industries. However, the pursuit of a zero risk policy is counterproductive globally and domestically. There is only one zero risk policy and that is total exclusion of all imports (Kellar 1993, MacDiarmid 1991a).

The benefits of the imports often accrue to a relatively small group of people only, usually the entrepreneurs, initial importers, and distributors of, say, new genetic material (Acree and Beal 1988). The risks, on the other hand, are borne by a much broader group which includes all livestock owners whose animals could be infected with an exotic disease, as well as the general public who may be expected to bear the cost of containing and eradicating an outbreak of exotic disease. For these reasons, a risk analysis may include a benefits–risk analysis of the proposed importation. However, importation nevertheless may be permitted, even in the absence of any demonstrable national benefit and, under the terms of membership in the WTO, lack of perceived national benefit is not seen as a valid reason for declining an import.

Import restrictions applied in the name of protecting animal health should be based on sound risk assessment methods and should not be used as disguised barriers to trade.

2. ANALYSIS OF RISK

Risk, in the context of the importation of animals or animal products, is a measure of the probability of the introduction of an exotic disease and the seriousness of such an outcome. Although based in science, risk analysis is not in itself science. It must be able to deal with situations as they arise and tolerate the mathematical limitations of the animal disease prevalence estimates or other such data on which it is based (Kellar 1993). Where the data are uncertain, the distinction between facts and the analyst's value judgments may blur. For this reason, it is important that the risk analysis be made as *transparent* as possible.

Risk analysis is a blend of inductive and deductive reasoning and judgment. It is considered to comprise *risk assessment*, *risk management,* and *risk communication* (Ahl et al. 1993).

In evaluating a proposal to import animals or animal products, the first step is to draw up a comprehensive list of all the pathogens (hazards) which could be associated with the species or commodity under consideration and then identify the possible routes by which these could come into contact with susceptible animals in the importing country. Once the hazards have been identified, *risk assessment* is the process estimating, as objectively as possible, the probability that the import would result in the entry of an exotic disease agent and that local livestock would be exposed to that agent. Risk assessment ought to examine the *effect* of the introduction of an exotic disease.

Risk management is the process of identifying and implementing measures which can be applied to reduce the risk to an acceptable level.

Risk communication is the process by which the results of risk assessment and risk management are communicated to decision makers and stakeholders. Adequate risk communication is essential in explaining official policies to stakeholders, such as livestock industry groups, who often perceive that they are exposed to the risks, but not the benefits of imports. Risk communication must also be a two-way process, with stakeholders' concerns being heard by officials and addressed adequately.

Risk assessment takes into account the prevalence of pathogens in the source population, the probability of their surviving in the animal or product during the process of importation, the probability of the pathogen coming into contact with local livestock after importation, and the seriousness of such contact. There is a substantial body of information on the survival of pathogens in many animal products (MacDiarmid 1991b, Morley 1993a). In theory, each of the other factors should be amenable to being quantified in a similar scientific fashion. In reality, it is often not possible to quantify them adequately, and much of the assessment is based on assumptions which are potentially controversial and open to challenge.

Risk management, on the other hand, is usually able to be quantified more objectively. For instance, there should be little debate over the sensitivity of a particular serological test, the efficacy of a particular embryo washing regimen for a specific pathogen on embryos of a given species, or the sterilizing effect of a given heat treatment.

3. MANAGING RISK

Consider, for example, a serological test applied to a group of animals to test for the presence of a particular disease. With some diseases, a policy decision may be made that a positive test result will disqualify only the individual animal which reacted positively to the test. However, with other diseases, usually those which have the potential to cause major economic damage, it may be decided that a positive test result in any one animal will disqualify the entire group intended for importation.

When only seropositive animals are rejected, the probability of including even one test-negative infected animal (c) in a group of n animals can be calculated thus (Marchevsky et al. 1989):

$$P(c \geq 1|N) = 1 - \left\{ \frac{(1-p)e}{(1-p)e + p(1-s)} \right\}^n \tag{1}$$

As the size of the group destined for import increases, so does the risk (Table 1).

Table 1 Probability That Test-Negative Infected Animal Be Included in Group Destined for Import

n	If only reactor animal excluded	If single reactor disqualifies group
100	4.92×10^{-2}	5.00×10^{-2}
200	9.61×10^{-2}	2.50×10^{-3}
300	1.41×10^{-1}	1.25×10^{-4}
400	1.83×10^{-1}	6.25×10^{-6}
500	2.23×10^{-1}	3.13×10^{-7}

Note: Prevalence = 0.01, sensitivity = 0.95, specificity = 1, entire group tested.

From MacDiarmid, S. C., Risk analysis and the international trade in animals and their products, in *Animal Health Economics — Principles and Applications,* Dijkhuizen, A., Morris, R. S., Eds., Post Graduate Foundation of the University of Sydney, Sydney, 1996, in press. With permission.

When a positive test result in any one animal will disqualify the entire group intended for importation, the probability of disqualifying an infected group increases as prevalence and/or the size of the group increases. The probability of a given test failing to detect at least one test-positive animal in an infected group, and thus failing to identify the group as infected, can be calculated thus (MacDiarmid 1988):

$$\beta = (1 - ts/n)^{pn} \tag{2}$$

Where t is the number of animals from the group which are tested.

The difference in risk between the two policies is illustrated in Table 1. It can be seen that where the presence of a single reactor animal disqualifies the entire group destined for export rather than just the reactor animal itself, the risks of an infected animal being imported are significantly reduced.

Whether a positive result to any one test disqualifies only the affected individual or the whole importation, the risks of importing unwanted disease can be further reduced by imposing a series of safeguards. When a series of safeguards is applied to an importation, it may be relatively easy to quantify the amount by which the risk is reduced, even if a consensus on the magnitude of the initial, unrestricted risk cannot be attained.

4. EXAMPLES

4.1 Surveillance Programs for Livestock Disease

Confronted with the need to reduce the cost of monitoring New Zealand's beef cattle for the disease brucellosis, we (MacDiarmid 1988) examined the risk of different surveillance programs failing to detect infected herds. Using Equation 2, we compared an automated complement fixation test conducted on serum samples collected on the farm, the same test conducted on samples collected at slaughter, and a delayed hypersensitivity skin test applied on the farm.

Considering the size of beef cattle herds in New Zealand, the percentage of animals from each herd slaughtered in a given period, and the number of infected animals recorded historically in herds identified as infected for the first time, the study concluded that the on-farm use of the relatively insensitive skin test had a greater probability of identifying infected herds than the more sensitive serological test applied to samples collected from animals at slaughter. While the application of the serological test to samples collected on the farm had a lower risk of failing to identify infected herds, the risk of failure from the application of the skin test was considered acceptable in light of its significantly lower cost.

4.2 The Risk of Introducing Anthrax by Importing Green Hides

A relatively simple example of risk assessment applied to the international trade in animal products is that which is used to estimate the risk of introducing anthrax into New Zealand through the importation of green hides from Australia.

Harkness (1991) considered that the annual probability (T) of anthrax introduction via the medium of unprocessed hides is related to the probability (p) that a hide contains anthrax spores and to the number of occasions (n) that susceptible animals are exposed to contact with those spores. The number of occasions that contact with spores causes infection follows a binomial distribution, so the chance of introduction of infection is

$$T = 1 - (1 - p)^n \quad \textbf{(3)}$$

However, when T is small, this approximates to

$$T = pn \quad \textbf{(4)}$$

To assess the probability of anthrax spores being present, the following assumption is also made.

$$p = ise \quad \textbf{(5)}$$

where:

i is the probability that an Australian animal was infected with anthrax at the time of slaughter. (The expected incidence of anthrax in Australia was calculated at 40 cases per year. The total number of sheep and cattle slaughtered each year in Australia was calculated to be about 40.23 million. The value of i was estimated therefore at $40 \div 40.23$ million $= 9.94 \times 10^{-7}$.)

s is the proportion of spore infectivity surviving preexport handling. (Since the spores of the anthrax bacillus are extremely resistant to environmental conditions, s was estimated as 0.9.)

e is the proportion of green Australian hides among all rawstock processed in New Zealand. (Approximately 38.4 million sheep and 3.1 million cattle are slaughtered in New Zealand annually, an estimated 31% of New

Zealand-produced hides and skins are processed in the country, amounting to 13.5 million pieces annually. The estimated annual import volume from Australia for green skins is 0.92 million. So, e was estimated at 0.92 m ÷ 13.5 m = 0.068.)

Thus, p was estimated as $0.068 \times 0.9 \times 9.94 \times 10^{-7} = 6.1 \times 10^{-8}$.

The number of occasions per year on which susceptible animals are likely to be exposed to contact with anthrax spores was calculated as follows:

$$n = gtvf \tag{7}$$

where:

- g is the number of approved tanneries in New Zealand ($g = 23$).
- t is the proportion of approved tanneries operating with a risk of contaminating pastureland by wastewater during flood periods. (No satisfactory information was available at the time the assessment was made. The average proportion presenting risk, t, was estimated to be 0.2.)
- v is the average number of days per year on which flooding occurs on pasture downstream of tanneries. (The value of v was estimated to be 25.)
- f is the probability of processing contaminated material during flood periods. (Calculated as average number of days of flooding divided by days worked per year, around 25 ÷ 235. Therefore, $f = 0.11$.)
- n was estimated as $23 \times 0.2 \times 25 \times 0.11 = 12.65$

Therefore, the calculations indicate that the probability of introducing anthrax in any one year is

$$T = 6.1 \times 10^{-8} \times 12.65 = 7.72 \times 10^{-7}$$

The risk is likely to be even lower when one considers that the probability of livestock encountering the anthrax organism on any contaminated pasture is less than 1 and that ante- and postmortem inspection at Australian abattoirs is highly effective in preventing anthrax cases from being processed for their hides.

Deterministic risk assessment models, such as the one just described, do not give the decision maker any estimate of the uncertainty of the risk estimate. A risk assessment method must include some estimation of the degree and source of **uncertainty** associated with predicting the likelihood of introducing an animal disease, as otherwise decision makers tend to focus on a single possible outcome. As most of the variables are only estimates of what is likely, the "real" risk estimate will be shrouded in uncertainty.

A Monte Carlo-type simulation model, using a PC software program such as @RISK (Palisade Corporation, Newfield, New York), allows each of the variables to be represented as a range of values and then, by a series of iterative calculations, presents the final risk estimate as a probability distribution.

4.3 Disease Transmission by Embryo Transfer

The development of embryo transfer techniques, whereby embryos may be recovered from donors in one country and frozen and transported internationally before being implanted into recipients in another country, has opened the way for the relatively low-risk introduction of new bloodlines. The importation of genetic material by embryo transfer carries considerably less risk of introducing exotic diseases than does the importation of live animals. However, while much evidence is available to indicate that many pathogens are unlikely to be transmitted along with embryo transfers, caution must be exercised because, in many cases, the number of experiments has been so small that the upper limit of a 95% confidence interval for the probability of disease transmission is still rather high.

To be 95% confident that transmission does not occur between viremic donor and susceptible recipient in, say, more than 1 transfer in 100, transfers would need to be carried out with negative results on 300 occasions. Such large-scale experiments are expensive and have been conducted for a relatively small number of pathogens only.

By taking into account factors such as sensitivity of the diagnostic test on the herd or flock of origin and on embryo-derived progeny, and the probability of the disease being transferred along with the embryo, one may calculate the risk of a particular disease entering a country through an importation based on an embryo transfer program and a policy of a single test positive disqualifying the entire shipment. In the approach adopted in New Zealand (MacDiarmid 1993), the first step was to calculate the risk that the disease of concern could be present in the source herd, but could escape detection by a serological screening test. Values for size of herd, test sensitivity, number of donors, average number of offspring/donor, and probability of transmitting disease by embryo transfer were incorporated into a Monte Carlo-type simulation model as distributions of values, using the PC software program @RISK.

Sutmoller and Wrathall (1995) used a more detailed model to assess the risk of foot-and-mouth disease transmission by transfer of bovine embryos imported from Brazil. Their model incorporated several more steps than that described by MacDiarmid (1993), but, like the latter, used triangular distributions incorporated into an @RISK Monte Carlo model. In the model of Sutmoller and Wrathall, the import risk was defined as the probability that one or more embryos carrying an infectious dose of foot-and-mouth disease virus would be included in a batch of 200 embryos. They concluded that the most likely risk of that happening with embryos from Brazil would be 1×10^{-11}, a very small risk indeed!

4.4 The Risk of Introducing Rabies Through the Importation of Dogs

As part of a review of policies regulating the importation of dogs, the New Zealand Ministry of Agriculture used a Monte Carlo-type simulation model to assess the risk of releasing a rabid dog from quarantine under each of a number of import policies based on quarantine periods of different duration, with or without verified vaccination status (Chief Veterinary Officer 1995).

The risk of selecting a rabies-infected animal, without any safeguard being in place, was estimated to be a function of the incidence of rabies among domestic dogs in the exporting country and the incubation period of the disease. The incidence and incubation period were described by triangular distributions.

Once the magnitude of the unrestricted risk of introducing a rabies-infected dog had been estimated for each country under consideration, the effects of vaccination and quarantine were assessed, again using a range of values in a simulation model. The effect of each safeguard is the product of the unrestricted risk and the estimate of failure of the safeguard.

The simulation model was run for 5000 iterations. Table 2 shows the estimated risks of introducing from France a dog incubating rabies following the application of different risk-reduction strategies. In 95% of iterations, the risk estimate was less than that shown in the table. The model also generated risk estimates for imports from several other countries proportional to the incidence of rabies in domestic dogs in those countries.

Table 2 Risks of Introducing from France a Dog Incubating Rabies

Safeguards	Infected dogs per million[a]
1 month quarantine	4.83
4 months quarantine	2.39
6 months quarantine	1.03
Vaccinated and 1 month quarantine	0.51
Vaccinated and 4 months quarantine	0.25
Vaccinated and 6 months quarantine	0.11

[a] 95% of the iterations of the @RISK simulation model produced estimates equal to, or less than the value shown.

On the basis of this risk assessment, it was concluded that vaccinated dogs imported without prolonged quarantine pose no greater risk of introducing rabies than dogs entering through a 6-month quarantine. By showing that prolonged quarantine could be replaced by vaccination without any reduction in security, risk assessment helped in developing a new policy which significantly reduced the cost of importation and eliminated the prolonged separation which was often so painful to pets and their owners.

4.5 The Risk of Introducing Fish Diseases in Table Salmon

For many years, the importation into New Zealand of ocean-caught Pacific salmon from Canada was prohibited because it was feared that such importations could introduce diseases into local fish stocks. To reevaluate this position, in 1994 the Ministry of Agriculture conducted an analysis of the risks of introducing exotic fish diseases through importation of salmon from Canada.

For table fish to serve as a vehicle for the introduction of fish disease, a chain of criteria must be met (MacDiarmid 1994). Taking these into account, a nonquantitative risk analysis led to the conclusion that of 23 diseases present in North American salmonids, furunculosis, caused by the bacterium *Aeromonas salmonicida*, is the disease which would be most likely to be carried in the type of commodity under consideration.

Having identified the disease most likely to be carried by the commodity, a quantitative risk assessment took into account what was known of the prevalence of *A. salmonicida* in wild Pacific salmon, the distribution and numbers of the organism found in infected Pacific salmon, the effect of processing on the number of *A. salmonicida* in the tissues of infected fish, the survival of the pathogen in the environment, the dose required to infect susceptible fish (of any species), and waste management practices in New Zealand.

The analysis concluded that there is negligible risk of introducing *A. salmonicida* through trade in chilled, headless, eviscerated salmon. The analysis concluded that there is a 95% probability that the risk is less than 1×10^{-7} per tonne imported. To put this into perspective, the analysis pointed out that the entire annual production of wild, ocean-caught Pacific salmon in British Columbia is no more than 100,000 t.

5. CONCLUSION

Even in situations where the risk from unrestricted entry can be quantified objectively, and little controversy surrounds the calculation of the extent to which safeguards reduce that risk, it may be difficult to attain agreement on what constitutes an acceptable risk.

In an assessment of the risk of various major diseases of livestock entering New Zealand through articles carried in the baggage of passengers or in the mail, Davidson (1991) considered a risk of less than one disease introduction per 100 years to be acceptable. However, using very different techniques, Forbes et al. (1994) estimated that the mean risk of foot-and-mouth disease entering New Zealand was about once in 50 years. That is, one regulator proposed a criterion for acceptability of risk that was less than what others actually estimated the risk to be. Observations such as this emphasize that the discipline of risk analysis, as it applies to the importation of animals and animal products, is still in its infancy. It will be some time before general agreement on what constitutes "acceptable" risk can be achieved.

Those regulating the importation of livestock and commodities derived from them must also strive for consistency in their decision making, and this is one of the obligations of membership in the WTO. If the disease risk posed by an importation from one country is considered acceptable, an importation from another country should not be declined if it poses no greater risk. An advantage of quantitative risk analysis is that it helps to ensure consistency. It also permits various risk management strategies to be compared.

While this chapter has concentrated on quantitative risk analysis, nonquantitative methods should not be forgotten. Until relatively recently, quarantine authorities tended to base decisions almost solely on nonquantitative risk analyses, and these still have a valuable role to play in the routine administration of imports, especially of animal products (Christensen 1994). Nonquantitative risk analyses can be objective, repeatable, and transparent and always take less time, and thus are less expensive, than quantitative analyses (Christensen 1994). Nevertheless, with increasing frequency, quarantine authorities are having to deal with import proposals which require a quantitative approach (Morley 1993b). While it has been possible in the past for regulators to avoid risk by refusing access, in the post-GATT environment this option is less acceptable, and so quarantine authorities around the world are beginning to adopt the discipline of quantitative risk analysis.

6. ACKNOWLEDGMENT

Sections of this chapter have been adapted, with permission of the publisher, from MacDiarmid, S. C., Risk analysis and the international trade in animals and their products, in *Animal Health Economics — Principles and Applications*, Dijkhuizen, A., Morris, R. S., Eds., Post Graduate Foundation of the University of Sydney, Sydney, 1996, in press.

REFERENCES

Acree, J. A., Beal, V. C., Animal health perspectives of international embryo exchange, *Animal and Human Health, 1*, 39, 1988.

Ahl, A. S., Acree, J. A., Gipson, P. S., McDowell, R. M., Miller, L., McElvaine, M. D., Standardization of nomenclature for animal health risk analysis, *Revue Scientifique et Technique de l'Office International des Epizooties, 12*, 1045, 1993.

Chief Veterinary Officer, Annual report 1994, *Surveillance, 22(3)*, 4, 1995.

Christensen, B., Non-quantitative risk analysis and the importation of animal products, *Surveillance, 21(4)*, 13, 1994.

Davidson, R. M., Assessment of risks posed by material of animal origin in mail and passengers' effects, *Surveillance, 19(4)*, 24, 1991.

Forbes, R. N., Sanson, R. L., Morris, R. S., Application of subjective methods to the determination of the likelihood and consequences of the entry of foot-and-mouth disease into New Zealand, *New Zealand Veterinary Journal, 42*, 81, 1994.

Harkness, J., Review of Conditions Applied to the Import of Hides and Skins into New Zealand, NASS Publication 91-3, Ministry of Agriculture and Fisheries, Wellington, N.Z., 1991.

Kellar, J. A., The application of risk analysis to international trade in animals and animal products, *Revue Scientifique et Technique de l'Office International des Epizooties, 12*, 1023, 1993.

MacDiarmid, S. C., Future options for brucelosis surveillance in New Zealand beef herds, *New Zealand Veterinary Journal, 36*, 39, 1988.

MacDiarmid, S. C., Risk analysis and the importation of animals, *Surveillance, 18(5)*, 8, 1991a.

MacDiarmid, S. C., The Importation into New Zealand of Meat and Meat Products; A Review of the Risks to Animal Health, NASS Publication 91-2, Ministry of Agriculture and Fisheries, Wellington, N.Z., 1991b.

MacDiarmid, S. C., Risk analysis and the importation of animals and animal products, *Revue Scientifique et Technique de l'Office International des Epizooties, 12*, 1093, 1993.

MacDiarmid, S. C., The Risk of Introducing Exotic Diseases of Fish into New Zealand through the Importation of Ocean-Caught Pacific Salmon from Canada, NASS Publication xx–x, Ministry of Agriculture and Fisheries, Wellington, N.Z., 1994.

Marchevsky, N., Held, J. R., Garcia-Carrillo, C., Probability of introducing diseases because of false negative test results, *American Journal of Epidemiology, 130*, 611, 1989.

Morley, R. S., A model for the assessment of the animal disease risks associated with the importation of animals and animal products, *Revue Scientifique et Technique de l'Office International des Epizooties, 12,* 1055, 1993a.

Morley, R. S. (Coordinator), Risk analysis, animal health and trade, *Revue Scientifique et Technique de l'Office International des Epizooties, 12*, 1000, 1993b.

Sutmoller, P., Wrathall, A. E., *Quantitative Assessment of the Risk of Disease Transmission by Bovine Embryo Transfer,* Scientific and Technical Monograph Series No. 17, Pan American Foot-and-Mouth Disease Center (PAHO/WHO), Rio de Janeiro, 1995.

QUESTIONS

1. Under what circumstances does GATT accept the imposition of measures restricting trade in animals or animal products?
2. When considering the importation of animals or animal products, what is meant by *risk assessment*?
3. Why are risk assessments on imports of animals or animal products sometimes controversial and open to challenge?
4. Why are the effects of risk management measures usually able to be quantified more objectively then risk assessments?
5. When a regulator adopts a policy of excluding only those animals which fail a test, what is the effect of increasing the size of a group of animals intended for import?
6. Where a single animal reacting to a test disqualifies the entire shipment, what is the effect of increasing the size of the group intended for import?
7. What is the main weakness of deterministic risk assessment models?
8. While there is evidence that many pathogens are unlikely to be transmitted by the practice or embryo transfer, why is it that one cannot be totally confident that disease will not be introduced along with imported embryos?
9. What are the advantages of quantitative vs. qualitative risk assessment?
10. Why do nonquantitative risk analysis methods still have a role in the regulation of imports, especially of animal products?

CHAPTER **IV.5**

Incorporating Tribal Cultural Interests and Treaty-Reserved Rights in Risk Management

Barbara L. Harper

SUMMARY

Risk assessment is increasingly being used as a primary analytical tool in risk-based decision making. It incorporates implicit and explicit values, biases, presumptions, and even, due to the specific parametrics selected for analysis, risk management goals themselves. Thus, both the technical methodology and the values basis of risk assessment must be examined for their adequacy in addressing different cultural perspectives in general and the rights and interests of sovereign American Indian nations in particular. Conventional risk assessment is especially inadequate for assessing unique tribal activity and exposure patterns and risks to tribal cultures, health, and identity. Further, the overall risk management framework frequently lacks holistic and coherent goals, as well as a process for ensuring equal access to the decision process. Specific examples are provided that relate to risk-based land-use planning and remediation.

Several solutions are presented here, including the comparative risk approach as a basis for evaluating a wide range of risks, evaluation of risks and impacts to the "ecocultural-landscape," and criteria used by the technical staff of the Confederated Tribes of the Umatilla Indian Reservation of northern Oregon for evaluating potential impacts to sovereignty and environmental, human, and cultural health.

Key Words: American Indian, cultural risk, comparative risk, landscape, equity, sovereignty, values

1-56670-130-9/97/$0.00+$.50

1. INTRODUCTION

Risk assessment is increasingly being applied to pollution control and remediation decisions, particularly in the context of cost-risk-benefit analysis and land-use planning. While there are certain advantages in using such methods to prioritize remedial actions and develop risk-reduction strategies, conventional assessment methods and decision processes are plagued by inherent limitations in their ability to incorporate unique cultural perspectives and the rights and interests of affected communities, particularly those of sovereign American Indian nations. Credible, technically defensible, and politically acceptable risk management strategies will result only if reformed risk assessment practices and open risk management processes fully embrace the perspectives and values of communities directly affected by such decisions.*

The issues described in this chapter have been identified as particular concerns to the technical staff of the Confederated Tribes of the Umatilla Indian Reservation (CTUIR 1993a,b, 1994b, 1995), but are likely to be applicable to many other community situations. Risk assessment increasingly comprises the principal technical decision tool for federal agency decisions about off-reservation activities that may have critical implications or impacts both on-reservation and in off-reservation ceded lands where tribes have sovereign rights reserved to them to use resources and pursue traditional activities. Major federal facilities within tribal ceded lands include the Hanford Nuclear Site in southeastern Washington (the most severely contaminated site in the Western Hemisphere) and the Umatilla Army Depot in northeastern Oregon (site of 12% of the nation's chemical and nerve agents stored under deteriorating conditions and slated for on-site incineration). The tribal reservation is downwind and downriver from both these facilities, putting at further risk the resources that tribal members have depended on for thousands of years.

Several major areas of deficiency have been identified in the overall risk assessment/risk management process: (1) lack of recognition of the range of risk information needed to provide a strong decisional information base; (2) growing recognition that conventional methods and metrics do not provide adequate details about impacts to tribal health and community well-being; (3) the need for a higher integrative perspective for combining diverse types of risk information into a format useful for both stakeholders and risk managers; and (4) growing recognition that personal values and (un)recognized biases of the assessor and manager are implicit or explicit throughout the risk assessment and management process (CTUIR 1995).

Conventional risk assessment is typically focused on "environmental safety and health" (ES&H) risks, overlooking much of what is actually at risk. Risk sources may directly impact not only human health and the environment — a particular

* This raises the point that Western science and indigenous science often have different criteria (rules of evidence, or ways of knowing) for establishing the validity of knowledge (Stoffel and Evans 1990), especially for impacts to tribal ecocultural human health. Risk assessment is exceptionally vulnerable to this conflict because it is inherently predictive, untestable, and value laden. Technical "experts" are often allowed to validate both the methods and the results, while those who have been risk assessed are limited to protesting this presumption of validity. Any resulting modifications in the methods, however, are likely to improve the accuracy of conventional (i.e., "approved") approaches by including factors that were heretofore overlooked.

concern to subsistence-dependent tribal families — but also tribal cultural values, traditional tribal lifestyles, and tribal cultures themselves for many generations to come. These risks are not often accounted for with existing methodologies, thus resulting in decisions which are "unstable" due to an inadequate information base. Impacts beyond ES&H risks are not just "considerations" to be used in risk management activities, and they are definitely different from conventional definitions of "perceived risk";* they are real risks that require an analysis that is just as rigorous and systematic as that for ES&H risks and that belong in the same quantitative risk framework (National Research Council 1994, Vermont Agency of Natural Resources 1991, California Environmental Protection Agency 1994).

There is also a more basic deficiency in the entire Western approach to environmental management, and this is also seen in toxics risk assessment and management. An indigenous worldview would seldom rely first or solely on a risk-based approach to either toxics management or land-use planning without first committing to principles such as sovereignty, protection, equity, and sustainability. In other words, the entire decision context must be framed using the worldview (especially views about sustainability, balance, cyclical time, and reciprocal relations) of the indigenous community, because it is logically inappropriate to use a Western context for evaluating impacts to indigenous values and cultures (Margolis 1993, Duran and Duran 1995, LaDuke 1993).

Several solutions are presented in this chapter, and they include suggestions for setting values-based integrated ecocultural risk management goals (particularly for complex remedial sites with multiple risk sources and multiple trustee resources); for redefining the risk information needs to include appropriate culture-specific parametrics; and for using concrete, but holistic evaluation criteria as "systems requirements." Whether the decision involves holistic conservation or prioritization ("cultural triage"; Stoffle and Evans 1990), these solutions should be useful.

Potential Tribal Risk Model Characteristics

1. Sovereignty and Treaty-Reserved Rights: CTUIR has sovereign authority to, among other things, protect treaty-reserved rights, to promote and enhance tribal self-determination and cultural integrity, and to protect tribal and individual rights to pursue traditional activities, including religious and cultural practices, both on-reservation and in off-reservation ceded areas and beyond.
2. Tribal, state, and federal governments, and their natural and cultural resource agencies, are responsible for protecting conditions and resources required for the above practices. Comanagement and codecision making by sovereign nations and other trustees is an absolute requirement for technically defensible and politically acceptable decisions.

(Table continued on next page)

* Conventional risk approaches tend to evaluate "human health, environmental impacts, and perception"; or "hazard (i.e., real risks) and outrage (i.e., unreal risks)"; or "cancer risk, ecological toxicity, and knowledge/dread" (see, for example, Morgan et al. 1994); or "human health, habitat disruption, and the social response to perceived risks" (see OSTP 1995). None of these approaches evaluates cultural risk correctly, because an evaluation of cultural risk bears little, if any, resemblance to an evaluation of potential health symptoms due to anxiety and fear which may arise, in part, from recognition of danger (even though neurophysiological symptoms are very real health effects and should be included in the portion of the analysis that addresses direct health risks).

Potential Tribal Risk Model Characteristics (continued)

3. The fundamental goal of strategic land-use planning should be long-term, culturally appropriate integrated ecocultural management. The fundamental principles of such plans are sovereignty, protection, equity, and sustainability.
4. Types of information that must form the risk information base after a principle-based mission plan is developed:
    a. Environmental/ecological integrity and quality
    b. Human health effects (including multigenerational)
    c. Individual and community sociocultural/religious well-being
    d. Temporal and spatial descriptors for each of the above

2. DEFICIENCIES IN CONVENTIONAL RISK-BASED DECISION MAKING FROM A TRIBAL PERSPECTIVE

Especially if a "course of action" at complex waste sites is composed of hundreds or thousands of individual decisions about risk, cost, and schedule, it is important to develop (and enforce) a set of risk principles that reflect the perspectives of the impacted communities. However, decision rules alone do not guarantee adequate participation of sovereign nations, nor do they guarantee that tribal perspectives are understood, much less used in the decision process. A truly open process will ensure that "interested and affected parties" are involved throughout the decision process and that their values, perspectives, rights, and goals frame and guide the decision process from policy development through problem formulation to decision implementation. It will necessarily shift some of the decision authority to tribal councils or other trustees/stakeholders and will require some initial investment of time and effort on the part of the responsible agencies to establish an open comanagement process. However, this will ultimately be more cost-effective over the long term than approaches such as "decide-announce-defend," "respond-to-comments," or "develop a utilitarian equation and let the computer optimize" (the "science tells us that . . ." approach).

2.1 Risk Management Goals of Achieving Affordable, Acceptable, or Allowable Risk Levels May Not Satisfy Principles of Equity, Protection, or Sustainability

Risk management goals and risk assessment assumptions generally reflect the perspective of the decision maker or risk manager. Risk management goals (e.g., achieving "acceptable risk," "allowable risk," or "affordable risk") are inherently value based, but are seldom developed democratically. A given level of risk may not be acceptable to stakeholders, but may be "allowable" under some statutes or "affordable" under others. Frequently, the terminology used to set risk management goals is confused, thus, for example, mistakenly equating safety or protection with available budget.

The basic problem statement of a decision process is often too narrow, and a coherent goal or mission plan is often lacking. It may not be clear whether the goal is to be health protective, cost-effective, or utilitarian (health-per-dollar-effective). This type of confusion may lead to questions such as "How little do I have to clean?" (also stated as "Don't clean up what doesn't make sense") or "What level of protection can I afford?" A narrowly focused risk manager may attempt to force a decision into a simplistic zero-sum format (for example, "More expensive remediation or less land use?"). This immediately creates competition among potential land users, especially between industrial users (who may tolerate "brownfield" cleanup standards) and prior-in-time-and-right users such as sovereign Indian nations for whom the land and its resources are supposed to be held in trust by the U.S. government for members to safely use "for as long as the grass should grow."

Risk management methods of "trading" one type of impact for another are also contrary to indigenous worldviews, because people and their culture are, in reality, inextricably intertwined with the natural environment (Figure 1), with no component being of greater or lesser intrinsic value than any other component. Failure to recognize this cultural dichotomy has resulted in a long history of paternalistic policies on the part of government and technology and paternalistic actions on the part of professional "experts" (Lowrance 1985).

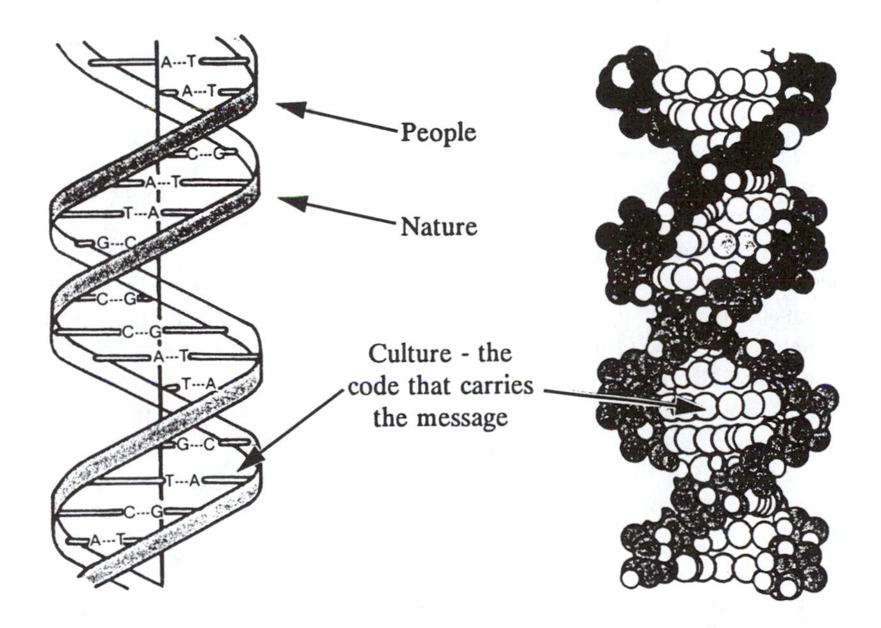

Figure 1 The "double helix" of risk assessment. People and nature are intimately linked by culture/religion, and an evaluation of all three is necessary in order to develop an appropriately comprehensive and holistic information base relevant to tribal health. (Modified from Office of Technology Assessment, 1986. "Technologies for Detecting Heritable Mutations in Human Beings." Washington D.C., p. 24.)

2.2 Ethical, Legal, and Social Issues Are Required Parts of the Information and Planning Base, Not Just a Final Clearance Step or Part of Postdecisional Stakeholder Acceptability

Values should guide the development of the overall problem statement; the selection of metrics; the collection, analysis, and integration of data; the construction of the information base; the selection of decision criteria; and the ultimate implementation of the decision. The evaluation of ecological and cultural risks is not a step to be postponed until the action is ready to be deployed in the field, because their evaluation encompasses much more than merely avoiding further harm (or minimizing future harm) to localized natural or cultural resources during implementation. This process actually begins with a values-based analysis of the available alternatives that will accomplish the mutually agreed upon goals. If protection of natural and cultural resources is perceived by managers solely as an end-of-process filter, this may result in, at best, project delay and stakeholder outrage and, at worst, project abandonment. Rather, the original mission statement should, at a minimum, include specific goals related to the ethical and sociocultural issues that will ultimately determine the degree of acceptability of the decision. This is particularly true when so many factors that affect "health" lie outside conventional Euro-industrial medical boundaries (Lowrance 1985) and exert a strong political or interpretive influence regardless of the weight of the technical evidence.

2.3 Particularly as Risk Results Are Presented as Point Estimates Within Risk Ranges, Uncertainty Must Also Be Managed

Technical uncertainty is sometimes considered analagous to stakeholder perception. The assessor typically addresses technical uncertainty by collecting more data, while the manager seeks to reduce the amount of perceived risk with more communication or education. Both data and communication are thought of as improving the accuracy of the risk estimates, but this is not entirely true for either case. The collection of more detailed data within the original restricted categories is less important than collecting the appropriate breadth of data at lower precision levels. Similarly, the education of risk assessors and managers about cross-cultural perspectives and about the need to modify "approved" risk assessment methods and presumptive risk management goals may be more difficult than ensuring that a community group (or its experts) has a sufficient level of technical understanding to participate meaningfully in the decision process (Silbergeld 1991, Shrader-Freschette 1991).

2.4 Principles of Environmental Justice Require Changes in the Fundamental Goals of Risk-Based Decision Making and the Practice of Risk Assessment

At least four factors tend to disproportionately increase risk to American Indian health from environmental contamination: (1) dose (potentially increased exposure due to cultural lifestyle activities), (2) response (potentially increased physiologic sensitivity due to genetic makeup, existing health conditions, or concurrent exposures), (3) mitigation (possible decreased access to health care, insurance compensation, and other forms of postharm amelioration), and (4) cultural well-being (potentially disproportionate impacts to individual and tribal community health and identity and cultural values). In addition, the responsibility of the present generation toward future generations (regarding long-term impacts of long-lived radioactive contaminants, for example) requires a description of the temporal risk profile and an evaluation of multigeneration and cumulative impacts. Conventional risk assessment addresses none of these systematically.

3. SPECIFIC DEFICIENCIES IN EVALUATING IMPACTS TO TRIBAL HEALTH AND IDENTITY

Narrowly scoped risk analysis methods tend to omit metrics related to unique use of treaty-reserved resources, unique (nonsurburban) lifestyle activities and exposure pathways, and ecocultural health and tribal identity. Omission of a data integration step and a description of the temporal risk profile may be compounded by other faulty assumptions to further distort the risk picture. Without correcting these deficiencies, it is not possible to evaluate the potential for a disproportionate burden of risks to fall on tribal communities through time. However, if these (and other) deficiencies are corrected, then risk assessment can indeed be one useful tool for risk management, but only after overall integrated, holistic goals and value-based decision criteria are established.

3.1 Unique Use of Treaty-Reserved Resources for Subsistence, Ceremonial, Cultural, or Religious Practices Must Be Evaluated with Tribal Guidance

Tribal members use numerous sources of food and other ceremonial, medicinal, and material resources that are not commonly used by the dominant society and are thus ignored in conventional risk assessments. Given the close relationship between nature and tribal peoples and their cultures, a complete understanding of contaminant exposure could only be obtained by charting whole ecosystems, as well as the cultural practices related to gathering and using many resources. Consideration of dietary factors alone includes a myriad of nonsuburban plants and animals (along with a variety of plant and animal parts not part of the suburban diet), seasonally fluctuating consumption rates that would cause peaks in contaminant intake rates, a variety of storage and preparation methods, and a higher proportion of locally obtained food than typical default exposure factors (U.S. EPA 1989) used in conventional assessments.

Further, many species serve multiple purposes (food, medicines, and materials). For example, the common cattail has many uses: in the spring the shoots are eaten, the roots are consumed, and the pollen is used in breads later in the season. The fibrous stalks are used in woven items such as baskets, in which other foods may be stored or cooked, or mats used for sleeping and shelter (Harris 1993, 1995). Thus, even describing multiple food uses does not necessarily describe all the ways people interact with even a single species. Further, even if it were possible (and only with tribal permission) to compile a catalog of dietary and medicinal species, biouptake and bioaccumulation factors are largely unknown for individual species. A more appropriate approach may be to start with an assumption that a given proportion (higher than the standard suburban default assumptions; U.S. EPA 1989) of the total diet is obtained locally, then "anchor" the assessment with key species for which contaminant uptake, contaminant bioaccumulation, foodchain transfer, and human ingestion rates are known.

In addition to the evaluation of direct and indirect foodchain exposures, part of an impact evaluation must include consideration of the loss of the traditional diet (including protein, vitamins, fiber, and so on) which is physiologically optimal for the people who have undergone millenia of genetic adaptation.

3.2 Unique (Nonsuburban) Lifestyle Activities and Exposure Pathways Can Be Assessed Only in Direct Consultation with Local Tribes

Cultural practices that are integral components of a traditional lifestyle may also result in increased exposure potential. Certain cultural, ceremonial, and spiritual practices, such as sweat lodges, are unique to tribal peoples and present multiple exposure pathways not addressed by conventional risk analyses. In addition, conventional parameters (such as the duration and frequency of time spent outdoors) may need to be increased to account for particular lifestyle practices. Again, a preferred approach begins with a recognition that exposure assumptions should be increased over suburban default levels, rather than attempting to catalog the myriad of individual, confidential, and tribal- or clan-specific activities. Activity patterns and therefore exposures may also differ substantially with age and gender, making it important to anchor generic parameters with local knowledge chosen by tribal members to represent particular lifestyles or activities of critical importance.*

3.3 Evaluations of Ecocultural Health and Cultural and Spiritual Values Are Core Elements in the Tribal Risk Information Base

The term "cultural risk" has been used in at least three ways. In the narrowest sense, it means impacts to cultural and historic sites and resources during project implementation. It may also include traditional activities and skills or knowledge, although this interpretation varies among applications. There are, in fact, significant issues relating to definitions of historic property, traditional cultural property, cultural

* As with specific exposure data, it should be recognized that all resulting information belongs to the affected tribe and can only be developed and used under their direction; the data do not belong to the assessor or ethnographer. At some point, too, it becomes ethically improper to pursue scholarly inquiry to the point of intrusion (Toelken 1995), especially if the degree of improvement in "data quality" does not provide a comensurate benefit to the people whose lifestyles are being publicly examined, possibly without their full knowledge or informed consent. In this context, "benefit" does not mean increased "accuracy" in toxicity/exposure data and, as a consequence, relaxed pollution controls and increased allowable exposure levels, but rather some real increase in protection or the provision of health services (using the broadest definition of health).

resources, cultural values and practices, and cultural risk. In addition, exactly what constitutes an adverse effect (physical, chemical/radiological, aesthetic, legal, institutional, and so on) is often an issue. In a broader sense, cultural risk also includes impacts to cultural values and to cultures themselves and is similar to quality-of-life definitions used in comparative risk projects. In some assessments, cultural risk is misused to mean solely a culture-specific social and behavioral response to risk — this reflects a perceptually limited understanding of non-Euro-American cultures (i.e., sociological imperialism; Duran and Duran 1995) that perpetuates cross-cultural communication problems and paternalism and can even exacerbate adverse effects on tribal health.

Tribal identity includes culture, religion, and place; if the link between the environment and the people is broken, the culture/religion is also broken (Figure 2). Tribal health includes personal well-being, which is derived from membership in a healthy community with strong traditional values, and the ability to follow traditional lifestyle, healing, religious, and educational practices in nondegraded surroundings. Since tribal culture/religion is inseparable from the place of origin, full and safe access to these places and their natural resources is required so that the cultural values of critical significance to the American Indian and her/his local community are preserved (Harris 1995).

Traditional tribal cultural practices evolved over long-term, sustainable associations between human and nonhuman species and their environment. The environmental landscape shapes modes of thinking, feeling, and behaving in a way that goes beyond mere survival. Language, culture, and religious symbols all coalesce together at particular locations in forms that reflect the unique local patterns of the naturospiritual realm. The people respond with a corresponding social organization and living religion that are unique to the area and inseparable from it and that follow the area's natural rhythms and demands. This not only provides a time-proven effective design for sustainable survival, but also represents a way of knowing that reinforces a feeling of real presence in the environment and a continual awareness of the harmonious coexistence of the material and spiritual realms that Euro-Americans seldom achieve (Jahner 1989, Bennett 1993).

3.4 Faulty Land-Use Assumptions in the Mental Model Bias the Outcome

Land-use and exposure assumptions can bias the outcome of the risk assessment tremendously. For instance, the (highly questionable) presumption that institutional controls and restricted access will be enforced for as long as contamination remains (thereby preventing exposure and risk) precludes the use of typical residential exposure scenarios and the evaluation of subsistence or other cultural-based activities and would likely lead to incorrect measures for evaluating progress in risk reduction. For instance, one might declare a site "safe for unrestricted surficial recreational use," while actually leaving in place a substantial amount of surface, subsurface, and groundwater and/or surface water contamination that could pose ecological and cultural risks and could also pose unacceptable human risk under reasonable tribal use scenarios, particularly over long time periods.

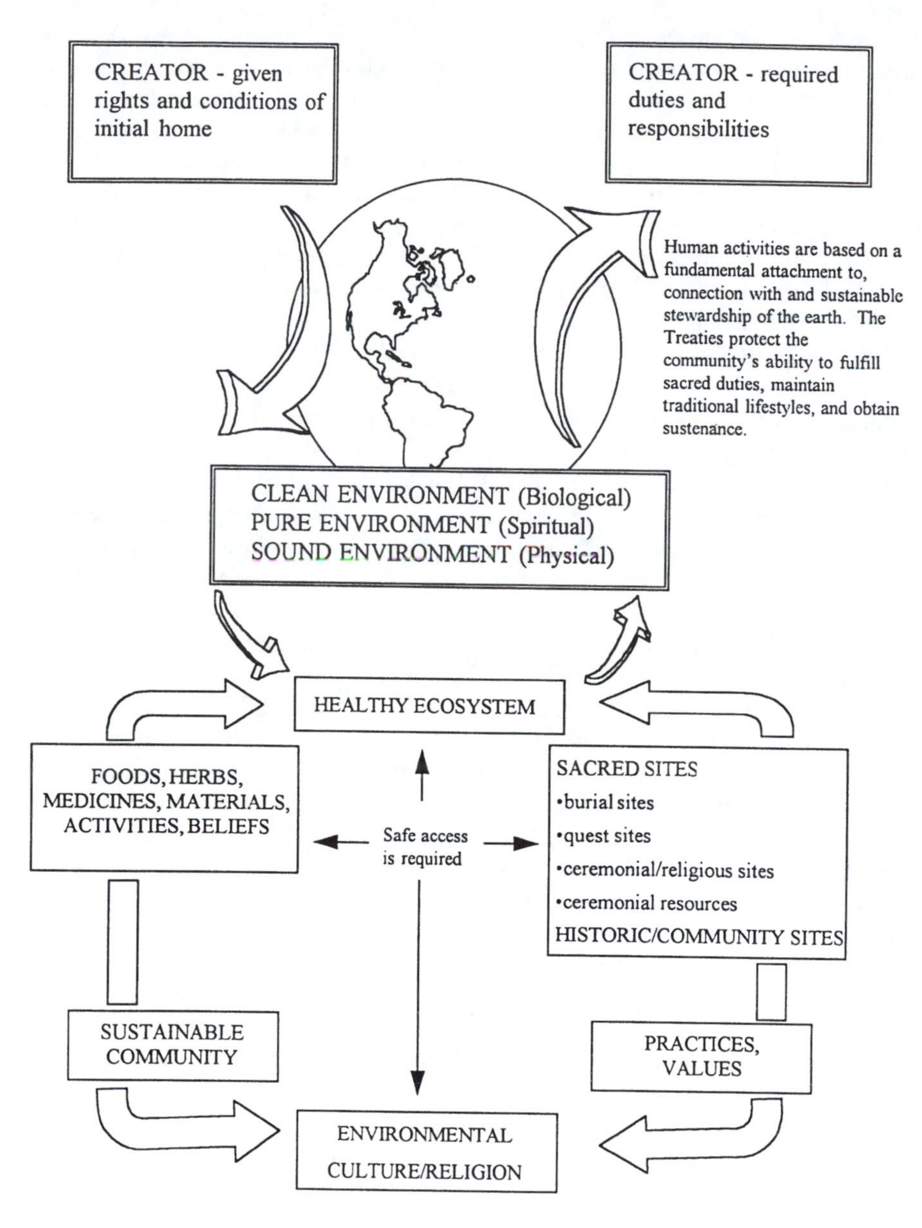

Figure 2 The Natural Law, illustrating why full and safe access to a healthy ecosystem is necessary for tribal cultural/spiritual health. The term "treaties" refers to the various treaties between Indian nations and the U.S. government, under which natural and cultural resources necessary for a healthy environment and traditional lifestyle will be protected by the U.S. government in perpetuity for tribal people. (With thanks to Russell Jim and Robert Cook, Yakama Indian Nation, and Stuart Harris, Confederated Tribes of the Umatilla Indian Reservation.)

Using a conventional narrow risk definition as justification for institutional controls, one could conclude that there is indeed no risk if there is no exposure. However, using the broader concept of risk, it is clear that such "mitigation" (i.e., breaking the exposure pathway) also breaks the land-culture link, which is both an immediate and a cumulative adverse effect on sovereign rights and the ability to safely follow traditional cultural practices. Risk managers may assume that this effect represents a zero-impact planning baseline, or that it is an "affordable" impact compared to other impacts, or even that preventing exposure by forbidding access to heritage lands provides a "net benefit." Similar arguments have been applied to natural resources (e.g., that contamination and restricted access may "protect" habitat from physical disturbance) and cultural resources (e.g., that contaminated gravesites are "protected" from looting). In at least one case, it has been proposed that "mitigation" of cultural impacts could occur through consultation with tribal members and payment for lost spiritual ceremonies on sites that are targeted for destruction through resource exploitation, to the abhorrence of traditional tribal peoples (Hall 1994). It should be recognized that loss of a natural/cultural resource is not a one-time impact borne by a single generation, but is a cultural harm that continues to accumulate over time.

4. SOLUTION: EVALUATE IMPACTS TO THE ECOCULTURAL LANDSCAPE

4.1 Whether the Decision Context Calls for Strategies to Prevent, Mitigate, Protect, Remediate, or Restore, Principles of Integrated Ecocultural Management Still Need to Be Followed

The basic premise of this approach to strategic planning and impact evaluation is that integrated environmental management must be combined with concepts of cultural landscapes and environmental justice into an integrated ecocultural management approach (Figure 3). The spatial dimensions include surface and subsurface ground, groundwater and surface water, and air and biota; due to influences from and on nearby geologic and natural features, these boundaries may extend beyond reservation, ceded, or traditional use boundaries. The temporal dimension includes cumulative past effects, present impacts (including future impacts deriving from present conditions), future impacts, and cumulative multigeneration effects. The ethical dimension may extend far beyond minimal legal requirements for trust resource protection and intergovernmental consultation.

Land-based decisions begin with a rigorous characterization of land and its cultural and natural resources and include the evaluation of current and potential impacts by stressors to environmental integrity and to human physical, sociocultural, and spiritual health associated with use of those resources. Stressors include physical, radiological, or chemical contamination and aesthetic impacts, including byproducts and side effects of actions or responses. With this wider evaluation, a different

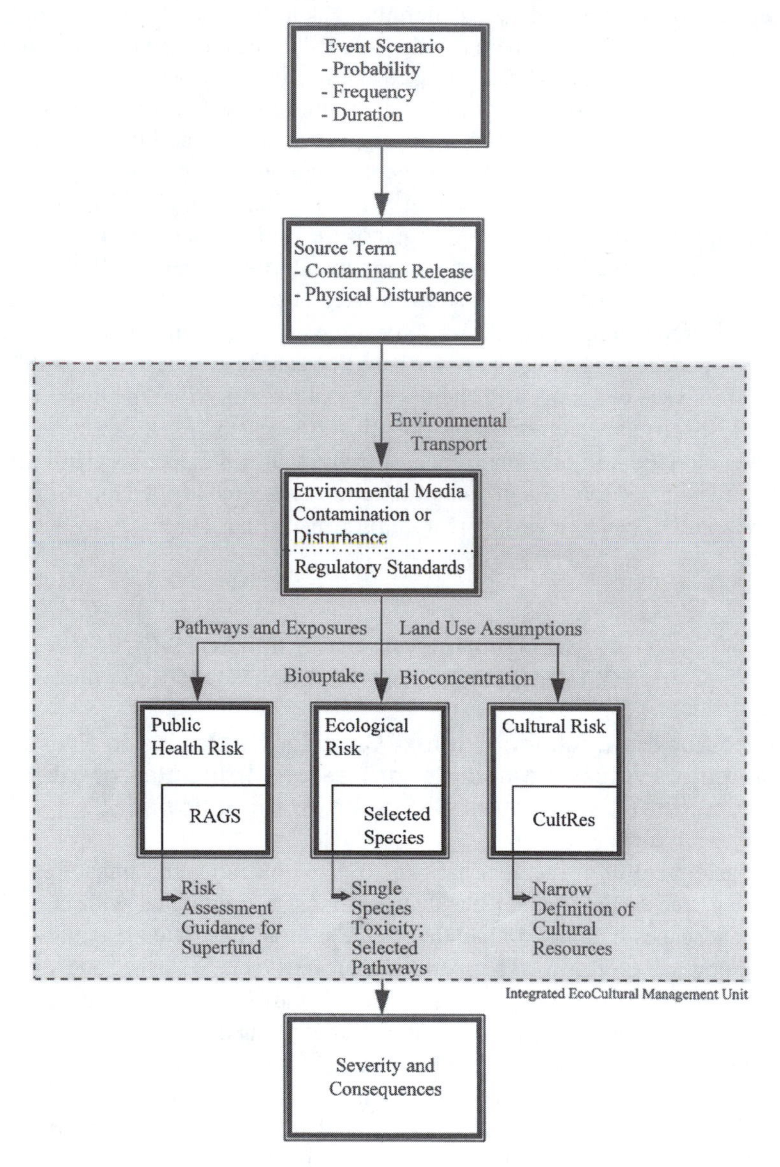

Figure 3 An ecocultural management unit. The shaded areas within the four components of the ecocultural unit indicate that, from a holistic tribal perspective, conventional methods or standards address only a portion of what is "at risk." Environmental impacts that are significant to tribal members may occur even when regulatory standards are not violated; RAGS Superfund guidance (USEPA 1989) is not appropriate for traditional lifestyles; single-species ecological toxicity does not address habitat and other landscape-scale impacts; a narrow legalistic definition of cultural resources ("stones and bones") does not reflect cultures and cultural values that may be at risk. Note that "severity" and "consequences" are not the same: severity is a (more or less) objective indicator of the level of harm that could occur to a given resource, while consequences measures severity plus the importance (weight) of the affected resource.

decision might be reached; for example, preservation or restoration of cultural/religious integrity may, in fact, be a key decision driver, and cleanup standards might be developed for ceremonial quality as well as for human health.

Principle: Temporary solutions to remedial actions may have lower short-term project costs, but higher cumulative natural resource and sociocultural compensation costs. Interim and final states of remediation, restoration, and disposal must be determined with trustees during the problem definition stage.

4.2 A Land-Use Plan Should Focus on Integrated Ecocultural Management Goals (Nonconflicting, Risk-Based Priorities and Remediation/Restoration Goals Then Can Be Established for Individual Risk Sources or Proposed Actions)

If mission statements are phrased in holistic ecocultural terms, then specific goals will be more coherent and integrated, regardless of the specific application. For instance, if the mission is to evaluate either prospective (e.g., under NEPA) or retrospective (e.g., under CERCLA) impacts, then information across the entire span of environmental/ecological/human/sociocultural risks would strengthen the information base. If the mission is to design remediation and restoration strategies, then the result would be a long-term integrated approach (some or all of which might be risk based), rather than piecemeal or project-by-project mitigation. If the mission is to choose among technical options, one would start with an "Alternatives Assessment" (O'Brien 1994) to reflect the full range of stakeholders' underlying goals and key issues (Keeney 1992) before developing risk-based standards and selecting a preferred alternative. Finally, if the mission is to develop land-use plans, then end state land uses might include risk-based criteria for an equitable and sustainable combination of restored treaty-reserved rights; long-term growth management; conservation/preservation; environmental resource use; economic development; and protection/enhancement of health, safety, and quality of life.

Neither "risk reduction" nor "land release" would be primary goals of a land-use plan — they are secondary to the primary goal of equitable and sustainable integrated ecocultural management. Only after value-based management principles have been established should risk-based evaluations (spanning the entire range of risk types) be used to prioritize actions for individual risk sources and to establish remedial and restorative goals relative to overall health protectiveness and cost-effectiveness.

Principle: In a land-use planning context (especially for complex sites), it is inappropriate to rely on a risk-based land-use approach without first developing an integrated, holistic, principle-based mission statement and site-wide plan. Temporally phased and spatially fragmented cleanup and land release actions should not proceed until comprehensive value-based goals are established. Tribal perspectives start with holistic goals and then move to specific objectives directed toward established goals and end states; they do not start with fragmented actions that are pieced together to construct some semblance of a whole plan.

5. SOLUTION: APPROACHES FOR HOLISTIC RISK EVALUATION

5.1 Comparative Risk Projects

Several comparative risk projects (U.S. EPA 1993) have evaluated impacts to quality of life, human health, and the environment. In particular, the Vermont (1991), California (1994), and Wisconsin Tribes (U.S. EPA 1992) projects stand out as examples where community values guided the selection of metrics for evaluating impacts ranging from human and environmental health to socioeconomic factors and aesthetics. The Wisconsin Tribes project modified conventional risk assessment concepts to accomodate unique tribal lifestyles and subsistence activities, overall tribal culture, natural resource use, cultural and religious values, and tribal priorities. Even so, the predetermined framework for the analysis perpetuated some of the limitations related to the difficulties in evaluating temporal factors, equitable distribution of risk, and long-term sustainability indicators. However, the Wisconsin Tribes project demonstrates that it is indeed possible to modify conventional parameters and develop additional ones that together provide a much more complete and satisfactory description of risk.

5.2 Specific Examples of Ecocultural Risk Evaluation: Map Based and Parameter Based

Two approaches are under development at the Pacific Northwest National Laboratory, Richland, WA, that attempt to accommodate tribal perspectives on human-ecocultural risk. One approach uses geographic information system (GIS) data layers relating to a variety of ecological resources (some of which may be threatened and endangered and some of which are not endangered, but are of critical importance to local tribal members) and identified cultural/historical resources. As work proceeds, human health risk "isopleths" using tribally developed exposure scenarios and modeled contaminant concentrations over time will be added. In addition, a "heritage" map indicating general areas of special importance to Hanford Site nations may also be developed. The philosophical issue here is that while it is necessary to relate impacts to tribal health, culture, and identity directly to the land, it may be improper to attempt to "map" cultural values at all, since any zonation implies a judgment as to relative importance of certain species or relative sacredness of different areas.

A more conventional approach has been to develop parameters reflecting ecocultural values expressed by local tribes, in addition to others modified from comparative risk projects. This approach also has limitations of being overly numerical and thus losing some of the cultural meaning behind the parameters, of inadvertently biasing the evaluation by the selection and wording of individual parameters, of including too little active participation by tribal staff, and of implying that one can prioritize some values over others. Both the map-based and parameter-based approaches do provide methodological starting points, however, and encourage the use of initial value statements to guide the development of parameters.

6. SOLUTION: THE LINK BETWEEN THEORY AND PRACTICE — "CTUIR CRITERIA" APPLIED WITHIN GEOGRAPHIC, GEOSPHERE, BIOSPHERE, AND ETHICSPHERE BOUNDARIES

The meaningful exercise of tribal treaty rights is entirely dependent on a healthy ecosystem; a right to fish or gather plants is hardly useful if the fish and plants themselves have vanished or become contaminated, or if the resources have been damaged to an extent that further exercise of rights will cause unacceptable injury to the resources (CTUIR 1993a).

An adequate evaluation of impacts to tribal sovereignty, environmental, cultural, and personal health requires a holistic and integrated approach that conventional risk assessment and management lack. As described previously, natural resources form the basis of traditional diets, ceremonies, material items, recreation, trade, and other cultural activities and practices. All indigenous plants and animals have religious significance to people who practice traditional Indian religion. People, culture, and nature evolved together and coadapted over many millenia; impacts to any one of these affects overall tribal health and identity, because impacts to a single resource may have ramifications for human health, environmental integrity, and religious use.

General criteria for evaluating impacts spanning the range of concerns discussed earlier are shown in the following table. Additional principles can be enumerated for specific proposed actions, such as "do not prejudice future options" through the choice or irretrievable waste forms or through the use of physical barriers between long-lived radioactive or chemical contaminants and the environment that must be replaced every 100 years for the next 10,000 years.

CTUIR Criteria for Evaluating the Impacts of Proposed Actions

1. Protection of tribal sovereignty including protection of tribal rights in ceded territory and areas over which CTUIR exercises off-reservation treaty rights in perpetuity
2. Protection and restoration of the environment, including the resources required for full and safe exercise of on- and off-reservation treaty rights
3. Protection of cultural, religious, and archeological resources; cultural integrity and heritage; the conditions necessary for traditional, subsistence, or religious activities (including aesthetic or spiritual qualities of an area or resource); tribal identity; and related tribal rights
4. Protection of the reservation and its members, including future generations, from hazards originating in off-reservation ceded lands or elsewhere

The spatial and temporal dimensions of such an evaluation may not stop at the boundary of the reservation or ceded territory, but may extend for as far distant as the resource (aquifers, habitat, and so on) and its buffer zones extend and for as far and as long as the impact persists on the land, natural resource, and human base of a whole and holistic community. It includes all environmental media (biotic and abiotic) and all uses, adaptations, and effects. It includes considerations of ancillary and cumulative impacts to ecocultural (including aesthetic) resources related to the exercise of treaty rights in either space or time. Finally, as recognition of a "global village" increases, an American Indian set of environmental ethics is required as the basis of a safe, healthy, equitable, and sustainable future for us all.

REFERENCES

JW Bennett, 1993. *"Human Ecology as Human Behavior. Essays in Environment and Development Anthropology."* Transaction Publishers, New Brunswick, NJ.

California Environmental Protection Agency, 1994. "Toward the 21st Century: Planning for the Protection of California's Environment." Final report by the California Comparative Risk Project, Sacramento, CA.

CTUIR, 1993a. "Criteria for Evaluation of Proposed Changes to the Hanford Federal Facility Agreement and Consent Order." CTUIR Department of Natural Resources, Pendleton, OR, July 21, 1993.

CTUIR, 1993b. "Comments on the Columbia River Impact Evaluation Plan." CTUIR Department of Natural Resources, Pendleton, OR, September 3, 1993.

CTUIR, 1994a. "Scoping Issues for the Environmental Restoration Disposal Facility." CTUIR Department of Natural Resources, Pendleton, OR, February 18, 1993.

CTUIR, 1994b. "Development of Draft Hanford Sitewide Groundwater Remediation Strategy and Perceived Impediments to Its Effective Implementation." CTUIR Department of Natural Resources, Pendleton, OR, May 11, 1994.

CTUIR, 1995. "Scoping Report: Nuclear Risks in Tribal Communities. A Report by the Confederated Tribes of the Umatilla Indian Reservation Outlining Concerns about Risk-Based Approaches to Environmental Management Decision-Making." CTUIR Department of Natural Resources, Pendleton, OR.

E Duran and B Duran, 1995. *"Native American Postcolonial Psychology."* SUNY Press, Albany, NY.

K Hall, 1994. "Impacts of the Energy Industry on the Navajo and Hopi." In: RD Bullard, ed. *"Unequal Protection. Environmental Justice and Communities of Color."* Sierra Club Books, San Francisco.

SG Harris, 1993. "The Nez Perce ERWM's Recommendations for Refinement of Risk Assessment Proposed by DOE's Columbia River Impact Evaluation Plan." Nez Perce Tribe Department of Environmental Restoration and Waste Management, Lapwai, ID, September 3, 1993.

SG Harris, 1995. "A Limited Sample of Concerns of the Confederated Tribes of the Umatilla Indian Reservation Community on Using An Appropriately Defined Risk Assessment Model."Appendix B of CTUIR, 1995 "Scoping Report: Nuclear Risks in Tribal Communities." CTUIR Department of Natural Resources, Pendleton, OR.

E Jahner, 1989. "The Spiritual Landscape." In: DM Dooling and P Jordan-Smith, eds. *"I Become Part of It. Sacred Dimensions in Native American Life."* HarperSan Francisco, HarperCollins Publishers, New York.

RL Keeney, 1992. *"Value Focused Thinking. A Path to Creative Decisionmaking."* Harvard University Press, Cambridge, MA.

W LaDuke, 1993. "A Society Based on Conquest Cannot Be Maintained. Native Peoples and the Environmental Crisis." In: R Hofrichter, ed. *"Toxic Struggles. The Theory and Practice of Environmental Justice."* New Society Publishers, Philadelphia, PA.

WW Lowrance, 1985. *"Modern Science and Human Values."* Oxford University Press, New York.

H Margolis, 1993. *"Paradigms Are Barriers: How Habits of Mind Govern Scientific Beliefs,"* The University of Chicago Press, Chicago, IL.

G Morgan, B Fischoff, L Lave, P Fischbeck, S Byram, K Jenni, G Louis, S McBride, L Painton, S Siegel and N Welch, 1994. "A Procedure for Risk Ranking for Federal Risk Management Agencies." unpublished report, Carnegie Mellon University, Pittsburgh, PA.

National Research Council, 1994. *"Building Consensus. Risk Assessment and Management."* National Academy Press, Washington, D.C.

M O'Brien, 1994. "A Proposal to Address, Rather than Rank, Environmental Problems." In: AM Finkel and D Golding, eds. *"Worst Things First? The Debate over Risk-Based National Environmental Priorities."* Resources for the Future, Washington, D.C.

Office of Technology Assessment, 1986. "Technologies for Detecting Heritable Mutations in Human Beings." Washington, D.C., p. 24.

OSTP, 1995. "Risk Assessment Research in the Federal Government." Draft discussion paper by the OSTP Committee on Environment and Natural Resources, Subcommittee on Risk Assessment; Mark Schaefer (lead author). Executive Office of the President, OSTP, Washington D.C., March 7, 1995.

KS Shrader-Freschette, 1991. *"Risk and Rationality. Philosophical Foundations for Populist Reforms."* University of California Press, Berkeley.

EK Silbergeld, 1991. "Risk Assessment and Risk Management: An Uneasy Divorce." In: DG Mayo and RD Hollander, eds. *"Acceptable Evidence. Science and Values in Risk Management."* Oxford University Press, New York.

RW Stoffle and MJ Evans, 1990. "Holistic conservation and cultural triage: American Indian perspectives on cultural resources." *Human Organization* 49(2): 91–99.

B Toelken, 1995. "Fieldwork enlightenment." *Parabola* 20(2): 28–35.

U.S. Environmental Protection Agency, 1989. "Risk Assessment Guidance for Superfund, Volume I: Human Health Evaluation Manual (Part A)." EPA/540/1-89/002.

U.S. Environmental Protection Agency, 1992. "Tribes at Risk. The Wisconsin Tribes Comparative Risk Project." Office of Policy, Planning and Evaluation, EPA/230-R-92-017.

U.S. Environmental Protection Agency, 1993. "A Guidebook to Comparing Risks and Setting Environmental Priorities." Office of Policy, Planning and Evaluation, EPA/230-B-93-003.

Vermont Agency of Natural Resources, 1991. "Environment 1991: Risks to Vermont and Vermonters." Vermont Agency of Natural Resources, Waterbury, VT.

QUESTIONS

1. What are the elements of comparative risk?
2. What is the difference between precision and accuracy?
3. How would you define health from different cultural perspectives (e.g., absence of clinical symptoms or holistically whole and well-balanced)?
4. How would you use a holistic definition of health to define "things at risk" from environmental contamination?
5. What are some risk management goals that might differ depending on the cultural perspective?
6. How might restricted access to traditional use areas cause harm to a culture?
7. Think of ways that would result in increased exposure to environmental contaminants for a person pursuing a traditional (subsistence) lifestyle.
8. Considering the various types of risk (human health, environmental, and socio-cultural quality of life), how effective is breaking the human exposure pathway in

reducing overall risk? How would your answer differ if you narrowly define risk solely as the probability that symptoms could occur after exposure to a given contaminant concentration (i.e., the old dogma that no exposure = no risk)?

9. Suppose a particular contaminant is at an environmental concentration that results in a 10^{-4} lifetime cancer risk to an individual, but the contaminant has a half-life of 1000 years. Describe how you would estimate risk and present risk information about (1) the risk to one person, (2) the risk to the current population, and (3) the risk to future generations (or everyone ever exposed to the contaminant). Does risk accumulate over time? If so, how would you determine when this risk becomes acceptable?
10. How would you characterize the validity (or accuracy) of qualitative data gathered from traditional experts (i.e., elders)?
11. Describe the difference between allowable risk (risk allowed under regulations), acceptable risk (acceptable to whom?), and affordable risk.
12. How might chronic (multigenerational) exposure to low levels of mutagens affect a small gene pool?

CHAPTER **IV.6**

Global Use of Risk Analysis for Sustainable Development

Vlasta Molak

SUMMARY

Sustainable development is defined as "integrated strategies that would halt and reverse the negative impact of human behavior on physical environment and allow for livable conditions for future generations on Earth" (UNCED 1992). The concept of sustainable development resulted in Agenda 21, developed at the Earth Summit in Rio de Janeiro in 1992, which presents a blueprint for development of humanity in the 21st century agreed upon by a majority of countries on earth. The majority of human societies (countries) are not sustainable, since they are highly dependent on fossil fuels and nonrenewable resources which are being rapidly depleted.

Environmentalists often dispute the causes of global environmental degradation. Some blame it on population growth [P] in developing countries (SOUTH), some on unbridled consumption [C] in developed countries (NORTH), and some on the polluting technology [T] and low energy efficiency (EAST and SOUTH) and/or large consumption of energy (NORTH) [E]. In reality, all of the four factors are important contributors, and modified Ehrlich' equation can be written as

$$\text{Environmental Impact (EI)} = P \times C \times T \times E$$

where P = population, C = consumption, T = technology, and E = energy consumption (E). (This is qualitative and not an exact mathematical expression.)

The importance of each factor in any locality varies. Western Europe and North America contribute mostly to consumption and energy use [C and E], while in

1-56670-130-9/97/$0.00+$.50

developing countries the major factors are population (P) and polluting technology (T) (in those countries that are attempting to industrialize). Also, often, the energy efficiency is very poor (China and the former USSR), even though useful energy consumption per person may be small. Countries of the former USSR and Eastern Europe have stable and limited populations, and consumption is also low. However, their outdated technology and poor energy efficiency (T and E) are major contributors to the tremendous environmental degradation of Eastern Europe uncovered after the fall of communism.

In order to deal with the environmental impacts of human activities and of development, one needs first to evaluate the problems. Risk analysis can help evaluate the problems resulting from improper industrial management, transport, and energy management; consider the proposed solutions; and establish better management strategies that would promote sustainable development. Decision- and policy-makers (businesses, governments) and the "doers" (engineers, architects, city planners, etc.) could perform risk analysis before they make investments and execute projects with a potentially high environmental impact. Comparing various options of risk management would result in the most reasonable solutions. In addition, properly performed and documented risk analysis may uncover hidden agendas of various interest groups and thus promote democratization within a frame of sustainable development. It is important that the decisions about such policies and developments are made at the local or national level where people will benefit (or be harmed) by such decisions, rather than by remote decision-makers whose sole motivation is short-term gain, and who will not suffer from the consequences of their decisions should anything go wrong.

Key Words: sustainable development, Agenda 21, energy, pollution, environmental impact, agriculture, efficiency, nutrition, transportation, population, wars

1. INTRODUCTION

Governments and the societies around the world are at the crossroads of environmental decision making which will determine the futures of the next few generations and possibly even the survival of humanity. They can proceed in an "as usual style," trying to get investments without determining the long-term environmental and economic impact of building new industries, transport, electric power plants, agriculture, etc., or they could learn from the mistakes and positive experiences of the western democracies and the local cultures in their environmental management. ***Better business through better environmental management is the new paradigm for 21st century, which enlightened governments and businesses are starting to understand.***

The number ONE environmental problem in the Western world is compartmentalization (Molak 1990). Compartmentalization along the lines of scientific disciplines and along the environmental media contributes to piecemeal solutions of environmental problems that often create bigger problems than those they are attempting to solve. Over the years, various disciplines have dealt with environmental problems from a very narrow scope. A chemical engineer would design a plant without worrying what to do with waste or how the emissions from the plant would affect the neighborhood. That problem was left for the environmental engineer, toxicologist, or lawyer years after the plant started operation. Toxicologists are studying mechanisms of action for various chemicals with little awareness of which chemicals are most likely to cause problems. Lawyers are demanding clear "yes" or "no" answers when scientific evidence can suggest only possibilities, and the answer is "maybe." To solve the problem of compartmentalization, "cultural change" is necessary both in government and society at large. An interdisciplinary scientific activity such as risk analysis may become a part of this cultural change.

When dealing with environmental problems such as global warming, acid rain, ozone hole, chemical contamination, resource depletion, deforestation, agricultural degradation, etc., we are attempting to obtain a complete picture using all available data to perform risk analysis or in some cases cost–benefit analysis. For example, a large interdisciplinary study on realistic mitigation options for global warming evaluated costs of various measures that could be undertaken in the United States to decrease amounts of greenhouse gases (CO_2, CH_4, NO_2, CFCs, and others) (Rubin et al. 1992). They evaluated electricity efficiency, industrial energy use, transport efficiency, and power plants. The authors calculated that with the current technology one could reduce the CO_2 equivalent production in the United States from 7900 Mt/year to less than 6000 Mt/year by just improving efficiency. The cost of such improvements would be actual savings, ranging anywhere from savings of $62/t conserved to spending $1/t conserved (in case of increase of the capacity of nuclear power plants). Generally, the largest saving could be achieved in residential and commercial energy use by improvements in efficiency. Measures to improve efficiency range from designing new household appliances such as refrigerators, furnaces, and efficient light bulbs to simple solutions such as planting trees around houses and having white roofs to deflect the sun in summer, or use of the sun to heat water or prepare food. Additional reduction in CO_2 emissions of 3600 Mt/year could be achieved by changes in transportation habits; developing new electric supply technology such as solar, wind, biomass, hydroelectric, and others; decrease in ruminant animals and nitrogen fertilizers; and reforestation. Some of these measures may require initial investment that will take longer to recover. Therefore, small alterations in the way we conduct our daily lives could have a doubly beneficial effect; decrease in fossil fuel consumption, decrease in greenhouse gases production, and saving in energy costs. A small investment would further substantially reduce our dependence on fossil fuels and decrease U.S. energy consumption by 75% without decreasing the quality of life. On the contrary, the quality of life would improve because of a decrease in pollutant emissions and an increase in exercise. This would be achieved without radical changes in transportation policies. With changes in transportation policy, where personal vehicles would be replaced by efficient public transportation systems, bicycles, and increased walking, the energy consumption could be further reduced. ***This study clearly demonstrates that energy efficiency is the most economic way of reducing greenhouse gas emissions and that the only barrier in their implementation are institutional barriers and an unwillingness to change (Rubin et al. 1992). A side benefit of decreased energy consumption would***

be also a decrease in acid rain, which is the result of the use of coal in electric power plants. Although this study does not specifically analyze risks of high energy consumption (which can be expressed in global warming and/or acid rain), it indicates that risk reduction would be a side benefit of energy management.

In former communist and developing countries, which are in the stage of forming new institutions, the implementation of energy efficiency in all types of activities is essential. During rebuilding or developing new infrastructures, one can implement all the findings on energy efficiency, rather than trying to imitate the Western way of high consumption. ***It may be more prudent to increase energy efficiency than to develop fast exploitation of natural resources in Russia and other East European countries and further contribute to CO_2 and other greenhouse gasses emissions. Risk analysis of various energy options in Eastern Europe can point toward most optimal developments of energy resources and their most efficient use.*** Based on Western experiences with transport problems, those countries may be able to avoid becoming dependent on cars and thus contribute to sustainability.

In an interesting way, it appears that humanity, in order to survive, has to adopt the problem-solving methods used by some native Americans: a "medicinal circle approach" in which various wise tribal people contribute to finding a solution and "considering the seventh generation" when making an important tribal decision. ***An interdisciplinary vs. simplistic approach and long-term vs. short-term reasoning seems to be the only way out of the environmental catastrophe toward which humanity seems to be heading. Risk analysis can provide a framework for this type of reasoning and contribute to the improvement of the quality of life in communities around the world.***

2. ENERGY AND POLLUTION

One of the most critical areas of human activities is the use of energy for heating, cooling, performance of mechanical work, and for transport (Molak 1990). Risk analysis can help to promote rational decision making in energy production and use policy most beneficial for the environmental quality (Molak B. 1990). For example, in order to have less sulfur dioxide pollution in Zagreb (Croatia), a decision was made to use oil (with low sulfur content) in the thermal plant. Individual apartment buildings are still heated by using high-sulfur coal. However, a risk analysis of the problem reveals that the bulk of pollution comes from individual houses, and the best cleanup strategy would be to continue burning a high-sulfur coal in the thermal plant and installing a scrubber for sulfur, but to use oil in the residential buildings. In that way, the cost of the scrubber would soon be made up with savings from fuel (coal is cheaper than oil). Additionally, the use of more expensive fuel (oil) in residential buildings would encourage energy efficiency because consumers would have an incentive to conserve energy. Similarly, locally based risk analysis could be used to determine the best energy production and use management. Energy production and use can be evaluated for its impact on the environment using risk analysis techniques, and choices should be made for the cleanest and most efficient energy production for a particular location (be it coal powered, solar, hydroelectric, wind, geothermal, nuclear, or other). Risk analysis could help make the most optimal choice by uncovering all the available data and putting them in the proper framework. Comparative risk analysis studies in the past pointed out the relative advantages of nuclear energy, based on number of deaths associated with unit energy production. However, the Chernobyl accident had altered those numbers, since thousands of deaths were attributed to radiation exposures in the aftermath of the accident.

World statistics on uses of coal, oil, gas, hydroelectric, nuclear, wind, and solar energy indicate some encouraging trends (*Vital Signs* 1994). Since 1990, oil production has leveled off and even decreased, as did coal production. Nuclear energy is becoming increasingly unpopular, with electricity production declining due to the age of many reactors. Natural gas production has increased, as has the use of wind energy and solar energy. These positive sustainable trends could be accelerated by proper use of risk analysis in evaluating available energy options in a given area.

Energy efficiency can be enhanced on a national level by developing more energy-efficient houses and household appliances. Instead of building light fixtures or other electrical appliance factories in the old way, technology of high-efficiency and long-duration products should be purchased from the best vendors. In addition, architects and builders could incorporate energy-efficient designs in mass production of both individual and apartment buildings. Reliance on local wisdom and materials at the site may be more prudent than attempting to imitate large, energy-demanding building development which has occurred in the last 50 years in the United States and other developed countries. The world lending institutions, which are responsible for many developmental projects, should incorporate long-term environmental and human health risk analysis in all proposed projects. Many of these projects in the past would not pass the test of risk analysis, since the long-term consequences of their implementation would be uncovered. For example, building large dams, such as Aswan or dams in Brazil, resulted in tremendous negative environmental impacts, while benefiting few people and only for a short time (Molak 1990).

Since the fall of communism, new opportunities toward sustainable development have presented themselves to the World Bank and other international lending institutions and private investors. Economic risk analysis may use methods of green accounting in analysis of trends rather than the GNP, which is generally a crude and often false measure of the well-being of the population (Gore 1994, *Vital Signs* 1994). Countries in tropical and subtropical climates may be able to directly capture the sun energy rather than depend on oil or other nonrenewable energy resources. Suggestions made in Rubin et al. (1992) could be implemented in each new energy initiative. Using risk analysis may provide the necessary rationale for a change (or maintenance of status quo). However, to derive all the necessary data to perform risk analysis, a free access to information must exist. Only in transparent democratic societies, with checks and balances, can possible hidden agendas of national and local decision makers be uncovered and the public be empowered to take the course of action that would benefit society as a whole rather than promoting short-term profits for small numbers of individuals, while destroying a country's resources and prosperity. Agenda 21 principles of sustainability could be incorporated into free-market mechanisms by creating policies that would encourage energy efficiency and renewable energy uses (UNCED 1992).

3. TRANSPORTATION

"Monsters are prowling our land. They are devouring irreplaceable resources at increasing rates. Their wastes are fouling our air and may even be changing the climate of our planet. And they may doom our children and grandchildren to a world of unimaginable problems" (Union of Concerned Scientists letter to members 1992). The monsters are cars, which are using almost 80% of the oil in the United States and in most cities account for over 60% of total air pollution (the rest comes from power plants and chemicals production and use!). Unfortunately, ~60 years ago in the United States, many efficient public transport systems were destroyed by car makers and oil companies that needed more consumers for their products. Tram lines were paved over, and tram cars were destroyed or retired into museums. Also, U.S. passenger railroad traffic was practically eliminated. Europe has fortunately maintained its public transportation systems, its trains, and the compactness of its cities. Eastern European countries could thus learn from the United States and European experience, either to maintain or develop an efficient public transportation system (including trains) and discourage use of personal vehicles or follow the fate of many U.S. cities where one cannot function without a car. ***Risk analysis can be used to point out the costs and environmental consequences of current road building trends in the United States and recently in Eastern Europe and developing countries, where valuable agricultural land is being lost in order to build roads, parking spaces, and shopping malls creating a nonsustainable society.***

Even if a nonfossil fuel car is invented, it would still contribute to traffic congestion in the cities. The only long-term solution to transportation problems is public transport and increased use of walking and bicycles. Bicycles are the most economical way of transport: per mile traveled a single passenger expands 50 times less CO_2 than when using a car (World Watch Institute 1991). Therefore, with developing of new streets or repaving old streets, bicycle and pedestrian lanes should be added, especially in flat areas. Gradually, a whole bicycle and pedestrian infrastructure could be built, during routine repaving of roads or during new developments, without large expenditures. ***Creating more livable communities, where the need for commuting to work is decreased either by making communities more compact or by use of computers in working from home is another way of dealing with transport problems.*** World statistics (*Vital Signs* 1994) point out some encouraging signs of change around the world. The car production has declined to 34 million (in 1993), while bicycle production is on the rise (108 million in 1993). In developing countries, most people cannot afford cars, while in developed countries the environmental awareness of citizens is gradually leading to policies away from the unbridled use of cars. In Japan and Denmark, combining bicycle and train commuting to work is very popular (in Japan 3 million commuters chose this option and in Denmark 25 to 30% of all train commuters come to the train by bicycle). ***Risk analysis could encourage the shift toward bicycles by illustrating the hidden costs of car use, such as costs of building and maintaining roads, hidden costs of oil, and costs of accidents on the roads (compared to those of public transport or bicycling).*** For example, the world death toll of car accidents is ~500,000 people per year, and in the United States alone more than 40,000 people die each year (*Vital Signs* 1994). Risk analysis indicates that with mandatory seatbelts and lower speed limits the number of deaths in the United States decreased (from 55,000 in 1970) even though the number of miles driven increased. Thus, any policy that deals with transportation could be evaluated in advance. For example, simple risk analysis indicates that seatbelt wearing and speed limits should not be abolished, unless one is prepared to deal with an increase of car accident deaths and injuries. If one also takes into account the high costs of injuries, one may make a compelling case for decreasing car use and finding alternate ways of moving people around or decreasing the necessity for commuting by building more compact and sustainable communities.

Based on analyses of CO_2 production and air pollution, one should consciously and energetically strive to remove "THE MONSTER" (cars) from the roads to the greatest extent possible and educate the public (which now accept cars as a status symbol) that cars lead to an overall degradation of the environment and dependency on foreign oil (Lowe 1994). ***Public transport by more efficient vehicles, walking, and bicycle use should be encouraged by developing national and local policies that would liberate free-market mechanisms in service of sustainability and building livable communities. At this point, most of the policies encourage the use of cars, and free-market mechanisms are tampered with by the policies in which municipalities and states bear the costs of new roads and infrastructure development and maintenance, while road builders and housing developers are the major recipients of the benefits.*** Our dependence on foreign oil increases the real cost of oil (since war machinery is necessary to maintain the access to oil fields, as was demonstrated in Desert Storm), although this cost is not reflected in prices, as it should in a free market. In addition, costs from air pollution attributed to cars were estimated to be $5 to 85 billion per year (Cannon 1990). Risk analysis can point out the hidden costs of car dependence and thus encourage municipalities to develop alternate transport means. Experts in risk communication could devise the best "persuasion scheme" to accomplish this goal. Since advertising is a tool companies use to increase their market share, municipalities could use it for promoting behaviors useful for sustainable development. The success of such an approach was demonstrated in Cincinnati in the summer of 1995, when ozone in the city air compelled city officials to request citizens to ride buses in order to save the city from consequences of ozone-nonattainment (Molak, personal communication). Within several days the bus ridership was increased by 15 to 20%. Decreases in the price of tickets also helped. ***Thus, equitable free-market mechanisms could be used as a driving force behind the sustainability principle implementation in transport (Colby 1990).***

Cars are one of the major causes of environmental crisis in the United States, resulting in dependence on foreign oil, air pollution in the cities, etc. Inefficient and short-sighted urban planning results in traffic jams and long-distance commuting to the suburbs, which are inaccessible without cars. Additionally, large areas of natural habitats are lost to suburban sprawl, roads, parking spaces, and shopping malls, leading to the disappearance of species. Lack of opportunity to walk and bicycle on the streets leads to a degeneration of American health due to nonuse of muscle, in addition to health effects of polluted air.

4. INDUSTRIAL POLLUTION

Traditionally, industrial pollution has been used as an indicator of environmental degradation. The amount of toxic waste generated in the United States is very large. According to the Community Right-to-Know data, companies release 100 lb of toxic chemicals per citizen every year (Molak 1989). This consists of chemicals released into the air and water and disposed as solid hazardous waste. However, according to the national Resource Conservation and Recovery Act (RCRA), the real amount of toxic waste is probably 2000 lb (950 kg) per year, since Toxic Release Inventories report only 5 to 10% of total releases because of various exclusions in the law. This pollution affects the quality of air in the vicinity of industrial facilities, surface and sometimes drinking water, and communities living near production plants and hazardous waste sites (Molak 1989). An effort is made to decrease industrial pollution in the United States. ***While in the past an emphasis was on pollution controls (filters and scrubbers at the end of the pipe or water filtering before releasing), current wisdom is to reduce pollution at the source (U.S. EPA 1990).***

Industrial production facilities can be built with closed circles of material flow, zero waste, and just-in-time delivery (see Chapter II.3). Comparative risk assessments of various products could be performed following "the cradle-to-grave" principle, and only those products should be chosen which have a minimal impact on the environment and fit into geobiochemical cycles of earth. The products could be designed for real needs rather then artificially created needs promoted by inaccurate and often false advertising (Fromm 1981). Such production would also decrease the need for consumption, which frequently serves as a substitute for lack of fulfillment of community and other social needs (Rifkin 1991). While tampering with market mechanisms is dangerous, it is essential to recognize that those mechanisms depend on government policies and thus promote policies that develop a more equitable and sustainable society (Hayek 1976). Only environmental laws established in the last 20 years have compelled free markets to encourage responsible corporate citizenship. Thus, dissolution of those environmental laws may lead us back to rivers bursting in flames (see Chapter III.5). Economic risk analysis can indicate that pollution prevention pays, and, thus, it also makes good business sense (Molak 1991). While investments into pollution prevention could, in some cases, be immediately recovered by savings in raw materials, longer-term policies are necessary for pollution prevention encouragement on a wider scale.

A possibility of acute toxic releases of toxic chemicals used in production of various products can be evaluated and every step taken to minimize the impact on the neighboring population and workers. A type of risk analysis called "hazard analysis" according to the Community Right-to-Know law could also help prevent accidents as it is doing in the United States (U.S. EPA 1979).

4.1 Pollution Prevention Pays

The law that made pollution prevention possible is the Community Right-to-Know law or SARA, Title III. Spurred by the 1984 accidental release of the toxic chemical methyl isocyanate in Bhophal, India, which killed over 3500 people,

SARA, Title III is intended to increase the protection of the American public from the adverse effects of hazardous chemicals used, produced, and released from factories and transported via highways, railroads, and waterways. In 1988, for the first time in history, it became known to the American public and officials how much toxic chemicals get thrown into the air, soil, and water of America. This awareness encouraged both private citizens and U.S. EPA officials to require industry to decrease the release of toxics. Reacting to the public outcry and realizing that waste reduction will also save money on raw chemicals, industry in America is evaluating its industrial processes and deriving more efficient engineering solutions for reduction of waste at the source. ***Since pollution can be viewed as resources distributed in the wrong places, pollution prevention equates to a decrease of resource waste, thus saving money.*** Numerous examples showed that pollution prevention leads to decreased production costs and increased profits (Molak 1990, 1991). Large companies in the United States, such as 3M, Dow Chemical, Monsanto, P&G, and others, all have pollution prevention programs that have already significantly reduced those companies' toxic wastes and releases. Further reduction is encouraged by various U.S. EPA programs.

5. URBAN PLANNING

The main cause of ecological impact and species extinction is destruction of natural habitats caused by development. It is essential that urban planning be carefully used before developing or enlarging a city. Current policies both on the local and federal level encourage urban sprawl, since larger profits are made by building on agricultural land than on existing city grids, a result of unwise policies where the municipalities bear the costs of transport to new developments. ***Thus, free market is only apparently free; in reality, current policies and subsidies promote artificial incentives toward nonsustainable development.*** (See Chapter I.7.) The size of a unit city can be small enough to encourage public transport, bicycle lanes, walking, car pooling, etc. Free-market mechanisms could be devised to limit the size of the community by a green belt around the city that would prevent developers from building new remote houses. The best example of such a city is Portland, Oregon (Lowe 1992). Planning of new building developments has been coordinated with increases of public transport activities; therefore, although Portland experienced a doubling of its work force downtown, the number of private cars entering downtown each day stayed the same. This is mostly attributed to improvement in city planning and public transportation. Bicycles could become vehicles of the future for 1- to 15-km trips, which are frequently the extent of commuting mileage of suburban dwellers. ***In addition, urban renewal of inner cities may decrease the need for new suburb development, which is harmful in the long run. Perhaps incentive for redevelopment of inner cities rather than new suburb development could be designed by the cities. In addition to planning the location of new housing units, urban planning could be used to promote energy-efficient materials and experimentation with the newest energy-efficient houses, rather than old-type architecture of energetically demanding high rises and houses. Local developmental***

policies can incorporate sustainable development concepts into each step of urban planning and provide incentives for sustainability using various market mechanisms rather than rigid uniform rules.

In order to minimize commuting from home to work, urban planning should be developed that takes into account the waste of time and energy in commuting. The factories, offices, and other workplaces can be located in areas most accessible to employees. Very sophisticated models in urban planning already exist in the United States and Western Europe and have been tried in several cities in the United States (for example, Portland, Oregon). World Watch Institute's report, "State of the World 1992," describes some of the progress made in urban planning and other activities that can lead to sustainable society.

6. AGRICULTURE

There is a movement among health-conscious consumers in the United States to buy "organically" grown fruits and vegetables, which has encouraged the development of many "organic" farms (Robbins 1990). These farms do not use pesticides, herbicides, or chemical fertilizers, but instead use organic fertilizers (manure), crop rotation, and mechanical weeding. The yield on these farms is somewhat smaller than at conventional farms, but this is made up by saving on chemicals (which are becoming more expansive, since the insects are becoming resistant to pesticides and larger amounts of pesticides are needed) (*Vital Signs* 1994). In this case, risk analysis may indicate the potential for worker and food contamination and make a stronger case for use of organic agriculture. Consumers are willing to pay more for certified organic products because they do not want to ingest pesticides and herbicides and because of higher nutritional value (Robbins 1990). Also, no-till agriculture is being implemented, which decreases soil loss, one of the major problems in the world (World Watch Institute 1990).

A gradual return to sustainable agriculture can be implemented, with a slow decrease in the use of pesticides and herbicides. More health-oriented diets could be developed, based on whole grains, vegetables, and fruits, with a decrease in meat consumption, which takes ten times more resources to produce than vegetables for grains (with the same nutritional value) (Robbins 1994). For every meat eater, three to four vegetarians could be fed, which would practically eliminate any food shortages currently existing (provided that the food is grown in the right places). One of the problems in current agriculture is that developing countries have become dependent for their food on Western nations, partially because of high population growth, but also because of developmental policies encouraged by lending institutions promoting cash crops rather than sustainable food supply. Such cash crops are energy intensive and lead to all the pitfalls of chemically based agriculture and soil loss. Again, a comprehensive risk analysis, both for human health effects and for environmental impacts, can provide a framework for more sensible policies to encourage sustainable development.

Numerous epidemiological studies and studies in animals have demonstrated a strong correlation of nutrition and various (especially degenerative) diseases

(Campbell 1992, O'Connor and Campbell 1987). ***Generally, a consumption of animal-derived foods (meat and dairy products) is associated with an increased incidence of degenerative diseases, while consumption of whole grains and vegetables decreases the likelihood of those diseases (Campbell 1992, Barnard 1993).*** For example, there is a strong correlation between nutrition and cancer (O'Connor 1987). Also, recent studies in heart-disease treatments have indicated clear benefits of eating nonanimal foods (grains, vegetables, beans, etc.) (Ornish 1993). Caloric restriction diets in experimental animals have been demonstrated to prolong the longevity and decrease the incidence of degenerative diseases such as heart disease and cancer. Prompted by this research, American recommendations for proper nutrition have changed: we do not teach our children "4 groups of food" any more, but a nutritional pyramid, where grains and vegetables are on the bottom (most recommended consumption) and meat and milk products are at the top (no or small consumption recommended).

Going one step further, one can make a link between eating from the high end of the food chain (animals and animal products) and environmental degradation. Eating meat and milk products is energetically wasteful: production of 1 lb of meat takes 10–15 lbs of grain, in addition to destroyed forests to provide pastures for cattle grazing, and ~35,000 gal of water necessary to produce each pound of meat (Robbins 1990). Also, one should consider the environmental impact of methane production from ruminants, water depletion associated with beef and other meat production, water pollution because of nitrates and phosphates and cow waste, air pollution around the feedlots, etc.

We are an "instant culture," where an abundance of material goods produced an impatience in achieving goals and a desire for instant gratification (Fromm 1981). Consumption (both industrial, energy, and food) in the United States is one of the major contributors to the overall negative environmental impact (Postel 1994). Food consumption, including a large consumption of meat, is one of the major contributors to agricultural degradation, since the production of meat requires large amounts of grain production. In the U.S., 90% of soy beans and 80% of all corn grown is for animal consumption (Robbins 1990). Long-term risk analysis of eating habits can help point the way to decrease the negative impact they have on our health and the environment.

7. POPULATION EXPLOSION

Finally, no responsible environmental policy can leave out of consideration the impact of population growth on the overall quality of life. Although European population is approaching zero growth, one of the major problems in developing countries (Africa and Latin America) is rapid population growth which is exceeding the earth's capacity. All the advances in environmental protection (clean air, water, and soil; easier commuting; energy efficiency; etc.), will be forgone if the population keeps growing. The finality of the earth's resources is a given, and, therefore, earth systems can support only a limited number of people. Thus, a prudent policy would be to encourage limits in the number of children per family, rather than encouraging large families. Although this policy may appear as a part of overall social policies, it is equally pertinent to pollution prevention and environmental protection in general. A goal of any enlightened community should be an overall zero-population growth and sustainable development that will enable a high quality of life to all its citizens. In a democratic society people would decide for themselves the meaning of the term "quality of life." As Thomas Jefferson stated: "People are inherently capable of making proper decisions when properly informed."

Recent trends in armed conflicts around the world (numerous local and regional conflicts) show increases in risks of death and injury from violence by other human beings. In addition, destruction of infrastructures and environmental risks are also increasing. Although the arms trade is rapidly decreasing (largely due to disintegration of the USSR, one of the biggest arms traders before the fall of communism) (Vital Signs 1994), currently, the world is filled with weapons, which often become sources of intimidation and increase risk of death. Risk analysis points out that in war situations all other risks combined pale in comparison with the risks caused by a deliberate destruction by the armed forces. While a cynic may argue that war is one of the ways of keeping a lid on the population explosion, it is incompatible with the foundations of democratic and humanistic culture and international foundations of human rights (UNCED 1992).

8. CONCLUSION

Humanity and the modern western culture as we know it may be in decline (Fromm 1981). Modern society is not sustainable, because of high consumption of nonrenewable resources such as fossil fuels and minerals and creation of waste which is very recalcitrant to enter the geobiochemical cycles of the earth. The Earth Summit in Rio de Janeiro (June 1992) has clearly expressed this concern. As Mr. Vaclav Havel, the first president of democratic Czechoslovakia, noted in his speech to the U.S. Congress in 1991, "Without a global revolution in the sphere of human consciousness, nothing will change for better in the sphere of our being as humans, and the catastrophe toward the world is headed — be it ecological, demographic, social, or general breakdown of civilization — will be unavoidable." Humanity is out of step with nature (Lovelock 1969) and the biosphere may respond by following the Le Chatelier principle of eliminating the species causing disturbances in the great scheme of things on earth. Risk analysis may contribute to a development of this "global consciousness" and avoidance of a gloomy prophecy.

On a smaller scale, each country can follow the sustainability principle, as expressed in the Bruntland report in 1989 or at the Earth Summit in 1992. The sustainability principle could be implemented at the local level even more diligently, thus promoting self-sufficiency. Developing countries and former communist countries (countries in transition) are in an ideal position to implement this principle, just as Germany and Japan were in a position to modernize their industrial production after their industry and infrastructure were destroyed in the World War II. The Marshall plan did wonders for both Germany and Japan, leading to their economic well-being and efficiency in production. There is an opportunity in former communist countries and developing countries to build a new infrastructure and an industrial system which will incorporate efficiency and clean technologies, while keeping material consumption relatively low.

Experience developed in the United States and Western Europe that could help in establishing sustainable societies and new clean and efficient technologies are already in existence. This experience and technologies could be adjusted to local conditions by local scientists and engineers who know the circumstances and needs of their particular country. In addition, local wisdom of sustainable living and adjusting to natural environments could be revived and applied. For example, rather than rebuilding houses, hospitals, kindergartens, and other buildings in an energy-wasting way, one should apply a combination of the old wisdom and the most modern energy research and build houses (components of which are already either commercially available or available locally) which use much less energy than that used by conventional buildings. The building industry should be given the incentives by local and regional governments to incorporate new technologies to provide durable housing and energy efficiency in its housing units. At the same time, one should develop local capacities for building and use of local knowledge and self-sufficiency, rather than making decisions far away from the localities where people have to live with those decisions.

In addition, technological solutions are only a part of achieving sustainability. The concept of development should not be tied to the industrial growth and the GNP, and a more sophisticated economic measure of development should reflect nonmaterial, less tangible parameters that increase the quality of life for the majority of people (Fromm 1981).

Risk analysis, in spite of its limitations, could be used on the regional and local level to evaluate alternatives in development and/or environmental management. It can indicate trends and patterns of environmental risks which could serve to establish more suitable environmental risk management strategies, taking into account all the available knowledge.

REFERENCES

Barnard N. 1993. *Food for Life.* Crown Publishers, Inc., New York.

Brundtland GH. 1987. *Our Common Future (Brundtland Report).* World Commission on Environment and Development. United Nations, New York.

Campbell C. et al. 1992. China: From Diseases of Poverty to Diseases of Affluence. Policy Implications of the Epidemiological Transition. *Ecology of Food and Nutrition,* 27: 133-144.

Cannon JS. 1990. *The Health Costs of Air Pollution.* A survey of studies published 1984–1989. American Lung Association. Washington, DC.

Colby ME. 1990. Economics and Environmental Management. Presentation at the International Society for Ecological Economics: Conference on "The Ecological Economics of Sustainability," Washington, DC, May 21–23, 1990.

Fromm E. 1981. *To Have or To Be?* Bantam Books, New York.

Gore A. 1993. *Earth in the Balance: Healing the Global Environment.* Houghton-Mifflin, Boston, MA.

Lovelock JE. 1979. *GAIA. A New Look at Life on Earth.* Oxford University Press. New York.

Lowe MD. 1992. *Shaping Cities: The Environmental and Human Dimensions.* World Watch Institute, Washington, DC.

Lowe MD. 1994. Reinventing Transport. in *State of the World 1994.* World Watch Institute, Washington, DC.

Molak B. 1990 Energy consumption structure and other factors important in environmental pollution and pollution decrease. Savjetovanje saveza energeticara Hrvatske, Zadar-Petrcane, 17–19 May, 1990, 41–54.

Molak V. 1989. Waste Reduction and Community Right to Know. *RISK Newsletter,* May: 8–9.

Molak V. 1990. Pollution as a Terminal Disease? *Risk Analysis,* 10: 605–607.

Molak V. 1991. Over 100 Billion/Year Wasted by Industry into the Air, Down the Drain, and into the Countryside of the United States. *J. Clean Technology and Environmental Sciences,* 1(2): 155–157.

O'Connor PO and Campbell TC. 1987. The Influence of Nutrition on Carcinogenesis. *Nutrition,* 3: 155–162.

Ornish D. 1990. *Dean Ornish's Program for Reverting Heart Disease.* Random House, New York.

Ornish D. 1993. *Eat More: Weigh Less.* Harper Collins, New York.

Postel S. 1994. Carrying Capacity: Earth's Bottom Line. in *State of the World 1994.* World Watch Institute, Washington, DC.

Robbins J. 1990. *Diet for a New America.* Avon Books, New York.

Robbins J. 1992. *May All Be Fed. Diet for a New World.* Avon Books, New York.

Rubin ES, Cooper RN, Frosch RA, Lee TH, Marland G, Rosenfeld AH, and Stine DD. 1992. Realistic Mitigation Options for Global Warming. *Science,* 257: 148–149, 261–266.

UNCED. 1992. AGENDA 21: Program of Action for Sustainable Development. United Nations Publication — Sales No. E.93.I.11.

Union of Concerned Scientists. 1992. Letter to members. Washington, DC.

U.S. EPA, FEMA, and DOT. 1987. Technical Guidance for Hazard Analysis. Government Printing Office, Washington, DC.

U.S. EPA. 1990. Reducing Risk: Setting Priorities and Strategies for Environmental Protection. SAB-EC-90-02.

Vital Signs. 1994.World Watch Institute, Washington, DC.

World Watch Institute. 1990. *State of the World 1990.* Washington, DC.

World Watch Institute. 1991. *State of the World 1991.* Washington, DC.

World Watch Institute. 1992. *State of the World 1992.* Washington, DC.

QUESTIONS

1. Define sustainable development.
2. What is Agenda 21?
3. Why is Western society (North) nonsustainable?
4. What is the modified Erlich equation? What are the most important factors in environmental impact?
5. What is the definition of South and North in the United Nations terminology? How does environmental impact differ in those two major earth regions? What factors are prevalent and where?

6. How can one deal with the problem of compartmentalization?
7. How does public transportation decrease environmental impact? What is the impact of personal cars?
8. What is the significance of urban planning on sustainable development?
9. What is the significance of pollution prevention in terms of efficiency?
10. How do our food choices impact (a) our health? (b) our environment?
11. What is the significance of population and growth on the caring capacity?

CONCLUSION

Vlasta Molak

The previous chapters should have convinced the reader that risk analysis is indeed a very complex process and that risk management is potentially an even more complex process, by which we are attempting to reach "optimal" decisions regarding risks (probability of an adverse effect). The honest practitioners of risk analysis are aware of this complexity and are struggling to develop more sophisticated methods to deal with it.

The concept of "risk" in our society has multiple definitions. Often the public interprets risk in terms of likelihood and outrage. One side issue in acceptability, then, is whether the adverse conditions are intentional or not. Although we choose to drive cars (certainly, based on actuarial data, a very risky activity), we do not want others to impose smaller risks upon us (such as a factory in our neighborhood that may pollute the water or air or add carcinogens to our food even in small, almost negligible, quantities). The choice of an "adverse effect" in our risk analysis is also dependent on public policy. The real issue in public health is quality of life as a function of time. The possibility that at some time in the future someone may get cancer may not be very important. Instead, perhaps we should be looking at life-shortening and decrement in life quality at earlier ages, not the chance of cancer during one's lifetime. A similar issue exists with noncancer toxicity: the type and severity of the effect is extremely important, not just whether the effect is adverse. The idea of unacceptable risk depends on the severity of effect. We would not close down a local pizza restaurant over a case of gastric upset, but we might do so for a case of hepatitis. However, these are not scientific issues but issues of societal values, which need to be dealt with in a democratic way, with all stakeholders having an opportunity to voice their preference.

Uncertainty in risk analysis also diminishes its predictive value. As previous chapters have indicated, there are many types of uncertainties in risk analysis. Some of them are consequences of lack of reliable data, some of models chosen, some

1-56670-130-9/97/$0.00+$.50

are consequences of our choice of studies included in risk analysis, and some come from our assumptions about exposures or hazards. For example, in nonprobabilistic risk analysis, usually performed by the U.S. EPA to derive criteria (for either reference doses or drinking water criteria), the choice of toxicological study determines the no-observable-effect level (NOEL) or lowest-observed-effect level (LOEL). Even if we use probabilistic risk analysis (PRA) for exposures and give a distribution rather than a single value for a particular human parameter of exposure, the choice of the study to determine LOEL or NOEL is a result of an expert's judgment, which varies between experts.

In application of PRA (for example, cancer or nuclear power plant failure), choice of a model can to a great extent affect the final probability value. Experts (whose honest scientific opinion is based on valid arguments) may disagree about which model is most applicable. Also, even if the model is carefully selected to fit perfectly the particular case of risk analysis, the lack of sufficient data, or even the distribution of data, may introduce new types of uncertainty in our final result: probability of an adverse effect from a particular hazard.

Risk analysis is almost inextricably "political." This is perhaps because the costs and benefits are of different types and they affect different individuals (e.g., cancer among the poor who may live near polluting plants or toxic sites and profits for those who own those plants and make profits while polluting). However, there is still a great deal of misunderstanding regarding the applications and limits of risk analysis in dealing with human condition, be it human health, ecological health, economic well-being, etc. There are those who believe that the application of risk analysis is a panacea that can rationally solve all our problems, and that if we rank the problems using risk analysis alone we can optimize our decision making. Others vehemently oppose any attempts to use risk analysis to evaluate and rank our problems, especially environmental problems. Opponents of risk analysis come from two diametrically opposite sides: environmentalists and conservative business proponents.

Many environmentalists have experienced risk analysis as a clever smokescreen used by industry to cover up the "real" damage that exposures to industrial pollution or technology in general have caused. In many court cases, environmentalists claim that risk analysis was brought in only to confuse the juries and sell them the "bill of health" story about their industry. On the other hand, the conservative probusiness institutions and individuals argue that risk analysis is used by "rabid environmentalists" to prevent legitimate business from making well-deserved profits, by scaring the public about risks that in their (conservative) opinions are nonexistent. Some examples are their claims that second-hand smoke does not cause cancer, that food choices do no affect degenerative diseases, and that asbestos is not as dangerous as environmentalists claim. Scientists and the government are accused of being overly conservative and worrying about nonexistent risks. The more politicized groups accuse the scientists from academia and governmental institutions of being puppets in the hands of the federal government's attempt to destroy U.S. industry by unreasonable regulations that are overly concerned with "nonexistent risks."

However, according to many scientists who study risks and legal ramifications of risk analysis, this claim that U.S. government-sponsored risk assessments invariably take the most conservative, pessimistic, or "safety first" positions on all risk assessment treatments is false. Many articles in the scientific and public policy literature demonstrate that federal risk assessment policies and practices are far less consistent and often far less conservative (see Chapter II.7). Some of the EPA approaches that are not conservative (i.e., do not embody a "safety first" or conservative risk assessment bias) are

1. A risk analysis method that produced "middle-of-the-road" results on some disputed issues
2. A risk analysis method that reflected a plurality of existing scientific views about the "best" scientific practices rather than deferring judgment pending acquisition of better data and testable hypotheses; this practice may be appropriate where the only objective is the pursuit of improved knowledge, but imposing it in actual risk analysis contexts promotes whatever status quo exists and may often allow continued toxic exposures for population groups
3. A risk analysis method based on administrative convenience rather than on "safety first" considerations; for example, the greatest preponderance of epidemiological data drawn from cohorts of healthy (often white) workers rather than potentially more sensitive subpopulations such as children; most federal health-based risk assessments study cancer risks and occasional chronic respiratory diseases while largely ignoring damage to immunological systems or other more subtle effects

The other group of people, represented in Congress and federal government agencies, is very much enamored with risk analysis and is trying to pass laws and establish regulations that would mandate risk–benefit analysis. In such proposals, risk analysis is treated as a panacea, as if the results of risk analysis are not subject to uncertainty inherent in complex social systems. The overuse and overtrust in numbers derived from such risk analysis could be very dangerous for general democratic processes and may gridlock the decision-making process.

As was previously noted, such blind reliance on technology (scientific fix) presents a danger to a free society. The best we can hope for in applying risk analysis to the complex problems we face today (such as environmental exposures to chemicals and radiation, ozone hole, resource depletion, soil loss, global warming, etc.) is to ascertain patterns that could be useful for risk management. The numbers derived by risk analysis are at best crude and often misleading, if the uncertainty associated with them is not clearly spelled out. We could compare risks of different cleaning methods at the hazardous waste sites, or risks of the use of different types of energy or different types of transportation with more certainty than we can predict the global warming phenomenon. Risk analysis could help us make some general predictions about the economic and human health impacts of certain decisions (e.g., to use either public transportation or personal cars; nuclear energy, coal-powered plants, or conservation), which could help develop more livable and equitable sustainable societies. Compared with the accurate predictions we can get in the physical sciences, this sort of mere pattern prediction is not satisfying. However, to pretend that we possess the knowledge and power to enable us to shape society entirely to our liking, knowledge which in the real world we do NOT have, is likely to harm us.

As Dr. Hayek pointed out, ***"The recognition of the insuperable limits to his knowledge ought indeed to teach the student of society a lesson in humility which***

should guard him against becoming an accomplice in man's fatal striving to control society — a striving which makes him not only a tyrant over his fellows, but may well make him destroy a civilization which no brain has designed but which has grown from the FREE efforts of millions of individuals." Risk analysis can become a valuable tool to bring all the available facts to the discussion table, where, in a proper democratic process, decisions about risk management can be made for a particular situation. ***It is important to be aware that we are not working with the "real" and complete truth; but as long as we may not know the complete "truth," a carefully performed risk analysis, with all the assumptions and uncertainties spelled out, provides the best available interpretation of the existing data.***

Answers

CHAPTER I.1

1. There are two types of noncancer chemical risk analysis uses: to derive criteria and standards for various environmental media and to characterize risks posed by a specific exposure scenario, e.g., at the Superfund site, by drinking contaminated water, by consuming contaminated food, by performing some manufacturing operations, by accidental or deliberate spill or release of chemicals, etc. Usually such exposure scenarios are complex and vary with each individual case; thus, methods in risk analysis must be modified to account for all possible exposures in a given situation.
2. Chemical risk analysis used for criteria development generally does not determine the probability of an adverse effect. Rather, it establishes concentrations of chemicals that could be tolerated by most people in their food, water, or air without experiencing adverse health effects either in short-term or long-term exposures (depending on the type of derived criterion). These levels (either concentrations of chemicals in environmental media or total intake of a chemical by one or all routes of exposure) are derived using point estimates of average consumption of food and drink, and body parameters, such as weight, skin surface, metabolic rate, etc.
3. Permissible Exposure Limits (PELs) are regulated by OSHA.
4. Exposure assessment is determining of the fate of the chemical in the environment and its consumption by humans. Ideally, by performing environmental fate and transport of chemicals, and by evaluating food intakes, inhalation and possible dermal contacts, one can assess total quantities of toxic chemicals in an exposed individual or population, which may cause adverse health effects. In criteria derivation, one uses either the worst-case exposure scenario or most-probable exposure scenario and point values for various human parameters.
5. Reference dose (RfD), previously known as daily acceptable intake (DAI) is defined as total daily dose of the chemical (in mg/kg of body weight) that would be unlikely to cause adverse health effects even after a lifetime exposure. An RfD for a chemical is "estimation (with uncertainty spanning perhaps an order of magnitude) of a daily or continuous exposure to the human population (including sensitive subgroups) that are likely to be without appreciable health risk." RfDs are established from all available toxicological data for several hundred chemicals, particularly those associated with Toxic Release Inventories (TRI). The RfDs and risk assessment methodologies used for their derivation are available on-line in the Integrated Risk Information System.

6. Uncertainty factor is a number (usually a multiple of 10) that is used as multiplier and dividend in order to derive more protective (conservative) criteria. Its function is to ensure safety when criteria are derived based on incomplete sets of data, or uncertainty in choosing the model for that particular risk assessment.
7. Criteria are calculated by using best toxicological studies on a particular toxic substance and models of average exposures in daily situations. Appropriate uncertainty factors are also used.
8. Since the contamination of groundwater is only half of a 10-day health advisory, it should not be recommended for regular drinking use unless the chemical can evaporate (by boiling or letting the water stand). Without knowledge of other properties of the given chemical XYZ and knowledge of its fate and transport in soil, as well as RfD, we cannot derive further recommendations about this chemical, except that the caution should be practiced in handling the contaminated soil.
9. If the bioaccumulation factor is 20, then the concentration of the chemical in a mature fish would be 100 mg/kg of fat in the fish. Since the RfD is only 2 mg/day, it would be unwise to eat the fish for an extended period of time even in very small quantities, since even 20 g of fish (fat) would contain 2 mg. Especially dangerous is the fact that the chemical can likely bioaccumulate also in human tissues.

CHAPTER I.2

1. National Research Council (NRC), *Risk Assessment in the Federal Government: Managing the Process,* National Academy Press, Washington, D.C., 1983.
   National Research Council (NRC), *Science and Judgment in Risk Assessment,* National Academy Press, Washington, D.C., 1994.
   Office of Science and Technology Policy, Chemical carcinogens: review of the science and its associated principles, 1985, *Federal Register* 50:10372–10442.
   U.S. Environmental Protection Agency. The risk assessment guidelines of 1986, *Federal Register* 51:33992–34005.
   U.S. Environmental Protection Agency. Proposed guidelines for carcinogen risk assessment, 1996, *Federal Register* 61(79):17960–18011.
2. Analytical and Descriptive. Analytical studies consider individual exposure. The two approaches to analytical studies are cohort and case-control methods. The cohort method identifies groups of exposed and nonexposed individuals and studies the difference in disease occurrence between the two groups. The case-control method compares persons with the disease and persons without the disease for differences in exposure and other factors. Descriptive studies are analyses of disease rates in groups of exposed and nonexposed persons. The primary difference between analytical and descriptive studies is that analytical studies consider individual exposure while descriptive studies consider measures of exposure for a whole group. An example of a descriptive study might be a correlation of esophageal cancer mortality rates among countries with the per capita alcohol consumption of those countries. Such a study might find a positive correlation, but it is unknown whether those who developed esophageal cancer actually consumed alcohol.
3. Meta-analysis is the comparing and synthesizing of studies dealing with similar health effects and risk factors. Its utility is that it can be used to formally examine sources of heterogeneity, clarify the relationship between environmental exposures

and health effects, and generate information beyond that provided by individual studies or a narrative review.

4. Biomarkers are generally considered to include:
   (1) biomarkers of effect (biologic evidence that damage has occurred).
   (2) biomarkers of susceptibility (biological evidence that the individual may have heightened disease susceptibility). Susceptibility could be inherited or acquired.
   (3) biomarker of internal dose (e.g., tissue level of a carcinogen).
   (4) biomarker of biologically effective dose (e.g., DNA adducts).
5. Questions include:
   Could the study have detected an increase in cancer risk, i.e., was the sample size large enough?
   Could the results of the study have been due to chance, bias, or confounding?
   Was cancer latency addressed?
   How was exposure determined?
   In a cohort study, was follow up of cohort members adequate?
6. Temporality — the disease has to occur within a biologically reasonable time after initial exposure.
   Consistency — the same result occurs in multiple studies.
   Magnitude of the association — the risk is large and precise.
   Biological gradient — the risk is found to increase as the exposure increases.
   Specificity of the association — the likelihood of a causal interpretation is increased if a particular form of cancer is related to exposure in several studies (e.g., asbestos exposure and mesothelioma, cigarette smoking and lung cancer).
   Biological plausibility — the association makes sense with respect to metabolism, pharmacokinetics, etc.
   Coherence — the cause and effect are in logical agreement with everything known about the agent, exposure to the agent, and the disease.
   (*Note:* None of the criteria are considered conclusive by themselves, and the only criterion that is essential is the temporal relationship.)

CHAPTER I.3

No exact answers; questions are creative thoughts.

CHAPTER I.4

1. The development of a risk analysis model offers several benefits, including: a greater awareness of uncertainty and risk for all those involved in the development of the model; the creation of a more blameless environment for the discussion and management of risk; the identification of "opportunities", i.e., events that may or may not occur but would accrue benefit to the group should they occur; and finally, the development of more informed and balanced decisions and risk reduction strategies.
2. Monte Carlo risk analysis modeling is superior to more traditional single point (deterministic) modeling as it can incorporate all identified uncertainties and risks and thus facilitate more informed decisions. The great advantages of Monte Carlo modeling over other risk analysis modeling techniques (e.g., algebraic solutions

and method of moments) are that building up a Monte Carlo model is very intuitive and this type of model allows for the inclusion of complex stochastic relationships between the model's variables with a minimum of effort.

3. 81.5%, zero, 97.5%.
4. Bounded distributions do not allow scenarios that extend a variable beyond the minimum to maximum range defined by the expert. Thus, if the expert underestimates the extent to which a value may vary (as very often happens) the model will have an unrealistically narrow range and fail to convey the extremes that the future might hold.
5. BetaPERT distributions have a smaller spread than triangle distributions. The distribution of a total project cost will therefore also have a smaller spread if BetaPERT distributions are used instead of triangles. Thus, the risk contingency will be smaller since the difference between the mean and 85 percentile will be smaller.
6. No specific answer; examples and correlations will vary according to workplace.
7. Mean: £, mode: £, median: £, standard deviation: £, variance: £^2, skewness: unitless; kurtosis: unitless.

CHAPTER I.5

1. Probabilistic risk analysis was developed to facilitate the quantification of risks associated with complex engineered systems. It is particularly well suited to analyzing the frequencies of extremely rare events, such as core melts in nuclear reactors or chemical plant accidents, for which little if any accident data will be available.
2. To simplify the difficult and complicated task of system design, system design is generally done by specifying the boundary conditions under which each subsystem is expected to operate (e.g., sources of electric power, cooling water, etc.), and performing detailed engineering design of each subsystem individually. Thus, the dominant sources of risk often arise from *interactions between subsystems* (e.g., situations in which one subsystem fails, and thus changes the environment faced by other subsystems), since such interactions may be overlooked in the ordinary engineering design process.
3. Quantitative estimates of accident frequencies or probabilities provide a more rigorous basis for evaluating the cost-effectiveness of alternative risk reduction actions, and for determining the relative importance of different risk contributors.
4. Without information on risk contributors, the only decisions available after the PRA is completed will be either to accept the status quo and continue operating, or to shut down the system (typically at great cost). Information on risk contributors can help facility owners and operators make good decisions about system design modifications, operations, and maintenance.
5. (1) What can go wrong?; (2) How likely is it to go wrong?; and (3) What will be the consequences if it does?
6. Hierarchical models provide a way of structuring the vast quantities of information that go into a risk analysis.
7. Event trees are well suited for displaying the *order of events*, displaying *dependencies between events* (e.g., the fact that the failure probability of subsystem B may depend on the status of subsystem A), and facilitating communication about

the assumptions made in the risk model (e.g., presenting a risk model to plant staff for review and discussion). However, because combinations of subsystem successes and failures are explicitly shown, event tree models can rapidly become extremely large, including literally billions of sequences. Fault trees provide a more compact way of representing large numbers of events, but can obscure dependencies and the chronological order of events.

8. In the large fault-tree approach, the sequences that are important to risk can be easily remembered and understood, thereby facilitating communication. By contrast, in the large event-tree approach, the individual split fraction models are relatively simple, and the failure probability of a particular subsystem will generally not depend on the specific *causes* of other subsystem failures or top events earlier in the event tree.
9. *Grouping redundant components or trains* under a single top event helps to ensure that the various top events in the event tree will be conditionally independent of each other, and makes it possible to model common-cause failure within an individual top event, rather than between different top events. Placing *causally dependent events* to the right of the events that influence them also helps to minimize complex dependencies between the various top events. Placing *more severe events* toward the left side of the event tree often makes it possible to *prune* the event tree by eliminating unnecessary branches, which can greatly reduce the number of accident sequences that must be represented in the event tree. Finally, putting the top events in *chronological order* can help to facilitate communication about the assumptions made in the risk model.
10. Data are needed on initiating event frequencies, component failure rates, common-cause failure rates, component maintenance frequencies and durations, component fragilities, and human error rates. *Success or exposure data* are also needed; for example, the number of hours of operation or experience over which the observed failures occurred, and the number of demands experienced by standby or by cycling components (such as thermostatically controlled heating or cooling systems). Possible data sources include maintenance requests, corrective action reports, significant event reports, anomaly reports, plant or mission logs, test results, and case histories, in addition to expert opinion and published or computerized databases.
11. The analyst must specify the *level of detail* of the analysis, the *components of interest*, the *database study period*, the relevant *failure modes* for each component, and the appropriate *units* for each failure mode (e.g., hours vs. demands). The analyst must also decide whether to pool information for similar components (which can increase the total amount of information that is available for the analysis, but can lead to misleading results if the components are not sufficiently similar), whether to use test data on partial failures, and how to account for corrective actions.
12. The application of PRA has been successful in risk management because PRA results, combined with engineering judgment, frequently make it possible to identify relatively inexpensive risk reduction options.
13. Every plant is unique, even nominally “identical” units on the same site. The influence of operating and maintenance practices can far outweigh the inherent design reliability of the equipment, so that even plants that start out as sister units can have very different risk and reliability profiles.

CHAPTER I.6

1. Ecological risk assessment is defined as "the process that evaluates the likelihood that adverse ecological effects are occurring, or may occur, as a result of exposure to one or more stressors." In contrast, ecological risk management is the process for making decisions or selecting options to manage the risk. Ecological risk *assessment* is one of several inputs into risk *management.*
2. Human health risk assessment defines adverse consequences in terms of effects on individual, classes, or groups of humans. Ecological risk assessment defines "adverse" in a variety of ways, but usually as the consequences to a nonhuman biological feature of the environment. For example, the main risk to converting native prairie to farmland may be the consequences on soil biological diversity.
3. Among the major reasons offered in support of ecological risk assessment are: it is an organized, systematic way to prioritize ecological policy problems; it allows policy makers to allocate scarce resources to solving the most important ecological problems; it clearly separates science from policy making; and it really only formalizes how decisions are actually made. Among the common objections to ecological risk assessment are: it is too easy for practitioners (technocrats) to insert their personal values and distort the results; in order to make ecological problems technically tractable for risk assessment, the problems must be simplified, which distorts their relevance to the policy problem; and the formulation of the risk problem defines the result and, in practice, technical experts often define the risk problem rather than the public.
4. In an ideal world personal values and priorities should be separated from science and the assessment process. However, in conducting ecological risk assessments this is nearly impossible because there are no clearly accepted values and priorities for ecological policy questions, so scientists and analysts must make many assumptions in order to carry out the analysis. The assumptions (values) are often the most important issues in public policy; unfortunately, they are often not explicitly stated and conveyed to risk managers.
5. The most common alternative to ecological risk assessment is to avoid defining adverse events (risk) initially by evaluating the ecological consequences and let policy makers or the public decide which of the alternatives are most desirable. Ecological consequences (or ecological change) can only be defined as adverse when a human value or criterion is applied. Another approach is to expand benefit/cost analysis to cover nonmonetary consequences (and costs). All of these approaches are subject to similar criticisms (and misuse) as is ecological risk assessment.
6. The most commonly held view is that ecological risk assessment needs to be closely linked to ecological risk management, but clearly separated. Risk assessment is an analytical process that provides input to decision making (management). It is separate and distinct, but should be a directly relevant policy question being addressed by risk managers.
7. The issue of which ecological changes or consequences are defined as adverse is one of the most difficult in ecological risk assessment. To label an ecological change as adverse requires the application of a human value or priority, which means that it is not a scientific or analytical choice. Societal involvement is required and this may be obtained through legislation, policy directives of elected or appointed officials, or direct stakeholder input.

8. Scientists and other technical experts play the dominant role in ecological risk assessment. In contrast, their role in risk management is one of providing input and technical council.

CHAPTER I.7

1. Opportunity Cost. At the least, a "free lunch" takes your time, and often the provider of this "free" meal wants more of your time or other resources.
2. While not necessarily "good" reasons, there are at least a couple of explanations for why businesses fail to undertake activities that will improve their profitability: (a) inefficient capital markets; and (b) habit. A small business with cash-flow problems may have difficulty obtaining a loan to purchase the needed equipment even though the equipment would improve cash flow and profitability. Bankers are not famous for their impartiality and objectivity. Just like the rest of us, business people are creatures of habit. Phrases like, "If it ain't broke, don't fix it," contain a rationale for inaction as well as folk wisdom.
3. Over time, technological advances may very well reduce (social) costs. Lunches might not be free, but some lunches might be much less expensive.
4. *Eliminating* the risks associated with automobile transportation means doing without it. Since much of America is built around automobile transportation, this is not a feasible goal for very many Americans. As individuals, we can *reduce* auto transportation safety risks by reducing commuting distance, making fewer unnecessary trips, and purchasing safer autos. Socially, we can reduce safety- and pollution-related risks by investing in alternative transportation, supporting policies to make highways safer and to make autos safer, longer-lasting, cleaner-burning, and more fuel efficient. We can also support investments in alternative transportation, long-range planning to reduce dependence on autos, and systems of penalties (taxes) and rewards (tax credits) to bring market prices closer to total social costs.
5. "Better economists" means behaving more like the "rational economic man" of neoclassical economics, where each person's actions are motivated only by his or her own narrow pecuniary self-interest. Many organizations, both religious and secular, organize and act on behalf of what they believe is "the right thing to do." In most work environments, some degree of cooperation and teamwork is necessary. Even basic civility requires some restraint on self-interest. If the "what's in it for me?" approach to life is becoming more widespread, then people are becoming "better economists" and one of the consequences will be poorer economic performance.
6. Yes, because in pure competition good information and an industry-wide standardized product combine with free entry and exit of sellers as well as buyers to ensure that profits earned have actually added value to the economy. If there is no market failure or opportunities for cost-shifting, competitive forces will "reward" profits in proportion to actual value added to the economy.
7. The "good economist" is unconcerned with others' rights, and is more likely to take advantage of an opportunity to shift costs.
8. The "right-to-know" provision of the 1986 Superfund reauthorization required companies to report all releases of toxic chemicals. This simple, low-cost statute forced companies to look at their own releases and contemplate the likely reactions of citizens and customers. Better and more complete information usually produces better decisions and better resource allocations. If the government reported eco-

nomic data with adjustments for increased risk or environmental degradation, many producers and consumers would begin to "rethink" their activities, gradually leading to much improved resource allocations.

9. Wider acceptance of the "intrinsic value" perspective would make us more cautious when it comes to irreversibilities. If value exists apart from what humans want, irreversible biological and ecosystem effects involve serious costs even where humans are not directly impacted. If all life is intrinsically valuable, many "value-creating" economic activities should be "trumped" to prevent irreversible species loss and ecosystem degradation.
10. In a democracy, every citizen's vote should carry approximately equal weight in the political process, which defines community needs and develops processes for meeting those needs. Increasing global interdependencies suggest that, if we value democracy highly, we need to develop international institutions that can "hear" the voices of all the world's inhabitants.
11. The point of this exercise is to stimulate thinking about the social consequences of what are (primarily) private choices. A decision to stop smoking has consequences for family and for the country's health care resources. Most private risk reduction decisions involve some social benefits or costs.
12. This question is intended to encourage the reader to approach risks from an economic perspective.
13. If "fairness" is defined to include consideration of differences in income, race, ethnicity, gender, and generation (assuming future generations are represented), and "most economical" is defined to take account of all costs, adjusting for market failures and factoring in irreversibilities, society might nevertheless be justified in "trumping" an economic risk reduction strategy on the grounds of community need.

CHAPTER II.1

1. A. dermal exposure during use of an antimicrobial soap
   B. VOC inhalation following offgasing from new furniture
   C. consumption of pesticide residues in agricultural commodities
   D. ingestion of microbial-contaminated water
2. Track contaminant into residence from outdoor soil/dust; exposure during resuspension of contaminated soil/dust.
3. Biological variation in body weight, skin surface area, and inhalation rate; body weight basis; for example, children would have a higher exposure (mg/kg) than adults.
4. Air exchange rate temperature; open windows increase air exchange rate and reduce inhalation exposures to air-borne chemicals.

CHAPTER II.2

1. FIFRA requires that any pesticide registered in the U.S. must perform its intended function without causing unreasonable adverse effects on the environment. This statute also requires that evaluations of potential risks to man or the environment must also take into account the economic, social, and environmental costs and benefits of the use of a given chemical. While the use of "unreasonable risk"

suggests that some risks will be tolerated under FIFRA, it is clearly expected that the anticipated benefits will outweigh the potential risks when a pesticide is used according to commonly recognized, good agricultural practices.

2. Scientific issues involved in the evaluation of potential dietary exposures and human health risks associated with pesticide residues in food include: the scientific and regulatory paradox created by the Delaney Clause in the FFDCA and analysis of uncertainty. The Delaney Clause specifically prohibits the presence of residues of materials found to "induce cancer in man or animal." This absolute standard is inconsistent with the "risk-benefit" statutes of FIFRA. Further, this creates a regulatory dilemma in that while residues of "carcinogenic" pesticides are allowed in RACs under Section 408 of FFDCA, they are not allowed under Section 409. The second issue, uncertainty analysis, underscores a fundamental issue for the practice of risk analysis in general. Quantitative evaluations of uncertainty, for example, the use of Monte Carlo simulation to develop distributions of dietary exposures (and risks), provide the most scientifically defensible approach for estimating potential exposures to pesticides. The NAS has recently recommended the use of distributions of consumption and pesticide residues in food rather than single-point data to characterize dietary exposures and risks. Uncertainty analysis, where appropriate data can support its use, provides much more information to risk management decision-makers.
3. Two examples of methods that can be used for monitoring potential occupational exposures to pesticides are (1) the use of dosimetry clothing or patches on workers for measuring dermal exposures and (2) the use of personal air sampling devices to measure breathing zone inhalation exposures.
4. Three exposure pathways that may be relevant to potential residential exposures to chemicals include: incidental ingestion of dislodgeable residues from treated surfaces following hand-to-mouth behavior in children; dermal exposure to dislodgeable residues from treated surfaces; and inhalation of air-borne chemicals during and post-application of spray (e.g., hand-held aerosols, total release foggers) products.
5. Benefits that result from the international harmonization of testing guidelines and protocols for studies for pesticide registration include: establishment of a uniform approach to data requirements and interpretation, minimization of regulatory staff resource duplication regarding study initiation and review, conservation of economic resources and prevention of trade barriers.

CHAPTER II.3

1. dose
2. International Committee for Radiation Protection
3. U.S. Environmental Protection Agency
4. The doses are considered acceptable in relation to background doses, but are as low as are technically feasible for that industry.
5. Whole-body dose is the dose received from uniformly irradiating the body with an external source of radiation.
6. Effective dose is the equivalent in risk to a unit of whole-body dose. If the body is not uniformly irradiated by an external source, the effective dose is a dose (usually lower) that is equivalent in risk that would be received by the whole body.

7. For the simple situation of whole-body dose from photons, energy deposited equals dose. For other types of radiation or other energies of photons, the dose may be higher or lower than the energy deposited.
8. 0.48 Sv. The doses are added without regard to source, age, or time interval between doses.
9. The dose is .04 Gy for two hours of exposure and dose equivalent is $1.5 \times .04 = .06$ from the graph in Figure 1.
10. Table 1 shows the weighting factor for the lung to be 0.12. The effective dose is 0.24 Sv.

CHAPTER II.4

1. Industrial pollution can be defined as the presence of toxic substances in air, water, or soil, often resulting from inefficiencies in production processes.
2. Pollution can also be regarded as resources distributed in wrong places. Therefore, pollution prevention at the source can be regarded as saving on resources.
3. Pollution prevention is a process of decreasing pollution at the source.
4. Waste minimization is a decrease of the amount of waste that must be shipped off-site. It is not necessary due to pollution prevention at the source.
5. Industrial pollution may present a health risk to humans or ecological systems as well as a risk to economic well-being. These risks can be estimated and compared using risk analysis methods. Increased new risks may come from nuclear plant accidents, radioactive waste, pesticides and other chemicals release, oil spills, chemical plant accidents, ozone depletion, acid rain, and global warming.
6. Mining, basic industry, chemical industry, other manufacturing, energy production, transportation, waste management, wastewater treatment.
7. Air pollution, water contamination, soil/land contamination, hazardous waste generation, radioactive waste generation, acid rain, ozone depletion, global warming.
8. Risk analysis can serve to establish a priority of pollution problems based on the magnitude of risk that they pose either to human health or ecological systems. Pollution can also be regarded as resources distributed in wrong places. Therefore, pollution prevention at the source can be regarded as saving on resources. Since economic risk analysis can indicate economic losses resulting from pollution, it can be used to encourage pollution prevention at the source as a means of improving the bottom line.

CHAPTER II.5

1. *Risk assessment*: evaluate undesirable outcomes and assign probabilities to their chance of occurrence (e.g., climate change and climate impact assessment).
   Risk management: involves political decisions concerning what can be done to control societal risks, e.g., response strategies.
2. Possible answers:
   - Concentration of CO_2 depends critically upon environmental sinks.
   - Atmospheric CO_2 concentrations do not match the total CO_2 emissions; roughly half the CO_2 is absorbed by terrestrial plants and the ocean.

- Scenarios of future emissions critically depend on the rates of population growth, energy consumption per capita, and the rate of penetration of the nonfossil energy sources (renewables and nuclear).

CHAPTER II.6

1. *Risk Analysis, Toxicology, Food and Chemical Toxicology, Journal of the American Medical Association,* and *Toxicology Modeling.*
2. One CD-ROM contains the equivalent of several hundred computer floppy disks and thousands of printed pages. The information can be searched and accessed very quickly. CD-ROM versions of printed documents often contain "extras" such as spoken text and other sounds, photographs, interactive maps, "movies," and animation.
3. A key drawback to CD-ROM databases is that they cannot be completely up-to-date, whereas online databases can be updated frequently and accessed as soon as they are updated. However, it is likely that the rather new online ability to update and revise the content of CD-ROMs via information downloaded to the user's computer hard drive will be used for at least some CD-ROMs of particular value to exposure assessors and risk assessors.
4. Created by the U.S. Department of Defense in the late 1960s, the Internet has evolved into a worldwide collection of computer networks used for many purposes. This includes sending electronic mail messages, accessing various databases, and as a way to share, publish, and distribute textbooks, journals, newsletters, and other sets of information.
5. A key evolving part of the Internet, the WWW includes a collection of documents (text, graphic, video, and audio files) that users can navigate through via use of browser software programs and "hypertext" links. Hypertext enables users to highlight certain pictures, words, or phrases, starting with information displayed on "Home Pages," and to then move to linked pictures or pages of information.
6. Risk assessors and risk managers can communicate via Internet e-mail messages and via information shared on WWW sites. There can also be use by risk assessors and risk managers of Usenet newsgroups and mailing lists to collaborate and to ask about or share information (e.g., research plans and results). The future is likely to see increased use of the Internet for teaching, group discussions, and scientific meetings held at "virtual facilities." Data can be shared and discussed, and one can even take a "walk" through the virtual facility being used.
7. *Intelligent Agent (or Smart Agent, Software Agent, Assistant Agent, Internet Agent, Good Virus, or Web Robot):* Software programs that are told or essentially learn what information the user likes to see, and which will then search through electronic mail, databases, networks, World Wide Web home pages, and Internet "Usenet" newsgroups on an ongoing basis to retrieve that information for the user. These types of programs can also be used to deliver data and messages to other users/systems. In a portable device, Intelligent Agents are used by "Intelligent Assistants" to sort incoming messages based on what the user has looked at first in the past.
   Intelligent Assistant (or Personal Intelligent Communicator or PIC): Small, portable electronic devices that can send and receive electronic mail and faxes, sort incoming messages, and handle other functions such as listings of addresses and phone and fax numbers.

Personal Assistant: Pocket-sized electronic devices that contain the contents of books or other information. The information can be searched using key words, with results displayed on the screen of the Personal Assistant.

CHAPTER III.1

1. Extensive news coverage of catastrophic accidents. Increased number of risk-assessment studies. Loss of trust in risk management.
2. Experts see risk as determined primarily by expected mortality whereas laypeople's perceptions include qualitative characteristics such as dread, control, catastrophic potential, etc.
3. The finding that white males see the world as much less risky than everyone else sees it.
4. Because of the public's greater sensitivity to adverse events amplified by a social and political system that highlights those events and keeps them in the "public eye."

CHAPTER III.2

1. Fire risks are normally independent of each other while earthquake risks are highly correlated. If insurers are risk averse they are likely to charge higher premiums for risks that are more highly correlated due to the increased variance in the losses.
2. The most important step would be to require a medical exam as a condition for insurance so the insurer is fully informed about the patient's health status before issuing a policy.
3. Catastrophic losses may cause insolvency of the insurer. The federal government has unlimited borrowing power so a severe flood will not create financial problems. In addition to the fear of insolvency, insurers are concerned with problems of adverse selection when insuring against risks, such as floods. Unless they inspect each house individually they may not be clear how safe it is in relation to the hazard.
4. The insurer would determine whether such a risk is insurable by utilizing epidemiological data to estimate the probability of a person contracting asbestosis when exposed to a certain number of particles of asbestos fibers in the air. The insurer also has the option to refuse to write coverage or cancel an existing policy if the number of asbestos fibers in the air exceeds a certain level.

CHAPTER III.3

1. Risk assessment is a process whereby the nature and size of a risk are assessed and characterized; it is one of two main parts of risk analysis.
2. Risk management, the second main part of risk analysis, is a process whereby the ways in which a risk may be abated or eliminated, or its consequences mitigated, are developed, and appropriate ways are chosen and implemented.
3. In comparative risk assessment, risks are characterized by being compared, qualitatively and/or quantitatively, to others; often risks are characterized comparatively by ranking them against each other, ordinally or categorically. The individual risks

associated with different problems or issues are not, and usually cannot be, calculated or characterized separately as in other forms of risk assessment.

4. Because data, theory, and calculations are usually not sufficient to characterize the risk, qualitative factors and judgments, and creative ideas, assumptions, and models, must be brought into play, as in any art.
5. Associated with an issue within the framework of a risk study, risk reduction methods may or may not be in place and functioning. Whatever risk still exists with such methods operating, or whatever risk exists if there are no methods operating, is the residual risk that is to be assessed before implementing any new, future risk reduction methods. Residual risk is, therefore, the risk that is to be characterized and, if necessary, abated or eliminated through further risk management efforts. Misinterpretation often occurs when participants in a comparative risk project forget that there are already risk management measures in place, when they are in place, and/or when they consider possible risk abatement strategies as though they have already been implemented. Policy makers should understand residual risk and that it has been used in a comparative risk assessment because, if they do not, they might prioritize specific issues for action incorrectly; a low residual risk may be the result of the risk being low even though no abatement measures are in place or, alternatively, it may be low because risk management programs are in place and operating well. In the latter case, priority needs to be given to maintaining such risk management measures.
6. Criteria, such as probability of exposure to stressors, extent of adverse effects within a population or among ecosystems, and others, clearly defined and consistently used in the course of a comparative risk study, make for clarity and ease of communication and better understanding among study participants and, thus, make for a better quality of comparative risk assessment. "Uncertainty" is sometimes suggested as a criterion for characterizing a risk; it is better thought of as a characterization of the uncertainty regarding the characterization of risk. Uncertainty, in this sense, arises from uncertainty and gaps in the data, in the theories used, and in judgments applied. It is useful to know, when using the results of a risk assessment, what the uncertainty in the assessment and characterization may be when planning risk management measures, for example.
7. A ranking cleanly and clearly based on comparative risk is one valuable input to a priority ranking. Other valuable and necessary inputs to a priority ranking are such considerations as, for example: available means for, and technical feasibility of, risk reduction; the costs of risk reduction; the benefits that might accrue from risk reduction; and social and political factors. A risk ranking and a priority ranking are, therefore, two very different things.
8. A single individual can carry out a comparative risk assessment, but the results would be highly suspect because no one person has the sum total of knowledge, skill, experience, and perceptions needed to produce well-rounded results of high quality, utility, and credibility.
9. Although risk is mathematically defined in principle, in practice, especially with the broad range of risks a comparative risk study must encompass, the information is not available for full, mathematically correct risk calculations to be made. As to public participation, among the factors that must enter into a comparative risk study are the perceptions and values of the risk takers and their views on consequences, such as impacts on their communities and quality of life, for example. By involving members of the public, and keeping them as well informed as possible

about the available scientific information, a better idea of the risks perceived by the public can be entered into a comparative risk assessment.

10. Inevitably, whether a comparative risk study is carried out only by technically trained individuals or by participants who are not necessarily scientists or scholars, individual perceptions and values must enter into the process, if only through judgments required to fill gaps in the available data. Mathematical definitions of risk deal with objective risk; risk as characterized even in a well-done comparative study is always largely perceived risk, informed by science and by risk principles. A comparative risk ranking such as discussed in this chapter may be described as a risk ranking if it is carefully based on residual risk and if a nonmathematical definition of risk, one including perceptions, is accepted.
11. Regional or national comparative risk rankings deal in perceived "averages" of residual risks and do not usually highlight specific, localized risks. Thus, risks associated with abandoned waste sites were not ranked highly in *Unfinished Business* although in a local community such a site may be the major concern. Where known, such local risks need to be mentioned specifically in the commentary that accompanies any regional or national comparative risk ranking.
12. Despite the enormous and growing base of sound scientific knowledge relevant to environmental risks, the base is still in no way sufficient to make comparative risk assessment possible by itself. Much as some scientists may not wish to speculate and make judgments in their fields, it is necessary to do so to help bridge the gaps caused by the absence of specifically applicable knowledge if comparative risk studies are to be done. Potential participants in a study must understand this need and commit themselves to making judgments if they are to become actual participants. The many studies that have been carried out with the help of distinguished scientists make it clear that there are individuals willing to participate in such important endeavors despite their possible misgivings.
13. Personal commitment of individuals to promoting a sound environment, the understanding that, since resources are not endless, priorities must be set and that risk is an important consideration in setting priorities, and the possibility of making a significant personal contribution to an important means for ensuring the continuation of a sound environment.
14. One would not expect there to be gross differences, but there would almost inevitably be differences in detail, such as the inversion of the rankings of two closely ranked issues, since two separate groups would not be likely to bring identical perceptions of the different risks to the table.
15. Comparative risk ranking involves many uncertainties. Examples of this are lack of pertinent information, uncertainties in the information and data that are available, honest differences of opinion on how to interpret available data or what judgments to use where data are not available, different views on how to weigh risks that differ in nature from each other, differences in perception among participants in a comparative risk assessment, and, even differences in the personalities and the abilities of participants to express their views during a consensus ranking process. For these kinds of reasons, expressing a final ranking by assigning issues to a limited number of risk categories more correctly brings the total uncertainty into the picture than expressing a final ranking in ordinal form — even though developing an ordinal ranking may be a useful step toward achieving a final ranking.
16. In setting priorities for action or in developing policies, risk managers and policy makers need to take the uncertainty in a risk assessment into account, and they must therefore be informed as to what the risk assessors think the uncertainties

are. How they do this is determined on a case-by-case basis and it can depend on how conservatively protective they may wish to be, or feel they must be, among other things. Thus, whether to consider an issue of more uncertain risk characterization to be of lower priority for action because it is not as certain in its ranking as another issue, or whether to give it a higher priority of action because, with its uncertainty, it could pose a much higher risk than its ranking would indicate, is a decision that must be made on a case-by-case basis.

17. There are alternatives to using risk, including comparative risk, and many have been used and are still used. Some alternatives are to prioritize on the basis of factors other than risk, such as feasibility of reduction once a risk is known or believed to exist, whatever its size, or, not to prioritize at all but to address risks as they become apparent or as they become politically evident ("squeaky wheel" prioritization). Although such approaches do reduce risk, they do not make for the best use of risk reduction resources because they do not, in principle, strive for maximizing the total reduction of risk with those resources. The author does not know of approaches that are better, in allocating resources, than those that make use of risk as an input to prioritization.

CHAPTER III.4

1. Comparative Risk Analysis (CRA) is a term used to describe a rapidly growing number of projects performed around the U.S. by state and local groups as a promised new cure for "irrational" environmental management.
2. CRA is supposed to combine the "science of risk analysis" with "community values" to derive an environmental problems priority list, which then could be used to rationally apply resources for risk management based on the magnitude of "real" risk rather than perception.
3. Conflict of interest of the participants, insufficient information for valid risk analysis, insufficient time and expertise to perform risk analysis in a competent manner by the participants.
4. Open question with no right answer! Good ideas should be sent to the author!
5. Peace of mind — safety, happiness, and health
   Mobility — ease of getting from one place to another
   Aesthetics — visibility, noise, odors, and any visual impacts
   Future generations — impact on our children, availability of alternatives, reversibility of effects
   Sense of community — neighborhoods and personal growth
   Economic impacts — maintaining a comfortable standard of living, achieving personal goals, costs
   Fairness — sense of equity, respect of individuals' or property owners' rights, number of affected persons, severity of effects on different groups
   Recreation — access to and quality of recreational lands, opportunities for solitude
6. Such a definition of risk totally defies the stated purpose of CRA, which is to combine the science of risk analysis and community values, rather than basing risk management on public perceptions of risk (as has often been the case in the past). Moreover, such a definition may miss some real risks which the general public may not be aware of and thus outrage would be nonexistent.

CHAPTER III.5

1. Environmental equity refers to how equally environmental risks and procedures to mitigate those risks are distributed across different sectors of the population. Environmental justice refers to a policy of affording subpopulations equal environmental and health protection. Environmental racism connotes "disproportionate environmental risks in racial minority communities."
2. The environmental justice movement originated as a convergence of the civil rights movement and grass roots environmental movements. It was believed to be precipitated by hazardous waste cases in Warren County, North Carolina, in the early 1980s, which signaled racial disparities in the location of waste sites.
3. Requirements to address environmental justice issues are currently embodied in Executive Order 12898, which requires each federal agency to consider environmental justice in its activities.
4. It is difficult to quantify health risks associated with hazardous waste sites for use in an environmental justice analysis, since those risks vary according to many factors. For example, the toxicity of individual contaminants varies, making an aggregate risk estimate difficult, and risk varies temporally as the presence and toxicity of those contaminants change.
5. Minority groups are often difficult to define, since people use different criteria in classifying themselves into groups and persons in more than one category (e.g., children of mixed marriages) find it difficult to classify themselves.
6. Spatial units used for data collection in environmental justice analyses include those defined by the U.S. Bureau of the Census, such as Blocks, Block Group, Tracts, Zip Codes, and Counties.
7. Circles of varying radii around specific sites have been used as one approach to aggregating data in the spatial units selected. One way to aggregate the data is to include all Census units within a certain radius in their entirety (actually the centroids of those data units are used to determine whether or not the Census unit falls within a certain distance of the site). Another way is to use Geographic Information Systems to intersect data units so that a desired geographic area, such as a circle, is obtained.
8. Once the data are extracted and aggregated, various methods and criteria are used to determine whether or not an environmental justice issue exists. A comparison area or areas are selected against which the population characteristics of a given area of interest are compared. One basis for concluding a justice issue may exist is if the interest area, having a potential environmental problem, has a greater proportion of minorities than the comparison area. Various numerical techniques are available for conducting the comparison.
9. Examples of areas of subjectiveness in environmental analyses include what particular threshold and/or difference between the area of interest and the comparison area is used to establish a disparity and what comparison area is selected.

CHAPTER III.6

There were no questions in this chapter.

CHAPTER III.7

There were no questions in this chapter.

CHAPTER IV.1

1. The distinguishing threats of nuclear power are radiation and the heat given off by the radiation decay process, which decreases with time, but requires cooling provisions for some time following reactor shutdown.
2. The evidence is strong that nuclear power is among the safest of the developed energy alternatives. This is in spite of the two serious accidents involving Three Mile Island and Chernobyl. No member of the public or the operating staff has been killed or injured from a nuclear power plant accident in the United States. This is extremely impressive considering that there are 109 operating nuclear plants in the U.S.
3. There have been two major accidents involving nuclear power plants: Three Mile Island (U.S.) and Chernobyl (Ukraine). Although the Three Mile Island accident did not result in any injuries or deaths, the Chernobyl accident did result in 30 fatalities from acute doses of radiation and the treatment of 300 people for radiation and burn injuries. The latent effects of the Chernobyl accident have yet to be quantified.
   Nuclear power suffered a severe setback from these two accidents, especially the Chernobyl accident. It is expected that it will take decades of safe operation of nuclear power plants to rebuild public confidence in spite of its many advantages over other energy alternatives.
4. Among the principal elements of managing nuclear power plant safety are an effective regulatory process, risk and safety assessment practices by industry that clearly reveal the safety performance of the plant with time, and the adoption of a quantitative risk assessment and management process based on the use of proven risk-based technologies.
5. There is tangible progress in moving toward risk-based regulation, but to most that progress is considered slow. Some of the reasons for the slow progress are institutional inertia in the government, concerns by industry that the cost may be too high, the lack of stability in the methods of analyses to support risk-based regulation, and continuing questions on how to control the quality of the supporting analyses.
6. The distinguishing feature of PRA is that it quantifies the uncertainty of how likely an event or a series of events is. Most other risk assessment methods deal only with questions concerning the occurrence of events and their consequences; they do not attempt to quantify the uncertainty in the results of the assessment. PRA addresses all three of the fundamental questions of risk: what can go wrong, what are the consequences, and how likely is it, including the uncertainties involved.

CHAPTER IV.2

1. Earthquakes also cause damage to buildings and their contents, as well as damage to lifelines (highways, power lines, gas distribution network, etc.). These forms of

damage cause homelessness, business interruption, unemployment, and other economic consequences. Furthermore, the earthquake causes a disaster by making all the damage occur simultaneously in one region. So the community's capacity to respond is diminished at the same time that it is called upon for support. All of this is too much to squeeze into a single chapter, so the chapter focuses on the most extreme consequence of an earthquake: the potential for death.

2. The buildings destroyed in the Northridge earthquake had withstood previous quakes, but that did not mean that they were immune. Earthquake damage depends on so many factors, such as frequency spectrum of ground motion, constructive or destructive interference of waves reflected under the earth, etc., that a vulnerable building can "luck out" most of the time.
3. Where there are many URM buildings, there are many owners. They get together and become a potent political force in opposition to a mandatory ordinance.
4. As long as the state has bond money set aside for this purpose, go for it now! However, if it takes a fight to get the money in either year, consider the differences between the two options. In terms of lives saved, the difference is that option 1 saves lives during the first 10 years and option 2 does not. (Both save lives after the first 10 years.) In terms of cost, option 1 costs full price now and option 2 costs half price 10 years from now. With a 7% discount rate, it follows that the present discounted value of the cost of option 2 is one-fourth the cost of option 1. So the cost of saving lives for the first 10 years is three-fourths the cost of retrofit now. To compare the value of 10 years of life saving with 30 years of life saving, calculate $[1 - \exp(-0.07 \times 10)]/0.07 = 7.2$ and $[1 - \exp(-0.07 \times 10)]/0.07 = 12.5$. The ratio is 0.57. The cost per life saved is thus $(3/4)/0.57 \times \$0.6$ million = \$0.8 million for a "typical" URM bearing wall building.
5. If the owner is willing to retrofit the building rather than demolish it, then the cost of life saving for a building lifetime of T years is the cost for 30 years divided by $[1 - \exp(-0.07 \times T)]/0.07/12.5$, as discussed in the answer to Question 4.
6. The answer depends on which quantity relating to retrofit you set equal to which quantity relating to WTP for risk reduction. If you set typical cost to \$3 million, you have to come down a factor of $9.6/3.0 = 3.2$, from \$25 to \$8/ft^2. If you look at a reasonable upper bound for a typical building (cost × uncertainty factor of 3) and equate it with \$7 million, you have to come down a factor of $3 \times 9.6/7.0 = 4.1$, from \$25 to \$6/ft^2. If you look at median cost for a high-cost building (cost × variability factor of 7) and set it to \$7 million, you have to come down a factor of 9.6, from \$25 to \$3/ft^2.
7. Neither earthquakes nor buildings are distributed in a random geographic fashion. A GIS enables the analyst to model realistic distributions of earthquake probabilities, soil types, etc., for buildings in a given location. Earthquake effects, and thus the cost of saving a life, do not really conform to the default lognormal distributions assumed in this chapter. A GIS is the best way to disaggregate the location-dependent features.

CHAPTER IV.3

1. Because of our need to feel in control of our lives, we tend to deny hazards which threaten our sense of control. Thus, unless we have suffered from a hazardous event, we will tend to deny that it will recur even though others might assure us

it will. Sometimes, even if we have lived through such an event, we will rationalize that it won't happen again.
Immediately after the event, almost everyone will recognize that it can happen. However, as people die, move out, or rationalize, fewer and fewer people will. The newcomers will not accept the hazard and so will not prepare, and the overall preparedness of the community will decline.

2. A disaster-management system requires the input from many cooperating agencies. If one agency does not perform as required, the effectiveness of the whole system is threatened.
   Unless the staff of each agency has experience working with the other agencies, there will tend to be tasks that fall "between the cracks" and are left undone. That is, each agency may think that the other is responsible.
   Moreover, key experienced staff in each agency will tend to be transferred to other positions or they may leave. This turnover may be every 4 to 5 years on an average. Since disasters may not happen more than once in twenty years or so, there will be plenty of time for the overall effectiveness of the disaster-management system to decline.
3. First, there is the tendency of communal preparedness to decline.
   Second, disaster-management systems will tend to become less effective as the time since the last event increases.
   Finally, developing (and developed) countries frequently run their disaster-management systems on hierarchical lines, so that potential victims will be led to assume that disaster management is the responsibility of the government and not their problem.
4. The most effective component of any disaster-mitigation system is the members of the public. Most people get their warnings from family and friends, and it is often members of the public who provide vital information to the disaster-management team. People can also substantially reduce their vulnerability to the next disaster by preparing for it. Finally, it is only by constant communal pressure that funding for disaster-management systems are sustained.
5. There is no unique answer to this. However, a key matter for attention is sustaining the preparedness of the community. Without this, funding for disaster management will tend to dry up.
6. A preparedness campaign using ideas set out in Appendix A may be a good start. Disaster-management systems can be categorized as falling into three groups:
   - controlling the event (e.g., levees or dykes to protect against floods),
   - avoiding the event (e.g., by planning regulations or keeping developments out of the way of the hazard),
   - mitigating the effects of the event (e.g., insurance, relief, etc.).

   What is the best mix of strategies could be assessed by using economic analysis, taking account of both monetary and nonmonetary risks.
7. Factors of advantage in many developed countries might be:
   - good communications,
   - good transport facilities,
   - high technology,
   - a tradition of individual initiative.

   Factors of advantage in many developing countries might be:
   - a tradition of communal cooperation,
   - a tradition of striving for consensus,
   - a tradition of avoiding assigning blame.

8. Perhaps one strategy might be to call a meeting to discuss how to maintain the safety of the village during the wet season, and steer the discussion toward matters of individual responsibility.

CHAPTER IV.4

1. Under the GATT measures restricting trade in animals or animal products may be imposed to protect animal or human health in the importing country.
2. When considering importation of animals or animal products, risk assessment is the process of identifying all the potential diseases that could be associated with the particular commodity and then estimating the probability of their being introduced through imports.
3. Risk assessments on imports of animals and animal products are sometimes controversial and open to challenge because many of the assumptions relating to disease prevalence in the source population, survival of pathogens in the commodity and exposure of local livestock to the pathogens have to be made on the basis of few hard data.
4. The effects of risk management measures are usually able to be quantified more objectively because more data are available.
5. When only those animals that fail a specific diagnostic test are excluded from a group intended for import, the risk of introducing the disease in question increases as the size of the group increases.
6. When a single animal reacting to the specific diagnostic test excludes the entire group, risk decreases with increasing group size.
7. The main weakness of deterministic models is that they do not give the decision maker any estimate of the uncertainty of the risk estimate.
8. With most diseases, one cannot be totally confident that embryo transfer is risk-free because insufficient studies have been conducted with most pathogens.
9. Quantitative risk analysis assists in obtaining consistency in decision making and also permits a comparison of the effects of different risk reduction measures.
10. Nonquantitative risk analysis methods are still useful, especially in the routine regulation of imports of animal products, because they can be objective, repeatable and transparent, and are always quicker, thus cheaper, than quantitative methods.

CHAPTER IV.5

1. Human health, environmental impacts, and quality of life.
2. "Precision" refers to the level of detail in the measurement of parameters, generally numerical in nature with a mean (or best estimate) with a range (e.g., standard deviation). Precision can also refer to qualitative data, such as elicitation of expert judgment, which can be either numerical or narrative in nature. In the latter case, precision refers to the level of resolution that an answer provides, and derives from the correctness of the question and how focused the answer is. "Accuracy" refers to whether the results capture the truth somewhere within the numerical bounds, regardless of the size of the error bars. Qualitative data can be accurate, even if it is not very precise.

3. Western medicine has largely focused on the disease rather than the patient, and defines health more in terms of clinically observable symptoms that can be measured by diagnostic tools. The public health disciplines tend to have a broader definition that includes whether a person's function or mobility is impaired or her/his activity is restricted. Only recently has western medicine begun to focus on the whole person, but still does not quite reach the level of holism that indigenous cultures have always followed. In indigenous cultures, some illnesses are seen as purely physical, and, at the other extreme, some illnesses are seen as being the manifestation of spiritual illness. An indigenous health care facility would include both a spiritual health care facility and a medical care facility. Some clinics for Native Americans are now including traditional spiritual healers as well as medical practitioners with better health outcomes for the patients.
4. If community well-being is in place (e.g., the ability to follow traditional activities and healing practices, psycho-spiritual well-being and so on) then contamination would not only affect the person by virtue of direct exposure to contaminants, but would also affect his/her health through the degradation of resources, loss of ceremonial resources, loss of community integrity through reduced social interactions and trade, and so on. A person's health can be adversely affected even if she/he is not directly exposed, but this must be measured as a degree of lost access or use, rather than by personal exposure or symptoms. Prospectively, impacts to community-wide health and personal health can be predicted from seeing adverse environmental or ecological effects. Similarly, the health of a community is, in some respects, a reflection of the degree of ecological health. Conventional risk assessment has yet to recognize this.
5. First and foremost is the long-term perspective that rejects short-term fixes or partial solutions that either postpone the final remedy by imposing it on future generations, or even prevent final cleanup by choosing interim states that preclude more cleanup later. Second, the total environmental contamination burden would be managed in addition to individual hotspots. Third, endstate management goals would tend to be expressed in positive language (such as "achieve holistic environmental stewardship") rather than in negative language (such as "avoid major adverse impacts"). Fourth, risk management decisions would abandon forced decisions between reducing human exposure at the expense of habitat and ecocultural resources through excavation, and would move to decisions about how to reduce contamination while protecting ecocultural resources by choosing less intrusive remediation technology.
6. Restricting access to traditional use areas and traditional cultural properties may violate treaty-reserved rights, it may result in lost community knowledge if access to specific sites is required for teaching, it may harm the spiritual well-being of the community if sacred or ceremonial sites are degraded or if access is denied altogether, it may result in language impacts (place-names, place-specific activity names), it may impair the gathering of specific foods and medicines the loss of which could cause a nutritional or medical decrement, and it may impose detrimental replacement costs on a community that already lacks sufficient funds for adequate health care and nutrition.
7. There are five aspects to this answer: direct food exposures, increased exposure due to food collecting practices (including hunting, fishing, and gathering), indirect exposures to materials that are used for household and cultural items (such as food storage baskets, cooking pots, etc.), unique cultural practices such as the use of

the sweat lodge (increased inhalation exposure if the water used to produce the steam is contaminated), and the wider exposure of the trade network (total community contaminant burden). Thus, a traditional subsistence exposure scenario would have to include more than just an individual's increased consumption of plants and animals that could potentially come from contaminated areas. It will need to reflect the many ways that people interact with the environment, and also the recognition that tribal communities are not exposed just one person at a time, but as whole extended families or communities at a time. It is also important to recognize that when persistent contaminants are present, exposure might extend for more than just one generation, thus resulting in another type of increased community exposure.

8. Breaking the human exposure pathway through the use of institutional controls reduces direct human exposure, but may not result in actual cleanup, and thus would not reduce environmental risk (exposure of biota or ecosystems). If the institutional controls result in lost access to traditional areas or specific ceremonial sites, the ability of the tribal community to exercise their culture and religion is diminished. If the institutional control limits the number of visits or types of activities in order to reduce exposure, this means that the people are being asked to accept exposure in return for being allowed access to their ancestral lands and resources. Using the narrow definition of risk (risk = probability of symptoms if excessive exposure occurs), then reducing risk would be measured solely by the level of human exposure, and success would be defined as preventing exposure in excess of regulatory standards or conventionally accepted risk levels. Similarly, the loss of habitat or cultural resources during remediation would not be valued as highly as human exposure, and those resources might be irretrievably lost if the remedial technology was not chosen specifically to be as least intrusive as possible.
9. The risk to one person would be assessed using an exposure scenario representing the maximum reasonable exposure for the lifestyle that we wish to protect. For tribal members this would be a subsistence lifestyle that includes comprehensive consideration of major pollutant sources to which she/he might be exposed. This might also include a child's exposure scenario, gender-specific activities, co-risk factors such as possible underlying health and nutritional deficits, and so on. The risk to the current population would include cumulative exposures with cancer risk summed over everyone exposed, an evaluation of the number of people exposed to additive hazards from noncarcinogens, and specific evaluation of target organ toxicity (such as neurotoxicity if neurotoxins are present. The risk to future generations would estimate the concentrations of the contaminant over time (10 half-lives, for instance) and evaluate how much cancer or noncancer risk this would result in. This would be expressed using the analogy of how much exposure a person would receive if he lived 1000 or 10,000 years. The number of people exposed at various exposure levels would also be evaluated. The determination of whether this cumulative risk is acceptable can only be made through a negotiation process involving the people whose future members would be impacted.
10. The elicitation of information from experts is an established procedure for developing technical information. Such a process can also be used with tribal elders to develop information about what are the appropriate risk measures to be evaluated, and whether there has been any adverse impact to them. There is no reason to think that this data is any less accurate than the information elicited from other experts, since it is just as verifiable as typical numerical data elicited from technical

experts. This data is no more "anecdotal" than best professional judgment is and should be regarded just as accurate.

11. The degree to which these are different will vary from situation to situation. There are some instances where multiple contaminants may be present, each slightly below its regulatory standard. This risk would be allowed under regulations even if the cumulative risk were above levels typically allowed during Superfund clean-ups. This is both because economic considerations can be part of the basis for developing the regulatory standard, and because contaminants are regulated individually. Regulation-based cleanup and risk-based cleanup might result in different cleanup levels. In other situations, the question is posed as how much risk reduction can we afford. The people whose children are being exposed, the polluters, and the people who must pay (the general taxpayer) may all be different, in which case there will be questions about whether society at large has a moral obligation to help protect someone else's children, and whether the federal government has a legal obligation as natural resource trustee (under NRDA) and guarantor of tribal health and safety (under the treaties). There may also be a disproportionate distribution of benefits versus risks, such as local communities receiving the benefit of jobs in an industry that causes the contamination of resources belonging to people who seldom receive any employment benefit. Thus the question of affordable risk may pit local jobs versus environmental cleanup, and can only be resolved through negotiation and education about respective rights and concerns.
12. This is a critical data gap that has had relatively little attention due to the presumption that if the concentration of a contaminant is low enough to be acceptable for one generation, it should be acceptable no matter how long it persists. For small confined gene pools, such as occurs with many tribes, the cumulative dose to the total DNA contained in the gene pool might be an appropriate unit of analysis. The accumulation of nonlethal detrimental mutations over time could be estimated, and perhaps verified by the examination of genetic polymorphisms. Any such research, however, must be carefully designed since the small numbers of people may preclude statistical significance. The ethics of such research must also be carefully considered, as we have seen with the Human Genome Project when it attempts to sequence the DNA of indigenous populations.

CHAPTER IV.6

1. Sustainable development is defined as: "integrated strategies that would halt and reverse the negative impact of human behavior on physical environment and allow for livable conditions for future generations on Earth" (UNCED 1992).
2. The Agenda 21, developed at the Earth Summit in Rio de Janeiro, 1992, presents a blueprint for development of humanity in the 21st century agreed upon by a majority of countries on Earth.
3. The majority of human societies (countries) are not sustainable, since they are highly dependent on fossil fuels and nonrenewable resources, which are being rapidly depleted.
4. Environmental Impact (EI) = P × C × T × E (this is qualitative and not an exact mathematical expression). The major factors are P = population, C = consumption, T = technology, E = energy consumption.
5. North is the term used to denote developed countries of North America, Europe, and East Asia, while South refers to developing countries located mostly in the

southern hemispheres (Latin America, Africa, South Asia). The importance of each factor in any locality varies. Western Europe and North America contribute mostly to the consumption and energy use (C and E), while in developing countries, the major factors are population (P) and polluting technology (T), in those countries that are attempting to industrialize. Also, often the energy efficiency is very poor (China and former USSR), even though useful energy consumption per person may be small. Countries of former USSR and Eastern Europe have stable and limited populations, and consumption is also low. However, their outdated technology and poor energy efficiency (T and E) are major contributors to tremendous environmental degradation of Eastern Europe uncovered after the fall of communism.

6. Compartmentalization can be overcome by an interdisciplinary and long-term approach to a problem.
7. Public transportation, use of bicycle and walking increases sustainability since the energy expanded by a mile traveled per person is decreased multiple times in comparison with use of private vehicles. Bicycles are the most efficient means of transportation per mile traveled, using about 60 times less energy than cars for the same distance.
8. Urban planning is one of the determinants of a necessity to commute to a work place or daily activities, thus having an impact on energy expenditures or conservation.
9. More efficient process leaves less waste, thus pollution prevention is equivalent to more efficient manufacturing.
10. (a) Food choices have a great impact on health and particularly on occurrence of degenerative diseases, such as heart disease, cancer, arteriosclerosis, diabetes, high blood pressure, and others, both in etiology of those diseases and their management. Meat and milk product consumption over a long time are associated with an increased incidence of degenerative disease, as compared with grains, vegetable and bean consumption.
    (b) Since each pound of meat takes 12–18 lbs of grain to produce, food choices have a direct impact on agricultural impact, which is one of the major factors in the earth-caring capacity; in addition, meat and milk consumption are associated with an increase of water needed for raising animals.
11. All the advances in environmental protection (clean air, water and soil, easier commuting, energy efficiency, etc.), will be foregone if the population keeps growing. The finality of the Earth's resources is a given fact, and therefore Earth systems can support only a limited number of people. Therefore, a prudent policy would be to encourage limits in the number of children per family, rather than encouraging large families.

GLOSSARY

Acute toxicity — the most obvious and easiest measure of toxicity for single dose or short periods of exposures, generally defined by the LD_{50} (lethal dose 50%). This is the dose expressed in mg/kg body weight, which causes death within 24 hours in 50% of exposed individuals after a single treatment, either orally or dermally. LD_{50} is usually derived from animal studies (mice and rats). Measure of acute toxicity for gases is LC_{50} (lethal concentration of chemical in air that causes death in 50% of animals if inhaled for a specified duration of time, usually 4 hours). Based on that definition, chemicals are divided to practically nontoxic, moderately toxic, very toxic, extremely toxic, and supertoxic.

Agenda 21 — presents a blueprint for development of humanity in the 21st century agreed upon by a majority of countries on Earth, developed at the Earth Summit in Rio de Janeiro, 1992.

Benefit-cost analysis (BCA) — a well-developed and highly refined technique used to evaluate many different types of projects and activities, attempting to produce comprehensive measures of benefits and costs over the life of a project, and proceeding to reduce them to a measure of the present value of net benefits.

Bioassay — a test to ascertain toxicity or harmfulness of an agent in a biological system, such as live cells or animals.

Biological diversity — having a multitude of species living harmoniously in a geographical area.

Browser — software programs that are used to move around and search for what is on the Internet's World Wide Web. Two widely used browsers are Mosaic and Netscape, although numerous other browsers are available and others are being developed.

Building safety — features of a building design to reduce casualties in case of earthquake or fire.

CD-ROM (Compact Disc - Read-Only Memory) — one CD-ROM contains the equivalent of several hundred computer floppy disks and thousands of printed pages. The information can be searched and accessed very quickly. CD-ROM versions of printed documents often contain "extras" such as spoken text and other sounds, photographs, interactive maps, "movies," and animation. Also being marketed are CD-Rs (Compact Disc - Recordable), which allow users to record data on compact discs.

Chronic toxicity — adverse effects caused by long-term exposures to relatively low concentrations of chemicals, usually associated with specific organ damage or cancer. Those effects could be systemic toxicity, cancer, or reproductive and developmental toxicity. Data are usually obtained from animal studies and sometimes from epidemiological studies.

Combustion appliances — appliances such as gas-operated water heaters, gas kitchen ranges, and gas/oil furnaces that may release combustive products into residential air.

Community values — societal values that the community considers important for their well-being.

Comparative risk analysis (CRA) — a term used to describe a rapidly growing number of projects performed around the U.S. by state and local groups as a promised new cure for "irrational" environmental management. CRA is supposed to combine the "science of risk analysis" with "community values".

Consensus building — a process by which a group of people could develop agreement about their reality and/or a particular course of action.

Conservation — preserving the current degree of natural habitats or increasing their size.

Consumer products — products used by consumers in and around a residence, e.g., laundry detergents, cosmetics, hard surface cleaners.

Cost shifting — occurs when accounting costs are reduced, not by employing production or management methods that actually reduce opportunity costs, but by burdening workers, the larger community, the environment, or the future with those costs. The product is "too cheap" because there are opportunity costs that are not counted in the price.

Decision analysis — a scientific discipline dedicated to study how people make decisions and to formula mathematical models of the decision-making process

Developing countries — countries that are not yet highly industrialized and where income per capita is much smaller than in Western Europe, North America or East Asia. Since many of those countries are located below the equator, they are also frequently referred to as the South.

Disaster — an event of catastrophic proportions affecting a large number of individuals; it could be caused by natural or technological accidents.

Discommodity — in the process of producing goods (commodities), nearly every economic activity also produces bads (discommodities). Everyone wants goods and no one wants bads, so it should come as no surprise that individuals looking out for their own interests, as they are supposed to do in capitalism, will devote more effort to "capturing" goods (benefits) than bads (costs).

Dose–response — determining toxicological properties dependent on amounts ingested, inhaled, or otherwise entering the human organism. These are usually determined from animal studies. Different end points of toxicity are observed depending on the targeted organ of a chemical. Severity of a particular effect is a function of dose.

Earthquake — a shaking or trembling of the earth that is volcanic or tectonic in origin.

Ecological health — overall quality of normal functioning in an ecological system.

Ecological risk assessment — risk assessment that studies adverse effects of a particular hazard on the entire ecological system.

E-mail (electronic mail) — messages sent via computer from one person to another person or a group, often using the Internet.

Embryo transfer — removing the embryo (usually in mammalian species) and transferring them to a surrogate mother for full development. This technique is used in domestic animal production and for endangered species research and preservation.

Environmental equity — a principle of enabling all individuals with the same healthy environment (air, water, and soil) irrespective of their socio-economic status in society.

Environmental impact assessment — broadly and qualitatively derived, reestablishing the impact on the total environment of a previous or future activity or agent. It is somewhat more than risk analysis.

Environmental justice — ensuring that all the consequences of human activities have equal impacts on population, irrespective of their socio-economic status.

Environmental racism — a situation where a particular race is suffering from a disproportionate, negative, environmental impact.

Environmental management — a process of managing potentially negative impacts of human activities on the environment.

Environmental priorities — a list of environmental issues that have been determined to require attention, based on some rational criteria.

Expert opinion — an opinion of an expert in his/her field, presumably well based on factual and state-of-the-art information in the field. (Could be wrong!)

Exposure assessment — determining the fate of a chemical in the environment and its consumption by humans. Ideally, by analyzing environmental fate and transport of chemicals, and by evaluating food intakes, inhalation, and possible dermal contacts, one can assess total quantities of toxic chemicals in an exposed individual or population, which may cause adverse health effects. In criteria derivation, one uses either worse-case exposure scenario or most-probable exposure scenario, and point values for various human parameters.

Externality — occurs whenever those who create foregone opportunities (costs) do not fully pay for them; that is, when actual costs are greater than accounting costs. If the Central American cattle rancher doesn't have to pay for all the lost opportunities he creates, his costs are lower than they should be, and so is the price of the fast-food burger. Most environmental problems involve externalities, and while some of them are accidental or unavoidable, most are the result of cost-shifting.

Global warming — a phenomenon observed in the last 20 years associated with the rise of the average temperature of Earth, presumably caused by greenhouse gasses (carbon dioxide, methane, and others), mostly resulting from burning fossil fuels, which are not returned back into fixed carbon but are found in increasing concentrations in the global atmosphere.

Hazard identification — identifying potentially toxic chemicals or materials that could cause physical harm.

Hazardous waste — waste that contains toxic chemicals or chemical mixtures, sometimes designated as such by the U.S. EPA.

Heating, ventilation, and air conditioning system (HVAC) — the mechanical systems used to heat, cool, and ventilate a residence.

Human Exposure Factors — important factors such as body weight, inhalation rate, and surface area of skin exposed, which are used in exposure assessments.

Insurance — payment to offset future losses from various adverse effects of our actions. Distribution of risks are to all the participants in such a process.

Insurability conditions — conditions necessary to establish an insurance policy. Condition 1 requires the insurer to set a pure premium by quantifying the frequency and magnitude of loss associated with specific events of the risk. Condition 2 specifies a set of factors, such as adverse selection, moral hazard, and degree of correlated risk that need to be taken into account when the insurer determines what premium and type of coverage (maximum limits, nature of deductible) it wants to offer. A risk is not insurable unless there is sufficient demand for the product at some price to cover the upfront costs of developing the product and the expenses associated with marketing policies.

Intelligent agent (or software agent or "good virus") — software programs that essentially learn what information the user likes to see, and which will then search through electronic mail, databases, and networks on an ongoing basis to retrieve that information for the user. These programs can also be used to deliver data and messages to other users/systems. In a portable device, intelligent agents are used by intelligent assistants (see below) to sort incoming messages based on what the user has previously looked at first.

Intelligent assistant (or personal intelligent communicator [PIC]) — Small, portable electronic devices that can send and receive electronic mail and faxes, sort incoming messages, and handle other functions such as listings of addresses and phone and fax numbers.

Internet (or The Internet) — created by the U.S. Department of Defense in the late 1960s to protect computer networks in the event of wars, the Internet has evolved into a worldwide collection of computer networks used for many purposes. This includes sending electronic mail messages, accessing various databases, and providing textbooks, journals, newsletters, and other sets of information. The result is a collection of computer networks, about 40,000 as of 1995, that appears to be one very large network. Each Internet site has a unique "domain" name, with domain names always having two or more parts separated by dots (e.g., the e-mail address lists the Internet site for the Procter & Gamble Company as "pg.com"). A glossary of Internet terms can be found in the Internet's World Wide Web at http://www.matisse.net/files/glossary.html

Market failure — occurs whenever markets do not live up to the unworldly conditions of perfect competition, because information cannot be perfect and adjustments cannot be instantaneous. When markets are "almost there," with insignificant barriers to the entry or exit of buyers/sellers, an industry-wide standardized product, a large number of small, autonomous sellers/buyers, and extensive accurate information, they are then described as purely competitive.

(Adjustment cannot be instantaneous, and information and foresight cannot be perfect.) Purely competitive markets are characterized by the absence of market failure. When markets fail, the prices of resources and products do a poor job of measuring opportunity costs. Market failure is usually discussed in terms of four general classes of problems: market structure, information, public goods, and externalities.

Microenvironment — immediate area surrounding the person(s) of interest in an exposure assessment, e.g., the shower stall and/or bathroom for a person maximally exposed to water-derived halomethanes during showering.

Modeling — representing a reality with a mathematical formulation.

Monte Carlo modeling — Monte Carlo risk analysis modeling encompasses a range of techniques to mathematically describe the impact of risk and uncertainty on a problem. Each uncertain parameter within the model is represented by a probability distribution. The shape and size of these distributions defines the range of values that the parameters may take and their relative probabilities.

Natural hazard — potential for events in nature, such as droughts, earthquakes, floods, hailstorms, hurricanes, tornadoes, and volcanoes that can threaten the health and welfare of people in vulnerable structures.

Online (or On Line/On-Line) — using a computer and modem to access the Internet, commercial services, and other sets of computer users or databases to send e-mail, download software programs, search for information, etc.

Opportunity cost — value of the next-best use of a resource, where value is measured as alternative benefits foregone. All costs are opportunity costs, and all opportunity costs are foregone benefits.

Personal Assistant — pocket-sized electronic devices that contain the contents of books or other information. The information can be searched using key words, with results displayed on the screen of the personal assistant.

Pollution — distribution of a chemical substance or a mixture of substances at an undesirable location (air, water, soil), where it can cause adverse environmental or health effects. Pollution can be also regarded as resources distributed at wrong places. Pollution could be caused by industrial production, transportation, agriculture, or runoff. Industrial pollution can be defined as the presence of toxic substances in air, water, or soil, often resulting from inefficiencies in production processes. The presence of these substances can present a health risk to humans or ecological systems.

Pollution prevention — methods that could be used to prevent pollution at the source (within a production process), rather than at the end-of-the-tailpipe, as contrasted with waste management and recycling.

Probabilistic risk analysis (PRA) — type of risk analysis that was developed to facilitate the quantification of risks associated with complex engineered systems. It is particularly appropriate for analyzing the frequencies of extremely rare events, such as core melts in nuclear reactors, for which little if any accident data will be available. PRA provides an integrated model of system response. PRA identifies the different types and levels of damage that could result from different system responses. PRA should answer three basic questions: (1) What can go wrong? (2) How likely is it to go wrong? and (3) What will be the consequences if it does? The first question is answered by a structured list of possible accident scenarios. The second question is answered by quantifying the likelihood of each scenario (including the uncertainty about that likelihood). Finally, the consequences of an accident can be assessed in terms of a variety of damage indices.

Public good — a state or result of an activity that is beneficial to most people in a community and constant with their community values.

Quarantine — separating a potentially diseased animal or individual for an interval of time in order to prevent spreading of a disease into an unaffected area.

Residential building factors — residential exposure assessment factors such as air exchange rate and house and/or room volume that are key determinants of the magnitude of potential exposures in a residence.

Risk — a measure of the probability of the introduction of an exotic disease and the seriousness of such an outcome in the context of the importation of animals or animal products.

Risk analysis — a body of knowledge (methodology) that evaluates and derives the probability of an adverse effect of an agent (chemical, physical, or other), an industrial process, a technology, or a natural process. Definition of an "adverse effect" is a value judgment. It could be defined as death or disease (in most cases of human health risk analysis); it could be failure of a nuclear power plant, or a chemical plant accident, or loss of invested money. In some recent cases of risk analysis even vaguely defined terms, such as "quality of life" or "sense of community", have been evaluated using risk analysis.

1. hazard (agent) identification
2. dose–response relationship (how is quantity, intensity or concentration of a hazard related to adverse effects)
3. exposure analysis (who is exposed? to what and how much? how long? other exposures?)
4. risk characterization (reviews all of the above and makes calculations based on data, with all the assumptions clearly stated

Risk perception and risk communication are also often considered as a part of the risk analysis field, and some practitioners consider risk management also within risk analysis. In the U.S., however, risk management is traditionally regarded as an independent process that is performed AFTER an independent risk analysis.

Risk assessment — risk analysis applied in a particular situation, although the term is sometimes used interchangeably with risk analysis.

Risk-benefit analysis — evaluation of risks and benefits of some activity or agent usually based on economic consideration.

Risk characterization — consists of evaluating and combining dose–response relationship data with an exposure assessment. For establishing criteria and standards, assumptions are made about "average exposures" and criteria are set at the concentration at which it is believed that no harm would occur. For example, reference dose (RfD), and health advisories (for 1-day, 10-day and subchronic exposures) are derived for many chemicals with use of safety (uncertainty) factors to protect most individuals. If an actual exposure to an environmental pollutant (or pollutants) exceeds limits set by the criteria, efforts should be made to decrease the concentrations of the pollutant. The magnitude of risk can be estimated by comparing the particular exposure to derived criteria or reference doses.

Risk communication — methods that explain the risks to lay people, so that their perceptions of the situation are not distorted.

Risk management — a set of methods to deal with a real or perceived risk.

Risk perception — how the risks are perceived by different groups of people. Frequently, risk perception is dependent on factors other than risks, such as unfamiliarity, acuteness, catastrophic image, etc. Risk communication is designed to make risk perceptions commensurable with risks derived by risk analysis.

Seismic retrofit — action taken to increase the earthquake resistance of an existing building.

Social cost — cost to the society from an individual or joint action process, which is usually not well characterized, either in amount of money or a long-term impact.

Source characteristics — the nature of the source of residential exposure, e.g., the concentration of a chemical of interest in a particular consumer product and how it might be released into the residence; such as aerosol or vapor at g/h).

Source reduction — pollution prevention at the source rather than at the end-of-the-tailpipe.

Standard — a numerical value established for a particular medium for a concentration of a pollutant. Examples are air standards, drinking water standards, and permissible exposure limits (PELs). Usually standards are derived from criteria (which are obtained using risk analysis) by applying factors other than health concerns.

Sustainability — ability of a system to continue its operation, rather than dying or disappearing.

Sustainable development — integrated strategies that would halt and reverse the negative impact of human behavior on physical environment and allow for livable conditions for future generations on Earth (UNCED 1992). The concept of sustainable development resulted in Agenda 21, developed at the Earth Summit in Rio de Janeiro, 1992, which presents a blueprint for development of humanity in the 21st century agreed upon by a majority of countries on Earth. The majority of human societies (countries) are presently not sustainable since they are highly dependent on fossil fuels and nonrenewable resources that are being rapidly depleted.

Total exposure assessment methodology (TEAM) — a series of human exposure assessment studies conducted by the U.S. EPA to learn about types and levels of chemicals and airborne particulate matter present in residential air and about their sources and relative contributions arising from outdoor (ambient) air.

Uncertainty — ignorance about the value or the probability distribution of a quantity that is useful in performing an analysis or making a decision.

Value of risk reduction — willingness-to-pay to save a statistical life; the incremental value of incrementally reducing the probability of death from some small level to another even smaller level.

Variability — a real-world complexity, in which the value of an important parameter is not the same for each case, but has some nontrivial distribution over the relevant population.

Volatile organic compounds (VOCs) — compounds entering a residence that can be readily volatilized, i.e., they have a fairly high vapor pressure and low boiling point.

World Wide Web (WWW) — a key evolving part of the Internet. It includes a collection of documents that users can navigate through via use of browser software programs and "hypertext" links. Hypertext enables users to highlight certain pictures, words, or phrases, starting with information displayed on "home pages", and then move to linked pictures or pages of information. The hypertext files are moved across the Internet by use of HyperText Transport Protocol (HTTP) software programs that are part of the user's computer and the WWW home page computer. The standard address for a home page or other resource accessible via the WWW is provided as a Uniform Resource Locator (URL), e.g., http://www.matisse.net/files/glossary.html for the Internet glossary.

INDEX

A

Accum, Frederick, 294
Acid rain, 8, 409, 410
ACPA. *see* American Crop Protection Association
Acts of nature, risk assessment of, 83
Actual uncertainty, 40
Actuaries, 4, 252
Acute toxicity
  definition, 15, 451
  ratings of chemicals, 15
Additive dose, 172
Adulterated food, 87
Adverse effect, 1, 2, 423
Agency for Toxic Substances and Disease Registry (ATSDR), 21
Agenda 21, 407, 411, 451
Agricola, 4
Agricultural degradation, 409
Ambient water quality criteria (AWQC), 19
Ambiguous probability, 256
American Crop Protection Association (ACPA), 152
American Indian nations. *see* Tribal interests and risk management
Analytical studies *vs.* descriptive, 25
Animal studies, 26
Animal studies, disadvantages of as opposed to human studies, 28
A posteriori risk analysis, 6
Asbestos, lawsuits resulting from, 301
At-risk criterion, 87
Availability bias, 358
Average parameter values, 14

B

Backward logic. *see* Inductive logic
Bayesian data analysis, 39, 78
BCA. *see* Benefit-cost analysis
Beck, Lewis C., 294
Benefit-cost analysis (BCA), 102–103, 119–120, 451
Benzene exposure, 134
BetaPERT distribution, 47, 48, 51–52, 56, 60
Bhopal, India, accident in, 1, 68, 234, 414
Bicycles, reduce transportation burden, 412, 415
Bioaccumulation, 17
Bioassay, definition, 451
Biological diversity, 91, 451
Biological gradient, 27
Biological plausibility, 27
Biomarkers, 26
Biospheric egalitarianism, 89
Bottomry, 4
Breast implants, lawsuits resulting from, 301
Bruntland report, 418
Buffet, Warren, 101
"Business *vs.* environment," 3
Buyer manipulation, 99

C

Cancer
  case-control studies, 25
  hazards, case reports used to identify, 26
  morbidity, 28
  mortality, 28
  potency slope, 7, 16, 19
  probability of developing, 7
  projected deaths, total, from radiation, 166
  risk
    OSHA definition of, 16
    three-parameter equation for, 38
    U.S. EPA definition of, 16
  screening, use of biomarkers in, 26
Cancer potency slope (q*), 19
Carbon cycle, 191
Carcinogen, 7, 147, 150, 164, 165, 204, 297, 298
  pesticides, 147, 157
  risk analysis, 6
  risk assessment of, 299–300

group types, 40
human data, use in, 23
U.S. EPA guidelines, 23–25
Carson, Rachel, 5, 144
Catastrophic miscalculations, 316–317
Causal dependence, 76
Causality, criteria for, 27
Causally dependent events, 76
"Cause unacceptable injury to the resources," 403–404
CCDF. *see* Complementary cumulative distribution function
CD-ROMs, use of, in risk assessment. *see* Computerized risk assessment
CEA. *see* Cost effectiveness analysis
Center for Epidemiology and Policy, 28
Center for Science in the Public Interest, 115
Central Limit Theorem, 49
CERCLA. *see* Comprehensive Environmental Response, Compensation and Liability Act
Challenger disaster, 234
Chemical contamination, 409
Chemical risk analysis
chronic toxicity, 16
dose–response relationship, 14
"end points" of toxicity, 14
exposure assessment, 14
exposure levels, methods for establishment of, 13
hazard identification, 14
long-term exposure criterion, 13
NIOSH model, 20
OSHA model, 20
risk characterization, 14
short-term exposure criterion, 13
subchronic toxicity, 16
U.S. EPA model, 18–20
Chemical transformation, 17
Chernobyl, accident in, 1, 7, 68, 182, 234, 329, 331–333, 338, 411
Chicago school of conservative economics, 114
Clean Water Act, 92
Code of Hamurabi, 4
Coherence, as criteria for causality, 27
Common mode failure, 39
"Common sense" risk analysis, 6
Community Right-to-Know Law, 20, 414
Comparative risk analysis (CRA), 3, 102–103, 118–119, 120
committees
composition of, as shortcoming, 275
expertise level, as shortcoming, 275
community values, influence of, 275–278
comparative risk assessment, stages of, 260–263, 265–266
comparative risk management, stages of, 260
credibility of, 279
critique of, 268–269
definition of, 99, 452
elements of, 274
environmental problem ranking, 275–278
environmental risk analysis, 260, 266–268, 279
history of, 259–261
opinion *vs.* objective science, 273
public committee, role of, 263–264
Quality-of-Life Subcommittee, 277
science of, 278
specialist committee, role of, 263–264
"straw" ranking, 267
U.S. Environmental Protection Agency's "Unfinished Business" report, 260, 265
Comparative risk assessment, 260, 261, 263, 265, 266
Comparative risk management, 260, 303
Compartmentalization, 408–409
Complementary cumulative distribution function (CCDF), 70
Comprehensive Environmental Response, Compensation and Liability Act (CERCLA), 18
Computerized risk assessment
CD-ROMs, 203, 205, 213, 228, 451
CD-ROMs, future uses, 223–225
database reviews of software and databases, 208–209
databases, 203, 204, 205
databases for toxicology and risk assessment data, 209–214
Toxline, 209–210, 217, 225
TOXNET, 210, 217, 225
DIALOG, 208–210, 228
global electronics communications, future uses, 215–223
"information appliances," 226
"intelligent agents," 226
International Chemical Regulatory Monitoring System, online, 213
Internet, 217–218, 227
Internet, books and other information on, 223
"Java," 226
magazine and catalog software and database reviews, 207–208
MiniDisc Data (MD Data), 225
National Library of Medicine, online, 209–211, 214, 217
online, software available, 203
online databases, 204

online services, 211
online services, commercial, 215–217
outdatedness of information, potential for, 204
predictions, future software and online applications, 227–228
publications available online, 203, 205–208
search engines, 218
society activities relating to software and databases, 214–215
software, 203, 204, 205
U.S. Environmental Protection Agency, online, 211–212, 217
U.S. Food and Drug Administration, online, 212
U.S. National Library of Medicine, online, 225
World Wide Web, 131, 203, 211, 215, 218–224, 226, 227, 458
Conditional branching, 53
Conditional probabilities, 74
Consequence evaluation, ecological, 95
Conservatives, economic, 111, 114
Consistency, as criteria for risk assessment, 27
Consumer Product Safety Commission, 301, 307
CONTAM model, 130
Correlation, in risk equations, 39
Cost-benefit analysis, 409
Cost-effectiveness analysis (CEA), 119, 120
Cost shifting, 104, 107, 452
Court of Appeals vs. scientific community, 313–314
CRA. *see* Comparative risk analysis
Crystal Ball, use in Monte Carlo spreadsheets, 46
CTUIR. *see* "Cause unacceptable injury to the resources"
CTUIR criteria, 403–404
Cultural risk, 396–397
Cumulative distribution, 48, 52–53
Customer, as core of TQM, 91

D

Dalcon Shield, lawsuits resulting from, 301
Dams, negative environmental impacts of, 411
Databases. *see* Computerized risk assessment
DCS. *see* Disappointed consumer syndrome
DDT, 16, 17, 145
Decision trees, 63
Decompose, dominant contributors in PRA analysis, 80
Deforestation, 409
Degradation, agricultural, 409
Delaney Clause, 148, 298, 299–300
de LaPlace, Pierre Simon, 4
Demand side economics, 110
Dependencies
expert opinion and, 59
rank order correlation and, 58–59
DES, 298, 299
Descriptive studies *vs.* analytical, 25
Desert Storm, as demonstration of oil dependence, 413
Deterministic risk analysis, 337
Developing countries, importance of implementing energy efficiency in, 410
Dichlorodiphenyltrichloroethane. *see* DDT
Dietary exposure, 150
Dietary exposure, in pesticide regulation, 148, 149
Diethylstilbestrol. *see* DES
Disaggregation, 47
Disappointed consumer syndrome (DCS), 106
Disaster-management systems
availability bias, 358
declining communal preparedness, 356
denial, problem of, 357–359
disaster-mitigation works, 356, 361–362
economic changes, 361
economic losses from disasters, 356
environmental changes caused by disasters, 361
improvement of
coercion *vs.* persuasion, 370
community control, 363–366
community preparedness, 369–370
disaster recovery systems, 370–371
disaster warning systems, 366–369
media, the, 371
involuntary *vs.* voluntary risky activities, 358
maintenance of disaster-mitigation works, 361–362
organizational readiness, decline of, 359–361
psychological barriers to, 355
psychological effects of disasters, 362
"Ready-Set-Go" warnings, 368
social effects of disasters, 362–363
strategies for, 372
voluntary *vs.* involuntary risky activities, 358
Discrete distribution, 48, 53
Dose estimate, in calculating cancer risk, 38
Dose–response assessment, 24, 28
Dose–response relationship, 2, 14, 16, 274, 278, 279

E

Earthquakes, risk assessment of. *see* Seismic risk and management, California
Earth Summit, 418
Ecocultural evaluation, 402
Ecocultural management, 399–401

Ecological alteration, 95
Ecological consequence analysis, 95
Ecological degradation, 95
Ecological health, 95, 96
Ecological risk analysis, 3
Ecological risk assessment
  bioassays
    limitations of, 92
    role of, 92
  biospheric egalitarianism, 89
  burden of proof, question of, 92
  "command and control" approach, 92
  computer simulation, 93
  defining the problem, 93–94
  definition of, 88
  ecological consequence analysis and, 95
  ecological health, concept of, 95, 96
  emotional reactions to, 88
  endangered species, 93
  environmental impact analysis and monitoring, 92
  environmental stressors, 88
  expert opinion, better use of, 95
  expert opinion, problems of, 93
  modeling, 93
  multiple-use model, 90
  policy perspective, 94
  proposed changes, 95–96
  public preference, 96
  risk paradigm and, 89, 90–92
  scientific perspective, 94
  selection of at-risk criterion, 87
  simplification of presentations, need for, 96
Economic risk analysis
  basic economic questions for, 100
  benefit-cost analysis, 119–120
  benefits and costs, 112–120
    consumer demand, 114
    preferences, 113–115
    preferences, individual, 115
  community need, 110–112
  comparative risk, 99
  comparative risk analysis (CRA), 118–119
  conservative viewpoint, 111, 114
  cost shifting, 104, 107
  costs of hazardous waste cleanup, 8
  extra-market concerns, 99
  individual benefits, 102
  individual preferences, 115
  marginal benefit, 101
  marginal cost, 101
  market
    failure, 101
    failure, types of
      externality (social cost), 107–108
      information failure, 105–106
      market structure, 105
      public good, 106
    prices, 101
    shortcomings, 109–110
    "voice," 101
  opportunity cost, 102–104
  risk benefit, 99
  risk benefit analysis, 116–118
  risk reduction, 100
Economics, demand side, 110
Economists, neoclassical, 105
Ecosystem management, 91
Ecosystems, 92
EIS. *see* Environmental impact statements
Elixir Sulfanilamide, 296, 297
Endangered Species Act, 92
"End points" of toxicity, 14
Energy efficiency, 410–411
Environmental risk analysis, 260
Environmental impact statements (EIS), 283
Environmental justice, 3, 394–395
  comparison areas, 288
  concept of, 284
  definition, 453
  environmental consciousness movement, 281–282
  environmental impact statements, 283
  equity, 283, 284–285
  ethnic controversies, 282
  fairness, 284
  future directions, 289
  grassroots nature of, 283
  locational issues, 286
  measuring risk, 284–285
  minority groups, definitions of, 285
  "Not-In-My-Backyard" concept, 282
  origins of, 281–283
  population selection, 286–287
  racial controversies, 282
  racism, 284
  risk, defining, 284
  statistical protocols, 288
  temporal dimension, 288
  U.S. Bureau of Census, database, 281, 285
Environmental management, 4, 453
Environmental management, Western approach, 391
Environmental problem ranking, 275–278
Environmental protection
  vs. business interests, 3
  vs. free markets, 3
Environmental risk analysis, 275–278
Environmental Risk Reduction Act, 88
Environmental tobacco smoke (ETS), 126

EPA. *see* U.S. Environmental Protection Agency
Epidemiological risk analysis, 3, 6
analytical, 25
case-control, 25
cohort, 25
causal relationships and, 7
descriptive, 25
evaluation of, 26–27
Errors of measurement, 36
ETS. *see* Environmental tobacco smoke
Evelyn, John, 4
Event trees, 67, 72
deductive logic, 72
dependencies between events, 74
guidelines for construction, 76
initiating event, 73
order of events, 74
pruning of, 76
top events, chronological order of, 76
Executive Order 12898, 282
Expert opinion, 48
brainstorming and, 56
combining two dissimilar views, 57
definition, 453
dependencies and, 59
interviewing techniques and, 56
risk analysis model, use in, 55
Exposure analysis, 2, 274
Exposure assessment, 17, 24, 278, 453
Extra-market concerns, 99

F

Failure causes, 80
Failure probability, 38
Fault tree, 72
analysis, 181
combinatorial numbers of events, 74
inductive logic and, 72
linking, 75
top event, 72
FDA. *see* U.S. Food and Drug Administration
Federal Executive Order 12898, 281
Federal Food, Drug, and Cosmetic Act, 146, 147, 296–297
Federal Food and Drugs Act of 1906, 295
Federal Insecticide, Fungicide and Rodenticide Act, 143, 144, 145, 152, 153
Federal Meat Inspection Act, 295
Federal Register, 299, 300
Federal risk assessment, evaluation, 305–309
FFDCA. *see* Federal Food, Drug and Cosmetic Act
FIFRA. *see* Federal Insecticide, Fungicide and Rodenticide Act
First National People-of-Color Environmental Leadership Summit, 283
Food and Drug Administration. *see* U.S. Food and Drug Administration
Forbes, Steve, 101
Formal risk analysis, classifications
a posteriori, 6
carcinogen, 6
epidemiological, 6
noncancer chemicals, 6
nonquantitative, 6
probabilistic, 6
Forward logic, 72

G

Gantt charts, 63
GATT. *see* General Agreement on Tariffs and Trade
Gauss, 36, 37
Gaussian distribution, 41, 42
General Agreement on Tariffs and Trade, 377–378, 386
General distribution, 48, 52
Global climate change risk assessment
acceptable risk, 193–196
lay *vs.* expert perception of, 193
aerosols, 194
atmospheric concentration of CO_2 as factor in, 188, 191
carbon cycle, 191
CO_2 emissions, 188–189, 191–199
CO_2 emissions per unit of energy as factor in, 188, 191
energy production *per capita* as factor in, 188, 190–191
global warming, 187–188
greenhouse gas, 188, 191
intergenerational equity, 197–198
International Panel on Climate Change, 195, 196
interregional equity, 197–198
population as factor in, 188, 190
probabilistic risk assessment, 193
risk-benefit equation, 198
sea level rise per unit as factor in, 188, 192–193
sequential decision-making strategies, 196
stochastic uncertainty of, 190
temperature rise as factor in, 188, 191–192
Global environmental degradation, causes of, 407–408
Global warming, 8, 187–188, 409, 453
"Good risk" *vs.* "bad risk," 253, 254
"Good science," 305, 309–314
Gore, Al, 193

Governmental control of risk
  adulterated food, concept of, 294
  carcinogens, risk assessment of, 299–300
  colonial America, laws protecting food and drugs, 294
  Delaney Clause, implementation of, 298–300
  DES crisis, 298, 299
  Elixir Sulfanilamide tragedy, 296–297
  Federal Food and Drugs Act (1906), 295
  Federal Food, Drug, and Cosmetic Act of 1938, 296–297
  Federal Meat Inspection Act, 295
  food and drug regulation, 293
  food preservatives, 295
  history of, 293
    pre-1900, 294–295
  pesticide residues, 300
  Select Committee to Investigate the Use of Chemicals in Food, 297–298
  U.S. Congress, role of, 295, 297
  U.S. Department of Agriculture, role of, 295, 298
  U.S. Food and Drug Administration, bans imposed by, 295, 300
  U.S. Food and Drug Administration, role of, 295–297, 299, 300
Government interventions, 4
Greenhouse gas emissions, 410
Groundwater contamination, 150

H

HA. *see* Health advisories
Halley, Edmond, life-expectancy tables of, 4
Hayek, Frederich, 8, 9–10
Hazard analysis, 414
Hazard identification, 2, 14, 24, 26, 274, 278, 453
Health advisories (HA), 19
Health Canada, 152
Hippocrates, 3
Historical data, use in developing risk analysis model, 54
Holistic risk evaluation, 402
Human data, 23
  future of, in risk assessment, 28–29
  use in exposure-response assessment, 28
Human studies *vs.* animal studies, 28
Hypergeometric distribution, 48

I

ICRP. *see* International Commission on Radiation Protection
Immediately dangerous to life and health (IDLH), 19
Independence, in risk equations, 38
Indigenous communities, approaches to environmental management, 391
Individual plant examination (IPE), 82
Inductive logic, 72
Industrial pollution risk assessment
  acid rain generation, 178
  chemical
    industry, 180
    informatics and, 184
    plant accidents, 178
    releases, 178
    risk analysis, 180, 183
  closed loop production systems, 178
  defining boundaries, 184
  definition of, 177
  economic risk analysis, 180
  energy production, 182
  environmental exposure assessment, 183–184
  fault-tree analysis, 181
  global warming, 178
  hazardous materials handling, 181–182
  incineration, 181
  Love Canal, 181
  manufacturing, other than chemical, 181
  modern risks of pollution, 178
  oil spills, 178, 182
  ozone depletion, 178
  probabilistic risk analysis and, 179, 181, 183
  public perception of risk, 178
  raw materials, 179, 180, 182
  reference dose, 180
  SARA, 179
  special interest groups, 178
  Superfund, 181
  Title III, 179
  toxic release inventory, 179–183
  transportation, 182–183
  U.S. Environmental Protection Agency and, 179
Influence diagrams, 63
Initiating event, 80
Insurability of risk, 3
  ambiguous *vs.* nonambiguous probability
  coinsurance, 255
  correlated risk, 256
  deductible, 255
  "good" risk *vs.* "bad" risk, 253, 254
  identifying risk and
    earthquakes, 249
    fire, 248
    underground storage tanks (USTs), 250
  insurer *vs.* the insured, 247
  loss ratio, 252
  marketability and effect on, 257

nonambiguous *vs.* ambiguous probability, 256
premiums, 247, 256
premium setting for specific risks
administrative costs and effect on, 257
adverse selection and effect on, 252–254
ambiguity of risk and effect on, 250–252
correlated risk and effect on, 255
moral hazard and effect on, 254–255
uncorrelated risk, 256
Insurance, origins of, 4
Integrated plant-specific model, 80
Interagency Regulatory Liaison Group (IRLG), 307–309
International animal trade risk analysis
anthrax risk, through import of green hides, 381–382
disease, spread of, through trade, 377, 378–379
disease transmission via embryo transfer, 383
disease transmission via table salmon, 384–385
fish diseases, 384–385
livestock disease surveillance programs, 380–381
Monte Carlo simulation to determine risk, 383
New Zealand beef industry, 380–381
policy decisions and risk management, 379–380
probability calculations, 381–382
rabies risk through dog importation, 383–384
serological tests and, 379–380
International Commission on Radiation Protection (ICRP), 166
International Panel on Climate Change (IPCC), 192, 195, 196
Internet, 3, 454. *see also* Computerized risk assessment
Ionizing radiation risk assessment
additive (cumulative) dose, 172
beta particles, 168
cancer deaths, total projected, 166
carcinogenic risk, 164, 165
chemical risk assessment vs., 163
conservatism, 174
cumulative (additive) dose, 172
dose standards, 166
effective dose, 163, 170
energetic electrons, 168
energy-dependent factor, 165
external dose, 163, 167–168, 171
calculation for, 173
external dose, combined with internal dose, 171
gamma photons, 168
H_{10}, 168, 171, 173
history of, 164–165
individual dose, 172
individual risk, 171
internal dose, 163, 169–171, 173
internal dose, combined with external dose, 171
mutagenic risk, 164
national and international standards, 166
neutron dosimetry, 168
photons and, 167, 170
population dose, 163, 164, 172
population risk, 173
probabilistic risk and, 164
radiation type, 163
radionuclides, 169, 170, 173
risk-benefit analysis, 166
stochastic risk and, 164
types of radiation
alpha, 165
beta, 165
gamma, 165
neutron, 165
X-ray, 165, 167
whole-body dose, 167, 169, 170–171
IPCC. *see* International Panel on Climate Change
IPE. *see* Individual plant examination
IRLG. *see* Interagency Regulatory Liaison Group

J

Justice, environmental. *see* Environmental justice

K

Keefe, Frank R., 297–298
Knight, Frank H., 114

L

Large event trees, 74–75
Large fault tree, 75–76
Lawn chemicals exposures, 134–136
Levels of concern, 18, 19
Life-expectancy tables, 4
LOCs. *see* Levels of concern
Loglogistic distribution, 48
Lognormal distribution, 48
Love Canal, 234, 282
Lowest-observable adverse-effect level (LOAEL), 18, 19
Lowest-observed-effect level, 424

M

MADD. *see* Mothers Against Drunk Driving
Magnitude of association, 27
Marginal cost and benefit, 101

Market
failure, 99, 101, 454
prices, 101
"voice," 101
Massengil Company, 296
MAVRIQ, 130
Maximum equilibrium yield, 90
Maximum sustainable yield, 90
Mechanistic risk analysis, 337
Mendel's law of heredity, 296
Meta-analysis, 25
MF. *see* Modification factor
Milken, Michael, 101
Modern risk analysis, 4
Modification factor (MF), 19
Monte Carlo risk analysis model, 3, 14, 17, 37, 39, 383
definition of, 45, 455
design of, 46–47
probabilty distributions of, 46
real-world distribution data and, 14
software, 46
stages of, 46
three-point estimate distributions, 47
Monte Carlo software, 63
Morbidity ratio, 7
Mortality ratio, 7
Mothers Against Drunk Driving (MADD), 102, 106
Mothers of East Los Angeles, 283
Multi-Chamber Concentration and Exposure Model, 130
Multiple-use model, 90
Multistage theory, 309
Mutagenic risk, 164

N

Nader, Ralph, 114
NAFTA. *see* North American Free Trade Agreement
National Academy of Sciences (NAS), 149, 192, 327
National Hydrometeorological System (NHO), 368
National Institute for Occupational Safety and Health (NIOSH)
criteria and standards for chemicals, 14, 21
establishment of, 5
National Research Council, 28, 29
risk assessment paradigm, 24
National Toxicology Program, 298
Native American tribal interests. *see* Tribal interests and risk management
NATO. *see* North Atlantic Treaty Organization
Natural disasters, 1
in relation to probabilistic risk analysis, 68
risk management. *see* Disaster-management systems
Natural risks, 3
Neoclassical economists, 105
NIMBY. *see* "Not-In-My-Backyard" concept
NIOSH. *see* National Institute for Occupational Safety and Health
NOAEL. *see* No-observable adverse-effect level
NOEL, *see* No-observable-effect level
Nonambiguous probability, 256
Noncancer chemicals risk analysis, 6
criteria and standards for, 7
exposure threshold and, 7
Noncarcinogenic chemicals, 7
Nonchemical cancer risk analysis
environmental media uses, 13
specific exposure scenario uses, 13
Nonparametric distribution, 48
cumulative, 48, 52
discrete, 48, 53
exceptions to using model expert opinion, 48
general, 48, 52
triangle, 48, 49
uniform, 48, 49–50
Nonprobabilistic risk analysis, 423
Nonquantitative risk analysis, 6
No-observable adverse-effect level (NOAEL), 18, 19
No-observable-effect level (NOEL), 16, 204, 297, 424
Normal distribution, 48
North, Douglas, 107
"North," United Nations definition of, 5
North American Free Trade Agreement (NAFTA), 145
North Atlantic Treaty Organization, 154
No-till agriculture, 416
"Not-In-My-Backyard" concept, 282
NRC. *see* National Research Council; Nuclear Regulatory Commission
Nuclear power industry, risk and
accident history, 331–333
industry groups, 337
Nuclear Regulatory Commission, 328, 336, 338
Reactor Safety Study, 327
regulatory practices, 334
risk-assessment practices, 334–335
risk-based regulation, 335–338
safety assessment practices, 334–335
safety of, 329–331
status of industry, 328
thermal hydraulics, 329

Nuclear power plants, 5
Nuclear Regulatory Commission, 81, 336, 338

O

Objective uncertainty *vs.* subjective, 40
Observational studies, 26
Occupational exposures, 4
Occupational Safety and Health Administration (OSHA), 301
  criteria and standards for chemicals, 14
  establishment of, 5
OECD. *see* Organization for Economic Cooperation and Development
Office of Science and Technology Policy, 24
Opportunity cost, 99, 102, 103–104, 455
Opportunity costs, measurements of, 108–109
Optimum sustained yield, 90
Organic farming, 416
Organization for Economic Cooperation and Development (OECD), 145, 154
OSHA. *see* Occupational Safety and Health Administration
OSTP. *see* Office of Science and Technology Policy
Out-of-spec operation, 79
Outrage factor, as applied to risk assessment, 193
Ozone hole, 409

P

Paracelsus, Philippus Aureolus, 15
Parameter-based ecocultural evaluation, 402
Parametric distribution
  beta, 48
  BetaPERT, 51–52
  hypergeometric, 48
  loglogistic, 48
  logonormal, 48
  normal, 48
  Pareto, 48
  Weibull, 48
Pareto distribution, 48
Pascal, Blaise, probability theory of, 4
Paternalism, in legislative initiatives, 114
PCBs. *see* Polychlorinated biphenyls
PELs. *see* Permissable exposure limits
People for the Community Recovery of Southside Chicago, 283
Perception of uncertainty, 40
Perchloroethylene exposure, 134
Permissable exposure limits, 20
Perot, Ross, 101
PERT networks, 51
Pesticide Handlers Exposure Database (PHED), 152–153
Petzoldt, Paul, 110
Pinchot, Gifford, 90
Plank, Max, 5
Plant-specific assessment, 335, 336
Policy decisions, 8
Pollution
  definition, 455
  historical example of risks from, 4
  industrial, *see also* Industrial pollution risk assessment, 414–415
  risk assessment of. *see* Industrial pollution risk assessment
Polychlorinated biphenyls (PCBs), 16
Population dose, 163, 164, 172
Population explosion, 417–418
Population risk, 173
PRA. *see* Probabilistic risk assessment
PRA quantification, 77–80
Premiums, 4
Probabilistic dependence, 76
Probabilistic risk analysis, 3, 6, 36, 336, 424
  accidents, quantification of risk of, 68
  definition, 455
  failure-tree analysis and, 7
  fault-tree analysis and, 7
  historical data, use of, in, 7
    human activities and, 7
    natural phenomena and, 7
  industrial process safety and, 7
  insurance industry, use by, 7–8
Probabilistic risk assessment (PRA), 164, 179, 181, 183, 193, 328
  applicability of test data, 79
  case histories and, 78
  causes of component failure, 78
  common cause failure rates, 77
  component
    failure rates, 77
    fragilities, 77
    maintenance durations, 77
    maintenance frequencies, 77
    -specific information, 77–78
    type, 78
  components of interest, 78
  corrective actions in reports, 80
  data base study period, 78
  development of, 67
  disturbances during operation and, 68
  environment and, 78
  expert opinion and, 78
  failure frequencies for similar components, 78
  failure modes, 78

human error rates, 77
incipient failures and, 79
in-depth assessment, need for, 82
initiating events, 77
integrated model of system response and, 68
interactions between subsystems and, 69
level of detail of analysis, 78
levels 1 and 2, 71
operations and maintenance staff and, 83
out-of-spec operation, 79
performance history and, 78
potential risk management actions and, 69
qualitative measures of risk and, 69
quantitative assessment, uncertainty in the results, 69
quantitative measures of risk and, 69
risk contributors and, 69
risk management using, 80–82
severe acts of nature and, 83
structure of model
fault trees, 72
level 3, 72
successful examples of risk management using, 81
support systems, failure of, 82
usage and, 78
Probabilistic safety assessment, 334
Probability judgements, 41
"Proposed and Interim Guidelines for Carcinogen Risk Assessment," 25
PTEAM, 128
Public perception of risks, 3
Pure public good, 106

Q

Q*. *see* Cancer potency slope
Quantitative risk analysis, 33, 45
Quantitative risk assessment, 299, 300

R

RAC. *see* Raw agricultural commodity
RACs, 132
Radiation risk assessment. *see* Ionizing radiation risk assessment
Radioactive waste, 7
Radionuclide, 163
Rank order correlation, 58
Raw agricultural commodity (RAC), 132, 146, 147
Rayleigh distribution, 54
RBA. *see* Risk benefit analysis
RCRA. *see* Resource Conservation and Recovery Act
Reactor Safety Study, 327
"Ready-Set-Go" warnings, in disasters, 368
Recommended exposure limits, 20
"Red Book," 24
Reference contractions (RfC), 20
Reference dose (RfD), 14, 18, 19, 180
general formula for, 19
Registry of Toxic Effects of Chemical Substance, 14
Regulation, pesticide
benefits *vs.* risks, 145–146
carcinogen(s), 147, 150
dietary exposure, 148–150
exposure, methods for regulating, 146
food safety, 146–151
groundwater contamination, 150
history of, 144–145
negligible risk, 147
occupational exposure, 151–155
monitoring history in U.S., 151
risk assessment, role in, 143
Reilly, William K., 260
RELs. *see* Recommended exposure limits
Reportable quantities, 18, 19
Report of the Sanitary Commission of Massachusetts, 294
Residential exposures assessment and analysis
data sources, 136–137
environmental chemicals and, 126
environmental tobacco smoke and, 126
human exposure factors, 125
incidental ingestion exposures, 132
indoor source cases, 132–134
inhalation exposures, 129–131
outdoor use cases, 134–136
PTEAM studies and, 128
residential exposure factors, 125
residential microenvironment, 126
"sinks," 127
software for, 130–131, 136
TEAM studies and, 126–128, 132, 134
VOCS and, 126, 127
Residential Exposures Assessment Project, 137
Resource Conservation and Recovery Act (RCRA), 18
Resource depletion, 409
RfC. *see* Reference contractions
RfD. *see* Reference dose
@RISK, 46
Risk analysis
animal trade. *see* International animal trade risk analysis
basic economics of, 3
chemical. *see* Chemical risk analysis
common elements of, 2

"common sense" type, 6
comparative. *see* Comparative risk analysis
complexity of, 2
complex problems, applied to, 9
computer software programs and, 3
conceptual development of, 5
deductive reasoning and, 378
definition of, 1–2, 456
eating habits and, 417
economics. *see* Economic risk analysis
economics of, 99
buyer manipulation, 99
market failure, 99
opportunity cost, 99
hazard analysis, 414
historical overview, 3–4
inductive reasoning and, 378
Internet, use of, and, 3
limitations of, 8–10
livestock, international trade. *see* International animal trade risk analysis
model
expert opinion, input of, 55, 453
historical data, use in, 54
test data, use in, 54
nonprobabilistic, 423
nuclear power plants and, 5
probabilistic, 424
quality of life defined in terms of, 1
residential chemical exposures. *see* Residential exposures assessment and analysis
results, presentation of, 61–63
sense of community defined in terms of, 1
sustainable development. *see* Sustainable development
uncertainty, 3
variability, 3
Risk Analysis for Superfund, 20
Risk assessment, 1
assumptions, 6
carcinogens, quantitative assessment, 299–300
CD-ROMs for. *see* Computerized risk assessment
classifications, 24
climate change. *see* Global climate change risk assessment
comparative. *see* Comparative risk analysis
computer software programs. *see* Computerized risk assessment
databases. *see* Computerized risk assessment
definition, 303, 456
earthquakes. *see* Seismic risk and management, California
ecological. *see* Ecological risk assessment
federal agencies and, 304, 307–309
global climate change. *see* Global climate change risk assessment
governmental control. *see* Governmental control of risk
guidelines, 29
industrial pollution of. *see* Industrial pollution risk assessment
insurability and. *see* Insurability of risk
Internet. *see* Computerized risk assessment
ionizing radiation. *see* Ionizing radiation risk assessment
law and. *see* Governmental control of risk
Monte Carlo method, 217
nuclear power industry. *see* Nuclear power industry, risk and
online systems, use of. *see* Computerized risk assessment
outrage factor, 193
pesticide regulation. *see* Regulation, pesticide
pollution. *see* Industrial pollution risk assessment
probablistic. *see* Probabilistic risk analysis; Probabilistic risk assessment
risk analysis, role in, 378–379
risk communication, role in, 378–379
risk management, as separate from, 24, 94, 303
risk management, role in, 378–379
software. *see* Computerized risk assessment
tribal interests. *see* Tribal interests and risk management
uncertainty in, 6, 33–36, 304
World Wide Web in. *see* Computerized risk assessment
Risk-based regulation, nuclear industry, 335–338
Risk-benefit analysis (RBA), 102–103, 116–118, 120, 166
definiton, 456
FDA, as used by, 116
Risk characterization, 2, 24, 274, 278, 456
Risk, definition of, 164, 423, 456
Risk distribution, 37
Risk estimates
historical trends in, 41
overconfidence in, 40–42
past projections in, 41
Risk judgement, 235
Risk management, 3, 293
animal trade. *see* International animal trade risk analysis
definition, 303, 457
disaster management. *see* Disaster-management systems
earthquakes. *see* Seismic risk and management, California

livestock. *see* International animal trade risk analysis
nuclear power industry. *see* Nuclear power industry, risk and
risk assessment, as separate from, 94
trade, livestock. *see* International animal trade risk analysis
tribal interests. *see* Tribal interests and risk management
Risk perception
cognitive maps, 234
definition, 457
"dread risk," 236
paradox of, 233
psychometric paradigm, 234
race, as factor in, 235
risk judgement, female scientists' *vs.* male scientists,' 235
sociopolitical factors, 236
trust, destruction of, 239
trust, importance of, 237–239
democracy, role in, 239
trust, psychological principles, 242–243
trust-building events, 239
trust-destroying events, 239, 240, 242
trust erosion in government, 243
"unknown risk," 235
Risk reduction, 116
RISKview Pro, use for comparing relative heights for values within a range, 56–57
Roosevelt, Franklin, 296
Roosevelt, Theodore, 295
RQs. *see* Reportable quantities

S

Safety, concept of, 294
Sanitary engineers, 33
SARA, 18, 20, 179, 282, 414, 415
Scalar quantities, 70
Scatter plots, 63
Science and social policy, integration of, 314–319
Scientific management, 90
SCIES. *see* Screening-Level Consumer Inhalation Exposure Software
Screening-Level Consumer Inhalation Exposure Software, 129
Seismic risk and management
mitigation programs
Long Beach, 350
Los Angeles, 350
retrofit, 350, 352
Seismic Safety Commission
standardized loss estimation methodology, 352
uncertainty, in probability estimate, 348
Unreinforced Masonry Building Law, 342, 350
unreinforced masonry (URM), 343, 344, 348–352
mandatory programs, 351
notification-only programs, 351
voluntary programs, 351
variability, in risk analysis, 349
Seismic risk and management, California, 341
building codes, 342
building types, as threats to safety, 344
discounting lives, 346–347
Loma Prieta quake, 343
Los Angeles, repair ordinance passed in, 344
Northridge quake, 342, 344
property damage, 342
pseudo-discounting lives, 346–347
retrofit, 341
cost of, 341
Long Beach, ordinance passed in, 341–342
Los Angeles, ordinance passed in, 342
risk analysis of, as strategy, 343
risk reduction, value of, 345–346
Unreinforced Masonry Building Law, 342
unreinforced masonry (URM), 343, 344
value-of-life calculations, 346
Select Committee to Investigate the Use of Chemicals in Food, 297–298
Series 875, 153
Shattuck, Lemuel, 294
"Signaling," by economic market, 103
Silent Spring, 144
"Sinks," in residential chemical exposure, 127
"South," United Nations definition of, 5
Southwest Network for Environmental and Economic Justice, 283
Specificity of association, 27
Split fractions, 74
Stiglitz, Joseph, 119
Stochastic risk assessment, 164, 190
Stochastic uncertainty, 33, 190
"Straw" ranking, 267
Subjective uncertainty *vs.* objective, 40
Superfund Amendments and Reauthorization Act. *see* SARA
Superfund assessments, 181, 282
Supply side economics, 110
Sustainable development
Agenda 21, 407, 411
agriculture, role of, 416
bicycles, use of, 412, 415
Bruntland report, 418
cars, as "monsters,' 412–413
compartmentalization and, 408
definition of, 407, 457
energy and pollution, 410–411

Earth Summit, 418
global environmental degradation, role in halting, 407
industrial pollution, role of, 414–415
population explosion, role of, 417–418
risk analysis and, 408, 419
urban planning, role of, 415–416
western culture, decline of, 418
System unavailability, 80

T

TEAM. *see* Total Exposure Assessment Methodology
Temporal relationship, 27
Test data, use in developing risk analysis model, 54
Theory of error, 36
THERMdbASE, 130–131
Three Mile Island, 1, 234, 242, 329, 331, 332, 338, 362
Threshold limit values (TLVs), 21
Title III, 18–20, 179, 414, 415
Top event, 76
Total Exposure Assessment Methodology (TEAM), 126, 127, 128, 132, 134, 457
Total Human Exposure Risk Database and Advanced Simulation Environment, 130–131
Total quality management (TQM), 91
Toulene exposure, 133
Toxic chemicals, exposure to, 13, 313
Toxic chemicals risk analysis, 3, 303
Toxicological Profiles, 21
Toxic release inventory, 19, 179, 180, 183
Toxic Substances Control Act (TSCA), 116
Toxline, 209–210, 217, 225
TOXNET, 210, 217, 225
Treatise on Adulterations of Food and Culinary Poisons, 294
Treaty-reserved rights in risk management. *see* Tribal interests and risk management
TRI. *see* Toxic release inventory
Triangle distribution, 48, 49, 56
Tribal interests and risk management
comparative risk projects, 402
conventional assessment and management *vs.*, 390–392
CTUIR criteria, 403–404
cultural risk and, 396–397
ecocultural management, 399–401
environmental justice and, 394–395
equity, protection, sustainability principles, 392–393
ethical, legal and social issues, 394
holistic risk evaluation, 402
land-use assumptions and, 397–399
map-based ecocultural evaluation, 402
treaty reserved resources, 395–396
tribal risk model characteristics, 391–392
uncertainty, management of, 394
unique lifestyle of tribes, impact of, 396
Trust principles in risk perception. *see* Risk perception
TSCA. *see* Toxic Substances Control Act
Turner, Ted, 101
Type A and B uncertainties, 35

U

UF. *see* Uncertainty factor
Uncertainty
actual, 40
group 2A carcinogens and, 40
group 2B carcinogens and, 40
group 1 carcinogens and, 40
group 3 carcinogens and, 40
influence in risk management, 39
objective, 40
perception of, 40
standard analysis of, 41
subjective, 40
types of
fact, uncertainties of, 33
objective, 33
stochastic, 33, 35
subjective, 33, 34
variabilty *vs.*, 35
Uncertainty factor (UF), 19
Uncorrelated risk, 256
Underwriters, 249, 251–252
Uniform distribution, 48, 49–50
Urban planning, 415–416
Urea-formaldehyde foam insulation, 313
U.S. Challenger disaster, 68
U.S. Department of Agriculture, 295, 298
U.S. Environmental Protection Agency, 21, 39, 126, 127, 130, 132, 137, 143–147, 152, 154–156, 179, 181, 190, 194, 211–212, 217, 233, 260, 261, 265, 275, 276, 279, 283, 301, 327, 347, 415, 423, 425
carcinogen guidelines, 307, 310–311
cost of preventing death, 347–348
criteria and standards for chemicals, 14
establishment of, 5
Guidelines for Exposure Assessment, 17
U.S. Food and Drug Administration, 21, 212, 295–300, 307

criteria and standards for chemicals, 14
saccharin, varied risk estimates, 304
U.S. Nuclear Regulatory Commission, 328

V

Valdez Air Health Study, 128–129
Value-of-life calculations, 346
Variability *vs.* uncertainty, 35
Veblen, Thorstein, 114
Viscusi, W. Kip, 119
Vitruvius, 4
Volatile organic compounds (VOC), 126, 127, 458

W

Weibull distribution, 48, 54
Weight of evidence, 40
Western approach to environmental management, 391
WHO. *see* World Health Organization
Whole-body dose, 167, 169, 170–171
Willingness to pay (WTP), 101
World Health Organization (WHO), 145
World Trade Organization, 377–378, 385
World Watch Institute, 416
World Wide Web (WWW). *see* Computerized risk assessment